BUSINESS METHOD PATENTS

BUSINESS METHOD PATENTS

Gregory A. Stobbs

Harness, Dickey & Pierce, P.L.C.

111 Eighth Avenue, New York, NY 10011
www.aspenpublishers.com

This publication is designed to provide accurate and authoritative information in regard to the subject matter covered. It is sold with the understanding that the publisher is not engaged in rendering legal, accounting, or other professional services. If legal advice or other professional assistance is required, the services of a competent professional person should be sought.

—From a *Declaration of Principles* jointly adopted
by a Committee of the American Bar Association
and a Committee of Publishers and Associations

© 2002 Aspen Publishers, Inc.
A Wolters Kluwer Company
www.aspenpublishers.com

Printed in the United States of America

2 3 4 5 6 7 8 9 0

Library of Congress Cataloging-in-Publication Data

Stobbs, Gregory A.
    Business method patents / Gregory A. Stobbs.
      p. cm.
    Includes index.
    ISBN 0-7355-2158-1
    1. Technological innovations—United States—Patents. 2. Industrial
management—United States—Patents. I. Title.

KF3133.B87 S76 2001
346.7304'86—dc21
                                                    2001053568

About Aspen Publishers

Aspen Publishers, headquartered in New York City, is a leading information provider for attorneys, business professionals, and law students. Written by preeminent authorities, our products consist of analytical and practical information covering both U.S. and international topics. We publish in the full range of formats, including updated manuals, books, periodicals, CDs, and online products.

Our proprietary content is complemented by 2,500 legal databases, containing over 11 million documents, available through our Loislaw division. Aspen Publishers also offers a wide range of topical legal and business databases linked to Loislaw's primary material. Our mission is to provide accurate, timely, and authoritative content in easily accessible formats, supported by unmatched customer care.

To order any Aspen Publishers title, go to *www.aspenpublishers.com* or call 1-800-638-8437.

To reinstate your manual update service, call 1-800-638-8437.

For more information on Loislaw products, go to *www.loislaw.com* or call 1-800-364-2512.

For Customer Care issues, e-mail CustomerCare@aspenpublishers.com; call 1-800-234-1660; or fax 1-800-901-9075.

Aspen Publishers
A Wolters Kluwer Company

SUBSCRIPTION NOTICE

This Aspen Publishers product is updated on a periodic basis with supplements to reflect important changes in the subject matter. If you purchased this product directly from Aspen Publishers, we have already recorded your subscription for the update service.

If, however, you purchased this product from a bookstore and wish to receive future updates and revised or related volumes billed separately with a 30-day examination review, please contact our Customer Service Department at 1-800-234-1660, or send your name, company name (if applicable), address, and the title of the product to:

ASPEN PUBLISHERS
7201 McKinney Circle
Frederick, MD 21704

ABOUT THE AUTHOR

Gregory A. Stobbs is a principal in the patent firm of Harness, Dickey & Pierce, P.L.C., where he specializes in computers and software. He holds B.S.E.E. summa cum laude (electrical engineering) and J.D. degrees from the Ohio State University. Before joining Harness, Dickey & Pierce, he practiced as a patent attorney for Sperry Corporation (UNIVAC) and as a prosecuting attorney in Ohio, conducting first chair civil and criminal trials.

To Beth and my parents.

SUMMARY OF CONTENTS

CONTENTS

Chapter 3
A PHILOSOPHY FOR BUSINESS MODEL PATENTS — THE STATUTORY SUBJECT MATTER ISSUE

CONTENTS

Chapter 4
THE ORIGINS OF COMMERCE 103

Chapter 5
THE NATURE OF COMMERCE TODAY — ELECTRONIC COMMERCE 137

CONTENTS

CONTENTS

Chapter 7
E-COMMERCE TECHNOLOGY **339**

Chapter 8
PURE BUSINESS MODEL PATENTS 395

Chapter 10
CLAIMING BUSINESS MODEL AND E-COMMERCE

Chapter 11
DRAWINGS FOR E-COMMERCE AND BUSINESS MODEL
PATENTS

Chapter 12
THE PATENT SPECIFICATION 593

PREFACE

This book owes everything to Judge Giles S. Rich, author of the watershed decision in *State Street Bank v. Signature Financial Group*. Decided in the senior years of his life, but while he was still very much keen of mind, *State Street* has changed the face of patent law. Make no mistake; Judge Rich knew what he was doing. He saw that countless judges before him had been blindly repeating a dictum that he could not accept: that business methods are not proper subject matter for patenting. He set out to fix that, and set patent law on a new course in the process.

In Hans Christian Andersen's tale of The Emperor's New Suit, an entire kingdom, afraid of ridicule, refused to see that the emperor walked naked. In a public act of blind conformity all agreed the emperor's new clothes were splendid indeed. It was finally a child, unencumbered by a lifetime of prejudices, who in loud voice proclaimed the unthinkable truth, "But he has nothing on at all." The child had the benefit of innocence. Judge Rich, on the other hand, had a long lifetime of experience. Yet both came to the same conclusion. The publicly accepted dogma had been wrong. The emperor had been wearing no clothes.

It is a tribute to Judge Rich's remarkable powers that he saw the issue fresh, even though he undoubtedly bore many scars of past judicial battles. Perhaps he wrote *State Street* to be his legacy; or perhaps he was simply deciding this case on the facts and law as he had decided countless others. In either case, the decision has made a profound impact.

This book has two main missions: to explore the controversy and implications surrounding the patenting of business methods and business models, and to provide inventors and practicing patent attorneys with tools needed to develop and exploit this interesting form of intellectual property protection. The following is a brief synopsis of its chapters.

Chapter 1 introduces the controversy surrounding business model and business method patents using a commonplace example: the drive-thru restaurant. Although the McDonald brothers did not seek business method patents on their fast food assembly line production techniques, or their drive-thru service methods, they could have. What would have happened if they had? Would the Patent Office of the 1950s have granted such patents? Most patent attorneys of the 1950s would probably have counseled the McDonald brothers not to even try, for it was widely accepted that business method innovations were off limits to patent protection. In this chapter you will find out how all of that changed when one experienced appellate judge decided the old way had been wrong.

Chapter 2 delves into the *State Street Bank* decision, with an in-depth analysis of the positions taken by the litigants and by the powerful lobbyist groups who saw this case as a watershed that would forever change the face commerce. What led Judge Rich to rule as he did? Was it the parties' arguments on the merits? Was it other economic factors urged by the special interest groups in their amicus briefs? Or was this Judge Rich's final word on a longstanding controversy over how far patent law should extend. Find out in Chapter 2.

Chapter 3 digs deeper into the philosophy behind business method and business model patents. Are they proper statutory subject matter for patent protection? Should they be? This chapter begins by presenting a philosophy by which the patentability of business inventions can be understood. The chapter then presents a case-by-case review of the seminal Supreme Court decisions that have shaped how statutory subject matter is defined today. The chapter concludes with a discussion of statutory subject matter issues with Internet patents.

Chapter 4 is perhaps the most unique chapter in this book. It seeks to understand the origins of commerce, through a study of the origins of money. From a business method patent standpoint, money is the quintessential business innovation. Money is the prior art. This chapter reminds us that the modern day marvels of e-commerce and the Internet are but the latest folds in a commercial cloth woven long ago.

Chapter 5 explores the nature of commerce today, with an emphasis on electronic commerce. If you are looking for ways to analyze and describe e-commerce inventions, this chapter is for you. The chapter presents several different models and methodologies for describing basic commerce, e-commerce and Internet inventions. The chapter explains how to identify and track the flow of value through the commerce invention. The chapter concludes with many examples of different electronic commerce innovations, showing how these innovations may be described and claimed.

Chapter 6 explores many of the judicial decisions that predate *State Street Bank*. Chapter 6 supports the conclusion Judge Rich reached in *State Street Bank* that the patent law has never officially prohibited business method patents. Yes, as it turns out, the prohibition against business method patents, which was once widely taught in law school, was a myth propagated by the early cases as dutifully repeated *dicta*. Now that Judge Rich has set the record straight, why study these old cases? There are many reasons. Perhaps most compelling today is that many of these early cases dealt with inventions that fell outside the technological arts (some could be practiced with simple paper and pencil). How the courts grappled with patentability of such nontechnological inventions is quite instructive, especially today.

Chapter 7 concentrates on the technology of e-commerce. Because many business model and business method inventions today use e-commerce or

Internet components, the well equipped patent attorney needs to have a good grasp of the technological underpinnings of the Internet. Therefore, this chapter describes such e-commerce building blocks as TCP/IP protocol, routers, bridges, hubs, gateways, the domain name system, client-server architecture, cookies, and streaming audio and video. A large portion of the chapter is devoted to how secure electronic commerce is effected using encryption and backed by supporting legislation. If you have been curious about how public key encryption works, this chapter is the place to begin.

Chapter 8 takes a look at those pure business model patents that often have little or no technological underpinnings. The chapter begins with a case study of the Toyota kanban system by which "just-in-time" manufacturing is achieved using simple, hand-printed cards, called kanban. Although Toyota did not seek patent protection for the early embodiments of the process, it did receive patents on subsequent improvements, which this chapter will examine. In this chapter you will learn how Taichi Ohno, inventor of the kanban system, received his inspiration — not from the Ford assembly line he was in the United States to observe — but from the fresh produce counter of the American supermarket. This chapter concludes with several examples of other pure business model patents. Each example illustrates techniques for describing and claiming the pure business model invention.

Chapter 9 is all about the prior art. It explains the importance of prior art to business method and business model patents, and describes how the United States Patent and Trademark Office classifies prior art patents in the business arts. The chapter gives a methodology for conducting prior art searches for business-related inventions. The chapter concludes with a discussion of the Internet as prior art.

Chapter 10 explains how to draft business model and e-commerce claims. In this chapter you will find a step-by-step methodology for analyzing inventions and drafting the claims. You will learn how to construct claim diagrams with which you and the inventor can test your claiming theories for proper scope before spending time crafting the precise claim language. The chapter contains numerous claim drafting templates to help get you started drafting some of the more sophisticated business model and e-commerce claims.

Chapter 11 is devoted to developing good patent drawings to illustrate business model and e-commerce inventions. Many patent drafters turn first to the common flowchart. Why be so limited? In this chapter you will find numerous drawing techniques to help explain your invention.

Chapter 12 discusses how to draft the business method, business model or e-commerce patent specification to meet the requirements of 35 U.S.C. § 112. The chapter discusses the leading cases concerning adequacy of disclosure. You will learn how to avoid having your application rejected by the Patent Office and how to immunize your patents from being invalidated by the courts. The chapter includes a handy checklist of topics to cover with the inventor when meeting prior to drafting and filing the application.

Chapter 13 concludes the book with a discussion on how business model, business method, and e-commerce patents can be exploited. Corporate patent portfolio managers and licensing specialists may find this chapter most helpful. The chapter begins with an in-depth treatment of the jurisdictional issues involving business model and e-commerce patents. This topic is particularly important where the invention may be practiced over the Internet, and where parts of the invention are deployed or practiced outside the United States. The remainder of the chapter focuses on the strategic patent portfolio. Find out how well your portfolio stacks up against the managed portfolios of others. Learn how to neutralize a computer's portfolio with a modest collection of well placed business model patents. Discover how you can organize your portfolio using the latest mapping techniques.

GREGORY STOBBS
Troy, Michigan
October 2001

BUSINESS METHOD PATENTS

CHAPTER 1

BUSINESS PATENTS—THE CONTROVERSY

§ 1.01   Drive-thru Windows

It is a lazy afternoon, like countless others. It is the summer of '49, dry and sunny. Beneath the buzz of cicadas and a stand of eucalyptus trees, the fast food industry is thriving. Eat in your car is the latest craze. Two families in Chevrolets, a group of teenagers crammed into a borrowed Buick, a little league team in a Ford station wagon, and a traveling salesman in a rented Oldsmobile have lined up to be waited on by youthful carhops delivering trays of burgers and fries. Later, when the sun goes down, the in-crowd will swarm beneath this same stand of eucalyptus trees to socialize, rev their engines, blast their AM radios loudly into the night and order burgers and soft drinks from the lighted menu.

The automobile has changed America. It has sped up commerce. Once dining out meant to walk into a restaurant, sit down at a table or slide into a comfortable booth, to study the menu carefully, to place your order and then seek entertainment in conversation or listen to juke box tunes while the food was being prepared. Now it is eat fast, in your car. The automobile has shaved minutes off the dining experience — no need to take the time to walk in, sit down and be waited on.

Thanks to the automobile, commerce of the 1940s has indeed sped up, but it is about to go even faster. In a similar hamburger drive-in restaurant in nearby San Bernardino, Maurice and Richard McDonald are experimenting with the very concept that Henry Ford used to spread the automobile across America: mass production. The McDonald brothers realize that they could make more money if they could increase the number of customers served per hour. This, they reason, will require changes in how orders are placed, how the food is prepared and how customers are served. Although carhop drive-ins represent a charming bit of Americana, the McDonald brothers realize that such drive-in restaurants have inherent limitations on throughput. Short order cooks are hard to come by and often temperamental. Youthful carhops attract crowds of teenagers who are often more interested in socializing than in ordering food.

After considerable thought, the McDonald brothers decided to perform some radical surgery on their San Bernardino hamburger restaurant. They fired all the carhops, reduced the menu to a few standardized items, served food in paper wrappers, eliminating dishes and dishwashers, and rebuilt the cooking area into an assembly line. Instead of carhops, counter attendants handed customers their orders, over the now familiar counter, minutes after the order was placed. It was fun to watch the food being prepared in assembly-line fashion, and the counter was low enough that children could place orders themselves, while their parents watched from their cars outside.

The assembly line mass production technique allowed the McDonald brothers to rely on cheaper labor. One person was trained to flip burgers; another to apply condiments; a third to wrap the finished product. Through this division of labor the McDonald brothers significantly reduced their manufacturing cost, allowing them to undercut their competitors' prices and still make a healthy profit. Their hamburgers sold for 15 cents when the going rate for a drive-in hamburgers was 30 to 45 cents.

Attracted by the low cost, the novel production technique, and the speed at which the product was served, customers flocked to the McDonald's counter in droves. Over time, fewer and fewer cars would pull under the stand of eucalyptus trees to be waited on by carhops of the neighboring restaurants.

However, one advantage the carhop restaurants still had over McDonald's mass production machine was that carhop customers did not have to get out of their cars to pick up their order. The McDonalds eliminated this advantage by installing drive-thru windows, through which customer orders could be delivered while the customers were still seated in their cars. That final innovation killed the carhop drive-in restaurant.

Today, McDonald's method of doing business is so familiar we no longer take notice; yet beneath the surface lies the mass production genius of Henry Ford at work. What many do not realize is that the McDonald's business method extends far beyond food preparation and delivery. The McDonald's restaurant business today employs a massive, worldwide machine that herds cattle, raises chickens, digs potatoes, and processes these "raw materials" into a homogeneous product destined for our stomachs.

Innovative in its day, and still quite commercially competitive, the McDonald's restaurant business can give only partial credit to the mass production vision of the McDonald brothers. Their vision was but the tip of an iceberg. Equal, if not greater, credit must be given to another innovation, far more profound than mass production, yet invisible to us today. That innovation is commerce. Commerce, the very yarn of which businesses are woven, has been spinning forth from the human mind for over the last two thousand years. If you want to understand business method patents, you need to understand commerce.

Two thousand years ago even the simple act of procuring a steer for slaughter involved an uncertain, often risky bartering process. There was no refrigeration, and only crude preservatives. Nearly every staple foodstuff had to be eaten fresh. Many things would have to be invented first, before the McDonald's restaurant business would even be remotely possible. Quite simply, the mass produced burger would have been impossible to prepare two thousand years ago.

[A]   Was the McDonald's Drive-thru Restaurant Patentable?

The McDonald brothers invented a new system of food delivery. Their way was more profitable, and the competition soon figured that out. Their drive-in produced $277,000 during 1951. That was nearly 40 percent better than they had been doing with the previous business model.[1] Should the McDonald brothers have patented their new method of doing business? In hindsight, yes, perhaps they should have. However, that was not the culture of their time.

Patents on business methods were considered unattainable. Business methods, it was said, were not the proper subject matter for patents. Indeed, the

---

[1] Love, J., *McDonald's Behind the Arches* (Bantam Books, 1995), p. 19.

United States Patent Office held this view, and expressed it in its official Manual of Patent Examining Procedure. The Manual of Patent Examining Procedure (or MPEP) is the patent examiner's bible that explains, in hundreds of pages of detail, exactly how an examiner must consider an application for patent and for what types of inventions a patent may be obtained. Quite simply, the Patent Office took a dim view of business method patents. Until the law was changed in 1996, this is how the Patent Office advised its examiners to respond:

> Though seemingly within the category of process or method, a method of doing business can be rejected as not being within the statutory classes. See *Hotel Security Checking Co. v. Lorraine Co.*, 160 F. 467 (2nd Cir. 1908) and *In Re Wait*, 73 F.2d 982, 24 U.S.P.Q. (BNA) 88, 22 C.C.P.A. 822 (1934).[2]

Thus, if the McDonald brothers had applied for a patent on their method of doing business, the examiner's official action would likely have read, "claim rejected as not being statutory subject matter. See *Hotel Security Checking v. Lorraine* and *In re Wait*."

If the McDonald brothers had decided to push ahead anyway and appeal the Examiner's ruling, they likely would have learned of the fate of Richard Hollingshead, the man who invented and patented the drive-in movie theater. Hollingshead lost his patent during litigation. The fate of his patent is interesting, and quite informative on the issue of whether a drive-in business model is patentable. Let us therefore step a bit further back into history, and learn what the McDonald brothers would have learned about Hollingshead's innovative drive-in business model, and his ill-fated patent.

[1] Hollingshead's Ill-fated Attempt to Patent the Drive-in Movie Theater

Shortly before the 1939 World's Fair, Richard M. Hollingshead, Jr. hit upon a novel way of cashing in on the motion picture craze. Motion pictures, popularized by Thomas Edison in the late 1800s, had become big business, evolving from silent films of the 1910s and early 1920s to "talkies" of the late 1920s to the Technicolor films of the 1930s.

In terms of historical development, the motion picture industry is quite unique. It captured — on film — an image of its very evolution. We can still relive today the very moment — as the first audiences did — when black and white film evolved into color before the viewers' very eyes. That happened in the 1939 classic, *The Wizard of Oz*.

By the early 1930s motion picture theaters had sprouted like poppies across the American landscape. Hollingshead conceived of a new type of theater in which patrons could drive in and watch the latest movie from the comfort of their cars. Hollingshead devised a special parking layout that would permit persons

---

[2] MPEP, § 706.03(a).

seated in either the front or back seat to see the screen without obstruction. He graded the parking lot slightly downward toward the screen, so that patrons seated in cars in the back row could see over the cars in the front rows.

Unlike automobiles of today, automobiles of the 1930s typically had tiny windshields, making it hard to see the screen when parked in the front rows. To address this problem, Hollingshead constructed his preferred parking area as a series of elevated terraces with individual ramps at the front of each parking space. By driving the front wheels partially or fully upon these ramps, patrons could adjust the viewing angle of their cars until all could see. Thus the ramps enabled each driver to adjust the viewing angle of his or her vehicle, to selectively accommodate individual vehicle occupants, without interfering with any other automobiles.

Hollingshead applied for a patent on his drive-in theater concept and was awarded United States Patent No. 1,909,537 on May 16, 1933. He assigned the patent to Park-In Theatres, Inc., which, in turn, licensed Loew's Drive-In Theatres, Inc. to commence operation of a drive-in theater in Providence, Rhode Island in June 1937.

Loew's Drive-In paid Park-In a percentage of its gross receipts as royalties until the close of the 1937 season in November of that year. The next spring Loew's again opened its theater, but refused to pay further royalties. Loew's Drive-In refused on advice of its legal counsel that the license was illegal and the Hollingshead patent was invalid for absence of patentable invention.

Hollingshead took Loew's to court and won in the district court. Loew's appealed to the First Circuit Court of Appeals. The appellate court reversed. It ruled that Hollingshead's patent was invalid.

Hollingshead's loss in the First Circuit Court of Appeals has many times been cited as rationale for why business methods are not patentable. What few people realize, however, is that Hollingshead's patent actually did not purport to cover the business method, per se. Instead, his patent covered the physical layout and construction details of the theater itself. His patent covered such features as the inclined risers that would allow patrons to view the movie screen through the tiny windshields of their 1930 Caddys. These features did not impress the Circuit Court as being particularly inventive:

> Making every allowance for viewing the patentee's contribution in the light of hindsight, it seems to us that grading the ground upon which an automobile is to be placed for the purpose of giving it the tilt desired would be the first expedient to occur to anyone who put his mind on the problem.[3]

Why Hollingshead did not seek to patent the broader concept of a new method of displaying movies, we can only hypothesize. Clearly there was something quite innovative in his idea to unite the American fondness for movies

---

[3] Loew's Drive-In Theaters, Inc. v. Park-In Theaters, Inc., 174 F.2d 547, 553 (1st Cir. 1949).

and its love affair with the automobile. The district court, which ruled in Hollingshead's favor, saw this clearly, finding his innovation to be:

> 'ingenious,' 'practical,' 'a good business investment,' 'a completely new idea of giving entertainment to people who could not ordinarily go to the theater,' and 'is a beneficial contribution to the art of exhibition of motion pictures.'[4]

Perhaps Hollingshead's patent attorneys failed to understand how to craft claims to the underlying business model; or perhaps his attorneys knew that the sentiment within the Patent Office weighed heavily against patenting such abstractions. There is considerable basis for the latter hypothesis, as Chapter 6 will explore in greater detail. Numerous cases from the 1800s and early 1900s show that patent law was simply not ready to accept the patentability of business methods and business models.

[2] Could McDonald's Have Patented the Drive-thru Window?

In the face of such precedent as Hollingshead's drive-in movie theater, it is understandable why the McDonald brothers did not seek patents on their new food preparation and delivery methods. However, what about the drive-thru window? Could the McDonald brothers have patented this key component of their business model and thereby blocked their competition from invading their newly developed business space?

In a word, the answer is no. Like Hollingshead, whose risers were considered to be obvious, the drive-thru window itself was already known in the art. Hard to believe, but true: the drive-thru window was invented long before the drive-thru craze began. In 1943 William Odenthal was awarded United States Patent No. 2,330,005 on what he termed a "Closure Operating Means," which was actually a hydraulically actuated mechanism for operating the service window of a drive-in restaurant. Odenthal conceived his invention during the days of the carhop drive-in. The hydraulically actuated window was devised to enable the waitress or cook to open and close the window, hands-free.

Why the need for such a window in the carhop days? Odenthal explained it was partly for hygiene and partly to improve the appearance of the restaurant. The following excerpt from Odenthal's patent paints a picture of those times:

> In viewing the many objects of the invention there stand out prominently among the leading ones, the following:
> The provision of a means for operating closures in "drive-in" eating establishments where it is the custom of the public, if desired, to eat in their automobiles and be serviced from the kitchens preparing the food and clearing the dishes wherein the food is served.
> Usually the "drive-in" restaurants have their patrons park in the rear, whether or not they are to be served without alighting. For this reason the

---

[4] *Id.* at 552.

customers face the building and have a broad view of the deliveries to and from the building.

Furthermore, the view must not be unsightly and should preferably be made as attractive as possible. The handling of garbage must be kept out of sight as the swarming of flies thereabout is obnoxious, and above all the spirit and practice of sanitation should prevail.

In order to provide for many objects whereby to offset the difficulties resulting from the several disturbing elements arising from the foregoing factors, the invention serves many useful purposes.

When the trays are returned to the kitchens, the invention makes possible the quick dispatch of the dishes to their respective automatic washers.

The invention provides for a quick automatic opening and closing of the closure.

Quickness of operation guards against the entrance of the flies and cold, the escape of kitchen fumes, and the chance to obtain an interior view.[5]

Figure 1-1 shows several views of Odenthal's drive-in window mechanism. The window opens and closes by sliding vertically under hydraulic assistance. The Odenthal drive-in window, while originally designed to obscure a patron's view of the short order cook, would certainly work fine in McDonald's new drive-thru restaurant concept. Indeed, the Odenthal patent was cited as prior art in several later drive-thru window patents that were clearly intended for the drive-thru restaurant business.

Thus, it is doubtful that the McDonald brothers could have cornered the drive-thru restaurant market with a patent on a window that slid horizontally to open, or one that swung outwardly. For the McDonald brothers to corner the market, it was clearly going to take a much more fundamental patent. It was going to take one that covered the very core of the McDonald's business idea, that of applying mass production techniques to the restaurant industry. Unfortunately for the McDonald brothers, business method patents were in shackles at the time. The shackles would not be removed for another 50 years.

§ 1.02   State Street Bank Unlocks the Shackles

Quite early in the development of their new restaurant concept, the McDonald brothers decided to cash in their chips. Wealthy from their early success, the McDonald brothers agreed to sell the franchise rights in their new restaurant concept to Ray Kroc, the traveling salesman who had sold them their first milk shake blenders. Through franchising, Kroc spread the McDonald's restaurant concept worldwide and turned the McDonald's Corporation into one of the most successful businesses on the planet.

Thanks to Ray Kroc, McDonald's became not only a ubiquitous restaurant chain, with golden arches, clown mascot and some of the world's hottest coffee,

---

[5] United States Patent No. 2,330,005, to Odenthal, issued December 27, 1940.

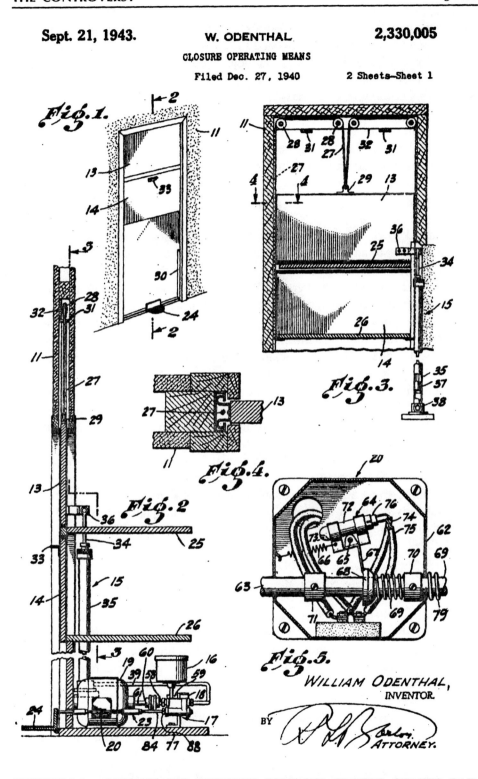

Sept. 21, 1943.   W. ODENTHAL   2,330,005

CLOSURE OPERATING MEANS

Filed Dec. 27, 1940   2 Sheets—Sheet 1

**FIGURE 1-1:   ODENTHAL'S DRIVE-IN WINDOW (UNITED STATES PAT-
ENT NO. 2,330,005)**

9

but a very popular publicly traded stock. Smelling the dollars even more than the coffee, mutual fund managers flocked to snatch up shares of McDonald's stock, paying for it with their investors' money. What would McDonald's have been worth if it had obtained patents on the drive-thru business model? One can only speculate. Certainly, in those days, patents on business methods were the furthest thing from your average mutual fund manager's mind. But those days would change. The irony is that it was the mutual fund industry that unwittingly unlocked the shackles on business method patents. That story begins with a massive institutional mutual fund manager known as State Street Bank.

[A] Introducing State Street Bank

Little known outside the financial community, State Street Bank & Trust is one of the oldest and largest institutional banks in the United States. State Street Bank began in 1792 as a simple bank in Boston. By the year 2001, it had grown to massive size, employing over 17,500 people worldwide, and commanding $5.3 trillion worth of assets under its custody and $500 billion under its management.

Unlike your local bank that issues ATM cards and handles checking accounts for the common citizen, State Street Bank is a bank for institutional investors. Its clients include pension funds, insurance companies, investment houses, labor unions, large corporations, and mutual fund companies. For these large and financially sophisticated clients, State Street Bank provides a broad range of services which include information services, custody, securities lending, investment management, performance and analytic measurement, cash management, and record keeping.

Until it tangled with Signature Financial Group over a rather curious patent involving the management of mutual funds, business method patents were the furthest thing from State Street Bank's institutional mind. Indeed, State Street Bank, already having a dominant position in the financial community,[6] would probably have preferred that the business method patent should not be allowed to exist.

[B] What Led State Street Bank to Unlock the Shackles?

In early 1991, R. Todd Boes filed a patent application for a data processing system capable of administering a new form of investment structure developed by Signature Financial Group. He assigned his invention to Signature Financial. Signature called its investment structure a Hub and Spoke structure, alluding to its main architectural feature. The Hub and Spoke structure works as explained in the following paragraphs.

Each of a plurality of Spoke funds sells shares to the public and invests all of its assets in a Hub portfolio. The Hub owns the pool of investment securities

---

[6] State Street Bank had custody of approximately 44 percent of the United States' mutual fund market at the time it clashed with Signature Financial Group.

while each Spoke fund owns a pro rata interest in the Hub portfolio. Unlike a typical standalone mutual fund, the Hub portfolio is not engaged in public offering of securities. Rather, it sells one beneficial interest to each Spoke investor. The Hub and Spoke structure is then treated as a partnership for tax purposes.

The Hub and Spoke structure gives each individual Spoke fund an economy of scale it would not otherwise enjoy. By pooling assets with other Spoke funds, all enjoy lower administrative costs. However, in order to comply with tax laws, the gains and losses must be passed daily from the Hub to each of the Spokes. The net asset value of each Spoke fund must be calculated at the close of every business day.

Signature Financial was awarded a patent on the Boes invention on March 9, 1993.[7] It offered a license to State Street Bank under the patent, but license negotiations were rocky and ultimately collapsed. Thereafter, Signature convinced a large customer of State Street Bank to switch four of its mutual funds to Signature for administration. Determined to put an end to this market erosion, State Street Bank developed its own system to compete with Signature's Hub and Spoke arrangement. It then boldly sued Signature Financial, seeking a declaratory judgment from the United States District Court for the District of Massachusetts that the patent was invalid.

The District Court granted the relief State Street Bank sought. It declared the Boes patent invalid for being directed to nonstatutory subject matter. Signature Financial appealed to the Court of Appeals for the Federal Circuit. Throughout the appeal, State Street Bank appeared confident that the District Court had properly ruled in its favor. State Street Bank urged that the Boes invention was like any other accounting method, an accounting method designed to manipulate and record numbers. It was, in State Street Bank's view, nothing but an unpatentable abstract idea.

The District Court had ruled in State Street Bank's favor by finding that the claimed invention performed no transformation or conversion of any subject matter representing or constituting physical activity or objects. The mathematical algorithm expressed in the Boes patent was, in the District Court's view, an abstraction. Although not its central theme, the District Court noted that its opinion comported with the doctrinal exclusion from patentability known as the "business methods exception." The Court cited a long string of cases for the proposition that business methods are unpatentable abstract ideas.

On appeal to the Federal Circuit, State Street Bank seemed to have little difficulty supporting the District Court's view of business method patents. It devoted only three pages in its brief to that issue. Whether confident that the issue needed little treatment, or wary that the string of cases relied upon by the District Court might not support the proposition urged, State Street Bank placed virtually no reliance on these cases.

---

[7] United States Patent No. 5,193,056, to Boes, issued March 9, 1993.

Instead, State Street Bank argued that the Boes invention was an improper attempt to preempt calculations necessary for financial service entities to be configured in a Hub and Spoke format, and calculations necessary for a Hub and Spoke configuration to comply with federal tax laws. All State Street Bank needed was for this view to hold in the Court of Appeals and the fate of business method patents would be sealed.

What State Street Bank had not bargained for was Judge Rich, the Federal Circuit Court's most experienced judge. Judge Rich was poised to become the tipping point for widespread acceptance of the business method patent. State Street Bank's declaratory judgment lawsuit was about to provide the final impetus.

§ 1.03   Judge Rich Is the Tipping Point

In his best selling book entitled, *The Tipping Point*,[8] Malcolm Gladwell looks at why major changes in society so often happen suddenly and unexpectedly, and may often be traced back to a single individual. For reasons we may truly never understand, certain people act as natural pollinators, spreading contagious new ideas in virus-like fashion. For instance, Gladwell traces an unexpected leap in the popularity of Hush Puppies shoes during 1995, to a handful of Manhattan hipsters, who chose to wear Hush Puppies precisely because no one else did. His evidence shows the tipping point occurred sometime between 1994 and 1995, when the Greenwich Village and Soho crowds started snatching the soft brown suede shoes from shelves of mom and pop stores that were still carrying the forgotten brand. By word of mouth, Hush Puppies became the "in thing" to wear and sales jumped from 30,000 pairs a year in 1994 to 430,000 pairs a year in 1995.

According to Gladwell, not all tipping points are triggered by an unnamed few. Sometimes a tipping point can be traced back to a single famous individual. Take Paul Revere, for example. The British were indeed coming to force the unruly colonists back into submission. That had been the rumor for weeks. Remarkably, news of an imminent attack leaked out to a livery boy who was working the stables where a British officer had checked his horse only hours before the attack was to begin. The livery boy ran and told Paul Revere, who in turn met with his friends and hastily organized a plot to warn the surrounding villages.

Paul Revere rode north and west to Lexington; his compatriot William Dawes rode west to Waltham. Although both delivered the identical message, "The British are coming!" Revere's message struck home, while Dawes' message went unheeded. The next day, when the British marched on Lexington, the colonists were well organized and fiercely resisted, defeating the British

---

[8] Gladwell, Malcolm, *The Tipping Point — How Little Things Can Make a Big Difference* (Little Brown & Co., 2000).

army. The citizens of Waltham were not so lucky. Dawes' message had made no impact and the British overwhelmed that village easily.

Gladwell uses the Paul Revere example to illustrate how many a tipping point can be traced back to a single individual with the unique gift: the ability to influence others. History now shows that Judge Rich was such a person.

In the mid-1990s, when State Street Bank filed suit to remove Signature Financial's patent from its path, hardly anyone would have seen the wave of business method patents coming. Thus when Judge Rich issued his landmark opinion in *State Street Bank v. Signature Financial Group*[9] on July 23, 1998, that the world had been wrong about business method patents, the business and finance world shook to the foundation.

Why did Judge Rich rule as he did? It would have been quite easy to duck the business method issue. The District Court's decision had turned primarily on a hyper-technical analysis of the Boes patent as a mathematical algorithm. The District Judge had applied a now outdated Freeman-Walter-Abele test to reach the conclusion that the Boes claim was a mere abstraction that wholly preempted others from using the basic mathematical algorithms needed to compute final share price. Under the Freeman-Walter-Abele test, the District Judge found that the claimed invention lacked any physical transformation that might demonstrate that the claimed invention achieved a useful, concrete, and tangible result.

Had he chosen to do so, Judge Rich could have simply reversed the District Judge, correcting that Judge's finding that the claimed invention was an abstraction. The Boes claims were directed to an apparatus having several processors. Thus the claimed invention was a machine, not an abstract mathematical algorithm. Because no claims were directed to a business method, per se, Judge Rich could have easily ducked the business method issue altogether. However, he chose not to do so. Why?

Judge Rich was no stranger to this esoteric question of what is proper statutory subject matter. Indeed many years earlier, sitting on the Court of Customs and Patent Appeals, he had authored a watershed decision on the patentability of computer software that the United States Supreme Court reversed in *Gottschalk v. Benson*. No judge likes to be reversed, particularly by the Supreme Court in a landmark decision. Although Judge Rich's views in the Benson case were partially vindicated by later Supreme Court decisions, he still must have lived with the scars. Perhaps, at age 94, Judge Rich saw the State Street Bank case as his opportunity to set the statutory subject matter record straight.

§1.04 Business Patent Filings Explode

Approximately one year after the State Street Bank decision, *Forbes* magazine ran a feature article on Jay Walker and his company, Walker Digital. *Forbes* described Walker as a modern day Thomas Edison, reporting that he

---

[9] State Street Bank v. Signature Financial Group, 149 F.3d 1368 (Fed. Cir. 1998).

already owned 12 business method patents, with 240 others pending.[10] Two of those patents protected Walker's name-your-own-price airline ticket business, priceline.com, which *Forbes* reported was valued at $18.5 billion.[11] Priceline.com, a Walker Digital spin-off company, went public at $16 a share, jumped to $88 a share by its second day, and at the time of the *Forbes* article was valued at $130 a share. On paper, Priceline.com was worth $18.5 billion. Walker Digital had retained a 49 percent interest in Priceline.com, which contributed $9 billion to Walker Digital's balance sheet.[12]

Walker's strategy for creating wealth was stunning. He set up Walker Digital as an information age clone of Thomas Edison's Menlo Park laboratory. With 25 employees, half in their early 30s with master's degrees, and the other half lawyers with patent experience, Walker set up his lab as a business model think tank. They would brainstorm once a week to identify internet-oriented solutions to problems that were still eligible for patent. They would then apply for the patent and either seek to license it or start a business based on it.

Walker had applied for his patent on the name-your-own-price scheme, along with two other co-inventors Bruce Schneier and James Jorasch, on September 4, 1996. That was two years before Judge Rich's State Street Bank decision. Were these men prescient, or just plain lucky? Probably neither. The title of their patent, "Method and apparatus for a cryptographically assisted commercial network system designed to facilitate buyer-driven conditional purchase offers," bespeaks of encryption, a subject that was core to the foundation of Walker Digital. *Forbes* reports that Jay Walker started Walker Digital in 1991 after being inspired by a trade journal article describing how the banking industry electronically traded billions of dollars daily using public key encryption. That being said, the patent issued one month after the State Street Bank decision, without a cryptographic limitation in its main claim. Thus if Walker was neither prescient or lucky, he certainly was astute.

Like Walker, thousands of other patent applicants flooded the Patent Office mail room with business method inventions of every description. The United States Patent Office classifies patents according to claimed subject matter. United States Patent Class 705 contains most of the business model and e-commerce patents. The Patent Office records show that patents issuing in Class 705 more than tripled in the era after the State Street Bank decision (see Figure 1-2).

§ 1.05   Business Method Patents Raise Concern

The rush to the Patent Office by countless would-be Jay Walkers, not to mention the thousands of dot-com start up companies, left many wondering whether business patents would bring prosperity or plague. Leveraging their business model patent filings, dot.com entrepreneurs became balance sheet

---

[10] Machan, Dyan, "An Edison for a New Age?" *Forbes*, May 17, 1999, p. 178.
[11] *Id.*
[12] *Id.*

Timeframe	Number of patents issued
pre-1976	293
1976-1989	184
1981-1985	300
1986-1990	673
1991-1995	1039
1996-2000	3753

FIGURE 1-2: BUSINESS PATENTS ISSUING IN CLASS 705

billionaires. Naturally, these people liked what they saw: get a patent, get funded, get rich. On the other hand, many others saw business model patents as creating a false wealth. It took only a pad of paper and a few hours to write down an idea. Unlike Edison who experimented for months before his light bulb would stay lit, business model applicants could file for patents on ideas they had never even tested. Some believed that these new business model patents would strangle the economy in a web of infringement lawsuits. Drawn to the controversy like so many sharks, journalists pounced on the issue. They openly denounced business model patents, just as they had done in the mid-1990s when software patents had been in the limelight.

[A] Amazon.com's One-click Patent Shocks the E-Commerce World

In September 1999, Amazon.com was awarded a patent on its one-click process for internet purchases. One month later it sued Barnesandnoble.com for allegedly infringing the patent. After a hearing, the district court issued a preliminary injunction against Barnesandnoble.com, prohibiting it from using the one-click process. The injunctive order was handed down in the midst of the Christmas season, causing Barnesandnoble.com to disrupt service while the one-click feature could be removed. The following are some of the patent claims that Amazon.com asserted against Barnesandnoble.com:

1. A method of placing an order for an item comprising:
        under control of a client system, displaying information identifying the item; and
        in response to only a single action being performed, sending a request to order the item along with an identifier of a purchaser of the item to a server system;
        under control of a single-action ordering component of the server system, receiving the request;
        retrieving additional information previously stored for the purchaser identified by the identifier in the received request; and

15

generating an order to purchase the requested item for the purchaser identified by the identifier in the received request using the retrieved additional information; and

fulfilling the generated order to complete purchase of the item whereby the item is ordered without using a shopping cart ordering model.

2. The method of claim 1 wherein the displaying of information includes displaying information indicating the single action.

3. The method of claim 1 wherein the single action is clicking a button.[13]

To many, it seemed absurd that the idea of purchasing with "one-click" could be considered a patentable invention. Longtime software patent opponent, Richard Stallman, called for a worldwide boycott of Amazon.com. Computer book publisher, Tim O'Reilly, captured public sentiment in a reply to Stallman's call for boycott. O'Reilly, whose books are sold by Amazon as well as Barnes and Noble, was unwilling to support the boycott, although he did support Stallman's view that the one-click patent had gone too far. O'Reilly wrote:

> In the first place, this patent should have never been allowed. It's a completely trivial application of cookies, a technology that was introduced several years before Amazon filed for their patent. To characterize "1-Click" as an "invention" is a parody. Like so many software patents, it is a land grab, an attempt to hoodwink a patent system that has not gotten up to speed on the state of the art in computer science. I'm not completely opposed to software patents, since there are some things that do in fact qualify as legitimate "inventions," but when I see people patenting obvious ideas, ideas that are already in wide use, it makes my blood boil.[14]

In addition, O'Reilly published an open letter, dated February 28, 2000, to Jeff Bezos, Amazon's chief executive officer, in which he urged Bezos to stop all attempts to enforce the one-click patent. The letter stated that the patent had been granted without proper review of the prior art, and that, even if shown to have been validly granted, it would only serve to hold back further innovation. In response to a request for public support of this open letter, O'Reilly received 10,000 signatures from internet readers in five days.

O'Reilly later interviewed Jeff Bezos and published notes of his interview on the O'Reilly website.[15] In response to O'Reilly's original point, that the one-click patent was a trivial application of existing internet cookie technology, Bezos made a cogent comeback. According to O'Reilly, Bezos agreed that the cookie implementation was fairly trivial. However, the true novelty of his invention, Bezos explained, was in how he had reframed the problem. When he

---

[13] United States Patent No. 5,960,411, "Method and system for placing a purchase order via a communications network," issued Sept. 28, 1999.

[14] O'Reilly, <http://www.oreilly.com/ask_tim/amazon_patent.html>, February 28, 2000.

[15] O'Reilly, My Conversations with Jeff Bezos, <http://www.oreilly.com/ask_tim/bezos_0300.html>, Mar. 2, 2000.

came up with the one-click concept, everyone was locked into the shopping cart metaphor: select an item, put it in your shopping cart, take it to the checkout counter and pay. In the physical, brick and mortar world, that is how we shop. Bezos realized that in the virtual, internet world, a different metaphor is possible: see an item, own it — one click.

The O'Reilly-Bezos exchange fanned the flames of controversy. Trade media journalists rekindled their anti-patent diatribe and soon Bezos and the Patent Office had a raging conflagration to contain. In March 2000, under fire for his company's lawsuit, Amazon's chief executive officer publicly called for broad changes in the patent system. He proposed that the patent term for internet-related patents be reduced from the current 20 year term to a much shorter three to five year term.[16]

[B] The Patent Office Reacts

The Patent Office responded to the public outcry as well. Although the Patent Office was hardly in a legislative position to effect the radical changes Bezos called for, it did propose its own solutions, much more in keeping with its regulatory posture. On March 29, 2000, the Patent Office announced that it was adopting a special procedure to give heightened scrutiny to business method patents, such as the Amazon.com one-click patent.

Under the plan, business-related applications would be initially examined in the usual way. As part of this process, the examiner assigns the application to a patent classification based on the claimed subject matter. If the allowed application falls within Class 705, the business model and e-commerce class, a different examiner conducts a second review, giving particular scrutiny to whether the proper search was conducted and whether the original examiner's reasons for allowance warrant the scope of protection granted.[17]

§ 1.06   Congress Reacts to *State Street Bank* Decision

Even before the controversy over the Amazon.com one-click patent ignited, Congress had been busy making adjustments to the patent laws to compensate for what some feared would be a flood of business method patent litigation in the wake of the *State Street Bank* decision. Specifically, in late 1999 Congress enacted special defense provisions to insulate a party from charges of business method patent infringement, if it is shown that the accused party was practicing the method before the patent was applied for.

The enacted provisions did not define what is meant by a business method, leaving open the question whether an apparatus claim involving a business system would be subject to this statutory defense. In any event, patent attorneys

---

[16] Bezos, "An open letter from Jeff Bezos on the subject of patents," <www.amazon.com/exec/obidos/subst/misc/patents.html>

[17] Electronic Law and Business Report, Vol. 5, No. 14, April 5, 2000, p. 340.

must now factor this defense into their claim drafting strategy. The statute is reproduced in *Chapter 13*.

Whether this special defense statute will get much play in court remains to be seen. Clearly, those drafting future business-related patent applications will be supplementing their business method claims with business apparatus or business system claims, in an effort to avoid the brunt of the statute.

§ 1.07   Some in Congress Urge Further Action

The Amazon.com one-click patent and the Priceline.com name-your-own-price patent motivated at least two congressmen to take the business method patent issue a legislative step further. On October 3, 2000, representatives Howard Berman (California Democrat) and Rich Boucher (Virginia Democrat) introduced into Congress, HR 5364, a bill captioned, the "Business Method Patent Improvement Act of 2000." In his speech to Congress, Representative Boucher had this to say about the bill:

> Two years ago, the U.S. Court of Appeals for the Federal Circuit ruled in the State Street Bank decision that a patent could be issued on a method of doing business. Since then, the Patent and Trademark Office has been deluged with applications for business method patents. Unfortunately, the PTO has granted some highly questionable ones. Last year, it awarded a patent to Amazon.com for its "one-click" method of shopping at a web site. The press recently reported that the PTO is now on the verge of awarding a patent covering any computer-to-computer international commercial transaction.
>
> Something is fundamentally wrong with a system that allows individuals to get patents for doing the seemingly obvious. The root of the problem is that the PTO does not have adequate information — what is called "prior art" — upon which to determine whether a business method is truly non-obvious and therefore entitled to patent protection. We're introducing this legislation in an effort to repair the system before the PTO awards more monopoly power to people doing the patently obvious.[18]

Representative Berman cited other patent examples, expressing similar concerns:

> It is clear from my conversations with those who are developing the Internet, those financing Internet ventures, individuals conducting business and those in the patent community — and the public at large — that the patenting of Internet and business strategies and techniques is controversial and deserves serious examination. Some believe that "business method" patents should simply not be allowed. They argue, by analogy, that a toaster should be patentable but the idea of toasting bread should not. Others argue that business methods should remain patentable, but the PTO should apply much

---

[18] Floor Statement of Representative Boucher before the United States House of Representatives, reported at <http://www.house.gov/boucher/docs/bmpiastatement.htm>.

greater scrutiny when it examines such patent applications. To extend the analogy: we have been toasting bread for a long time and if you are going to patent a method of doing so, the PTO better make sure that it has never been done in just that way before. Some note that people have received patents on activities that have been undertaken for decades and even centuries, and argue that merely placing an activity on the Internet does not make for novelty. Finally, there are a number of strange examples that lend themselves to questions about whether such common human activities deserve patent protection at all. Surely, the patent system is functioning in a curious manner when patents have been issued on a technique for measuring a breast with a tape to determine bra size (Pat. No. 5,965,809), methods of executing a tennis stroke (Pat. No. 5,993,366) and swinging a golf club (Pat. No. 5,616,089), an architect's method of eliminating hallways by placing staircases on the outside of buildings (Pat. No. 5,761,857), and a method for teaching custodial staff basic cleaning tasks (Pat. No. 5,851,117). Others have noted with suspicion the patent for a method of exercising a cat using a laser light as a tease (Pat. No. 5,443,036).

Other patents, granted to more serious endeavors, have also have been roundly criticized. With regard to patenting Internet adaptations of brick-and-mortar businesses, questions have arisen about patents granted for a method of selling music and movies in electronic form over the Internet (Pat. No. 5,191,573), a method of developing a statistical "fantasy" football game using a computer (Pat. No. 4,918,603), a method of allowing car purchasers to select options for cars ordered over the Internet (Pat. No. 5,825,651), a method of rewarding online shoppers with frequent flyer miles (Pat. No. 5,774,870), and an arguably very broad patent on managing secure online orders and payments using an "electronic shopping cart" to purchase goods on the Internet (5,745,681).

In lay terms, the basic question in each case is whether the patent owner merely adapted a well known business activity to the Internet in a straight forward manner. In patent parlance, the question is whether any of these activities are truly new and would not be obvious to one skilled in the relevant art. Other questions that may be relevant are whether others in the U.S. had known of the invention or had used it, and whether the invention was used or sold in public prior to the filing for a patent.[19]

The full text of HR 5364 has been reproduced in Appendix B. Congress did not pass the Bill during the session in which it was introduced. Whether the text of the Bill, as reproduce in Appendix B, will someday be read as part of the legislative history of enacted patent law, or simply as the forgotten high water mark of the business method patent controversy of 2000 remains to be seen. In summary, the Bill contains these provisions:

- Definition of "business method": Section 2 defines "business method" as
  "(1) a method of: (A) administering, managing, or otherwise operating an

---

[19] Floor Statement of Representative Berman before the United States House of Representatives, reported at <http://www.house.gov/berman/floor100300.htm>.

enterprise or organization, including a technique used in doing or conducting business; or (B) processing financial data; (2) any technique used in athletics, instruction, or personal skills; and (3) any computer-assisted implementation of a systematic means described [in (1)] or a technique described in [(2)]." A "business method invention" is defined as "(1) any invention which is a business method (including any software or other apparatus); and (2) any invention which is comprised of any claim that is a business method."

- Determining an application is for a business method and Public Participation: section 3 provides that within 12 months from the date of filing of an application, the Director shall determine whether an application is for a patent on a business method invention. The Director shall provide notice to the applicant of the Director's determination, and provide 60 days for the applicant to respond. The Director shall publish the application 18 months after filing. The public will have the opportunity to submit for the record any prior art, including, but not limited to, evidence of whether the invention was known or used, or was in public use or on sale (pursuant to section 102), file a protest, or petition the Director to conduct a proceeding to determine whether the invention was known, used or was in public use or on sale under section 102. The Director is required to conduct such a proceeding upon receipt of a valid petition. Information submitted during this period shall be considered during the examination of the patent application. A patent applicant has provisional rights during the pre-issuance period.

- Post-grant opposition: section 3 further provides that within 9 months after the granting of a patent on a business method invention, any third-party can request the Director to order an opposition to a patent on a business method invention on the basis of sections 101 (subject matter is new and useful, i.e., utility), 102 (novelty), 103 (non-obviousness over prior art), and 112 (disclosure, description and enablement, with definite claims to distinguish over prior art). Director, within one year of the date of enactment, shall establish an Administrative Opposition Panel comprised of 18 administrative opposition judges and publish notice of the date of the establishment of the Panel. Each opposition is to be heard by one judge who may, pursuant to the APA, take evidence by oral testimony (including exhibits) in direct or cross examination, or in any deposition, affidavit, whether voluntary or compelled. The Federal Rules of Evidence apply. During the opposition, a patent applicant may propose amendments to their patent. Within 18 months after the filing of a request for an opposition, the judge shall determine the patentability of the subject matter. The determination may be appealed by any party to the Board of Patent Appeals and Interferences, and subsequently to the Court of Appeals for the Federal Circuit.

An opposition proceeding shall not prejudice a party's right to pursue civil remedies and court (other than the court hearing an appeal on the opposition) may, in its discretion, consider any matter related to the patent independently of an opposition proceeding. However, a party to an opposition may obtain a stay in any related pending civil litigation. Where there is a final decision against a party challenging a patent in a civil action, or there is a favorable determination for a patent owner in an opposition or interference, the party challenging the validity of the patent may not thereafter bring a civil action, opposition or interference on any issues that the party raised or could have raised in the first instance of challenge. A party may bring a civil action, or initiate an opposition or interference as to newly discovered evidence.

- Fees for filing a petition to the Director to conduct a proceeding to determine whether the invention was known, used or was in public use or on sale under section 102 is set at $35. Fees for filing an opposition on the basis of prior art is $200, and on any other basis, $5000.

- Burden of Proof: In the case of reexamination, interference, opposition or other legal challenge including civil litigation to a patent (or an application for a patent) on a business method invention, the party producing evidence of invalidity or ineligibility of the patent, has the burden of making their showing by a preponderance of the evidence. (section 3)

- Nonobviousness of a computer implemented business method (i.e., Internet application of a previously known business method): If the subject matter within the scope of a claim addressed to a business method would be obtained by combining or modifying one or more prior art references, and any of those prior art references disclose a business method which differs from what is claimed in the application only in that the claim requires a computer to implement the invention, the invention is presumed obvious. The presumption of obviousness may be rebutted by a showing by a preponderance of the evidence that the invention is not obvious to persons of ordinary skill in all relevant arts. For the purposes of this subparagraph, the areas of art which are relevant include the fields of the business method and the computer implementation. (section 4)

- Requirement to disclose search: An applicant for a patent on a business method invention must disclose in the application the extent to which the applicant searched for prior art. (section 5)

- Effective dates: Provisions of this Act apply to any patent application that is pending on, or filed on or after the date of enactment. As to patents issued on or after the date of enactment, but before the establishment of the Administrative Opposition Panel, a request for an opposition must be filed within 9 months of the date the Director publishes notice of the establishment of the Administrative Opposition Panel.[20]

§ 1.08   Will Business Model Patents Lead to Downfall of Civilization?

The *Wall Street Journal* reported that Priceline.com, once valued in the stratosphere, has since seen its stock prices plummet. After being unsuccessful in using its patents to stop Microsoft's competing name-your-own-price Expedia service, Priceline.com's stock dropped from $150 a share high to prices in the single digits.[21]

---

[20] Source of summary: <http://www.house.gov/berman/bmpia_summary.htm>.

[21] "Intangibles are Tough to Value, But the Payoff Matters in Dot-Com Era," *Wall Street Journal*, May 14, 2001, p. B1.

So, is the world in the grips of a business model patent epidemic that will eventually destroy civilization? If so, then what happened to Priceline.com? The answer is quite simple. Patents represent an important component of modern day business strategy, but they do not represent the only component. Business relationships, strategic alliances, customer loyalty, and just plain good management are equally important to the successful business. In Priceline.com's case, its suit against Microsoft resulted in a settlement whereby Microsoft agreed to pay a royalty. Specific details of the settlement remain confidential.[22]

If Priceline.com's stock value dropped because investors lost confidence after Priceline.com's patent failed to stop Microsoft in its tracks, that simply shows how stock prices are frequently tied to perception rather than reality. There are many, many variables in patent infringement litigation, only one of which is the inherent strength of the patent. Equally important is the financial staying power of the respective litigants. In this regard, Microsoft had a huge advantage.

The point is that civilization as we know it did not end when the Patent Office granted Priceline.com its business model patent on the name-your-own-price concept. The patent gave Priceline.com a ticket to the court house door, which it used to bring suit against Microsoft. It did not give Priceline.com a broad monopoly over offer-and-acceptance.

The remaining chapters will explore business methods, business models and e-commerce patents in much more detail. After reading this book you may conclude that business model patents are nothing to be especially concerned over. There will undoubtedly be good ones and bad ones, just as there are good patents and bad patents in any other field of the human endeavor. If, however, after reading this book, you conclude that business model patents are a plague on the patent system and bound to be our undoing, then perhaps you can sympathize with the ancient Romans, whose civilization slid into ruin upon a Lydian invention called money. In the big scheme of things, money is, after all, the root of all evil. Business model patents don't even finish a close second.

---

[22] TheStreet.com, "Priceline.com, Expedia Settle Lawsuit," <aol.thestreet.com/brknews/internet/1248402.html>, Jan. 9, 2001.

CHAPTER 2

STATE STREET BANK v. SIGNATURE FINANCIAL—JUDGE RICH'S LEGACY

§ 2.01   Judge Rich's Legacy

The *State Street Bank* decision jolted the United States legal system, deflecting it onto a new evolutionary path. If there had been any lingering doubt about the reach of patent law, *State Street Bank* decimated it. Patent law's reach now extends to all practical applications of the human intellect. That is the legacy of Judge Giles Sutherland Rich. How did the *State Street Bank* decision happen? What arguments did Judge Rich hear that persuaded him? Did he have any doubts about how far he should go? Was this a case of first impression for Judge Rich, or did he already have a preconceived plan on where he would take the patent law? These are questions that this chapter will attempt to answer.

§ 2.02   State Street Bank Litigation in the District Court

The State Street Bank suit was initiated as a declaratory judgment action brought by State Street Bank & Trust Company, seeking a declaration that U.S. Patent No. 5,193,056 owned by Signature Financial Group, Inc. is invalid. Signature's patent, entitled "Data Processing System for Hub and Spoke Financial Services Configuration," relates to a data processing system for monitoring, calculating, and recording information involved in a financial investment vehicle termed a "Hub and Spoke" configuration.

State Street Bank prevailed in the District Court. The district judge granted State Street Bank's motion for partial summary judgment, ruling that Signature's patent was invalid under 35 U.S.C. § 101. The issue, in the words of the District Court, was the patentability of mathemataical accounting functions.

> The core issue on summary judgment is whether computer software that essentially performs mathematical accounting functions and is configured to run on a general purpose (i.e., personal) computer is patentable under 35 U.S.C. § 101.[1]

The District Court applied the *Freeman-Walter-Abele* test,[2] which prescribes a two-step analysis: is a mathematical algorithm recited directly or indirectly in the claim; and if so, is the algorithm applied in any manner to physical elements or process steps. The District Court found that the claim did recite a mathematical algorithm and that it lacked a physical transformation and hence was invalid. In reaching its conclusion, the District Court drew a strong parallel between this invention and the invention in *In re Schrader*.[3]

In *Schrader* the claimed invention involved a way of conducting auctions by specifying a method of competitive bidding on a plurality of related items,

---

[1] State Street Bank v. Signature Financial Group, 927 F. Supp. 502 (D. Mass. 1996).

[2] *In re* Freeman, 573 F.2d 1237 (C.C.P.A. 1978); *In re* Walter, 618 F.2d 758 (C.C.P.A. 1980); *In re* Abele, 684 F.2d 902 (C.C.P.A. 1982).

[3] *In re* Schrader, 22 F.3d 290 (Fed. Cir. 1994).

such as contiguous tracts of land or the like, so as to maximize sales revenue or profit to the seller. Those claims had been held non-statutory.

The District Court reasoned that Signature Financial's claims were also non-statutory, because the element of physicality was lacking:

> [T]he '056 Patent claims an invention that essentially performs mathematical calculations on data gleaned from pre-solution activity and stores and displays the results. As with Schrader's bids, the fact that those numbers represent financial constructs, such as the Hub and Spoke configuration, does not save Signature's patent. The claims do not recite any significant pre- or post-solution activity. Neither does the invention measure physical objects or phenomena as in Arrythmia or Abele nor does it physically convert data into a different form as in Alappat.[4]

The pre- and post-solution activity referred to by the District Judge are what the Patent Office calls the "Safe Harbors" of statutory subject matter. Although not required to render a claim statutory, the Patent Office Manual of Patent Examining Procedure gives two templates that can render a process claim statutory. Called "safe harbors," the first template shows how to recite "post solution activity" so that the claim crosses the statutory line. The second template shows how to recite "data gathering" steps so that the claim crosses the statutory line. Both templates focus on deriving data from some physical structure or change in data that corresponds to a physical structure.

[A] The Subject Matter of the *Signature Financial* Hub and Spoke Patent

The claimed invention involves software that is closely tied to a business model. To understand the *State Street Bank* holding you may wish to distinguish between the business model and the data processing system that Signature devised to facilitate doing business under the business model. In this case, the patent claims were drawn to the data processing system.

Under the business model facilitated by the claimed invention, mutual funds (called Spokes) pool their assets in an investment portfolio (called Hub) organized as a partnership. This investment configuration provides the administrator of a mutual fund with the advantageous combination of economies of scale in administering investments, coupled with the tax advantages of a partnership.

To facilitate administration of the Hub and Spokes investment vehicle, Signature devised a data processing system with the following components:

- a personal computer;

- a data disk for storing information;

---

[4] *State Street Bank*, 927 F. Supp. at 515.

- an arithmetic logic circuit configured to prepare the data disk to magnetically store selected data; and

- four additional arithmetic logic circuits, each configured to retrieve information from a specific file, calculate incremental increases or decreases based on specific input, allocate the results on a percentage basis, and store the output in a separate file.

The additional four arithmetic logic circuits, described above, respectively process the following data:

- assets in the portfolio of each of the funds from the previous day, including increases or decreases in each of the funds' assets and how to allocate the percentage share that each fund holds in the portfolio;

- daily incremental income, expenses, and net realized gain or loss for the portfolio and how to allocate such data among each fund;

- daily net unrealized gain or loss for the portfolio and how to allocate such data among each fund; and

- aggregate year-end income, expenses, and capital gain or loss for the portfolio and each of the funds.

[B] Claims at Issue in *State Street Bank*

Claim 1, the only independent claim in issue, reads as follows:

A data processing system for managing a financial services configuration of a portfolio established as a partnership, each partner being one of a plurality of funds, comprising:
    (a) computer processor means for processing data;
    (b) storage means for storing data on a storage medium;
    (c) first means for initializing the storage medium;
    (d) second means for processing data regarding assets in the portfolio and each of the funds from a previous day and data regarding increases or decreases in each of the funds' assets and for allocating the percentage share that each fund holds in the portfolio;
    (e) third means for processing data regarding daily incremental income, expenses, and net realized gain or loss for the portfolio and for allocating such data among each fund;
    (f) fourth means for processing data regarding daily net unrealized gain or loss for the portfolio and for allocating such data among each fund; and
    (g) fifth means for processing data regarding aggregate year-end income, expenses, and capital gain or loss for the portfolio and each of the funds.

Note the claim is drafted in the means-plus-function format. Thus, it is to be evaluated under 35 U.S.C. § 112, paragraph 6. Under a § 112, paragraph 6 analysis the means-for terms are given meaning by reference to the patent specification. State Street Bank was highly critical of this claim. State Street sued to have the claim declared invalid because, according to State Street, the claim cast a monopoly over all compliance with certain federal tax laws. Signature Financial was equally adamant that the claim was properly drafted to cover a statutory business *machine*.

The next section will present some of the arguments State Street Bank made to the Court of Appeals in seeking to have the District Judge's decision affirmed. The following section will then present the arguments that Signature Financial made to have the lower court decision overturned. The first portion of this chapter will then conclude with a discussion of some of the *amicus* arguments presented by special interest groups and how Judge Rich ultimately ruled and why.

§ 2.03   State Street Bank Argues the Signature Claims Exalt Form Over Substance

State Street Bank saw the matter in black and white: Signature's patent was a blatant attempt to monopolize the safe harbor provisions of the federal tax laws pertaining to mutual fund partnerships.[5] The hub and spoke legal structure placed the mutual fund participants into a certain safe tax harbor that ensured their tax allocation schemes would not be criticized by the Internal Revenue Service (IRS).[6] The patent, State Street Bank maintained, was drawn to nothing more than an abstract idea — a partnership accounting method. If allowed to stand, the Bank argued, the patent "subjects the ability to comply with the tax laws' substantial economic effect requirements to a private monopoly."[7]

According to State Street Bank, the magic word "computer," gratuitously placed in the language of the claim had no legal impact. State Street argued that to allow Signature Financial to characterize the invention as a computer "machine" or "apparatus," merely because the claim language was drafted in means-plus-function apparatus claim format, exalted form over substance. State Street Bank saw Signature's claims as using the word "computer" as a linguistic tool to transform an unpatentable abstraction into a statutory invention. Referring to the patent's only independent claim, reproduced in subsection [B] above, it is easy to see why State Street Bank took this position

The first three elements of the claim (processor, memory, initialization) are found in any computer system. Moreover, nothing in these first three elements had anything specific to do with the Signature hub and spoke system. Every computer has a processor, memory and suitable initialization or boot-up software that gets the computer started when the power switch is turned on. These are conventional components, which were arguably included in the claim to provide antecedent basis for the remaining data processing elements. As State Street Bank

---

[5] State Street Bank's Brief of Appellee, 28.
[6] 26 U.S.C. §§ 701-761; 26 C.F.R. §§ 1.701-1 to 1.771-1.
[7] State Street Bank's Brief of Appellee, 30.

characterized it, "though computer processor means and data storage means are recited as elements of the patent's only independent claim, those means are simply thrown in and are not interrelated by function or otherwise with any of the other operationally significant claimed elements."[8]

The remaining data processing elements of the claim appear to float, unrelated and unconnected to each other, except that the processing steps are arguably carried out using the previously recited processor and memory. State Street Bank seized upon this shortcoming in its argument:

> The '056 patent lacks any specific structural limitations and any sense of interrelatedness. For example, none of the elements (d) through (g) of the first claim that purport to take the claim beyond an unprogrammed general purpose computer is recited as being interrelated to any of the other elements in the claim. Every clause in the claim dealing with these elements simply states part of a procedure for solving a mathematical problem.[9]

State Street Bank knew, of course, that Signature Financial would be arguing that the invention was not an abstract solution to a mathematical problem, but a statutory machine. State Street sought to defuse this argument by criticizing the way the claim elements had been drafted. Because these claim elements were not structurally interrelated, State Street Bank urged, the claim did not recite a machine. The argument is a bit esoteric, but basically says that to be a machine the elements must be linked together. The processor elements in Signature Financial's claim were not so limited. State Street bolstered this reasoning with excerpts from several prior Federal Circuit Court of Appeal decisions:

> [T]he claim examined in *Alappat* as a whole was directed to a *"combination of interrelated* elements which combine[d] to form a machine." 33 F.3d at 1544 (emphasis added); *see also Arrhythmia*, 958 F.2d at 1060; *In re Iwahashi*, 888 F.2d 1370, 1375 (Fed. Cir. 1989). To combine to form a machine, such an apparatus claim must have "specific structural limitations" in its interrelated elements. *Iwahashi*, 888 F.2d at 1375.[10]

State Street painted Signature's position as that of an extremist. Allowing one to convert an abstract idea into a patentable invention merely reciting a general purpose computer in the claim exalted form over substance, it argued:

> Signature argues, the Court should pronounce that any general purpose computer programmed with any software is patentable subject matter. This is an extreme position that finds no support in the law. Indeed, its consequence would be to elevate form over substance and render any and all computer implementable algorithms patentable. Courts have consistently rejected

---

[8] State Street Bank's Brief of Appellee, 2.
[9] State Street Bank's Brief of Appellee, 17.
[10] State Street Bank's Brief of Appellee, 16.

similar attempts to evade § 101's limits, and Signature's efforts here should be treated no differently.[11]

§ 2.04   Signature Financial Counters — Arguing that the *Alappat* Decision Compels a Finding that the Patent Is Valid

Signature argued that its invention was a new and useful *machine*, capable of administering a fund configured in the Hub and Spoke structure. It argued the statutory patentability could be demonstrated through direct analogy to *In re Alappat*.[12] In *Alappat* the Court of Appeals, sitting *en banc*, had ruled that a programmed general purpose computer could constitute a new machine for § 101 purposes. Judge Rich wrote the majority opinion.

Specifically, the invention in *Alappat* involved an algorithm for modulating the electron beam of a digital oscilloscope to fool the eye into seeing a smooth waveform instead of a digitally jagged one. A programmed microprocessor performed the algorithm. Faced with a Patent Office rejection of his apparatus claims, Alappat appealed to the Federal Circuit Court, and the Court held the programmed microprocessor (computer) was a statutory machine — a rasterizer — serving the practical and useful purpose of smoothing out digital artifacts in an oscilloscope display.

Signature reasoned that if the Alappat invention was a statutory machine, expressed in apparatus claims, then its financial data processing machine was also statutory. In *Alappat*, the Court had analyzed the Alappat claims under § 101 by looking to the specification for underlying support that the claimed elements were, in fact, structural elements of a claimed machine. The Court demonstrated compliance by grafting words of the specification onto the claim. The grafted words appear in brackets in the Alappat Claim below:

A rasterizer [a ''machine''] for converting vector list data representing sample magnitudes of an input waveform into anti-aliased pixel illumination intensity data to be displayed on a display means comprising:

(a) [an arithmetic logic circuit configured to perform an absolute value function, or an equivalent thereof] for determining the vertical distance between the endpoints of each of the vectors in the data list;

(b) [an arithmetic logic circuit configured to perform an absolute value function, or an equivalent thereof] for determining the elevation of a row of pixels that is spanned by the vector;

(c) [a pair of barrel shifters, or equivalents thereof] for normalizing the vertical distance and elevation; and

(d) [a read only memory (ROM) containing illumination intensity data, or an equivalent thereof] for outputting illumination intensity data as a predetermined function of the normalized vertical distance and elevation.[13]

---

[11] State Street Bank's Brief of Appellee, 7.

[12] *In re* Alappat, 33 F.3d 1526 (Fed. Cir. 1994).

[13] State Street Bank v. Signature Financial Group, 33 F.3d 1526, 1540 (Fed. Cir. 1994).

What worked in *Alappat* should work here; so Signature sought to do the same thing. It grafted language onto its claim and sought to show how the invention was, in fact, a claim to a statutory machine. Here is how Signature rewrote its claim to demonstrate that the subject matter was statutory:

A data processing system [a machine] for managing a financial services configuration of a portfolio established as a partnership, each partner being one of a plurality of funds, comprising:

(a) [a personal computer including a central processing unit, or an equivalent thereof] for processing data;

(b) [a data disk formatted magnetically and implicitly understood apparatus for writing information thereto and reading information therefrom, including, *inter alia*, motors, transducers, interface circuits, and RAM circuits, or an equivalent thereof] for storing data on a storage medium;

(c) [an arithmetic logic circuit configured to prepare the data disk to magnetically store selected data, or an equivalent thereof] for initializing the storage medium;

(d) [an arithmetic logic circuit configured to retrieve information from a specific file, calculate incremental increases or decreases based on specific input, allocate the results on a percentage basis, and store the output in a separate file, or an equivalent thereof] for processing data regarding assets in the portfolio and each of the funds from a previous day and data regarding increases or decreases in each of the fund[s'] assets and for allocating the percentage share that each fund holds in the portfolio;

(e) [an arithmetic logic circuit configured to retrieve information from a specific file, calculate incremental increases or decreases based on specific input, allocate the results on a percentage basis, and store the output in a separate file, or an equivalent thereof] for processing data regarding daily incremental income, expenses, and net realized gain or loss for the portfolio and for allocating such data among each fund;

(f) [an arithmetic logic circuit configured to retrieve information from a specific file, calculate incremental increases or decreases based on specific input, allocate the results on a percentage basis, and store the output in a separate file, or an equivalent thereof] for processing data regarding daily net unrealized gain or loss for the portfolio and for allocating such data among each fund; and

(g) [an arithmetic logic circuit configured to retrieve information from a specific file, calculate incremental increases or decreases based on specific input, allocate the results on a percentage basis, and store the output in a separate file, or an equivalent thereof] for processing data regarding aggregate year-end income, expenses and capital gain or loss for the portfolio and each of the funds.[14]

One slight shortcoming in Signature's argument, which State Street Bank was quick to point out, was that the language Signature grafted onto its claim was

---

[14] Signature Financial Corp.'s Brief of Appellant, 16-17.

nowhere to be found in the Signature Financial patent specification. State Street Bank scathed Signature for this tactic with these words:

> Purporting to rely on *Donaldson* and *Alappat*, Signature has concocted a reading of the claim that makes it appear to have all sorts of supporting structure. Yet one could spend a lifetime searching the '056 patent specification for that structure and still never find it — for a simple reason: it is not there. Nowhere in the '056 patent specification is there mention of the "arithemetic logic circuits" that appear in Signature's brief. Nowhere in the '056 patent's specification is there mention of the "motors" that appear in Signature's brief. And nowhere in the '056 patent's specification is there mention of the "interface circuits" or "RAM circuits" that appear in Signature's brief.
>
> Signature instead "borrowed" those types of phrases from the *Alappat* specification and edited them into the '056 patent claims. Then, having incorporated the *Alappat* specification into the '056 patent through this sleight of hand, Signature pronounces that "the claims in *Alappat* and the claims of the '056 Patent are *scientifically and legally identical at a structural level* and both are directed to true machines for section 101 purposes." Signature Brief 18 (emphasis added). Just the opposite is true.[15]

§ 2.05   Special Interest Groups Attempt to Influence Outcome

The District Court decision in *State Street Bank v. Signature Financial* sent shock waves throughout both the e-commerce and banking communities, awakening lobbyist forces from both sides. The e-commerce community saw the decision as a troublesome nettle that threatened to choke out software patent protection for internet commerce. They wanted this District Court decision pulled up by its roots before it spread. Some powerful players within the banking community saw things differently. Among them were Visa International Service Association and MasterCard International, Inc. They applauded the District Court for putting patent law in check. They feared that if the District Court's ruling did not stand, there would be nothing to stop patent law from extending its reach into their pot of gold. Both interest groups submitted *amicus curiae* briefs to the Federal Circuit Court of Appeals, each seeking to sway the Court to their way of thinking.

As it turns out, the banking community's worst fears came true. The Court of Appeals did reverse the District Court. Patent law did extend its reach into the pockets of the banking community. Bankers and financiers found themselves now needing to juggle patent law considerations with their already complicated array of banking laws, federal regulations and international economics.

It is instructive to analyze how these special interest groups sought to persuade the Court. Unlike the named litigants, who drew more traditional, point-counterpoint battle lines, these special interest groups each approached the issue

---

[15] State Street Bank's Brief of Appellee, 14-15.

and crafted arguments from their unique perspectives. The e-commerce community argued economics. They played up the importance of software to the U.S.
economy and urged that a strong patent law was essential if the United States was
to maintain its lead. The banking community argued Constitutional law. They
urged that the Framers of the Constitution had never intended that patent law
should reach beyond the technological arts.

[A] The Information Technology Industry Supports Computer-Related Business Patents

Representing the e-commerce industry was a lobbyist trade organization
known as the Information Technology Industry Council (ITI). Its members
comprise an impressive list:

> AMP Incorporated
>
> Apple Computer, Inc.
>
> AT&T
>
> Compaq Computer Corporation
>
> Dell Computer Corporation
>
> Digital Equipment Corporation
>
> Eastman Kodak Company
>
> Gateway 2000, Inc.
>
> Hewlett-Packard Company
>
> Hitachi Computer Products (America), Inc.
>
> IBM Corporation
>
> Information Handling Services
>
> Intel Corporation
>
> Lexmark International, Inc.
>
> Lucent Technologies, Inc.
>
> Mitsubishi Electric America, Inc.
>
> Motorola, Inc.
>
> NCR
>
> Panasonic Communications and Systems Company
>
> Philips Key Modules/USA
>
> Silicon Graphics, Inc.
>
> Sony Electronics, Inc.
>
> Storage Technology Corporation

Sun Microsystems, Inc.

Symbol Technologies Inc.

Tektronix, Inc.

Texas Instruments Inc.

Xerox Corporation

The ITI played the economics card in its *amicus* brief. Its members accounted for over 3 percent of the U.S. gross domestic product. They supported strong patent protection for computer-related inventions, urging that it was crucial to maintain the U.S.'s competitive advantage in the world software market. Interestingly, several of ITI's members were U.S. subsidiaries of major Japanese and European electronics corporations.

In support of its economic claims, the ITI cited impressive facts and figures. The United States, ITI noted, is the world's leading innovator and producer of computer software.[16] Global revenue from the sale of software was more than $90 billion in 1994, with U.S. software sales accounting for more than 56 percent of such sales.[17] The sales of personal computer application software in North America for 1995 reached $7.53 billion, a 12 percent increase from 1994.[18] The U.S. software industry's contribution to the cross domestic product grew 21.2 percent per year between 1989 and 1994.[19]

ITI played to U.S. fears that its software industry would be lost to Asian and European competition if the United States let down its guard, as many believed it had done in the consumer electronics and semiconductor markets. The argument needed no elaboration—perhaps because none could be given. ITI stayed far away from such metaphysical questions as whether business methods were "art" within the Constitutional sense. Instead it began its legal argument with a frontal attack of the District Court's finding that "physical transformation" was required for patentability. ITI tied its argument to economic considerations, urging that the District Court's reasoning—if taken to its logical conclusion—would threaten America's lead in the software industry:

> Under the District Court's reasoning, virtually no software program in the financial services area would be entitled to patent protection, since they all involve calculations of financial statements and do not involve the transformation of physical objects. This position would threaten one of the

---

[16] *See* U.S. Congress, Office of Technology Assessment, OTA_TCT-527, *Computer Software, Intellectual Property and the Challenge of Technological Change*, 12 (U.S. Gov't Printing Office 1992).

[17] *See* Office of Computer Staff, Int'l Trade Comm'n, *Research Study 21*, "Global Competitiveness of U.S. Computer Software and Services Industries" (June 1995).

[18] "Personal Computer Application Software Sales Reach $7.53 Billion in North America in 1995," *Software Publishers Association News Release*, March 25, 1996.

[19] "Economic Contribution of the Packaged Software Industry to the U.S. Economy," *Software Publishers Association News Release*, August 1994.

most dynamic aspects of the American economy, and one in which this country currently holds a comparative advantage against the rest of the world.[20]

ITI's basic argument went like this:

1. Congress enacted § 101 of the patent law in response to the expansive prescription given by the Constitution.
2. The legislative history of § 101 and several significant Supreme Court decisions establish that Congress intended the patentable subject matter to "include anything under the sun that is made by man."[21]
3. The patent law drives innovation. Prudent investors demand patent protection for new computer-related inventions before placing their risk capital into a new venture.
4. Only abstract ideas, laws of nature, and natural phenomena are excluded from patentability — there is no exclusion for machines that do not cause an external transformation.
5. The District Court improperly treated the machine claims as a process — to reach the erroneous conclusion that a "transformation" was required — and failed to consider whether the claims were limited to a practical application in the technological arts.
6. The Patent Office is already issuing patents on banking and business transactions (1,331 to date). The validity of many of these will be drawn into question if the District Court's ruling is upheld.

As the above demonstrates, ITI's main thrust was to attack the District Court's requirement of a "physical transformation" on economic policy grounds. On the business method issue, ITI distinguished the Signature Financial invention as being a machine and not a method. ITI argued:

> Thus, if the method of doing business exclusion is to be applied at all, it should be applied to method claims only, not to machines. It is also submitted that the focus in any such inquiry regarding a method claim should be on the technological operations performed or lack thereof, not on the product or result obtained from those operations.[22]

[B] Visa and MasterCard Oppose Business Method Patents

Visa and MasterCard urged the District Court decision be affirmed. They addressed the business method issue head on and argued that business methods

---

[20] Brief *amicus curiae*, Information Technology Industry Council, State Street Bank v. Signature Financial Group.

[21] S. Rep. No. 1979, 82nd Cong., 2d Sess. 5 (1952); H.R. Rep. No. 1923, 82 Cong., 2d Sess. 6 (1952). Diamond v. Chakrabarty, 447 U.S. 303, 309 (1980), and Diamond v. Diehr, 450 U.S. 175, 182 (1981).

[22] 447 U.S. at 309 (1980), and Diamond v. Diehr, *id.* at 182.

are not patentable subject matter. Why did these financial institutions take such a hard stand? Their reason, they admitted in their Amicus Curiae brief. Both were parties to another pending suit, in which patentability under 35 U.S.C. § 101 was at issue.[23] The Court's decision in *State Street Bank v. Signature* was likely to impact the outcome of that suit, for they wrote, "As patent holders (and as accused infringers), Visa and MasterCard believe that this Court's decision in *State Street v. Signature* may have far-reaching implications in the financial services industry."[24]

The District Court held Signature Financial's patent 5,193,056 was invalid because the invention was, in the court's view, an abstract idea — a mathematical algorithm and a method of doing business. Visa and MasterCard concurred in the ruling that the claimed invention was an unpatentable mathematical algorithm. However, they filed their *amicus* brief to urge the other point, namely that the invention was an unpatentable business method.

They began their argument focusing on the Constitutional reference to the "useful arts." Article 8, Section 8, states:

> Congress shall have the power . . . To promote the Progress of Science and useful Arts, by securing for limited Times to Authors and Inventors the exclusive Right to their respective Writings and Discoveries.

Like many, Visa and MasterCard assumed that business methods were beyond the reach of patent law. They believed that the infusion of computer technology into the business world was confusing the issue, causing the Patent Office to issue patents on things that did not fall within the "useful arts."

> Visa and MasterCard, and their members, are increasingly confronted with patents that involve business methods, bookkeeping techniques, and marketing schemes that traditionally have been considered beyond the scope of 35 U.S.C. § 101 and the Constitutional reference to the "useful arts." Because of the widespread reliance on computers in all facets of business, these otherwise unpatentable business concepts are increasingly claimed in terms of computer programs, systems, or apparatus. This approach lends to the claims a superficial air of technology, but often the only "art" involved in the purported invention is that of business.[25]

Visa and MasterCard said they had no problem with patents on business machines, such as cash registers. However, the underlying business methods made possible by those machines, Visa and MasterCard believed were off limits to patent protection. That may have been Visa and MasterCard's philosophical position, however they also had wisely hedged their bets. Each had applied for

---

[23] MasterCard v. Meridian Enterprises Corp., No. C94-4105 DRD (consolidated case) (D.N.J., filed August 24, 1994).

[24] Brief *amicus curiae*, Visa and MasterCard, State Street Bank v. Signature Financial Group.

[25] *Id.*

and received business method patents themselves, as the two examples below demonstrate.

[1] Visa Patent 5,500,513 to *Langhans, et al.*

Visa owned a patent entitled Automated Purchasing Control System, which it had applied for on May 11, 1994. The patent issued March 19, 1996, well prior to the *State Street Bank v. Signature Financial* appeal. The patent deals with how to authorize banking transactions by manipulating account numbers. Exemplary claim:

1.   A method for authorizing transactions for distributing currency or purchasing goods and services, comprising the following steps:

generating a plurality of card numbers, each card number including an account number and a bank identification number (BIN), corresponding to card numbers encoded on a plurality of cards;

creating a database on a central computer having at least a first field for said BIN and a second field for said account number;

assigning a billing hierarchical position to each of said account numbers in said database;

assigning a reporting hierarchical position to each of said account numbers in said database, said reporting hierarchical position being independent of said billing hierarchical position;

assigning an authorization hierarchical position to each of said account numbers in said database, said authorization hierarchical position being independent of said billing and reporting hierarchical positions;

creating a plurality of distinct authorization tests, each of said account numbers being subject to at least one of said authorization tests, said authorization tests varying according to the hierarchical position of an individual account number;

subsequently receiving, from a remote terminal, a transmitted card number and a debit amount; determining, from said BIN in said transmitted card number, a location of said database;

determining, from the transmitted account number in said transmitted card number, the hierarchical authorization tests applicable to said transmitted account number; applying said hierarchical tests to said debit amount; transmitting to said remote terminal an authorization message if said hierarchical tests are passed;

transmitting to said remote terminal a message in accordance with a predetermined failure response option if said hierarchical tests are failed;

generating reports of said transmitting steps organized according to said reporting hierarchy; and generating invoices for debit amounts corresponding to authorization messages organized according to said billing hierarchy.

[2]   MasterCard Patent 5,557,516 to *Hogan*

MasterCard, too, had business method patents in its portfolio. An example of which is a patent entitled, "System And Method For Conducting Cashless Transactions." MasterCard applied for the patent on February 4, 1994. The MasterCard patent was issued September 17, 1996. It covers a method for conducting cashless transactions.

> Exemplary claim:
>   44. A method for conducting a cashless transaction with a user card apparatus comprising the steps of:
>
>> providing data in said user card apparatus representing at least a number of renewals and an available fund for said cashless transaction;
>>
>> increasing said available fund in response to at least said data;
>>
>> causing said number of renewals to be changed when said available fund is increased; and
>>
>> processing said cashless transaction.

[C]   Visa and MasterCard Cite a Judge Rich Law Review Article as Secondary Precedent

In advancing their argument, that business methods were not patentable, Visa and MasterCard cited interesting authority, one notable example being a law review article written by Giles S. Rich, *Principles of Patentability* and published in 1960. Author, Giles S. Rich, was, of course, the Judge Rich who wrote the *State Street Bank* decision that is the subject of this chapter.

Judge Rich penned the law review article primarily to dispel what he characterized as the unsound notion that to be patentable an invention must be better than the prior art. In developing his thesis, Judge Rich made the following observation regarding 35 U.S.C. § 101, the statutory section proclaiming what is patentable subject matter:

> Section 101, entitled "Inventions patentable," enumerates the categories of inventions subject to patenting. Of course, not every kind of an invention can be patented. Invaluable though it may be to individuals, the public, and national defense, the invention of a more effective organization of the materials in, and the techniques of teaching a course in physics, chemistry, or Russian is not a patentable invention because it is outside of the enumerated categories of "process, machine, manufacture, or composition of matter, or

any new and useful improvement thereof." Also outside that group is one of the greatest inventions of our times, the diaper service.[26]

Visa and MasterCard's basic argument went like this.

1. The Constitution restricts patent law to the useful arts.
2. Useful arts are those, and only those, that involve technological innovation.
3. You'll know a useful art (that is, a technological art) because it puts to practical use the forces and materials of nature.
4. Business methods, like fine arts, music, drama, sporting performance techniques, are not within the technological arts.
5. Patent law heralds substance over form — do not be fooled by an unpatentable business concept assuming a superficial air of technology because computers are used.
6. To distinguish between true technological inventions and business methods merely masquerading as such, ask this key question: "What did the applicant invent?"
7. Clues to the answer may be found in (a) whether general purpose or specific claim elements are recited, (b) what the written specification focuses upon (c) the nature of the problem confronting the inventor, and (d) whether the subject matter represents or constitutes physical activity or objects.
8. Signature Financial's invention is, in essence, a bookkeeping system that falls outside the technological arts and is therefore not statutory subject matter.

The above argument hinges on the premise that business methods are not patentable. The Amici argument strips away the technological dressing to reveal a business method at the core.

§ 2.06   Holding in *State Street Bank*

The Federal Circuit Court of Appeals reversed the District Court, ruling that Signature's claims do represent statutory subject matter. The Court based its ruling, not on the presence or absence of physicality as the District Court had done, but on the "practical utility" of the claimed invention. Judge Rich summarized the Court's holding as follows:

Today, we hold that the transformation of data, representing discrete dollar amounts, by a machine through a series of mathematical calculations into a final share price, constitutes a practical application of a mathematical algorithm, formula, or calculation, because it produces "a useful, concrete and tangible result" — a final share price momentarily fixed for recording

---

[26] Rich, G., "Principles of Patentability," 28 George Washington L. Rev., 393-407.

and reporting purposes and even accepted and relied upon by regulatory authorities and in subsequent trades.

Judge Rich noted that the statutory language of 35 U.S.C. § 101 lists four categories of statutory subject matter: process, machine, manufacture, and composition of matter. He indicated, however, that the question of whether a claim encompasses statutory subject matter "should not focus on which of the four categories of subject matter a claim is directed to . . . but rather on the essential characteristics of the subject matter, in particular, its practical utility."

Concluding that the claimed invention was a statutory machine, Judge Rich adopted Signature Financial's analytical technique to bolster his finding that this invention has practical utility. After each means-plus-function recitation, Judge Rich inserted, in brackets, language purportedly from the patent specification to add hardware substance and emphasize practical utility. Judge Rich was on sound legal footing for doing this because the claims were drafted in means-plus-function format and thus required reference to the specification to properly construe them. Thus Judge Rich rewrote claim 1 to include the bracketed material as follows:

(a) a computer processor means [a personal computer including a CPU] for processing data;

(b) a storage means [a data disk] for storing data on a storage medium;

(c) first means [an arithmetic logic circuit configured to prepare the data disk to magnetically store selected data] for initializing the storage medium;

(d) second means [an arithmetic logic circuit configured to retrieve information from a specific file, calculate incremental increases or decreases based on specific input, allocate the results on a percentage basis, and store the output in a separate file] for processing data regarding assets in the portfolio and each of the funds from a previous day and data regarding increases or decreases in each of the funds' assets and for allocating the percentage share that each fund holds in the portfolio; . . .

(e) . . .

(f) . . .

(g) fifth means [an arithmetic logic circuit configured to retrieve information from specific files, calculate that information on an aggregate basis and store the output in a separate file] for processing data regarding aggregate year-end income, expenses, and capital gain or loss for the portfolio and each of the funds.

It is interesting that much of the added bracketed language is not found in the patent specification, as State Street Bank had pointed out. The specification does recite use of a personal computer and a data disk, but specific mention of the Central Processing Unit (CPU) and the arithmetic logic circuits are nowhere to be found. Apparently Judge Rich took judicial notice that a personal computer has at least one CPU and that a CPU includes at least one arithmetic logic unit.

Unlike the District Court, which had viewed the invention as essentially a non-statutory business method, the Court of Appeals for the Federal Circuit viewed the invention as a machine, albeit one used to facilitate a business relationship.

[A] Judge Rich's Claim that Construction Went Hand-in-Hand with his Interpretation of 35 U.S.C. § 101

The Federal Circuit Court of Appeals is not bound to give deference to a District Court's grant of summary judgment, or to a District Court's claim construction or statutory construction. The Federal Circuit considers these issues *de novo*, and the reversal of a District Court on these issues is certainly not noteworthy. What is noteworthy about *State Street Bank* is how the Federal Circuit reached its result.

First, the Federal Circuit began its analysis by characterizing the invention as a machine. Judge Rich thus wrote, "When independent claim 1 is properly construed in accordance with § 112, paragraph 6, it is directed to a machine." Contrast this with the District Court's analysis, which characterized the invention as "an accounting system for a certain type of financial investment vehicle claimed as means for performing a series of mathematical functions."

Whereas the Federal Circuit looked to the claim language, and the underlying language in the specification, and found a machine, the District Court looked behind the claim language to answer the question, "What did the applicant invent?" In the District Court's view the applicant's invention was not a combination of CPU, data disk, and arithmetic logic circuits, but rather it was the underlying series of computations performed by those conventional circuits.

A similar disagreement in claim construction may be seen in the majority and dissenting opinions of the Federal Circuit Court of Appeals in *In re Alappat*.[27] The majority in that case viewed the invention as a rasterizer — an apparatus, unquestionably an artifact of human design. The minority viewed the invention as the conversion of one set of numbers (input waveform magnitudes) into another set of numbers (illumination intensity data) — a non-statutory abstract idea.

At the heart of this disagreement over claim construction, in both *Alappat* and *State Street Bank*, lies the statutory language of 35 U.S.C. § 101. That language reads, "Whoever invents or discovers any *new* and useful process, machine, manufacture, or composition of matter, or any new and useful improvement thereof, may obtain a patent therefor, subject to the conditions and requirements of this title (emphasis added)." There are two ways to construe this statutory language.

One statutory construction all but ignores the word new. The applicant's chosen claim language defines the invention or discovery for § 101 analysis, and

---

[27] *In re* Alappat, 33 F.3d 1526 (Fed. Cir. 1994).

whether the invention or discovery is new or not is irrelevant to the § 101 determination. Newness comes into play only in connection with §§ 102 and 103.

The other statutory construction looks at the applicant's chosen claim language — through the lens of the prior art — to identify what *new* thing the applicant claims to have invented or discovered. The claim as a whole is still considered, but only that which is new and useful bears on the statutory subject matter determination.

The Court in *State Street Bank*, however, flatly rejected the latter interpretation, citing its predecessor court's decision in *In re Bergy*:

> If the invention, as the inventor defines it in his claims (pursuant to § 112, second paragraph), falls into any one of the named categories [process, machine, article of manufacture, or composition of matter], he is allowed to pass [through the § 101 door] . . . Notwithstanding the words "new and useful" in § 101, the invention is not examined under that statute for novelty because that is not the statutory scheme of things or the long-established administrative practice.[28]

Although the Federal Circuit has chosen to ignore the word "new" in construing § 101, there remains at least one Supreme Court Justice's interpretation that seems to question this construction. Justice Stevens, writing for the 5-4 minority in *Diamond v. Diehr*,[29] questions the wisdom of ignoring the prior art in assessing the applicant's claim under § 101:

> The Court . . . fails to understand or completely disregards the distinction between the subject matter of what the inventor *claims* to have discovered — the § 101 issue — and the question whether that claimed discovery is in fact novel — the § 102 issue.[30]

[1] The Approach to Claim Construction Determines the Outcome

Like so many of the hotly contested statutory subject matter cases, the outcome usually turns on how you choose to construe the claim. If you read the claim language literally — taking the applicant at his or her word regarding the nature of the claimed subject matter — then any claim reciting a machine recites statutory subject matter, provided the specification supports the claim. On the other hand, if you look behind the literal claim language — to seek what the applicant is actually claiming to have discovered — then claims seeking coverage for abstract algorithms can be seen as failing to recite statutory subject matter.

In *Benson*, for example, the claims recited "re-entrant shift registers" to practice the claimed method. Re-entrant shift registers are clearly components of

---

[28] *In re* Bergy, 596 F.2d 952, 960, 201 U.S.P.Q. (BNA) 352, 360 (C.C.P.A. 1979).
[29] Diamond v. Diehr, 450 U.S. 175 (1981).
[30] *Id.* at 211 (emphasis in original).

a machine. Yet the Supreme Court, in that decision, looked behind the literal claim language and found that the applicant had not claimed to have discovered a machine, but rather an abstract mathematical algorithm for converting binary coded decimal numbers to binary numbers.

Contrast that decision with the holding in *State Street Bank*, where the Court looked no further than the applicant's apparatus claim and determined that the subject matter was a machine and therefore statutory. The Court grafted language from the specification onto the means-plus-function elements of the claim, but that was done mostly to demonstrate that the means clauses should not be viewed as process claims. (" '[M]achine' claims having 'means' clauses may only be reasonably viewed as process claims if there is no supporting structure in the written description that corresponds to the claimed 'means' elements.'')[31]

The net effect of the Court's holding in *State Street Bank* is to grant the patent applicant considerable power to protect business models, while avoiding the § 101 rejection. The Federal Circuit, at least for now, is willing to take the applicant at his or her word regarding the subject matter being claimed. If the claim recites a machine, and the specification supports such a reading, then the subject matter is statutory, period. Thus, while a business model, in the pure abstract sense, may not be patentable subject matter, patent applicants can obtain patent protection for the computer "machine" for carrying out that business model, by simply drafting the appropriate claim. Such a claim can be quite potent, as use of a computer may be the only effective way of carrying out the business model.

Two other things are worth noting about the decision in *State Street Bank*. First, the Federal Circuit has all but pulled the *Freeman-Walter-Abele* test up by its roots. Although not eradicating the doctrine outright, the Court had this to say about it:

> After Diehr and Chakrabarty, the Freeman-Walter-Abele test has little, if any, applicability to determining the presence of statutory subject matter. As we pointed out in *Alappat*, 33 F.3d 1543, application of the test could be misleading, because a process, machine, manufacture, or composition of matter employing a law of nature, natural phenomenon, or abstract idea is patentable subject matter even though a law of nature, natural phenomenon, or abstract idea would not, by itself, be entitled to such protection.[32]

Second, the Federal Circuit has even more forcibly wiped out the "business method" exception to statutory subject matter:

> We take this opportunity to lay this ill-conceived [business method] exception to rest. Since its inception, the "business method" exception has merely represented the application of some general, but no longer applicable legal principle, perhaps arising out of the "requirement for invention" —

---

[31] State Street Bank v. Signature Financial Group, 149 F.3d 1368, 1371 (Fed. Cir. 1998).
[32] *Id.* at 1374.

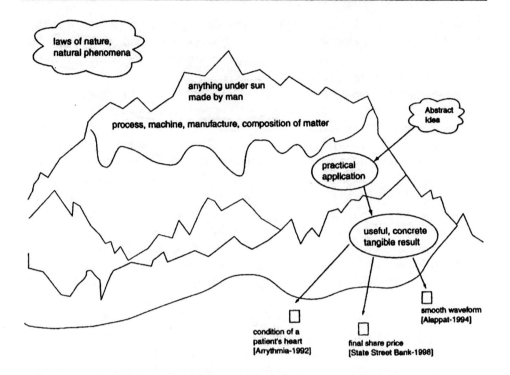

FIGURE 2-1:   THE METAPHYSICAL HIGH COUNTRY

which was eliminated by § 103. Since the 1952 Patent Act [which added § 103], business methods have been, and should have been, subject to the same legal requirements for patentability as applied to any other process or method.[33]

[B]   Putting the *State Street Bank* Holding in Context

The *State Street Bank* decision can be understood by looking down upon the decision from the metaphysical high country, as depicted in Figure 2-1. Begin in the clouds with the Supreme Court's ruling that laws of nature, natural phenomena, and abstract ideas can never be the subjects of patents.

The subject matter of patents begins with the words of Congress, repeated by the Supreme Court in *Chakrabarty*, that "anything under the sun that is made by man" may be the subject of patent, and in the statute drafted by Congress, § 101, that statutory subject matter comprises one of the four categories: process, machine, manufacture, and composition of matter.

When an invention is placed in this statutory high country the question may arise: is the invention grounded in one of the four man-made creations — process, machine, manufacture, and composition of matter — or does it remain an abstract

---

[33] 149 F.3d 31 at 1375.

idea floating in the clouds? The Supreme Court has guided us several times on this question.

Benson held that with a process, the transformation and reduction of an article to a different state or thing is one clue to determining if the subject matter is statutorily grounded. If such transformation and reduction is lacking, you may be dealing with a non-statutory abstract idea. *Benson*, however, involved a method claim and did not directly address the other three statutory categories.

Diehr held that although abstract ideas shall forever remain in the non-statutory clouds, practical applications of abstract ideas would represent statutory subject matter. Given that "everything under the sun that is made by man" lies on statutory ground, then practical applications equate to man-made creations. Practical applications are man-made creations that lie in the foothills on statutory soil. So far, so good, but you are still not down to the valley floor.

What is a practical application? The Federal Circuit in *Alappat* answered that question. Practical applications provide "useful, concrete and tangible results."[34] Useful, concrete and tangible results are individual settlements nestled in the lowlands, each founded by a different Federal Circuit decision.

The Alappat village, founded in 1994, is built on the premise that a "smooth waveform" is a "useful, concrete and tangible result." The Arrythmia village, founded in 1992, is built on the premise that the "condition of a patient's heart" is a "useful, concrete and tangible result." In this context, the State Street Bank village, founded in 1998, is built on the foundation that "a final share price momentarily fixed for recording and reporting purposes" is a "useful, concrete and tangible result."

Judge Rich was 94 when he wrote the *State Street Bank* decision. With it, he changed the course of patent law. Had *State Street Bank* been his only legacy, Judge Rich still would have earned his place in history. However, that is not Judge Rich's story. In fact, he was active his entire professional life in shaping what patent law is today. Indeed, if you look behind each of the seminal decisions on statutory subject matter you will find Judge Rich's mind at work. It is only fitting, therefore, that the following section should be dedicated to his memory.

§ 2.07 A Short Biography of Judge Giles S. Rich

Giles Sutherland Rich was born May 30, 1904, in Rochester, New York, the very same year that St. Louis hosted the World's Fair. The Louisiana territory had become part of the United States just one year before, and visitors to the fair were welcomed to view the actual desk at which purchase had been consummated. Like other World's Fairs before and after, the 1904 World's Fair featured exhibits of high technology. The high technology of 1904 was a turntable on which a locomotive and its tender of coal could be rotated 360 degrees under electric power provided by unseen motors beneath the floor.

---

[34] 33 F.3d at 1544.

Giles Rich would not have seen the Fair. Although growing up he undoubtedly saw many steam powered locomotives criss-crossing the landscape and occasionally pulling into a roundhouse to be rotated in a different direction by a turntable like the one that had been on display in the 1904 World's Fair.

Son of a lawyer, Giles Rich graduated from Harvard College in 1926 and from Columbia Law School in 1929. He began his professional career as a patent lawyer in his father's New York law firm just as the stock market crashed in 1929. He practiced as a patent lawyer for the first 27 years of his career. Rich was intensely interested in patent law. In addition to his law practice, Rich lectured at Columbia Law School from 1941 until 1956. In 1950-1951 he served as President of the New York Patent Law Association, and was one of the key drafters of the Patent Act of 1952.

In 1956, President Dwight D. Eisenhower appointed Rich to the bench, to sit on what was then known as the Court of Customs and Patent Appeals. Judge Rich was the first patent lawyer to be appointed to that position. When the U.S. Court of Appeals for the Federal Circuit replaced the Court of Customs and Patent Appeals in 1982, Judge Rich became one of the founding members of that court, the court which now hears all patent appeals.

While sitting on the bench, Judge Rich remained active as a lecturer. After a modest respite from his lecturing duties at Columbia, he joined Georgetown University Law School as an adjunct professor, where he taught from 1963 to 1969. Currently Judge Rich holds the record for being the oldest active federal judge in the nation's history. He never took senior status, but worked a full caseload up until shortly before his death on June 9, 1999. Judge Rich died at age 95.

Judge Rich's hand appears on five seminal decisions that have unquestionably shaped the patent law of biotechnology, computer software and business models:

- *In re Benson*[35]

- *In re Chakrabarty*[36]

- *In re Diehr*[37]

- *In re Alappat*[38]

- *State Street Bank v. Signature Financial*[39]

---

[35] *In re* Benson, 441 F.2d 682 (1971).
[36] *In re* Chakrabarty, 571 F.2d 44 (C.C.P.A. 1978).
[37] *In re* Diehr, 602 F.2d 912 (C.C.P.A. 1979).
[38] 33 F.3d 1526 (Fed. Cir. 1994).
[39] 149 F.3d 1368 (Fed. Cir. 1998).

Judge Rich was also the genesis of much of the Supreme Court's landmark patent opinion in *Graham v. John Deere*.[40] The Supreme Court in that case relied heavily on Judge Rich's article, "The Vague Concept of Invention as Replaced by Section 103 of the 1952 Patent Act."[41]

Why Judge Rich ruled the way he did in *State Street Bank* is easier to understand in light of some of his earlier decisions. In particular, his decision in *In re Benson*[42] is quite enlightening.

[A] How Judge Rich Had Ruled in *Benson*

In November of 1970, when he was 66 years old, Judge Rich sat on a panel with Judges Almond, Baldwin Lane and Re to hear the appeal of Gary Benson and Arthur Tabbot. At issue were two independent claims covering a method for converting binary numbers into binary coded decimal (BCD) numbers. Independent claim 8 recited hardware "reentrant shift registers" as being required elements of the method. Independent claim 13 did not restrict the method to any particular hardware elements. The patent examiner and the Board of Appeals had both ruled that the claims were drawn to non-statutory subject matter under 35 U.S.C. § 101 because the claimed method involved mental or mathematical steps.

Six months later, on May 6, 1971, Judge Rich published the decision of the Court, reversing the Board of Appeals and finding the claims were statutory under 35 U.S.C. § 101. As it would turn out, Judge Rich's ruling would not stand.

Of the two independent claims, claim 8 was perhaps less troublesome for Judge Rich. This was a method claim that contained an express recitation of "reentrant shift registers" to be used in performing the claimed method. A reentrant shift register is a hardware structure found within computers of Benson's day. In general, registers are hardware memory circuits in which data are stored. Shift registers have the ability to shift the stored digits to the right or left. A reentrant shift register has the added ability to carry digits from the left-most digit position to the right-most digit position, and vice versa.

Claim 8 read as follows:

> 8. The method of converting signals from binary coded decimal form into binary which comprises the steps of
> > (1) storing the binary coded decimal signals in a reentrant shift register,
> > (2) shifting the signals to the right by at least three places, until there is a binary "1" in the second position of said register,
> > (3) masking out said binary "1" in said second position of said register,
> > (4) adding a binary "1" to the first position of said register,
> > (5) shifting the signals to the left by two positions,
> > (6) adding a "1" to said first position, and

---

[40] Graham v. John Deere, 383 U.S. 1 (1966).

[41] 46 J.P.O.S. 844 (1964).

[42] 441 F.2d 682 (1971).

(7) shifting the signals to the right by at least three positions in preparation for a succeeding binary "1" in the second position of said register.

The Commissioner of the Patent Office had argued that the above claim 8 was non-statutory because it was directed to mental processes or mathematical steps. To reach this conclusion the Commissioner urged that the programmable computer, containing the reentrant shift registers, was merely a tool of the mind, and that the method was basically mental in character because the workstuff of the method was numbers.

Judge Rich had little difficulty rejecting this argument. Using language quoted from Judge Baldwin's decision in *In re Musgrave*,[43] Judge Rich asked the question, "Would a reasonable interpretation of the claims include coverage of the process implemented by the human mind?"[44] Judge Rich answered the question, "No." In his words, "Claim 8 is for a method to be practiced in part on particular apparatus specified to be a 'reentrant shift register'."[45]

Claim 13 was more difficult to rationalize. It contained no express recitation to particular hardware:

13. A data processing method for converting binary coded decimal number representations into binary number representations comprising the steps of
(1) testing each binary digit position i, beginning with the least significant binary digit position, of the most significant decimal digit representation for a binary "0" or a binary "1";
(2) if a binary "0" is detected, repeating step (1) for the next least significant binary digit position of said most significant decimal digit representation;
(3) if a binary "1" is detected, adding a binary "1" at the (i + 1)th and (i + 3)th least significant binary digit positions of the next lesser significant decimal digit representation, and repeating step (1) for the next least significant binary digit position of said most significant decimal digit representation;
(4) upon exhausting the binary digit positions of said most significant decimal digit representation, repeating steps (1) through (3) for the next lesser significant decimal digit representation as modified by the previous execution of steps (1) through (3); and
(5) repeating steps (1) through (4) until the second least significant decimal digit representation has been so processed.

Judge Rich approached this claim by first conceding its lack of "reentrant shift register" recitation, and pointing out that the "signals" terminology used in claim 8 was replaced by seemingly broader "representations" terminology in

---

[43] *In re* Musgrave, 431 F.2d 882 (1970).
[44] *In re* Benson, 441 F.2d at 687.
[45] *Id.* at 687.

claim 13. The claimed method was one for converting binary coded decimal number "representations" into binary number "representations."

The Patent Commissioner pointed out that claim 13 could arguably be infringed by a human using nothing more than pencil and paper, or even red and blue poker chips. In the Patent Office view, claim 13 recited a series of mental steps in the purest sense.

Judge Rich believed otherwise. He analyzed the recited steps and concluded that no human judgment or decision was required to practice the claimed method. The only mental activity required was that of making sure the steps of the claim were being performed as recited. Judge Rich injected a practical approach into his judicial reasoning. While the method could be practiced with pencil and paper, or poker chips, the only practical way to do so was with a digital computer. In Judge Rich's words:

> The only argument put forward by the Patent Office for holding claim 13 non-statutory under section 101 is that the method is basically "mental" in character. Looking at the present case in the light of all its circumstances, we observe in claim 13 a process consisting of a sequence of steps which can be carried out by machine implementation as disclosed in the specification, by still another machine as disclosed during the prosecution, and even manually although in actual practice it seems improbable anyone would ever do that, speed measured in milli- or even micro-seconds being essential in the practical utilization of such a process. Only in the manual performance would it require the operator even to think and then only to the extent necessary to assure that he is doing what the claim tells him to do. In no case is the exercise of judgment required or even the making of a decision as between alternatives.[46]

Judge Rich refused to view Benson claim 13 as mental steps. He saw the invention as a method for more efficiently operating a digital computer. Because computers were within the statutory art, methods for operating them should be also. As Judge Rich put it:

> Realistically, the process of claim 13 has no practical use other than the more effective operation and utilization of a machine known as a digital computer. It seems beyond question that the machines — the computers — are in the technological field, are a part of one of our best-known technologies, and are in the "useful arts" rather than the "liberal arts," as are all other types of "business machines," regardless of the uses to which their users may put them. How can it be said that a process having no practical value other than enhancing the internal operation of those machines is not likewise in the technological or useful arts? We conclude that the Patent Office has put forth no sound reason why the claims in this case should be held to be non-statutory.[47]

---

[46] *Id.* at 688.
[47] *Id.* at 688.

[B] The Supreme Court Reverses Judge Rich

In what became the landmark decision that would crush the advance of software patents for almost 20 years, the Supreme Court reversed Judge Rich in *Gottschalk v. Benson*.[48] *Gottschalk v. Benson* is discussed more fully in **Chapter 3**. Essentially the Supreme Court accepted the Patent Commissioner's position that Benson's claims were wholly preempting the mathematical formula underlying Benson's method. The Supreme Court thus chose not to focus on the fact that at least claim 8 was directed to the operation of a computing machine as Judge Rich had. Instead it focused on the algorithm, which the Court seemed to treat as synonymous with an abstract idea. In the words of the Supreme Court:

> It is conceded that one may not patent an idea. But in practical effect that would be the result if the formula for converting BCD numerals to pure binary numerals were patented in this case. The mathematical formula involved here has no substantial practical application except in connection with a digital computer, which means that if the judgment below is affirmed, the patent would wholly pre-empt the mathematical formula and in practical effect would be a patent on the algorithm itself.[49]

[C] Judge Rich's Reaction to the Supreme Court Reversal

No judge likes to be reversed, and Judge Rich was no exception. Shortly after the decision in *Gottschalk v. Benson*, Judge Rich had the opportunity in *In re Johnson*[50] to express his disagreement with the Supreme Court's views, even though he acquiesced to the higher court's position. The opportunity arose in an appeal over the patentability of certain automatic data processing equipment that would empower banks to create reports for its customers, categorizing checking account expenditures into different customer-defined categories, much like the reports today's personal financial software programs can do. One of the issues in the case was whether the invention was statutory under 35 U.S.C. § 101.

The majority of the panel on which Judge Rich sat saw the invention as a recordkeeping machine and ruled that such machines were clearly within the technological arts, and were therefore statutory. Judge Rich, undoubtedly still licking the wounds over his recent reversal in *Benson*, took exception to the majority view. His words reveal much about his firmly rooted views on the statutory subject matter issue. They also reveal his equally firmly rooted belief in the judicial system and in his duty to conform with what the Supreme Court had staked out in *Benson*, even though he did not agree with it:

> As the author of the opinion of this court in *Benson*, which was wholly reversed, I have not been persuaded by anything the Supreme Court said that

---

[48] Gottschalk v. Benson, 409 U.S. 63 (1972).

[49] Gottschalk v. Benson, *id.* at 71-72.

[50] *In re* Johnson, 502 F.2d 765 (C.C.P.A. 1974).

we made a "wrong" decision and I therefore do not agree with the Supreme Court's decision. But that is entirely beside the point. Under our judicial system, it is the duty of a judge of a lower court to try to follow in spirit decisions of the Supreme Court — that is to say, their "thrust." I do not deem it to be my province as a judge to assume an advocate's role and argue the rightness or wrongness of what the Court has decided or to participate in what I regard as the inconsistent decision here, supported by a bare majority which tries in vain, and only briefly, to distinguish *Benson* by discussing form rather than substance and various irrelevancies like pre-Benson decisions of this court, the banking business, social science, and the liberal arts.[51]

Section 6.23 discusses the *In re Johnson* case in more depth. The case is interesting from several standpoints. Perhaps even more than *Benson*, *In re Johnson* helped mold what business method patents are all about today. Like *Benson*, *In re Johnson* also rose to the Supreme Court. Unlike *Benson*, in which the Supreme Court took the statutory subject matter issue head on, *In re Johnson* was reversed by the Court on obvious grounds.

The statutory subject matter issue that lies at the core of *In re Johnson*, *Gottschalk v. Benson*, and *State Street Bank* presents an opportunity to explore what fundamental philosophy is at the core of patent law. That is the subject of the next chapter.

---

[51] *Id.* at 774.

CHAPTER 3

A PHILOSOPHY FOR BUSINESS MODEL PATENTS— THE STATUTORY SUBJECT MATTER ISSUE

§ 3.01   Is a Bucket Brigade Statutory Subject Matter?

Many of the e-commerce and business model patents described in this book contain a fairly heavy dose of technology. Virtually all of the e-commerce patents, by their very nature, involve electronic communication technology, or Internet technology. In addition, many of the business models being patented today exploit computer technology, database technology or other technology components. This then begs the question: is technology a requirement? Stated differently, are business inventions that employ *no* technology, or extremely *low* technology proper subject matter for patent under the United States Patent Laws?

If you are leaning towards the technology-is-required side of the argument, it may surprise you to learn that the U.S. Patent Office does issue patents on business inventions that involve technology that is so basic (like pencils and note paper) that few would call it technology at all. While these patents are relatively few in number, in comparison to the over six million patents issued, it would be doing the inventors and the Patent Office a disservice to conclude that these are simply mistakes.

Perhaps we should begin by defining terms. "Technology" is an ancient word, coming from the Greek word "techne," meaning skill one needs to practice an art and "art" means made by the hand. If we adopt a definition of technology that remains true to its ancient origins, then anything made by hand (including pencils and note paper) can constitute technology. There has been considerable judicial debate over just how far Congress intended the patent grant to extend. We will be exploring some of the more important Supreme Court opinions on this issue in the latter half of this chapter.

Although it took years to reach this conclusion, today it is well accepted that "anything under the sun that is made by man" will constitute statutory subject matter.[1] Thus, clearly a pencil, or a note pad, is statutory subject matter. The reason a pencil or a note pad is not patentable today is because both were invented long ago and can hardly be deemed novel today. One should not, however, confuse novelty with statutory subject matter. Novelty is the subject of 35 U.S.C. § 102 (and its companion requirement, non-obviousness, is the subject of 35 U.S.C. § 103). Statutory subject matter is the subject of the 35 U.S.C. § 101.

The business model patents that trouble most people are the ones where the technology component is commonplace or trivial. These are the ones where the ingenuity, if any, lies in the human interaction, not in the hardware. To explore this issue, let us consider whether a certain method for extinguishing a fire — the bucket brigade — will constitute statutory subject matter. The bucket brigade is, of course, no longer new. Forget that point for this analysis. The issue we wish to explore is whether the bucket brigade method of putting out a fire constitutes statutory subject matter.

---

[1] Diamond v. Chakrabarty, 447 U.S. 303, 309 (1980).

[A] The Bucket Brigade Scenario

Fire breaks out in the town hall. An alarm goes out. The citizenry is called to action. The town fire chief produces one wooden bucket and proceeds to shout orders comprising the following sequence of method steps:

1. lay out a reasonably straight line between the burning town hall and Walden pond;
2. align yourselves along that line, spacing yourselves approximately an arm's length apart;
3. one of you grab the wooden bucket;
4. look at the contents of the bucket and behave accordingly;
5. if the bucket is empty of water, pass it to your immediate neighbor who is closer to Walden pond;
6. if the bucket is empty and there is no one closer to the pond than you, dip the bucket into Walden pond so that it fills with water;
7. if the bucket is full of water pass it to your immediate neighbor who is closer to the fire;
8. if the bucket is full of water and there is no one closer to the fire than you, throw the contents of the bucket on the fire;
9. repeat steps 4.-8. until the fire is extinguished or Walden pond is dry.

Do these steps constitute a statutory method of doing business, in this case, a method of extinguishing a fire? If you are convinced that the method is statutory because of the technological component — the bucket — then please consider whether the method would be statutory if the bucket is eliminated and the citizens instead transport the water using their cupped hands.

§ 3.02   A Philosophy for Understanding Statutory Subject Matter

The patent law concept of what shall constitute statutory subject matter for a patent comes from ancient history, from the U.S. Constitution, from the Supreme Court and most recently from the Court of Appeals for the Federal Circuit. Section 3.07 below gives a brief overview of how patent law notions of statutory subject matter, under 35 U.S.C. § 101 have evolved. The present section will momentarily jump past the historical development to present a philosophical basis by which the statutory subject matter concept may be understood.

[A] Gaul Is Divided into Three Parts

Gaul is divided into three parts. That is how Julius Caesar began his Memoirs of the Gaelic wars. Caesar knew instinctively that to rule one must understand, and that to understand the whole one must understand its parts. Although Caesar was probably too busy fighting wars to contemplate the wisdom of Plato, had he done so, he would have understood that the world can also be subdivided into three parts: the natural world, the artificial world and the abstract world. Existing in the natural world is the earth itself, the land, the oceans, the flora and fauna, and, of course, human beings. In the artificial world are all the physical concoctions of man. Houses, boats, plows and swords, these are all artificial creations of the human species. Well apart from the concrete domain of the natural world and the artificial world floats the abstract world. Within the abstract world lies such intangible concepts, as truth, beauty, and the value of pi. The barrier between tangible and intangible cannot be crossed.

The Supreme Court has told us that statutory subject matter under the patent laws can arise only in the artificial world. Laws of nature and natural phenomena, both creatures of the natural world, are off limits to patent protection. Likewise, abstract ideas, those pure creatures of the abstract world, are also off limits. The only domain for which our patent law recognizes a property right of ownership is the artificial world. It is entirely within the artificial world that "everything under the sun that is made by man" exists.

[B] There Must Be Change Before Statutory Subject Matter Can Arise

Having restricted statutory subject matter to the artificial world, what else must be present before a patent property right may be granted? First, the inventor must have discovered or invented a way to effect a *change*, either to a naturally existing entity, or to an artificially existing entity. The process of tanning leather effects a change to a naturally existing entity, namely an animal hide. The process of tanning leather is statutory subject matter for patent. A machine made of gears and pulleys (which may in turn be made of metal extracted from the ground) effects a change to artificial entities, namely the gears and pulleys. The machine is likewise statutory subject matter for patent. By changing either a natural entity or an artificial entity, a new artificial entity is created. The tanned leather is artificial — it does not exist in nature; the machine is artificial also — it does not exist in nature, neither did its component parts.

[C] The Change Must Result from an Artificial Agency — Human Interaction

Thus, in assessing whether an invention is statutory subject matter, one should first be able to identify a change in either a natural entity or an artificial one. However, not all change is the result of human innovation. A lightning bolt will start a forest fire; the sun will dry up the rain; and the earth will nourish the willow trees, all without any human intervention. Thus, statutory subject matter has another requirement. The change in natural or artificial state must be brought about by artificial means. In other words, statutory subject matter arises when some artificial agency (human ingenuity) effects a change in either a naturally occurring entity or an artificial, human-created entity.

[D] Conclusion: The Bucket Brigade Is Statutory Subject Matter

Using this analysis, the bucket brigade is statutory subject matter. It operates upon elements in the natural world, namely fire and water. The water is transported from the pond to the fire by operation of an artificial agency, namely the line of humans working in concert to pass the bucket (or their cupped hands) from station to station until the fire is extinguished.

But wait. Humans are not artificial beings; they are natural beings. Does this not mean that the change (in the position of the water and the ultimate state of the fire) is brought about by *natural* means, as opposed to *artificial* means? Yes, humans are creatures of nature, just as lightning and sunshine are phenomena of nature. However, the *line of people* strategically arranged and spaced between pond and fire, and organized to perform the water passing function is not a phenomenon of nature. People don't just line up and start handing buckets of water to their neighbors as part of their natural or instinctive behavior. The bucket brigade line, and the bucket brigade procedure are *artificial* creations, a unique human innovation.

If the foregoing philosophy seems a bit too abstract, perhaps a few examples of actual issued patents will help.

§ 3.03 The Patent on a Method of Putting

The philosophy described above can be used to explain how the Patent Office was able to issue a rather surprising patent on a method of holding a golf club for putting. The patent, entitled simply, ''Method of Putting,'' was issued to inventor Dale D. Miller as U.S. Patent No. 5,616,089 on April 1, 1997. Here are the abstract and claims from that patent:

[A] Abstract of Method of Putting

A method of putting features the golfer's dominant hand so that the golfer can improve control over putting speed and direction. The golfer's non-dominant hand stabilizes the dominant hand and the orientation of the putter blade, but does not otherwise substantially interfere with the putting stroke. In particular, a right-handed golfer grips the putter grip with his or her right hand in a conventional manner so that the thumb on the right hand is placed straight down the top surface of the putter grip. The golfer addresses the ball as if to stroke the putter using only the right hand. Then, the golfer takes the left hand and uses it to stabilize the right hand and the putter. To do this, the golfer places his or her left hand over the interior wrist portion of the right hand behind the thumb of the right hand with the middle finger of the left hand resting on the styloid process of the right hand. The golfer presses the ring finger and the little finger of his or her left hand against the back of the right hand. The golfer also presses the palm of the left hand against the putter grip and squeezes the right hand with the left hand. The golfer then takes a full putting stroke with the above described grip.

[B] Claims of Method of Putting

1. A method of gripping a putter comprising the steps:

> gripping a putter grip with the dominant hand;

> placing the non-dominant hand over the interior wrist portion of the dominant hand behind the thumb of the dominant hand;

> resting the middle finger of the non-dominant hand on the styloid process of the dominant hand;

> pressing the ring finger and the little finger of the non-dominant hand against the back of the dominant hand;

> pressing the palm of the non-dominant hand against a forward surface of the putter grip as the non-dominant hand squeezes the dominant hand.

2. A method of gripping a putter as recited in claim 1 further comprising the steps of:

> placing the thumb on the non-dominant hand on the forearm of the dominant hand; and

> adjusting the orientation of a putter blade by moving the thumb on the non-dominant hand clockwise or counter-clockwise with respect to the forearm of the dominant hand.

3. A method as recited in claim 1 wherein the putter grip has a front surface that is flat, and the palm of the non-dominant hand presses against the flat forward surface of the putter grip.

4. A method of gripping a putter as recited in claim 1 wherein the dominant hand is positioned between two to four inches below a top end of the putter grip.

5. A method of putting comprising the steps of:

> gripping a putter having a shaft length of at least 32 inches by gripping a grip of the putter with the dominant hand;

> placing the non-dominant hand over the interior wrist portion of the dominant hand behind the thumb of the dominant hand;

> resting the middle finger on the non-dominant hand on the styloid process of the dominant hand;

> pressing the ring finger and the little finger of the non-dominant hand against the back of the dominant hand;

> pressing the palm of the non-dominant hand against a forward surface of the putter grip as the non-dominant hand squeezes the dominant hand; and

> stroking the putter to impact a golf ball with a blade of the putter.

6. A method of gripping a putter comprising the steps:

    gripping a putter grip with the right hand;

    placing the left hand over the interior wrist portion of the right hand behind the thumb of the right hand;

    resting the middle finger of the left hand on the styloid process of the right hand;

    pressing the ring finger and the little finger of the left hand against the back of the right hand;

    pressing the palm of the left hand against a forward surface of the putter grip as the left hand squeezes the right hand.

7. A method of gripping a putter as recited in claim 6 further comprising the steps of:

    placing the thumb on the left hand on the forearm of the right hand; and

    adjusting the orientation of a putter blade by moving the thumb on the left hand clockwise or counter-clockwise with respect to the forearm of the right hand.

8. A method as recited in claim 6 wherein the putter grip has a front surface that is flat, and the palm of the left hand presses against the flat forward surface of the putter grip.

9. A method of gripping a putter as recited in claim 6 wherein the right hand is positioned between two to four inches below a top end of the putter grip.

10. A method of gripping a putter comprising the steps: gripping a putter grip with the left hand;

    placing the right hand over the interior wrist portion of the left hand behind the thumb of the left hand;

    resting the middle finger of the right hand on the styloid process of the left hand;

    pressing the ring finger and the little finger of the right hand against the back of the left hand;

    pressing the palm of the right hand against a forward surface of the putter grip as the right hand squeezes the left hand.

11. A method of gripping a putter as recited in claim 10 further comprising the steps of:

    placing the thumb on the right hand on the forearm of the left hand; and adjusting the orientation of a putter blade by moving the thumb on the right hand clockwise or counter-clockwise with respect to the forearm of the left hand.

12. A method as recited in claim 10 wherein the putter grip has a front surface that is flat, and the palm of the right hand presses against the flat forward surface of the putter grip.

13. A method of gripping a putter as recited in claim 10 wherein the left hand is positioned between two to four inches below a top end of the putter grip.

[C]   Is a Method of Putting Statutory Subject Matter?

As the claims of Miller's Method of Putting patent demonstrate, the invention lies in the proper placement of the hands with respect to the putter grip. On the one hand, so to speak, gripping something is a natural act that humans have been able to do since the dawn of time. Miller didn't give us our ability to grip. All of us were born with that. On the other hand, the Patent Office did issue Miller a patent for his inventive way of gripping a putter. So can the philosophy of statutory subject matter be used to show why the Patent Office was right in granting this patent, or wrong? Let us see.

To begin, does the Miller invention effect a change on something in the natural or artificial world? The answer is yes. His innovative way of gripping the club changes the way the golf ball is putted, presumably for the better. Golf is a creation of man. As Mark Twain put it, "Golf is a good walk spoiled." Thus, golf ball putting is within the artificial world. So far, so good.

Does the Miller invention effect its change on putting through an artificial agency? That is where the analysis gets a bit tricky. In one respect, the human hands and their ability to grip are part of the natural world. Thus, one might conclude that the change in putting is brought about by a natural agency, the hands. Changes brought about by natural means do not qualify for patenting. However, to say that the hands alone are responsible for the change in putting would be to ignore the golf club shaft. When the physics of the putting stroke are analyzed, the fulcrum and lever moments produced by the interaction between the hands, arms, and golf club shaft are what make this putting stroke unique. The hands and club, when considered together as they must be, form an artificial device. It is this artificial agency that effects the change in the putting stroke. Thus, Miller's method of putting appears to qualify as statutory subject matter for patent.

§ 3.04   Application of the Philosophy to Business Methods

It would be a rather hollow philosophy if only golf strokes and bucket brigades could be analyzed with it. Most assuredly that is not the case here. If you think about it, many business inventions are similar to the golf stroke, in that they involve a series of steps, or a model for behavior to be carried out by people. To illustrate, consider the following IBM business method patent. The patent is entitled, "Method and system for accommodating electronic commerce in the semiconductor manufacturing industry." It was issued May 29, 2001, as U.S. Patent No. 6,240,400. Although the abstract and the preamble of the claim both make reference to the use of a computer, make careful note of the method steps that are recited. These steps could all be performed by people simply talking to one another.

[A] Abstract of IBM Business Method Patent

The method comprises the steps of identifying a plurality of players in the semiconductor manufacturing capacity market, each of which players can solicit capacity in a semiconductor manufacturing capacity market; providing a neutral third-party, the neutral third party and the plurality of players configured in a hub arrangement for communicating with each of the plurality of players in semiconductor manufacturing capacity trades; and realizing an open market conditionality between each of the plurality of players and the neutral third party so that the semiconductor manufacturing capacity supplied by one or more of the players can be bought and sold among the players; and, the neutral third party can preserve anonymity of each of the plurality of players soliciting semiconductor manufacturing capacity.

[B] Sample Claim from IBM Business Method Patent

1. A computer implemented method for accommodating electronic commerce in a semiconductor manufacturing capacity market, the method comprising:

1) identifying a plurality of players in the semiconductor manufacturing capacity market, each of which players can solicit capacity in a semiconductor manufacturing capacity market;

2) providing a neutral third-party, the neutral third party and the plurality of players configured in a hub arrangement for communicating with each of the plurality of players in semiconductor manufacturing capacity trades; and

3) realizing an open market condition between each of the plurality of players and the neutral third party so that:

i) the semiconductor manufacturing capacity supplied by one or more of the players can be bought and sold among the players; and

ii) each of the players may individually select at any time during said market condition, one of the following regardless of selection of the other players: identifying themselves to the other players and preserving their anonymity among each of the plurality of players soliciting semiconductor manufacturing capacity.

[C] Is the IBM Business Method Patent Statutory Subject Matter?

Approaching this patent in the same fashion as we did with the golf putting patent, we first ask, does the invention effect a change in either the natural or the artificial world, or is the invention strictly operating within the abstract world? If it is the latter, then the subject matter would not be statutory. Although many of the recited steps themselves involve manipulation of abstract ideas — such as identifying players and providing a neutral third-party — the claim specifies that these parties are "configured in a hub for communication." That configuration effects a change in the natural world because the human players and human third-party are not arranged in this way in the natural world.

Having identified a change in the natural world, the next step is to determine to what we can attribute this change. If the change is effected by an artificial (human-made) agency, then the subject matter is statutory. If the change is effected by a natural agency (like sunshine, or lightning, or gravity) then the subject matter is not statutory. Here, the change — the configuration of players into a hub for communication — is presumably effected by computers. We presume this because the preamble of the claim states that this is a computer-implemented method for accommodating electronic commerce. Thus, the agency effecting change is the computer, a member of the artificial world. The method is therefore statutory subject matter.

The IBM business method patent example featured here gives us an opportunity to test the statutory subject matter envelope by making a few simple modifications to the claim. Instead of the recited computer-implemented method, what if the invention were disclosed and claimed as relying solely on face-to-face human dialog to form the "hub for communication." That might significantly change the statutory nature of the invention.

[1] IBM Business Method — Modified to Rely on Human Dialogue

Here is the IBM business method claim, which has been modified to remove all vestige of computer technology, replacing it with simple, face-to-face human dialogue. Do the claims recite statutory subject matter now?

1. A method for accommodating commerce in a semiconductor manufacturing capacity market, the method comprising:

    1) identifying a plurality of players in the semiconductor manufacturing capacity market, each of which players can solicit capacity in a semiconductor manufacturing capacity market;

    2) providing a neutral third-party, the neutral third party and the plurality of players configured in a hub arrangement for face-to-face communication with each of the plurality of players in semiconductor manufacturing capacity trades; and

    3) realizing an open market condition between each of the plurality of players and the neutral third party so that:

i) the semiconductor manufacturing capacity supplied by one or more of the players can be bought and sold among the players; and

ii) each of the players may individually select at any time during said market condition, one of the following regardless of selection of the other players: identifying themselves to the other players and preserving their anonymity among each of the plurality of players soliciting semiconductor manufacturing capacity.

[2] Analysis of the Modified Claim

The modified claim pushes the statutory subject matter envelope to the limit. The claim in so many respects is just people talking to one another about the semiconductor industry and making deals among themselves with the aid of a neutral third party. In this light, the claimed invention hardly seems statutory. However, let us do the analysis to make sure. First, does the invention effect a change in the natural or artificial world? The answer to that question hinges on whether people naturally arrange themselves in a hub for face-to-face communication. That, it would seem, is a natural form of human behavior. Picture the tribal council meeting, or the campfire. A fairly strong case could be made that humans will naturally arrange themselves in a hub configuration when wishing to collectively discuss a matter of importance. Thus, the invention seems to fail the first test.

If you are not convinced that the hub configuration comprises natural human behavior, let us say *arguendo* that it is not. Thus, if we treat the hub configuration as a change in the natural world, then we must consider the second test. Is the change effected by artificial means? In this case, without introducing some artificial element — such as tin cans and string for communicating — it is difficult to conclude that any change has been brought on by artificial means. Thus, the invention also seems to fail the second test.

We could inject an artificial component that might turn the tide in favor of being statutory. The claim recites that the players have the option of preserving their anonymity while soliciting capacity. This might be done by simply whispering in the ear of the neutral third party. Whispering is a natural act and thus adds nothing to raise the invention to a statutory level. However, to prevent others from *seeing* a party whispering, all players may be provided with blindfolds. The introduction of such blindfolds would inject an artificial component into the invention that would render the subject matter statutory.

§ 3.05 Another Example to Test the Statutory Subject Matter Philosophy

On August 19, 1991, the Toyota Motor Company was awarded U.S. Patent No. 5,278,750, entitled "Production Schedule Making Method." The patent describes and claims a production schedule method that uses a simple cardboard note card, called a *Kanban*, to keep everyone on schedule in a just-in-time production operation. The abstract and a sample claim from this patent are reproduced below.

[A] Abstract of Kanban Patent

A method for making a production schedule of an instant process which produces a plurality of types of products, and supplies the products to a plurality of following processes by trucks. A truck delivery schedule including a number of truck deliveries and times is taken into account when the production schedule of the instant process is made. On the production schedule, a production order of the products is determined, and a stocking schedule also is made on the production order schedule.

[B] Sample Claim from Kanban Patent

1. A method for making a production schedule for an instant process that produces a plurality of types of products and supplies the products to a plurality of second processes by trucks, the method comprising the steps of:
   recording a second process schedule which includes numbers of products needed by the second process with respect to respective product types and respective days;
   recording a truck delivery schedule from the instant process to the second process and information included in a card called a *KANBAN* which is carried by each truck between the instant process and the second processes and which includes information about delivery types;
   determining a product shipment schedule of the instant process based on the second process schedule, the truck delivery schedule, and the *KANBAN* information;
   recording a working condition and a production condition of the instant process;
   determining a production schedule of the instant process based on the product shipment schedule, the working condition of the instant process, and the production condition of the instant process; and
   determining a production order of the products to be produced at the instant process based on the production schedule of the instant process.

[C] Analysis of the Kanban Patent

It is possible to dream up a schedule for something that remains entirely within the abstract domain. A schedule is inherently abstract. However, a schedule enters the concrete world of the artificial domain when it is written down or recorded. Such is the case here. The Kanban patent claims recite several recording steps that result in creation of an artificial entity — a written shipment schedule.

The writing down of the schedule itself effects a change in the artificial world because now something exists in the artificial world that did not exist before, namely a written shipping schedule. The agency responsible for effecting this change is the act of writing on the kanban card. Until the schedule was written on this card it was within the abstract domain. The card, an artificial element, gave substance to the schedule. Under the statutory subject matter philosophy presented here, the Kanban invention is clearly statutory subject matter.

What is interesting about the Kanban patent is its relatively "low-tech" nature. Unlike the IBM business method patent, with its high-tech, computer-implemented communication hub, the Kanban patent relies on a simple note card to keep the shipping schedule in play. The Kanban patent illustrates how a business method patent can be crafted, without reliance on high-tech, computer or Internet components. While you might dismiss the golf stroke patent as interesting, but commercially unimportant, you cannot dismiss the Kanban patent in this fashion. Low-tech, Kanban scheduling systems are used around the world to maintain just-in-time production schedules. Chapter 8 discusses this technology in greater detail. Although the Kanban card is a simple artifact, its use in mediating lean production schedules yields profound benefits.

§ 3.06   Should a Patent System Reward Inventors of Low-Tech Business Inventions?

After the decision in *State Street Bank v. Signature Financial* was announced, some criticized it as a serious erosion of the fundamental principles upon which the patent law is based. The critics reasoned that if business methods devoid of any technological innovation could be patented, it was only a matter of time before the whole of commerce would implode in a mass of patent infringement litigation. So far, that has not happened. If history is any indication, it is not likely to happen. A similar cry was made in the mid-1990s, when software patents were beginning to get popular. Today, the software industry is a healthy as ever. Software patents have simply not brought about the ruination that critics had predicted.

The analogy to software patents aside, should a patent system reward inventors of low-tech business methods and business models? There is reason to answer this question, yes. When considering a patent system and its value to society, many focus shortsightedly on the rights granted to the patentee. The U.S. patent laws give the patent owner the right to exclude others from making, using, selling and importing the patented invention in the United States. That can be a valuable right, and in certain instances it can force others to redesign their products and services to avoid violation of the patent owner's right. However, there is another side to the patent system that, from society's standpoint, is far more important. The patent system is a massive knowledge base into which inventors voluntarily pour their ideas so that others may learn from them.

[A] The Patent System Is the Modern-Day Library of Alexandria

Throughout history, humans have strived to preserve knowledge. Knowledge is the information upon which human societies are built. We differ today from our ancient ancestors more in our store of knowledge than in our genetic makeup. Whereas the genetic code that controls how our bodies form and grow is stored inside our DNA, the knowledge upon which our societies are based is stored outside, in our libraries.

The great libraries of ancient Egypt, Greece, and Rome all consisted of collections of papyrus scrolls. These libraries disappeared long ago, but the memory of their existence remains in the scholarly literature. The literature tells us that the Egyptians maintained a library at Amarna in 1300 B.C. and at Thebes in 1200 B.C. One thousand years later, in 330 B.C., Alexander the Great founded a great library at Alexandria, Egypt. The library is said to have contained over 400,000 scrolls. Like the earlier libraries at Amarna and Thebes, it disappeared without a trace.

While there is no way to ensure longevity, we certainly hope that the knowledge maintained in the U.S. patent records will survive for eternity. The currently available digital media certainly make it more likely that the knowledge will be preserved because digital media is far easier to copy and propagate than papyrus scrolls. The U.S. patent records, namely the millions of issued U.S. patents, represent a massive warehouse of knowledge, with far more breadth and depth than Alexandria's papyrus scroll collection could possibly have contained. In the U.S. patent library there is recorded how to generate electricity, how to fabricate electric lights, how to fly airplanes, how to program computers, and how to make telephone calls. Someday, certainly the cure for cancer will also be stored there.

[B] Business Method and Business Model Innovations Belong in the Patent Library

The patent system is designed to collect new ideas of all descriptions. It does not make value judgments. As long as an idea has practical utility, and is novel and non-obvious, the patent system welcomes it into the fold.

In this light, it seems highly doubtful that we would *not* want to record new ideas touching upon business and commerce. Who is to say that some novel business ideas should be preserved and others rejected? From society's standpoint, the far safer approach is to accept all new ideas and let the marketplace decide which are valuable and which are worthless.

In their book, *Natural Capitalism*,[2] authors, Hawken *et al.*, make a strong case that business model innovations may be the only way humankind will solve some of its largest problems, like global warming. Clearly if we keep burning fossil fuels and polluting the atmosphere with damaging hydrocarbons and CO_2,

---

[2] P. Hawken, A. Lovins, L. Lovins, *Natural Capitalism* (Little, Brown and Co., 1999).

our life on this planet could be choked out altogether. Then our libraries will no longer matter. Perhaps it will be a business method or business model innovation that will ultimately allow us to solve this problem, by altering economic conditions so that we no longer find it beneficial to burn fossil fuels or expel other harmful chemicals into the atmosphere. Would it not be worth it to society to grant a 20 year patent right in exchange for a solution to global warming? If the business method or business model solution that abates global warming happens to exploit innovations in developing human capital, would that make the invention any less valuable to society? The answer is clearly no.

Thus, when confronting business method and business model innovations that seem to lack the high-tech flavor of today's Internet e-commerce technology, do not reject the patent system out of hand. Even though the technology may be low, the importance to society could be quite high. Even if it is not, that is a matter for the marketplace and not the patent system to decide.

§ 3.07   Where Did Our Concept of Statutory Subject Matter Come From?

The origins of our patent law date far back into Middle Age England. From its very beginning, the patent has been designed as an inducement, an incentive, a reward. In medieval times, a patent was an inducement designed to lure goods and technologies from foreign lands into England. Be the first to travel abroad and bring back new goods or technology, and get rewarded with letters patent, the exclusive right to sell the new goods or to profit by the new technology — that was the way it worked.

The medieval system of patents had a bright side. It lit the fires of ingenuity and helped lift the standard of living in England, which at that time was lagging far behind the rest of civilized Europe and Asia. The system had a dark side as well. Rooted in greed, the dark side cast the shadow of monopoly over many necessities of life (to the benefit of the Crown and the Crown's close friends).

There has always been a bright side and a dark side to the patent system. The sides are inseparable, just as two sides of the human personality are inseparable. The legal system of patents has this one simple goal: to allow the bright side to shine on innovation, and to keep the dark side in check.

In this regard, the United States patent system takes the following tack:

> Whoever invents or discovers any new and useful process, machine, manufacture, or composition of matter, or any new and useful improvement thereof, may obtain a patent therefor, subject to the conditions and requirements of this title.[3]

The above statutory language, commonly referred to as § 101, states, in positive terms, what subject matter is eligible to earn a patent. Looking on the bright side, the statutory language lists broad areas of technology: processes,

---

[3] 35 U.S.C. § 101.

machines, manufactured articles, compositions of matter, for which innovation is rewarded. To keep the dark side in check, only the enumerated subject matter is eligible to earn a patent. Elsewhere in the patent statute, the term of the patent is fixed to roughly two decades. The patent term is far shorter than the copyright term, currently 75 years for a corporation, for example.

It is instructive to compare 35 U.S.C. § 101 to the comparable statutory language in the European Patent Convention:

> (1) European patents shall be granted for any inventions which are susceptible of industrial application, which are new and which involve an inventive step.

> (2) The following in particular shall not be regarded as inventions within the meaning of paragraph 1:

>> (a) discoveries, scientific theories and mathematical methods;

>> (b) aesthetic creations;

>> (c) schemes, rules and methods for performing mental acts, playing games or doing business, and programs for computers;

>> (d) presentations of information.[4]

Note how the European law goes further than United States law by expressly listing certain things that are not patentable subject matter. Under European law, the list is not considered exhaustive, and software inventions may be patentable, despite the statutory exclusion of computer programs.

The issue of what is patentable subject matter is central to many business patents. Much can be learned about this issue by studying history, and in particular Supreme Court history. There are several important precomputer Supreme Court decisions that have shaped the law on this issue, and three Supreme Court decisions that have expressly ruled on software patentability, from which many business and e-commerce inventions derive. These decisions are discussed in the next sections.

§ 3.08   Precomputer Era Supreme Court Decisions

The Supreme Court interpretation of what is patentable subject matter dates back long before the computer era. The Supreme Court first addressed a software patent in 1972, in *Gottschalk v. Benson*.[5] This section reviews those pre-*Benson* Supreme Court decisions that developed the law on which *Benson* and the later cases rely. If you are primarily interested in what the Supreme Court's position has been on patentability of software, you may want to turn immediately to § 3.09.

---

[4] Art. 52(2), Eur. Pat. Convention, Patentable Inventions (1973).
[5] 409 U.S. 63 (1972).

To many, the law in this area is confusing. What is patentable subject matter and what is not? Is computer software or a business model patentable subject matter or not? Are business models patentable subject matter or not? These have been controversial issues. One objective of this chapter is to unify the diverse decisions in this area, and to uncover the common thread woven through the fabric of these decisions. Believe it or not, there is a common thread to all the § 101 patentable subject matter cases.

[A] The Common Thread to Statutory Subject Matter Cases

So what is the common thread? What does the patentable invention have that the unpatentable discovery does not? The answer is human control. Human control is the common thread. *Diamond v. Chakrabarty*[6] maintains that ''anything under the sun that is made by man,'' qualifies as patentable subject matter. To make something is to control its existence. Without control there is no possession and no ownership.

It is the old adage that possession is nine-tenths of the law. Case in point is the 1805 property law case of *Pierson v. Post*.[7] Post did not own the fox he was chasing because he did not have possession of it. Thus when Pierson jumped out of the bushes and shot the fox before Post could get to it, Post had no cause of action, even though he had been cheated out of his trophy (or perhaps his bounty). You cannot own a wild animal you cannot catch.

In translating *Pierson v. Post* to intellectual property law, what does it mean to have control? Some examples may help. You can draw a circle, but you cannot control pi. You can have control over how big the circle is, but you cannot change pi — the relationship between your circle and its diameter. Pi is a principle or law of nature. You can exploit a law of nature, but you cannot change it. You can build and control a nuclear reactor, but you cannot change nuclear energy or make nuclear energy cease to exist. A law of nature is a wild animal that can never be caught. Push an anvil off a cliff, and it falls, irreversibly. You control the push; gravity does the rest.

Think about what Newton and Einstein did. They each changed the way we understand nature. They did not change the way we think. Our thoughts are our own. No matter how brilliant their discoveries were, Newton and Einstein gave us understanding, but not control. Newton taught us calculus to explain why an apple falls from a tree. Calculus or no calculus, the apple always falls. Einstein taught us curved space-time, to explain why an apple falls from a tree. But the apple still always falls. Gravity remains a wild untamed animal, and we are powerless to change it.

Therefore, when confronted with the dilemma of patentable invention or unpatentable principle, ask yourself about the claimed invention, ''Can humanity control this thing? Does humanity have the power to change this thing, or make

---

[6] 447 U.S. 303 (1980).
[7] 3 Caines 175 (N.Y.S. Ct. 1805).

this thing cease to be true or cease to exist?'' If so, then this thing, if new, may be patented, for it qualifies as ''anything under the sun that is made by man.'' If the answer is no, then you have discovered an untamable wild animal, beautiful to look at, perhaps, profoundly empowering, perhaps, but something that no one can own.

[B] Neilson v. Harford

Neilson v. Harford[8] is not a United States Supreme Court decision. It is an important 1841 decision of an English court. It is included here because several U.S. Supreme Court decisions cite it. The invention, the blast furnace, was described as follows: ''a blast or current of air must be produced by bellows or other blowing apparatus, and is to be passed from the bellows, &c, into an air-vessel or receptacle, by means of a tube, pipe, or aperture, into the fire . . . .''[9]

The defendant asserted that the alleged invention is not patentable subject matter because it claims a principle.

The English court upheld the patent, but not without admitting to being troubled by the issue of patentable subject matter. The problem the court faced was that the specification stated that the ''size of the receptacle will depend on the blast necessary.''[10] This description played to the defendant's argument that the invention was not a machine or apparatus, but sought to wholly preempt all use of the air blast principle. ''The only question,'' the court stated, ''is, whether on the whole the inventor has given sufficient information to the public by which the invention can, on the expiration of the term for which the patent is granted, be brought into public use without experiments or expense.''[11]

This case has been cited by the United States Supreme Court in several decisions. It is important because it shows that what you disclose (or fail to disclose) in the specification will bear on whether the invention is patentable subject matter. In other words the concepts of § 112 and § 101 are intimately entwined. This case also demonstrates how overclaiming can get you into trouble. Here the patentee invented a blast furnace, but almost lost the patent by being accused of trying to patent the scientific facts that fire needs oxygen to burn and that hot air heats better than cold air.

Evidently *Neilson v. Harford*[12] created quite a stir within the U.S. patent bar of that era, for the U.S. Supreme Court analyzes this decision carefully in *O'Reilly v. Morse*[13] and *Tilghman v. Proctor.*[14] These cases are discussed in § 3.08[D] and § 3.08[G] respectively.

---

[8] 151 Eng. Rep. 1266 (1841).
[9] *Id.* at 1266.
[10] *Id.* at 1274.
[11] *Id.* at 1272.
[12] 151 Eng. Rep. 1266 (1841).
[13] 56 U.S. 62 (1854).
[14] 102 U.S. 707 (1881).

[C] Le Roy v. Tatham

In *Le Roy v. Tatham*,[15] the patent was for an improvement in the making of pipe. The patent disclosed an apparatus for making pipe, but the essential improvement was that these pipes were made of lead. The lead was formed under heat and pressure, so that it first melted and then cooled, resulting in a superior product. Simply, it was the melting property of lead that made the invention work.

The Supreme Court, with a lengthy dissent by Justice Nelson, held that the invention did not constitute patentable subject matter. The majority emphasized the premise that a principle, in the abstract, is not patentable. According to the Court, "A principle, in the abstract, is a fundamental truth; an original cause; a motive."[16] Explaining the dividing line between patentable inventions and forbidden monopolies, the Court cited the example of a newly discovered form of energy or power:

> Through the agency of machinery a new steam power may be said to have been generated. But no one can appropriate this power exclusively to himself, under the patent laws . . . The same may be said of electricity, and of any other power in nature . . . In all such cases, the processes used to extract, modify, and concentrate natural agencies, constitute the invention. The elements of the power exist; the invention is not in discovering them, but in applying them to useful objects.[17]

Again, the Court looked to the patent specification, to see if the claimed invention was patentable subject matter.

A new property discovered in matter, when practically applied, in the construction of a useful article of commerce or manufacture, is patentable; but the process through which the new property is developed and applied, must be stated, with such precision as to enable an ordinary mechanic to construct and apply the necessary process.[18]

This case reveals an important concept. To be a patentable invention, there must be more than merely the discovery of a property existing in nature. The *disclosure* must show that the inventor is able to exert physical control over the discovered property. The invention must extract, modify, and concentrate a property of nature, not merely identify or name that property of nature.

---

[15] 55 U.S. 156 (1852).
[16] *Id.* at 173.
[17] *Id.*
[18] *Id.*

[D] *O'Reilly v. Morse*

O'Reilly v. Morse[19] involved a Copenhagen scientist named Oersted, who discovered electromagnetism in the winter of 1819-1820. During the decade that followed, several prominent inventors saw Oersted's miraculous new form of energy as a possible way to communicate at a distance. Many tried, but Samuel Morse is credited as the one who first made it happen.[20] Morse's telegraph device solved the problem of how to put electromagnetic energy into one end of a long wire and receive the minuscule amount of energy remaining after the long journey at the other end so that it could be used to print intelligible characters. The U.S. Patent Office granted Morse a patent in 1840 and reissued it in 1948.

Morse brought suit on his patent, and the case rose to the Supreme Court, primarily because of patent claim 8, which read:

> I do not propose to limit myself to the specific machinery or parts of machinery described in the foregoing specification and claims; the essence of my invention being the use of the motive power of the electric or galvanic current, which I call electromagnetism, however developed, for marking or printing intelligible characters, signs or letters, at any distances, being a new application of that power of which I claim to be the first inventor or discoverer.[21]

The Supreme Court upheld the first seven claims of Morse's patent (including one that might be the first software patent); however, claim 8 was simply too broad, and the Supreme Court struck it down. The Court cited, with approval, the English case, *Neilson v. Harford*[22] (discussed in § 3.08[B]) and questioned whether Morse had sought claim 8 because he had misconstrued that case.[23] The Supreme Court contrasted the Neilson's claim with Morse's:

> Undoubtedly, the principle that hot air will promote the ignition of fuel better than cold, was embodied in this machine [of Neilson]. But the patent was not supported because this principle was embodied in it . . . his [Neilson's] patent was supported because he had invented a mechanical apparatus, by which a current of hot air, instead of cold, could be thrown in. . . .
>
> * * *
>
> But Professor Morse has not discovered that the electric or galvanic current will always print at a distance, no matter what may be the form of the machinery or mechanical contrivances through which it passes.[24]

---

[19] 56 U.S. 62 (1854).

[20] European readers may not agree that Morse was the first. Steinheil, Wheatstone, and Davy also lay claim to being first.

[21] 56 U.S. at 112.

[22] 151 Eng. Rep. 1266 (1841).

[23] 56 U.S. at 117.

[24] *Id.* at 117-118.

In essence, the Supreme Court struck down Morse's claim 8 because it claimed more than Morse had truly invented. To make this point, the Court noted:

> If this claim can be maintained, it matters not by what process or machinery the result is accomplished. For ought that we now know some future inventor, in the onward march of science, may discover by means of the electric or galvanic current, without using any part of the process or combination set forth in the plaintiff's specification. His invention may be less complicated — less liable to get out of order — less expensive in construction, and in its operation. But yet if it is covered by this patent the inventor could not use it, nor the public have the benefit of it without the permission of this patentee.[25]

[E] *Rubber Tip Pencil Co. v. Howard*

You are no doubt familiar with the rubber eraser that slips over the end of a pencil. James E. Blair got a patent on that in 1869. Blair had not discovered that rubber will erase pencil marks. That discovery had already been made. Blair's invention was claimed as "a new article of manufacture, an elastic, erasive pencil head, made substantially as in the manner described."[26]

The problem was that Blair's claim could only be interpreted in light of the teachings in his specification; and his specification left everything open to design choice. For instance, the eraser could be of rubber or rubber and some other material. Its shape could be any convenient external form. It could have a socket to receive one end of a pencil, and that socket could be of any shape. And so forth.

In the patent infringement litigation, the defense offered into evidence a rubber nipple, not designed to erase pencil marks, but a device which did fit onto the end of a pencil and that one witness testified he had used to erase pencil marks.

The Court took little time in ruling Blair's patent invalid, stating "[t]he idea of this patentee was a good one, but his device to give it effect, though useful, was not new. Consequently, he took nothing by his patent."[27]

Today the Patent Office will not allow omnibus claims, directed as Blair's claim was, to a device "made substantially as described."[28] Nevertheless, the error Blair made is still made today. Trying to cover every possible embodiment of his idea, he wrote his specification like a broad, open-ended claim. The result: "He took nothing by his patent."[29]

---

[25] *Id.* at 112-113.

[26] Rubber Tip Pencil Co. v. Howard, 87 U.S. 489 (1874).

[27] *Id.* at 507.

[28] *Id.* Biotechnology claims are a possible exception.

[29] *Id.* at 507.

[F] *Cochrane v. Deener*

In *Cochrane v. Deener*,[30] the claimed invention was an improved process for making flour consisting of separating from the meal, first the superfine flour and then middlings meal. The middlings meal, when reground, was mixed with the superfine flour to produce a product of improved quality. During a patent infringement lawsuit, the defendants argued that the claimed invention was not patentable subject matter, because it was simply a process and was not limited to any arrangement of machinery.

The Supreme Court held that "a process may be patentable, irrespective of the particular form of the instrumentality's used."[31] This case is important because it decides that processes may be patentable subject matter. The law prior to this decision was not clear on that point. Many later cases build on this decision.

[G] *Tilghman v. Proctor*

In *Tilghman v. Proctor*,[32] the invention was a treatment of fats and oils used in such products as margarine. The invention was a process of separating the fats and oils into their component parts using water action at high temperature and pressure. In the patent infringement litigation, the defendants argued that they were not using the apparatus or the high temperatures described in the patent. Defendant's process used heated water under pressure, but at a lower temperature and mixed with lime (lime was used in another prior art technique). The defendants argued that to construe the patent to cover them gave the plaintiff a patent on the principle of using heated water under pressure.

The Supreme Court addressed the issue head on, asking, "What did Tilghman discover?" The Court reasoned that:

> [H]ad the process been known and used before and not been Tilghman's invention, he could not then have claimed anything more than the particular apparatus described in his patent; but being the inventor of the process, as we are satisfied was the fact, he was entitled to claim it in the manner he did.[33]

Thus, although in his patent Tilghman recommends a high degree of heat, the Court did not limit him to the temperature stated and defendants were found to infringe.

In reaching that decision, the Supreme Court had to decide whether Tilghman's discovery was a patentable process or an unpatentable principle. The Court resolved that question by looking at the scope of Tilghman's claim. The Court pointing out that Tilghman "only claims to have invented a particular

---

[30] 94 U.S. 780 (1877).
[31] *Id.* at 787.
[32] 102 U.S. 707 (1881).
[33] *Id.* at 721-722.

mode of bringing about the desired chemical union between the fatty elements and water. He does not claim every mode of accomplishing this result."[34] To bolster its reasoning, the Court pointed out several other known ways of accomplishing the result of Tilghman's process, for example, lime saponification, and sulfuric acid distillation.

Tilghman contains a valuable practice tip for drafting business process patents (or any patent involving a process). When drafting broad claims, keep the claim scope in context with other prior art techniques. Then you will be able to demonstrate, as Tilghman did, that the claim does not cover an unpatentable principle because other prior technologies also employ the principle.

[H] *Expanded Metal Co. v. Bradford*

If you frequent building supply stores you have probably seen expanded sheet metal screen or lath with diamond-shaped openings made from a single sheet of metal. If you have not, expanded sheet metal screen may be generally described as metal openwork, held together by uncut portions of the metal, and constructed by making cuts or slashes in the metal and then opening or expanding the cuts to form a latticework. The claimed invention in *Expanded Metal Co. v. Bradford*[35] was a method of making expanded sheet metal screens.

The case rose to the Supreme Court on two conflicting *writs of certiorari*. The Third Circuit had ruled the patent invalid; the Sixth Circuit had ruled the patent valid. At issue was the fundamental question of whether mechanical processes are patentable subject matter. In prior decisions the patentable processes had involved chemical action. Unlike the patent statute today, the patent statute of that era did not include express language that a process was patentable subject matter.

The Supreme Court explained, "the word 'process' has been brought into the decisions because it is supposedly an equivalent form of expression, or included in the statutory designation of a new and useful art."[36] The Court had little difficulty holding that processes of all types, not just chemical processes, are patentable subject matter. As part of its rationale, the Court cited the following language from *Tilghman:*

> A machine is a thing. A process is an act, or a mode of acting. The one is visible to the eye, — an object of perpetual observation. The other is a conception of the mind, — seen only by its effects when being executed or performed. Either may be the means of producing a useful result.[37]

---

[34] *Id.* at 729.
[35] 214 U.S. 366 (1909).
[36] *Id.* at 384.
[37] *Id.* at 382 (citing Tilghman v. Proctor, 102 U.S. 707, 728 (1881)).

[I] *MacKay Radio & Telegraph Co. v. Radio Corp. of America*

In *MacKay Radio & Telegraph Co. v. Radio Corp. of America*,[38] the invention was a radio antenna. The antenna was shaped in the form of a ''V,'' at an angle described by a mathematical formula. The V-shape gave the antenna a desirable directional property. The patentee did not invent the formula. It had been discovered and published in a scientific journal 30 years before. The Court stated that the antenna was patentable subject matter, notwithstanding the mathematical language stating:

> While a scientific truth, or the mathematical expression of it, is not patentable invention, a novel and useful structure created with the aid of knowledge of scientific truth may be.[39]

The Court narrowly construed the claim and found that the patent was not infringed. This case is occasionally cited for the proposition that a mathematical expression is not patentable. More correctly, it is a scientific truth, or the mathematical expression of a scientific truth, that is not patentable. Keep in mind that mathematics is simply a very precise language or means of describing something. It can be used to describe scientific truths and features of patentable inventions alike. As this case demonstrates, if a mathematical expression is used to describe a useful structure created with the aid of knowledge of the scientific truth, the useful structure is patentable subject matter.

[J] *Funk Bros. v. Kalo Co.*

In this case the Supreme Court declared the patent was invalid. The patentee had discovered that the nitrogen fixing ability of leguminous plants could be improved by mixing certain naturally occurring strains of different species of root-nodule bacteria and inoculating the plants with the mixture. The strains were selected because they could coexist, without inhibiting the effect of each other as other strains would do.

The Court held the patent invalid, stating:

> He who discovers a hitherto unknown phenomenon of nature has no claim to a monopoly of it which the law recognizes. If there is to be invention from such a discovery, it must come from the application of the law of nature to a new and useful end.[40]

Addressing whether this discovery qualified as an application of the law of nature, the Court ruled against the patentee:

---

[38] 306 U.S. 86 (1939).
[39] *Id.* at 94.
[40] Funk Bros. v. Kalo Co., 333 U.S. 127, 130 (1948).

But however ingenious the discovery of that natural principle may have been, the application of it is hardly more than an advance in the packaging of the inoculates. . . .

* * *

The combination of species produces no new bacteria, no change in the six species of bacteria, and no enlargement of the range of their utility. Each species has the same effect it always had.[41]

§ 3.09   Computer Age Supreme Court Decisions

By the time of the *Funk Bros.* case, the Supreme Court had fairly well resolved that processes would be patentable, which certainly opened the door for business process patents. What remained to be resolved, however, was to what extent information, such as computer software information, could be the subject of a patent. As we shall see, the legal system struggled with this issue for years, several times asking the Supreme Court for guidance.

[A]   *Gottschalk v. Benson*

Gottschalk v. Benson[42] is the first Supreme Court decision to address the patentability of computer software, and the patent was struck down as being unpatentable subject matter. Decided in 1972, during the reign of the IBM 360/375 mainframe computer, many viewed this decision as a condemnation of all software patents. It was not. The Supreme Court stated, ''It is said that the decision precludes a patent for any program servicing a computer. We do not so hold.''[43] However few took this statement to heart, for during the decade that followed *Gottschalk v. Benson* virtually no one sought software patents. No doubt the anti-software patent attitude within the Patent Office at the time contributed.

The *Benson* case found itself in the Supreme Court because Acting Commissioner of Patents, Robert Gottschalk, petitioned on behalf of the U.S. Patent Office for a *writ of certiorari*. The Patent Office had rejected claims 8 and 13 of Benson's application and the Court of Customs and Patent Appeals reversed, Judge Rich writing the opinion.[44]

[1]   Numbering Systems in *Gottschalk v. Benson*

The invention in *Gottschalk v. Benson*[45] is a process for converting binary coded decimal numbers (BCD) into pure binary numbers. Why the big fight over this rather esoteric subject? One possible explanation is this. Computers use binary numbers. Many electronic devices use BCD. For example, most of the

---

[41] *Id.* at 131.
[42] 409 U.S. 63 (1972).
[43] *Id.* at 71.
[44] *In re* Benson, 441 F.2d 682 (C.C.P.A. 1971).
[45] 409 U.S. 63 (1972).

digital readouts found in calculators, digital clocks, microwave ovens, and VCRs internally use BCD numbers. Thus for a computer or microprocessor to read these devices, it needs to be able to convert BCD numbers into binary numbers. Here is what you need to know about these numbering systems in order to follow *Benson.*

People count in base ten. The Arabic numbering system that the entire world uses is a base ten numbering system. Why base ten? Probably because people have ten digits on their two hands and naturally learned to count using these digits.

Computers, on the other hand, count in base two. The binary numbering system is a base two numbering system. Why base two? Simply, computers do not have ten fingers. Computers have only electrical impulses, which can be either turned ON or turned OFF. It is as if computers have only two fingers — ON and OFF. Thus computers must count using only these two digits.

Binary numbers, while easy for computers, are difficult for people to read. The number 53 (in base ten) looks like 110101 (in base two). Imagine trying to balance your checkbook using the binary numbering system. It would be next to impossible. Thus, it is clear that you need a way to convert back and forth between decimal (base ten) numbers and binary (base two) numbers. That conversion, however, is not what *Benson* is about. *Benson* involves something more. It involves converting from "binary coded decimal" numbers to binary numbers. Thus you need to know about the binary coded decimal numbering system.

The binary coded decimal (or BCD) numbering system is a base ten system. Only it uses *binary digits* as replacements for the standard Arabic digits with which you are familiar. The best way to understand BCD is to see an example. The number 53 (in decimal numbers) is represented as 0101 0011 (in BCD). To understand what is going on here, note that in a decimal system the number 53 is made up of two side-by-side digits, a 5 and a 3. If you replace the Arabic digit 5 with its binary equivalent 0101, and if you replace the Arabic digit 3 with its binary equivalent 0011, and if you place those substituted digits side-by-side, you get 0101 0011, which is the BCD equivalent for the number 53.

[2] Claims at Issue in *Gottschalk v. Benson*

At issue in *Gottschalk v. Benson*[46] were two claims, differently phrased, but covering essentially the same thing:

8. The method of converting signals from binary coded decimal form into binary which comprises the steps of
    (1) storing the binary coded decimal signals in a re-entrant shift register,

---

[46] *Id.*

(2) shifting the signals to the right by at least three places, until there is a binary '1' in the second position of said register,

(3) masking out said binary '1' in said second position of said register,

(4) adding a binary '1' to the first position of said register,

(5) shifting the signals to the left by two positions,

(6) adding a '1' to said first position, and

(7) shifting the signals to the right by at least three positions in preparation for a succeeding binary '1' in the second position of said register.

13. A data processing method for converting binary coded decimal number representations into binary number representations comprising the steps of

(1) testing each binary digit position '1,' beginning with the least significant binary digit position, of the most significant decimal digit representation for a binary '0' or a binary '1';

(2) if a binary '0' is detected, repeating step (1) for the next least significant binary digit position of said most significant decimal digit representation;

(3) if a binary '1' is detected, adding a binary '1' at the $(i + 1)$th and $(i + 3)$th least significant binary digit positions of the next lesser significant decimal digit representation, and repeating step (1) for the next least significant binary digit position of said most significant decimal digit representation;

(4) upon exhausting the binary digit positions of said most significant decimal digit representation, repeating steps (1) through (3) for the next lesser significant decimal digit representation as modified by the previous execution of steps (1) through (3); and

(5) repeating steps (1) through (4) until the second least significant decimal digit representation has been so processed.[47]

Although claim 8 recites use of a ''re-entrant shift register'' and claim 13 recites a ''data processing'' method, you can practice both methods with pencil and paper. Try it yourself. Try following claim 13 to convert BCD equivalent of 53 into pure binary:

0101 0011; BCD for 53

0110 101; binary for 53

[3] Holding in *Gottschalk v. Benson*

Justice Douglas wrote the opinion of the Court. He frames the issue as ''whether the method described and claimed is a 'process' within the meaning of

---

[47] *Id.* at 73.

the Patent Act."[48] Previous Supreme Court decisions held that the discovery of a new and useful process was patentable subject matter; whereas the discovery of a new and useful principle was not.[49] For background on some of these decisions, see preceding § 3.08[C], [D], and [I].

In contrast the statutory backdrop to some of the earlier Supreme Court decisions on this issue, the Patent Act of 1952 contains an express definition of the term process. "The term 'process' means process, art or method, and includes a new use of a known process, machine, manufacture, composition of matter, or material."[50] Under prior patent acts, the term "process" was not defined, but was nevertheless treated as patentable subject matter as an equivalent form of expression, or included in the statutory designation of a new and useful art."[51]

Reviewing several of the Supreme Court's prior decisions on the patentability issue, Justice Douglas, in a single paragraph, found the BCD conversion routine did not pass muster:

> It is conceded that one may not patent an idea. But in practical effect that would be the result if the formula for converting BCD numerals to pure binary numerals were patented in this case. The mathematical formula involved here has no substantial practical application except in connection with a digital computer, which means that if the judgment below is affirmed, the patent would wholly pre-empt the mathematical formula and in practical effect would be a patent on the algorithm itself.[52]

The Court thus concluded that the claimed invention was not a "process" within the meaning of the Patent Act. The claimed method was so abstract as to cover both known and unknown uses of the BCD to binary conversion.

The decision was read by many as a prohibition of software patents. Certainly the Court's decrying that this is an attempt to obtain "a patent on the algorithm itself"[53] promoted this reading. After all, reasoned many, the Court has in effect held that an algorithm is not patentable, and what is software but an algorithm? This reading was made further plausible by the Court's closing remarks:

> It may be that the patent laws should be extended to cover these programs, a policy matter to which we are not competent to speak. The President's Commission on the Patent System rejected the proposal that these programs be patentable.[54]

---

[48] *Id.* at 64.

[49] MacKay Radio & Tel. Co. v. Radio Corp. of Am., 306 U.S. 86 (1939); Rubber Tip Pencil Co. v. Howard, 87 U.S. 498 (1874); Le Roy v. Tatham, 55 U.S. 156 (1852).

[50] 35 U.S.C. § 100(b).

[51] Expanded Metal Co. v. Bradford, 214 U.S. 366, 382 (1909).

[52] 409 U.S. at 71.

[53] *Id.*

[54] *Id.* at 72 (citing To Promote the Progress of . . . Useful Arts, Report on the President's Comm'n on the Pat. Sys. (1966)).

The President's Commission Report, to which the Court refers, was drafted by a commission appointed by President Lyndon B. Johnson to study all aspects of the patent system. It makes interesting reading on a variety of patent law issues — recommending massive changes in numerous areas. The primary reason the President's Commission did not advocate patentability of software was that the Patent Office, in the Commission's view, could not examine applications because it lacked a reliable classification and searching technique.[55]

[4] Analyzing *Gottschalk v. Benson*

Benson is often misunderstood. Many cite it for the proposition that a mathematical algorithm cannot be patented, and then jump to the erroneous conclusion that anything claimed in the language of mathematics is therefore unpatentable. *Benson* does not stand for this. Rather, *Benson* simply follows existing Supreme Court precedent that:

> Phenomena of nature, though just discovered, mental processes, and abstract intellectual concepts are not patentable, as they are the basic tools of scientific and technological work.[56]

If you will accept the notion that numbers and numbering systems have different properties (for example, commutative and distributive properties) and different innate characteristics (for example, rational, irrational, prime), then you are on your way to understanding *Gottschalk v. Benson*. These properties and characteristics of numbers and numbering systems exist, and have always existed, like any other law of nature. You can discover these properties and characteristics, and teach them to your children, but you are powerless to change them. For instance, try changing pi to a rational number and see if your circles still have 360 degrees.

When properties and characteristics of numbers and numbering systems are seen as phenomena of nature, *Benson* makes perfect sense. Nobody can claim a monopoly over all uses, now known or later discovered, of a phenomenon of nature. Viewing the decision in this way demonstrates that *Benson* is consistent with later decisions upholding the patentability of software.

[B] *Parker v. Flook*

Parker v. Flook[57] is the second software patent decision of the Supreme Court. Decided in 1978, six years after *Benson*, *Parker v. Flook* represented, at the time, another blow to the viability of the software patent. However, as it had

---

[55] See Ch. 9 for a discussion of how the Patent Office currently classifies and searches business-related prior art.

[56] 409 U.S. at 67.

[57] 437 U.S. 584 (1978).

in *Benson*, the Court made it clear that it was not condemning all software patents:

> Neither the dearth of precedent, nor this decision, should therefore be interpreted as reflecting a judgment that patent protection of certain novel and useful computer programs will not promote the progress of science and the useful arts, or that such protection is undesirable as a matter of policy.[58]

However, while the Court did not condemn all software patents, its decision in *Parker* did go quite far in that direction:

> Very simply, our holding today is that a claim for an improved method of calculation, even when tied to a specific end use, is unpatentable subject matter under § 101.[59]

If the "method of calculation" in the above language refers to a mathematical principle of nature, then *Parker* forbids patentability of no new territory. Mathematical principles of nature, like abstract ideas, cannot be patented. However, if the "method of calculation" refers more broadly to any data processing method, involving Boolean logic calculations for example, then the viability of the software patent is severely threatened. For example, what if the invention is a method of parallel processing, involving collections of calculating steps designed to allow multiple processors to work together as one? Is this method of facilitating multiple processors to act as one unpatentable because it is a mere "method of calculation"?

[1] Technology in *Parker v. Flook*

The catalytic chemical conversion of hydrocarbons is a process widely used in the petrochemical industry. With the aid of acidic catalysts, the catalytic chemical conversion process decomposes large hydrocarbons, originally from crude oil, into smaller hydrocarbons for use as gasoline. During the conversion process a number of process variables (temperature, pressure, and flow rate) must be closely monitored. If these variables stray too far from proper values, an alarm must sound. Importantly, the alarm must sound if conditions are becoming dangerous (for example, pressure or temperature too high). The alarm must also sound if the process is operating far below acceptable efficiency (for example, pressure or temperature too low).

Monitoring the catalytic conversion process would seem a rather simple task of comparing the measured temperature, pressure, and flow rates with predetermined alarm limit values. If the measured parameters stray too far from the alarm limit values, the alarm is sounded. Naturally, there would be a set of

---

[58] *Id.* at 595.
[59] *Id.* at n. 19.

upper alarm limits to signal danger, and a set of lower alarm limits to signal inefficiency.

However, the task is more complicated than this. Because petrochemical processes cannot be guaranteed to operate in a steady state, the value of the proper alarm limit itself may fluctuate. In other words, one fixed set of alarm limits, (predetermined in advance for use under all operating conditions) is not good enough. Rather, a range of alarm limits is needed, one set for steady state operation, and another set for transient or start-up operation. To work this way, petrochemical engineers must update the alarm limits, in effect, choosing the proper set of alarm limits for the current mode of operation.

Flook's patent application described a method of updating the alarm limits, consisting essentially of three steps: an initial step which merely measures the present value of the process variable (for example, temperature), an intermediate step which uses an equation or algorithm to calculate an updated alarm-limit value, and a final step which adjusts the actual alarm limit to the updated value.

The Patent Examiner rejected Flook's application. He found that the mathematical formula was the only difference between Flook's claims and the prior art and therefore a patent on this method "would be in practical effect a patent on the formula or mathematics itself."[60] The Board of Appeals agreed with the Examiner and sustained the Examiner's rejection. The Court of Customs and Patent Appeals saw the matter differently and reversed.

Acting Commissioner of Patents, Lutrelle F. Parker, filed a petition for a *writ of certiorari*, urging that "the decision of the Court of Customs and Patent Appeals will have a debilitating effect on the rapidly expanding computer 'software' industry, and will require him to process thousands of additional patent applications."[61] The Supreme Court granted *cert*.

[2] Claims at Issue in *Parker v. Flook*

At issue, claim 1 of the patent describes the method as follows:

1. A method for updating the value of at least one alarm limit on at least one process variable involved in a process comprising the catalytic chemical conversion of hydrocarbons wherein said alarm limit has a current value of
$$Bo + K$$
wherein Bo is the current alarm base and K is a predetermined alarm offset which comprises:

   (1) Determining the present value of said process variable, said present value being defined as PVL;

   (2) Determining a new alarm base B_1, using the following equation:
$$B_1 = Bo(1.0 - F) + PVL(F)$$

---

[60] *Id.* at 587.
[61] *Id.* at 587-588.

where F is a predetermined number greater than zero and less than 1.0;
(3)   Determining an updated alarm limit which is defined as $B_1 + K$; and thereafter
(4)   Adjusting said alarm limit to said updated alarm limit value.[62]

Flook's method can be practiced using pencil and paper. In order to use Flook's method for computing a new limit, the operator must make four decisions. Based on individual knowledge of normal operating conditions, the operator first selects the original "alarm base" (Bo); if a temperature of 400 degrees is normal, that may be the alarm base. The operator must next decide on an appropriate margin of safety, perhaps 50 degrees; that is the "alarm offset" (K). The sum of the alarm base and the alarm offset equals the alarm limit.

Then the operator decides on the time interval that will elapse between each updating; that interval has no effect on the computation although it may, of course, be of great practical importance. Finally, the operator selects a weighting factor (F), which may be any number between 99 percent and 1 percent (a number greater than 0, but less than 1), and that is used in the updating calculation.

If the operator has decided in advance to use an original alarm base (Bo) of 400 degrees, a constant alarm offset (K) of 50 degrees, and a weighting factor (F) of 80 percent, the only additional information needed in order to compute an updated alarm limit value (UAV), is the present value of the process variable (PVL). Flook's application does not explain how to select the appropriate margin of safety, the weighting factor, or any of the other variables.

The computation of the updated alarm limit according to Flook's method involves these three steps:

1.   First, at the predetermined interval, the process variable is measured; assuming the temperature is then 425 degrees, PVL will then equal 425.
2.   Second, the solution of Flook's formula will produce a new alarm base (B_1) that will be a weighted average of the preceding alarm base (Bo) of 400 degrees and the current temperature (PVL) of 425. It will be closer to one or the other depending on the value of the weighting factor (F) selected by the operator. If F is 80 percent, that percentage of 425 (340) plus 20 percent (1 − F) of 400 (80) will produce a new alarm base of 420 degrees.
3.   Third, the alarm offset (K) of 50 degrees is then added to the new alarm base (B_1) of 420 to produce the updated alarm limit (UAV) of 470.

The process is repeated at the selected time intervals. In each updating computation, the most recently calculated alarm base and the current measurement of the process variable will be substituted for the corresponding numbers in

---

[62] *Id.* at 596-597.

the original calculation, but the alarm offset and the weighting factor will remain constant.

[3] Holding of *Parker v. Flook*

In *Benson*, the Supreme Court held a mathematical formula is like a law of nature; the patent law cannot allow anyone to wholly preempt a mathematical formula or law of nature. In ruling on Flook's application, the Court of Customs and Patent Appeals read *Benson* to apply only where *all* uses of a mathematical formula or law of nature are preempted. Only then is the subject matter nonstatutory. Flook's claim does not seek to preempt all uses of his mathematical formula — there are unclaimed uses of Flook's formula outside the petrochemical industry that remain in the public domain. Hence, the Court of Customs and Patent Appeals reasoned that Flook's claims did recite patentable subject matter. The Supreme Court did not agree.

Justice Stevens delivered the opinion of the Court. In his opinion, he quickly acknowledges that *Flook* does not seek to cover every conceivable application of the formula. However, that does not put an end to the matter, as the Court of Customs and Patent Appeals had assumed. The only novel feature of Flook's method, Justice Stevens notes, is the mathematical formula. Hence, the claim is still tantamount to a patent on the mathematical formula.

Flook had argued that the presence of specific "post solution" activity — the adjustment of the alarm limit — distinguishes the case from *Benson* and makes his process patentable. Justice Stevens rejected this argument, writing:

> The notion that post-solution activity, no matter how conventional or obvious in itself, can transform an unpatentable principle into a patentable process exalts form over substance.[63]

The analysis Justice Stevens uses is interesting, although somewhat confusing. He writes at the outset that the case turns entirely on the proper construction of § 101 of the Patent Act, which describes the subject matter that is eligible for patent protection. He writes that the case does not involve issues of novelty and obviousness that arise under §§ 102 and 103. He further writes that the Court's approach "is not at all inconsistent with the view that a patent claim must be considered as a whole."[64]

Justice Stevens proceeds to focus on Flook's mathematical formula (which he assumes to be novel) but nevertheless treats it as though it were a familiar part of the prior art — hence incapable of imparting novelty to the claimed invention. Treating the formula as prior art, he disposes of the entire claim almost as if performing an obviousness analysis under 35 U.S.C. § 103.

---

[63] *Id.* at 590.
[64] *Id.* at 594.

The question is whether the discovery of this feature [the mathematical formula] makes an otherwise conventional method eligible for patent protection. . . .[65]

* * *

Respondent's process is unpatentable under § 101, not because it contains a mathematical algorithm as one component, but because once that algorithm is assumed to be within the prior art, the application, considered as a whole, contains no patentable invention.[66]

The analysis used by Justice Stevens — treating the mathematical formula as prior art — is not original to *Parker v. Flook.* The analysis is derived from the decision in *O'Reilly v. Morse,*[67] which is in turn borrowed from the English decision *Neilson v. Harford,* in which the court said, "we think the case must be considered as if, the principle being well known, the plaintiff had first invented a mode of applying it . . . ."[68]

[4] Analyzing *Parker v. Flook*

Parker stands for three propositions.[69] First, the *Benson* rule of unpatentable subject matter is not limited to claims that *wholly* preempt a principle of nature, such as a principle of nature expressed as a mathematical formula. Form over substance distinctions, such as insignificant post-solution activity, will not save a claim when the only novelty lies in the unpatentable principle. Second, an improved method of calculation, even when employed as part of a physical process, is not patentable subject matter under § 101. This proposition is best understood when you consider a method of calculation to be a mathematical calculation expressing a principle of nature. Care must be taken, however, not to equate the term calculation with the result of all operations performed by a computer. There are numerous computer operations that can result in solutions to problems that do not involve preempting a principle of nature. For example, a software air traffic control system may employ numerous computer operations. Simply characterizing the air traffic control system as a calculation does not render the subject matter unpatentable.

Third, a principle of nature or mathematical formula is treated for § 101 purposes as though it were a familiar part of the prior art; the claim is then examined to determine whether it discloses some other inventive concept. This is an analysis technique. It is a way of testing whether the claim as a whole recites a patentable invention. The technique originates from the 1841 English case

---

[65] *Id.* at 588.

[66] *Id.*

[67] 56 U.S. 62, 115 (1854).

[68] 151 Eng. Rep. 1266 (1841).

[69] According to J. Stevers in dissenting opinion in Diamond v. Diehr, 450 U.S. 175, 204 (1981).

Neilson v. Harford.[70] In that case the principle "that hot air will promote the ignition of fuel better than cold" was treated as well known. The court in that case found the invention was not this well-known principle, but rather a mechanical mode of applying the principle to furnaces.

In this context Flook's invention is different from Neilson's blast furnace. *Parker v. Flook* teaches only the formula and suggests its possible application to the catalytic conversion process. He does not teach how to apply it to the catalytic conversion process, leaving to the human operator the problem of figuring out how to measure the variables, and how to select the appropriate margin of safety, the weighting factor, or any of the other variables.

Parker v. Flook gives you an important practice tip. Make sure the specification contains a full disclosure, not only of the principle or formula involved, but of how to apply the formula or principle to a specific problem. When a mathematical formula is involved, be sure to disclose how to measure the variables and how to select appropriate values of any constants, multiplier factors, and so forth.

Another lesson of *Parker v. Flook* is that, when possible, avoid claiming a software process or algorithm such that the result of the process is simply a number. In *Parker v. Flook* the "alarm limit" is simply a number, which does not in any way alter or control a physical property, or transform or reduce an article to a different state or thing. Flook's attempt to render patentable the computation of this number, by restricting the claim to a field of use, did him no good. Had *Parker v. Flook* tied the calculation of the alarm limit to a step that controls a petrochemical process, the claim would have recited statutory subject matter. See for example, *Diamond v. Diehr*,[71] discussed in § 3.09[D].

[C] *Diamond v. Chakrabarty*

Decided by the Supreme Court on June 16, 1980, *Diamond v. Chakrabarty*[72] holds that a human-made microorganism is patentable subject matter, constituting a "manufacture" or "composition of matter." You may wonder what this case has to do with patentability of software. Admittedly it does not have to do with software, but it does have everything to do with what is patentable subject matter. Chief Justice Burger, delivering the opinion of the Court, writes, "Congress intended statutory subject matter to 'include anything under the sun that is made by man.'"[73] Certainly, computer software, like human-made microorganisms, basks under the same sun.

---

[70] 151 Eng. Rep. 1266 (1841).

[71] 450 U.S. 175 (1981).

[72] 447 U.S. 303 (1980).

[73] *Id.* at 309.

[1] Biotechnology in *Chakrabarty*

Chakrabarty, a microbiologist, discovered how to genetically alter the Pseudomonas bacterium with plasmids to digest oil spills. Plasmids are hereditary units physically separate from the chromosomes of the cell. In prior research, Chakrabarty and an associate discovered that plasmids control the oil degradation abilities of certain bacteria. In particular, the two researchers discovered plasmids capable of degrading camphor and octane, two components of crude oil. In the work represented by Chakrabarty's patent application, Chakrabarty discovered a process by which four different plasmids, capable of degrading four different oil components, could be transferred to and maintained stably in a single Pseudomonas bacterium, which itself has no capacity for degrading oil.

Prior to Chakrabarty's discovery, the biological control of oil spills required the use of a mixture of naturally occurring bacteria, each capable of degrading one component of the oil complex. In this way, oil is decomposed into simpler substances that can serve as food for aquatic life. However, for various reasons, only a portion of any such mixed culture survives to attract the oil spill. By breaking down multiple components of oil, Chakrabarty's genetically altered microorganism promises more efficient and rapid oil spill control.

[2] Claims at Issue in *Chakrabarty*

Chakrabarty presented three types of claims: process claims for the method of producing the bacteria, claims for an inoculum comprised of a carrier material floating on water (such as straw) and the new bacteria, and claims to the bacteria themselves. The bacteria claims read:

> [A] bacterium from the genus Pseudomonas containing therein at least two stable energy-generating plasmids, each of said plasmids providing a separate hydrocarbon degradative pathway.[74]

The Patent Examiner allowed the claims in the first two categories, but rejected the claims for the bacteria. The examiner reasoned that the bacteria are "products of nature" and "living things," and for both reasons they are not patentable subject matter. The Board of Appeals affirmed; the Court of Customs and Patent Appeals, by a divided vote, reversed; and acting Commissioner of Patents, Sidney A. Diamond, petitioned for *certiorari*.

[3] Holding in *Chakrabarty*

Chief Justice Burger delivered the opinion of the Court, which affirmed the Court of Customs and Patent Appeals, in a 5 to 4 decision. "The question before us in this case," Chief Justice Burger writes, "is a narrow one of statutory

---

[74] *Id.* at 305.

interpretation requiring us to construe 35 USC § 101."[75] He writes, "Specifically we must determine whether respondent's micro-organism constitutes a 'manufacture' or 'composition of matter' within the meaning of the statute."[76]

Chief Justice Burger draws on the legislative history of the patent law of the United States, referring to the Patent Act of 1793, authored by Thomas Jefferson. In that early statute, Burger notes, Jefferson defined statutory subject matter as "any new and useful art, machine, manufacture, or composition of matter, or any new or useful improvement [thereof]."[77] The Act embodied Jefferson's philosophy that "ingenuity should receive a liberal encouragement."[78]

When the patent laws were revised in 1836, in 1870, and in 1874, Jefferson's language remained intact. In 1952, when Congress enacted the present Patent Act, Jefferson's language was changed only slightly — the word "process" replaced the word "art." By citing Thomas Jefferson, the Court clearly adopts an expansive view of what is patentable subject matter. However, the quote most often attributed to *Diamond v. Chakrabarty* comes not from Jefferson, but from P. J. Federico, the principal draftsperson of the 1952 Act:

> [U]nder section 101 a person may have invented a machine or a manufacture, which may include anything under the sun that is made by man . . .[79]

Chief Justice Burger notes that § 101 is not without its limits: "[t]he laws of nature, physical phenomena, and abstract ideas have been held not patentable."[80] However, Chakrabarty's genetically altered microorganism does not exist in nature. Hence Chakrabarty's claim, the Chief Justice reasoned, "is not to a hitherto unknown natural phenomenon, but to a non-naturally occurring manufacture or composition of matter — a product of human ingenuity."[81]

[4] *Chakrabarty's* Impact on *Benson* and *Parker v. Flook*

The decision in *Chakrabarty* does not overrule *Gottschalk v. Benson* or *Parker v. Flook*. To the contrary, *Chakrabarty* states that both of these prior precedents still stand, defining the limits of § 101. These prior cases both stand for the proposition that laws of nature, physical phenomena, and abstract ideas are not patentable. Nevertheless, the expansive view that "anything under the sun" that is made by woman or man is patentable subject matter, clearly removes the stigma implied by *Benson* and *Parker v. Flook* that software may not be patentable.

---

[75] *Id.* at 307.

[76] *Id.*

[77] Act of Feb. 21, 1793, § 1, 1 Stat. 319 (1793).

[78] 5 *Writings of Thomas Jefferson* 75-76 (1871).

[79] *Hearings on H.R. 3760 before Subcomm. No. 3 of the House Comm. on the Judiciary*, 82d Cong. 1st Sess. 37 (1951) (statement of P. J. Federico).

[80] 447 U.S. at 309.

[81] *Id.*

Very little must be said to square the decision in *Chakrabarty* with that in *Gottschalk v. Benson*, which simply held that things existing in nature (even newly discovered things), including mathematical formulas, are not patentable.

Parker v. Flook requires more explaining because in that decision Justice Stevens states that the judiciary "must proceed cautiously when . . . asked to extend patent rights into areas wholly unforeseen by Congress."[82] Chief Justice Burger, in the *Chakrabarty* decision, considers this cautionary statement, but does not read it to mean that inventions in areas not contemplated by Congress when the patent laws were enacted are unpatentable per se.[83] However, *Chakrabarty* was a 5 to 4 decision, and it is clear that the dissent (Justices Brennan, White, Marshall, and Powell) would have left the patentability of human-engineered lifeforms to Congress.

All in all, *Diamond v. Chakrabarty* breaks important new ground, for it opens the patent system to the cutting edge technology of the coming decade. Just as the English decision, *Neilson v. Harford*,[84] first shocked the patent world and then spread throughout it, so too will *Diamond v. Chakrabarty*. As seen by the Supreme Court's decision in *Diamond v. Diehr*,[85] discussed in § 3.09[D], the software community has already felt *Chakrabarty*'s impact.

[D] *Diamond v. Diehr*

Diamond v. Diehr[86] is the first Supreme Court decision to clearly open the door for software patents, although prior Supreme Court decisions (*Benson* and *Flook*) did not expressly close that door. Also, by the time *Diamond v. Diehr* was decided in 1981, the Court of Customs and Patent Appeals had already made it clear that it was prepared to accept software patents. Nevertheless, *Diamond v. Diehr* is an important software patent case, following directly in the footsteps of *Diamond v. Chakrabarty* that "anything under the sun that is made by man" is patentable subject matter.

[1] Technology in *Diamond v. Diehr*

The patent involves a process for curing synthetic rubber. When molding synthetic rubber into a precision product, raw, uncured synthetic rubber is placed in a mold, heat and pressure are applied, and the mold is then opened when the rubber has "cured" for the proper length of time. Once cured, the rubber retains the shape of the mold that produced it.

The cure is obtained by mixing curing agents into the uncured rubber polymer before molding and by then applying heat for the right length of time

---

[82] 437 U.S. 584, 596 (1978).

[83] 447 U.S. at 315.

[84] 151 Eng. Rep. 1266 (1841), which held patentable the applied principle of using hot air to feed a blast furnace fire.

[85] 450 U.S. 175 (1981).

[86] *Id.*

while the mixture is in the mold. The curing time must be neither too long nor too short. Curing time is a function of mold temperature, according to the relationship described by the well-known Arrhenius equation. Discovered by Svante Arrhenius, the relationship expresses the total cure time (v) as the following nonlinear function of activation constant (C), mold temperature (Z), and mold geometry-determined constant (x):

$$\ln v = CZ + x$$

The activation constant (C) is a unique figure for each batch of each compound being molded. It is determined by making viscosity measurements of the rubber mixture, using an instrument known as a rheometer. The geometry-dependent constant (x) takes into account the fact that the cure time will depend on the thickness and shape of the article to be molded.

Prior to Diehr and Lutton's invention the temperature variable (Z) was viewed as uncontrollable. This was because every time the mold press was opened to load the uncured rubber mixture, and again to remove the cured rubber product, the mold temperature cooled by an unknown amount. The industry practice was to calculate the cure time as the shortest time in which all parts of the product would definitely be cured, assuming a reasonable amount of mold-opening time during loading of the uncured rubber mixture and subsequent unloading of the cured rubber product.

Diehr and Lutton's invention eliminates the need to make assumptions about how long the mold has been opened and how much it has cooled. Diehr and Lutton use temperature sensors to continuously read the temperature inside the mold cavity. These sensors automatically feed the mold cavity temperature information to a digital computer which constantly recalculates the cure time, based on the Arrhenius equation. The computer then signals precisely when to open the mold.

[2] Claims at Issue in *Diamond v. Diehr*

Diehr and Lutton's application contains eleven different claims. Three examples are claims 1, 2, and 11 that are reproduced below:

1. A method of operating a rubber-molding press for precision molded compounds with the aid of a digital computer, comprising:

    providing said computer with a database for said press including at least,

    natural logarithm conversion data (ln),

    the activation energy constant (C) unique to each batch of said compound being molded, and

    a constant (x) dependent upon the geometry of the particular mold of the press, initiating an interval timer in said computer upon the closure of the press for monitoring the elapsed time of said closure,

constantly determining the temperature (Z) of the mold at a location closely adjacent to the mold cavity in the press during molding,

constantly providing the computer with the temperature (Z),

repetitively calculating in the computer, at frequent intervals during each cure, the Arrhenius equation for reaction time during the cure, which is

$$\ln v = CZ + x$$

where v is the total required cure time,

repetitively comparing in the computer at said frequent intervals during the cure each said calculation of the total required cure time calculated with the Arrhenius equation and said elapsed time, and

opening the press automatically when a said comparison indicates equivalence.

2. The method of Claim 1 including measuring the activation energy constant for the compound being molded in the press with a rheometer and automatically updating said database within the computer in the event of changes in the compound being molded in said press as measured by said rheometer.

11. A method of manufacturing precision molded articles from selected synthetic rubber compounds in an openable rubber molding press having at least one heated precision mold, comprising:

   (a) heating said mold to a temperature range approximating a predetermined rubber curing temperature,

   (b) installing prepared unmolded synthetic rubber of a known compound in a molding cavity of predetermined geometry as defined by said mold,

   (c) closing said press to mold said rubber to occupy said cavity in conformance with the contour of said mold and to cure said rubber by transfer of heat thereto from said mold,

   (d) initiating an interval timer upon the closure of said press for monitoring the elapsed time of said closure,

   (e) heating said mold during said closure to maintain the temperature thereof within said range approximating said rubber curing temperature,

   (f) constantly determining the temperature of said mold at a location closely adjacent said cavity thereof throughout closure of said press,

   (g) repetitively calculating at frequent periodic intervals throughout closure of said press the Arrhenius equation for reaction time of said rubber to determine total required cure time v as follows:

$$\ln v = cz + x$$

93

wherein c is an activation energy constant determined for said rubber being molded and cured in said press, z is the temperature of said mold at the time of each calculation of said Arrhenius equation, and x is a constant which is a function of said predetermined geometry of said mold,

(h) for each repetition of calculation of said Arrhenius equation herein, comparing the resultant calculated total required cure time with the monitored elapsed time measured by said interval timer,

(i) opening said press when a said comparison of calculated total required cure time and monitored elapsed time indicates equivalence, and

(j) removing from said mold the resultant precision molded an cured rubber article.[87]

The Patent Examiner rejected these claims on the sole ground that they were drawn to nonstatutory subject matter under 35 U.S.C. § 101. The examiner determined that those steps in the claims that are carried out by a computer were nonstatutory subject matter under *Gottschalk v. Benson*.[88] The remaining steps — installing rubber in the press and the subsequent closing of the press — were, the examiner found, "conventional and necessary to the process and cannot be the basis of patentability."[89]

The Patent and Trademark Office Board of Appeals agreed with the examiner. The Court of Customs and Patent Appeals disagreed and reversed.[90] The Court of Customs and Patent Appeals noted that a claim drawn to subject matter otherwise statutory does not become nonstatutory because a computer is involved. The Appeals Court saw Diehr and Lutton's invention as an improved process for molding rubber articles. In a now familiar pattern, the Patent Office appealed. Commissioner, Sidney A. Diamond, filed a petition for *certiorari*, arguing that the decision of the Court of Customs and Patent Appeals was inconsistent with prior decisions of the Supreme Court.

[3] Holding in *Diamond v. Diehr*

Justice Rehnquist delivered the opinion of the Court, which in a 5 to 4 decision sided with the Court of Customs and Patent Appeals and found the claimed invention was patentable subject matter. Justice Rehnquist begins his analysis with *Diamond v. Chakrabarty*,[91] reiterating the premise relied upon in that case that "anything under the sun that is made by man" is statutory subject

---

[87] *Id.* at 175, n. 5.
[88] 409 U.S. 63 (1972).
[89] *Diamond v. Diehr*, 450 U.S. at 181.
[90] *In re* Diehr, 602 F.2d 982 (C.C.P.A. 1979).
[91] Diamond v. Chakrabarty, 447 U.S. 303 (1980).

matter under § 101.[92] Justice Rehnquist acknowledges, however, that § 101 does have its limits:

> This Court has undoubtedly recognized limits to § 101 and every discovery is not embraced within the statutory terms. Excluded from such patent protection are laws of nature, natural phenomena, and abstract ideas.[93]

As to the Court's prior computer-related cases, *Gottschalk v. Benson* and *Parker v. Flook*, Justice Rehnquist treats these as simply standing for the following "long-established principles."

1. Excluded from patent protection are laws of nature, natural phenomena, and abstract ideas.[94]
2. An idea of itself is not patentable.[95]
3. A principle, in the abstract, is a fundamental truth; an original cause; a motive; these cannot be patented.[96]
4. A law of nature, like a mineral discovered in the earth or a new plant found in the wild, is not patentable.[97]

Finally, distinguishing the Diehr and Lutton invention from the invention in *Flook*, Justice Rehnquist finds that whereas in *Flook* the claims were drawn to a method for computing a number (an alarm limit), in Diehr and Lutton the claims seek protection for a process of curing synthetic rubber. Flook sought to foreclose all use of a mathematical formula in the petrochemical industry, but taught nothing of how the process variables are obtained; Diehr and Lutton, the majority found, "seek only to foreclose from others the use of that equation [the Arrhenius equation] in conjunction with all of the other steps in their claimed process."[98]

[4] Analyzing *Diamond v. Diehr*

It would seem that *Diamond v. Diehr* and *Parker v. Flook* lie only inches away from one another but on opposite sides of the judicial line in the sand. Each includes a significant calculation step. The calculation step in each involves solving a mathematical formula. In each the mathematical formula represents some principle of nature. Each purports to include at least one additional method step, such that neither wholly preempts its mathematical formula. Yet, *Diehr* is patentable and *Flook* is not. Why?

---

[92] *Diamond v. Diehr*, 450 U.S. at 182.

[93] *Id.* at 185.

[94] Parker v. Flook, 437 U.S. 584 (1978); Gottschalk v. Benson, 409 U.S. 63 (1970); Funk Bros. v. Kalo Co., 333 U.S. 127 (1948).

[95] Rubber Tip Pencil Co. v. Howard, 87 U.S. 498 (1874).

[96] Le Roy v. Tatham, 55 U.S. 156 (1852).

[97] Diamond v. Chakrabarty, 447 U.S. 303 (1980); Funk Bros. v. Kalo Co., 333 U.S. 127 (1948).

[98] *Diamond v. Diehr*, 450 U.S. at 187.

One reason is that the invention in *Diehr* effects a change in state of a physical thing — a mold is opened. In *Flook* the invention effects a change in state of a nonphysical thing — a number is updated. Citing the 1877 decision in *Cochrane v. Deener*[99] and the Court's subsequent adoption of the principle in *Gottschalk v. Benson*,[100] Justice Rehnquist offers the following guidance on what distinguishes a patentable process:

> Transformation and reduction of an article "to a different state or thing" is the clue to the patentability of a process claim that does not include particular machines.[101]

This clue is a helpful one because a computerized process running on a general purpose digital computer qualifies as a process that "does not include particular machines."

Another reason for the different outcome between *Diehr* and *Flook* may be a change in attitude of certain members of the Supreme Court. Figure 3-1 shows how each Justice voted in *Benson, Flook, Chakrabarty*, and *Diehr*. Note how certain Justices have voted consistently for an expansive reading of § 101 whereas others have voted consistently for a restricted reading; note also how other Justices have changed their stance on this issue.

This is not to say that any of the Justices are pro-software patent or anti-software patent, per se. Rather, the difference of opinion seems to be whether the claim language alone should dictate whether an invention qualifies as statutory subject matter, or whether the claim language should be construed first in light of the prior art to determine what the applicant has actually discovered. *Diamond v. Diehr* best illustrates this difference of opinion.

Diamond v. Diehr is a 5 to 4 decision. The principal difference of opinion between the majority and the minority is not over the construction of § 101, but over the construction of the claims. The majority construes the claims to cover a method of constantly measuring the actual temperature in a rubber molding press and determining when to open the mold. The minority construes the claims to cover a method of calculating the time that the mold should remain closed during the curing process. In effect, the majority places more emphasis on the claim language — taking the applicant's word that this is the applicant's discovery; the minority looks throughout the claim language — in light of the prior art — to identify what the inventor claims to have discovered.

This can be a bit confusing because ordinarily the prior art comes into play in determining novelty under § 102 and obviousness under § 103. Here, however, the minority uses the prior art to assess what the applicant claims to have discovered. The justification for this discovery requirement comes from the

---

[99] 94 U.S. 780 (1877).
[100] 409 U.S. at 70.
[101] Diamond v. Diehr, 450 U.S. 175, 184 (1981).

Case/Year	Majority	Dissent	No Part
Gottschalk v. *Benson* 1972	Douglas Burger Rehnquist Marshall Brennan White		Blackmun Stewart Powell
Parker v. *Flook* 1978	Stevens Marshall Brennan White Powell Blackmun	Burger Rehnquist Stewart	
Diamond v. *Chakrabarty* 1980	Burger Rehnquist Stewart Stevens Blackmun	Marshall Brennan White Powell	
Diamond v. *Diehr* 1981	Rehnquist Burger Stewart White Powell	Marshall Brennan Stevens Blackmun	

FIGURE 3-1: HOW THE JUSTICES VOTED

language of § 101, itself: "Whoever invents or *discovers* any new and useful process . . ."

If you are having difficulty understanding how the prior art has any bearing on what is patentable subject matter, you are not alone. Justice Stevens, writing for the minority in *Diehr*, accuses the majority of having the same difficulty:

> The Court . . . fails to understand or completely disregards the distinction between the subject matter of what the inventor *claims* to have discovered — the § 101 issue — and the question whether that claimed discovery is in fact novel — the § 102 issue.[102]

---

[102] *Id.* at 211 (emphasis in original).

§ 3.10   Statutory Subject Matter Issues with Internet Patents

The statutory subject matter question is largely settled. Software patents can constitute statutory subject matter[103] as can business methods.[104] Nevertheless, some courts still have statutory subject matter concerns, particularly where the claims have broad-reaching implications. A case in point is *AT&T v. Excel Communications.*

In 1982, AT&T agreed to divest itself of local telephone companies as part of the consent decree in *United States v. AT&T.*[105] Since that time, local phone companies (local exchange carriers) and AT&T have operated independently. Under the "equal access" directive of the decree, callers must have the ability to pre-subscribe their telephones to a long distance carrier (interexchange carrier) other than AT&T. Before the advent of "equal access," all interstate long-distance calls dialed on a 1+ basis were routed by the local exchange carrier to AT&T. A caller could reach another interexchange carrier only by dialing a seven-digit phone number, as well as a lengthy identification code, prior to dialing the called number. "Equal access" enabled callers to select a carrier other than AT&T to provide them long-distance phone service on a simple 1+ dialing basis. Thus, any selected carrier could become the caller's "primary interexchange carrier" or PIC.

On May 6, 1992, AT&T inventors, Doherty, Lanzillotti and Paulus, applied for a patent on an invention they termed, a "Call Message Recording For Telephone Systems." The system allowed AT&T to offer lower long-distance billing rates if both the calling party and the called party subscribed to AT&T as their primary interexchange carrier. The system worked by adding a PIC indicator to the standard exchange message. The PIC indicator identified which primary interexchange carrier the called party had selected. Using this PIC indicator information, and knowing the PIC of the calling party, AT&T could determine if both parties used AT&T as their primary long-distance carrier and could provide favorable billing treatment.

The patent issued July 26, 1994, with 41 method claims. Claim 1 is exemplary:

1. A method for use in a telecommunications system in which interexchange calls initiated by each subscriber are automatically routed over the facilities of a particular one of a plurality of interexchange carriers associated with that subscriber, said method comprising the steps of:

   generating a message record for an interexchange call between an originating subscriber and a terminating subscriber; and

   including, in said message record, a primary interexchange carrier (PIC) indicator having a value which is a function of whether or not the

---

[103] *In re* Alappat, 33 F.3d 1526 (Fed. Cir. 1994).
[104] State Street Bank & Trust, Co. v. Signature Financial Group, 149 F.3d 1368 (Fed. Cir. 1998).
[105] United States v. AT&T, 552 F. Supp. 131 (D.D.C. 1982).

interexchange carrier associated with said terminating subscriber is a predetermined one of said interexchange carriers.

Excel Communications is one of a number of resale carriers that contracts with facility owners to route their subscribers' calls through the facility-owners' switches and transmission lines. Excel began offering services with reduced rates if both parties used Excel as their primary interexchange carrier. AT&T sued Excel Communications for infringement in the U.S. District Court for Delaware. Excel moved for summary judgment on the grounds that the patent did not recite statutory subject matter.

Judge Robinson of the District Court of Delaware ruled in Excel's favor. She found that the claims implicitly recited a mathematical algorithm. Judge Robinson was mindful of the holding in *Diamond v. Diehr*,[106] that a claim drawn to statutory subject matter does not become nonstatutory simply because it uses a mathematical formula, computer program or digital computer. However, AT&T's claims, in Judge Robinson's view, lacked substance. Quoting from the *Diehr* decision, Judge Robinson wrote:

> Although the first element of these claims, generating a message record, is not "new" [fn omitted] but a standard practice in the industry, the court is directed to look at the claims as a whole to determine whether the process claimed "is performing a function which the patent laws were designed to protect (e.g., transforming or reducing an article to a different state or thing) . . ."
>
> In this regard, the '184 patent can be described as claiming an invention whereby certain information that is already known within a telecommunications system (the PICs of the originating and terminating subscribers) is simply retrieved for an allegedly new use in billing. Characterized as such, the §101 inquiry employed, e.g., in In re Grams, would dictate a finding of nonpatentability: Where "the only physical step [in the claim] involves merely gathering data for the algorithm," the subject matter is unpatentable.[107]

Judge Robinson thus granted Excel's motion for summary judgment, holding the patent invalid under 35 U.S.C. § 101.

AT&T appealed to the Federal Circuit Court of Appeals and that court reversed. Judge Plager delivered the opinion. In it, he drew upon the court's prior decisions in *Alappat*[108] and *State Street Bank*[109] which had both reached the conclusion that the presence of a mathematical algorithm or mathematical formula in a claim did not *ipso facto* render the claim nonstatutory under § 101. However, both *Alappat* and *State Street Bank* involved apparatus claims. As such, the claims recited physical structure. However, the claims at issue in this

---

[106] Diamond v. Diehr, 450 U.S. 185 (1981).

[107] AT&T Corp v. Excel Communications, Inc. 1998 WL 175878 (D. Del. 1998).

[108] 33 F.3d 1526 (Fed. Cir. 1994).

[109] 149 F.3d 1368 (Fed. Cir. 1998).

case were method claims. No physical structure was directly recited. Thus, arguably, the claims did not require the transforming or reducing of an article to a different state or thing. Was Judge Robinson correct? Did the lack of a physical transformation render the claims nonstatutory?

Judge Plager answered this question, no. First, he made it clear that § 101 is to be applied the same, no matter whether apparatus or method claims are involved:

> In both *Alappat* and *State Street Bank*, the claim was for a machine that achieved certain results. In the case before us, because Excel does not own or operate the facilities over which its calls are placed, AT&T did not charge Excel with infringement of its apparatus claims, but limited its infringement charge to the specified method or process claims. Whether stated implicitly or explicitly, we consider the scope of § 101 to be the same regardless of the form — machine or process — in which a particular claim is drafted.[110]

As to the physical transformation which District Judge Robinson had found lacking in AT&T's claims, Judge Plager clarified that physical transformation is not a requirement under 35 U.S.C. § 101. He reached this conclusion by carefully analyzing the words of the Supreme Court in *Diamond v. Diehr*:[111]

> Excel argues that method claims containing mathematical algorithms are patentable subject matter only if there is "physical transformation" or conversion of subject matter from one state into another. The physical transformation language appears in *Diehr*. . . and has been echoed by the court in *Schrader* [22 F.3d 290, 294 (Fed. Cir. 1994)] . . .
>
> The notion of "physical transformation" can be misunderstood. In the first place, it is not an invariable requirement, but merely one example of how a mathematical algorithm may bring about a useful application. As the Supreme Court itself noted, "when [a claimed invention] is performing a function which the patent laws were designed to protect (*e.g.*, transforming or reducing an article to a different state or thing), then the claim satisfies the requirements of § 101". . . (emphasis added) The "e.g." signal denotes an example, not an exclusive requirement.[112]

If a physical transformation is but one indicia that statutory subject matter is present, are there any others? Judge Plager answered this question by reference to the court's earlier decision in *Arrhythmia*[113] in which the presence of "transformed" data could be an indicia of statutory subject matter, if the data had a specific meaning that gave a useful, concrete and tangible result, as opposed to a mere mathematical abstraction.

---

[110] AT&T Corp. v. Excel Communications, Inc., 172 F.3d 1352, 1357 (Fed. Cir. 1999).

[111] 450 U.S. 175 (1981).

[112] *Excel Communications, Inc., 172 F.3d at 1358-1359.*

[113] Arrhythmia Research Technology, Inc. v. Corazonix Corp., 958 F.2d 1053 (Fed. Cir. 1992).

> The finding [in *Arrhythmia*] that the claimed process "transformed" data from one "form" to another simply confirmed that Arrhythmia's method claims satisfied § 101 because the mathematical algorithm included within the process was applied to produce a number which had specific meaning — a useful, concrete, tangible result — not a mathematical abstraction.[114]

The result reached by the court in this case left hanging the prior decisions in *In re Schrader*[115] and *In re Grams*.[116] In both of those cases the court had relied upon the *Freeman-Walter-Abele* test, which involved the two-step process of, first, analyzing the claims to determine if a mathematical algorithm was recited and, second, assessing whether the claims as a whole preempted all use of the mathematical algorithm. The *Freeman-Walter-Abele* test thus focused sharply on the mathematical algorithm and did not require a court to assess whether a useful, concrete and tangible result ensued. Because the *Schrader* and *Grams* courts had applied the *Freeman-Walter-Abele* test, and had not assessed whether a useful, concrete and tangible result ensued, Judge Plager rejected the *Schrader* and *Grams* cases as "unhelpful."

It would seem that any computer program that "transforms" data, which arguably all programs do, would constitute statutory subject matter under Judge Plager's analysis. To prevent the jump to that conclusion, Judge Plager revisited the court's decision in *In re Warmerdam*,[117] a decision which he authored. In that case, the court had rejected as nonstatutory Warmerdam's method claims for generating bubble hierarchies through the use of a particular mathematical procedure. The court found that the claimed process did nothing more than manipulate basic mathematical constructs, and concluded that "taking several abstract ideas and manipulating them together adds nothing to the basic equation."[118]

The decision in *Warmerdam* is troubling because the Warmerdam "bubble" data structure was described as being useful for controlling the motion of objects and machines to avoid collision with other moving or fixed objects. That would seem to place the subject matter within the domain of the useful, concrete and tangible result. However, as Judge Plager noted, the Warmerdam court concluded on the facts that the claims did not encompass such useful subject matter but rather the claims encompassed the manipulation of basic mathematical constructs. About *Warmerdam* Judge Plager wrote:

> The court found that the claimed processes did nothing more than manipulate basic mathematical constructs and concluded that "taking several abstract ideas and manipulating them together adds nothing to the basic equation"; hence, the court held that the claims were properly rejected under § 101 . . .

---

[114] 172 F.3d at 1359.

[115] *In re* Schrader, 22 F.3d 290 (Fed. Cir. 1994).

[116] 888 F.2d 835 (Fed. Cir. 1989).

[117] *In re* Warmerdam, 33 F.3d 1354 (Fed. Cir. 1994).

[118] *Id.* at 1360.

Whether one agrees with the court's conclusion on the facts, the holding of
the case is a straightforward application of the basic principle that mere laws
of nature, natural phenomena, and abstract ideas are not within the categories
of inventions or discoveries that may be patented under § 101.[119]

Warmerdam involved a computer data structure that represented basic relationships among objects. To the extent the claims called for a transformation of data,
the transformation, according to Judge Plager's analysis, did not involve a useful,
concrete, and tangible result. This suggests that computer data structures that
represent laws of nature, natural phenomena and abstract ideas remain off limits
to patent protection. Warmerdam's method claim read as follows:

1. A method for generating a data structure which represents the shape of
    [sic] physical object in a position and/or motion control machine as a
    hierarchy of bubbles, comprising the steps of:

    first locating the medial axis of the object and then creating a
hierarchy of bubbles on the medial axis.

Warmerdam's apparatus claim was of similar scope:

5. A machine having a memory which contains data representing a bubble
    hierarchy generated by the method of any of Claims 1 through 4.

Both of these claims make superficial reference to a position or motion control
machine. Nevertheless, the court found that Warmerdam was actually claiming
an abstract relationship. For a more detailed discussion on how to avoid claiming
an abstract relationship, see Chapter 10.

---

[119] *Id.* at 1360.

CHAPTER 4

THE ORIGINS OF COMMERCE

§ 4.01   The Internet and Money

The Internet mesmerizes the current business world. Why shouldn't it? According to a report in *Forbes*, describing a study conducted by the University of Texas at Austin, U.S. companies generated $301 billion in revenue from Internet-related activities in 1998. The Internet has thus surpassed the energy sector ($223 billion), the telecommunications sector ($270 billion) and is about to overtake the U.S. automobile sector ($350 billion). These are easy numbers to become mesmerized with. From a present-day perspective, it is easy to get caught up in the Internet excitement, and thereby lose sight of the bigger picture. While the Internet is indeed an important advance, it is only the most recent chapter in the book of commerce whose genesis chapter was written well over five thousand years ago.

Yes, you might say, humankind has engaged in commerce for eons, but never before has a single invention made such a profound, society-transforming impact. In that you would be partially right. The Internet is indeed transforming the very way we do business. It joins buyer and seller who otherwise would have no opportunity to meet. It affords a means of advertising goods and services. It hosts the marketplace where offers are made and accepted. It delivers goods and services. It supports the transfer of money.

This is all heady stuff; and yet, in terms of humanity's development as commercial beings, there is a far more fundamental business invention that must be considered first. That invention is money. Money is power. Money is the sea upon which modern commerce floats. Money was invented four thousand years ago. Yes, if the concept of electronic cash continues to develop as predicted, the Internet may one day *become* money. It therefore pays to have a better understanding of money, so that we can be prepared to patent it and its use.

§ 4.02   Barter and Proto-money

The year is 3000 B.C. Resting momentarily in the spotty shade beneath a scraggy olive tree, Agga stops to gaze far across the sun-baked hillside to the green valley below. His son sits beside him tightening the leather thong on his sandal. With a skillful eye Agga quickly surveys each of his small herd of cattle. Agga does not need to count. He knows every member of his herd by name. All is well. Heat rises in eddies from clay and rocks below, causing the Mesopotamian valley to shimmer like a thousand green jewels in the sun.

Agga and his son will stay on the hillside tonight with his herd. Tomorrow morning Agga will travel to the marketplace in the valley below, where he hopes to barter one or two of his calves for olive oil, wheat, and several goatskins of wine. Agga is prepared to be flexible. Although he knows what his family needs, he cannot be certain that he will find a farmer or merchant willing to trade him the commodities he needs in exchange for his calves. Because a single calf is the smallest unit of value Agga is able to trade, he must find a merchant or farmer who can give him equal value in olive oil, wheat, and wine. Agga is a skillful trader. He has made this trip many times. However, Agga knows that if no one wants his calves, he will have to return from his shopping trip empty-handed.

On previous visits to the valley, Agga had seen some of the wealthiest merchants trading entire warehouses of wheat for carefully weighed ingots of gold and silver. Gold and silver were extremely scarce and had great value. Like most traders in the marketplace, Agga had no access to this system of gold and silver ingots. If he had, Agga could have traded his calves for ingots and then spent those ingots purchasing the goods he needed. Unfortunately for Agga, even a single ingot was valued much higher than his entire herd. Wealthy merchants had no interest in subdividing an ingot to equal the value of a calf.

The gold and silver ingots used by wealthy Mesopotamian grain merchants were actually a feature of the early banking system. Grain was an important commodity that required storage. Grain was stored in the temple.

The marketplace that Agga would visit tomorrow was a patchwork of colorful tents, tables, and awnings of every description. Immediately behind the marketplace towers a gleaming white alabaster marble temple bearing carved bas relief images depicting the exploits of a now forgotten antediluvian king. Two immense bull-like guardian statues stand at the entrance of the temple. Beneath the guardian statues, temple officials watch over tables where records of the contents of the temple are maintained in cuneiform.

Deep within the temple, vast grain deposits are stored. Grain merchants wishing to deposit their grain harvest for storage brought their wheat to the temple in measured baskets. Priests would receive and store these baskets in exchange for carefully weighed ingots of gold or silver. The ingots served as a receipt for the wheat deposit made.

The exchange that Agga hoped to make in the morning required careful planning. The calves that Agga would trade were fat and healthy, thanks to the ample rain that had fallen. Last year, due to a drought, his herd was scrawny and fetched a poor exchange. Agga timed his trip to the valley when his calves would be at their prime. Although he would like to take his son into the city so that he could learn how to trade, that would not be possible. His son would need to remain behind on a nearby hillside to watch the herd.

The barter system that Agga and other common people used left much to be desired by today's standards. Glyn Davies, in his comprehensive treatise, *A History of Money From Ancient Times to the Present Day*, enumerated the many drawbacks of a barter system.[1] They are fairly apparent, but you might not think of them all. First, barter has no generalized or common standard of values; it supports no unified price system. The exchange rate between cows and wheat is completely different than the exchange rate between wheat and wine, and so forth. To shop at the local marketplace, Agga needed to retain a matrix of relative exchange rates in his head. With a large number of goods available, the number

---

[1] Davies, Glyn, *A History of Money From Ancient Times to the Present Day* (Cardiff: Univ. of Wales Press, 1994), 15-18.

of exchange rates get geometrically staggering. The mathematics is fairly straightforward.

In a simple marketplace of only three commodities, only three exchange rates are required. Things progress geometrically from there. In a four commodity market, six rates are required; in a five commodity market, ten rates are required; in a six commodity market, fifteen rates are required and in a ten commodity market, forty-five rates are required. Of course, Agga faced considerably higher permutations. In a Mesopotamian marketplace where, say, one hundred different commodities were offered, the number of exchange rates explodes to 4,950.

A second significant drawback of the barter system was that an exchange could take place only if the one trading partner, at that moment, wanted what the other trading partner had to offer, and vice versa. Agga could not trade calves for wheat, if the wheat merchant wanted only olive oil. Agga could, of course trade his calves for olive oil and then exchange the oil for wheat, provided he could find an olive oil merchant interested in cattle.

A third drawback involved the bulky and perishable nature of many commodities themselves. Often great effort had to be expended to protect a commodity's value. Grain had to be stored in a dry, vermin-free place; wine had to be stored in flasks or goatskin bags so it would not evaporate; cattle had to be fed and watered. While commodities such as sheep and cattle were self transporting, other commodities such as grain were not so easily transported in large quantity. There was little choice but to keep such commodities stored until use. Indeed, the earliest known wheel was constructed in Agga's time in Mesopotamia, around 3500 to 3000 B.C. No doubt the wheel was put to early use transporting grain, olive oil, and wine behind the hooves of oxen.

Goods could be stored. Services could not. Thus a forth drawback of the barter system was the inherent difficulty with trading in futures. In ancient times, this difficulty carried an exceedingly costly human price — slavery. As Davies reports,

> Services, by their nature cannot be stored, so that bartering for future services, necessarily involving an agreement to pay specific commodities or other specific services in exchange, weakens even the supposed normal superiority of current barter, namely its ability to enable direct and exactly measurable comparisons to be made between the items being exchanged. In the absence of money, or given the limited range of monetary uses in certain ancient civilizations, it is little wonder the completion of large-scale and long-term contracts was usually based on slavery. Thus the building of the Great Pyramid of Ghiza, the work of 100,000 men, and a logistical problem commensurate with its immense size, was made possible at that time only by the existence of slavery (even though these slaves enjoyed higher standards of living than others).[2]

---

[2] *Id.* at 17.

The gold and silver ingots used in the ancient Mesopotamian river valley were the precursor to perhaps the most important commercial invention of all time: money. In actuality, these ingots lacked some of the properties that we now ascribe to money. Thus, these ancient ingots may best be referred to as proto-money. Davies lists the functions of money as follows:

Specific functions (mostly micro-economic)
1. Unit of account
2. Common measure of value
3. Medium of exchange
4. Means of payment
5. Standard for deferred payments
6. Store of value

General functions (mostly macro-economic)
7. Liquid asset
8. Framework of the market allocative system
9. A causative factor in the economy
10. Controller of the economy[3]

In ancient Mesopotamian times, gold and silver ingots served wealthy merchants as a means of payment and a store of value for wheat they placed in temple storehouses. These ingots did not, however, serve as a medium of exchange or common measure of value, as most commodities could not be purchased using the ingots. From our modern day view we know that gold and silver can serve as the backbone of a monetary system; however, that discovery had not been made in ancient Mesopotamia. Gold and silver were still simply commodities — albeit ones that could not be eaten to sustain life — that merchants traded in the marketplace.

§ 4.03   Lydians Invent Coins

Sometime around 2000 B.C. a people known as the Lydians took up residency in Asia Minor, or Anatolia, roughly where Turkey lies today. The Lydians settled in the valley between the Hermus and Cayster rivers (now the Gediz and Buyukmenderes rivers). Like the Mesopotamian river valley, the Lydians' valley was green and fertile. But it had one additional feature. Gold and silver nuggets tumbled down the riverbeds and lay scattered along the banks. Most frequently the gold and silver was amalgamated in a naturally occurring mixture called electrum. Electrum was an amber-colored metal. Perhaps the name was chosen because the metal resembled amber, a substance that when rubbed would exhibit properties of electrostatic attraction. The Lydians learned to pan this gold and silver electrum from the rivers. It was prized for making jewelry.

Not far to the east of Lydia was the kingdom of Phrygia. The capital of Phrygia was the city of Gordion, home of the famous Gordion knot. It was said

---

[3] *Id.* at 27.

that whoever could untie the knot would gain the key to Anatolia. All who tried, failed, until Alexander the Great. On his march along the King road, bent on conquering all of Anatolia, he simply cut the famous knot in half with his sword. Aside from its famous knot, Phrygia also boasts of being ruled by King Midas, the one with the golden touch. Midas ruled Phrygia near the end of the 7th Century B.C. He died around 700 B.C. and was buried in Gordion along with the unwashed dishes of his funeral feast.[4] Greek legend has it that the rich deposits of electrum in the river which flowed through Lydia were the result of King Midas' bathing upstream, trying to wash away the curse of his golden touch.[5]

No doubt, at first the Lydians traded electrum as any other commodity, just as the ancient Mesopotamian people had traded cattle and wheat a thousand years before. However, sometime between 640 and 630 B.C. an enterprising Lydian king made the connection that electrum could be made into coins that could serve as a medium of exchange for his people. The genius of this now forgotten king was that each coin should have the value of no more than a few days' labor. That made the electrum coins accessible to the common citizen.

The coins produced a powerful effect. Freed of the many shortcomings of barter, the Lydian economy flourished. The first electrum coins were nothing more than crude, lima bean sized blobs of electrum, stamped on one side with an emblem of a lion's head to ensure authenticity. As metallurgical skills of the Lydians improved, the coins became flattened and round.

By making the coins the same size and weight, the Lydian king eliminated one of the most time consuming steps of early commerce — that of weighing the precious metal prior to consummating the transaction. Instead, merchants could assess the adequacy of payment by *counting*. In comparison with ancient barter standards, Lydian commerce could now be conducted at high speed.

The Lydians built a magnificent capital city, Sardis, on the banks of the Pactolus River (now Baguli River) at the north base of Mount Timolus (now Boz Dag). As the Lydians prospered, Sardis became the seat of a powerful Lydian empire where many came to engage in the efficient Lydian commerce. Fueled by the highly efficient coinage system, wealth poured into Sardis. Because Sardis was the place to trade, buyers and sellers flocked to Sardis. Goods and services of all description could be had in Sardis. The abundance of goods and services lead to another Lydian innovation — the retail market. Unlike the ancient marketplace of the Mesopotamian civilization, where the grain farmer sold the grain he harvested and the wine maker sold the wine he produced, the Lydian marketplace was run by shopkeepers who *purchased* goods for resale instead of raising them.

In 560 B.C., Croesus ascended to the throne. Croesus understood the Lydian economy well. He instructed his metallurgists to use their newly discovered skills to melt down the electrum coins so that gold and silver could be separated. He then ordered his mint to begin producing separate gold and silver coins of different, standardized monetary values. The added monetary flexibility and

---

[4] Gee, Henry, "Phrygian Funeral Feast," *Nature*, Dec. 23, 1999.
[5] *Supra*, note 1 at 61.

heightened quality control standards increased the speed of the economy further. Commerce that began as a means to fulfill the necessities of life became big business. During the reign of Croesus, one could shop in Lydian markets using gold and silver coins to purchase necessities, such as grain, oil, beer, wine, leather, pottery, and wood, as well as luxury items, such as perfume, cosmetics, jewelry, musical instruments, glazed ceramics, bronze figurines, mohair, purple cloth, marble, and ivory.[6]

Unfortunately for the Lydians, while their commercial innovations would spread throughout the world, their empire was not last. The wealth the Lydians accumulated would prove to be their downfall. Croesus used his wealth to build armies, which he sent to conquer neighboring lands. He conquered most of the Greek city-states in Asia Minor, spreading the Lydian monetary and retail market innovations to the skillful Greeks. Croesus then attacked Persia, which proved to be his downfall. Croesus' mercenary army met fierce opposition in the Persian army, lead by Persian king, Cyrus. The Persian army quickly defeated the Lydians and then marched into Sardis itself, conquering the Lydian capital and adding Lydia to the Persian Empire.

It is interesting to reflect upon the Lydian innovations: the creation of coins and the establishment of retail markets. The effect of these innovations was two-fold. The Lydian monetary system greatly increased the speed at which commerce could be conducted and the associated retail markets significantly expanded the variety of goods and services the common person had access to. These are, of course, some of the same benefits now attributed to the Internet and e-commerce. As we shall explore later in this chapter, some believe that the world is currently undergoing a transition — that we cannot fully see but historians will someday write about — in which hard currency — our coins of Croesus — will disappear, to be replaced by electronic money built of digital technology.

§ 4.04   Chinese Invent Coins Independently

Many historians attribute to the Lydians the invention of coins because it was their ideas about money and commerce that spread to the Greeks, then to the Romans and ultimately to the entire Western Civilization. However, like so many innovations that we trace to the Cradle of Civilization in Asia Minor, the ancient Chinese may have had the idea first.

The Stone Age ended when the humans discovered how to work metal. At the time the Stone Age ended in China, the people there were using cowrie shells as a primitive form of money. The cowrie shell was actually a quite popular form of primitive money throughout much of the primitive world. In addition to China, cowrie shells were used as primitive money in Africa, the Middle East, and Oceania. Cowrie shells probably represented small increments of value,

---

[6] Weatherford, Jack, *The History of Money* (New York: Three Rivers Press, 1997), 31.

commensurate with the fact that they could be gathered in great numbers along the seashores of the Indian and Pacific oceans.

When the Chinese learned to work metal, they cast cowrie shells of bronze and copper to serve as an additional source of money. Initially these metal cowrie shells probably had greater value in the ancient Chinese economy. Eventually, as metalworking skills advanced, the Chinese began using their highly prized metal implements, such as knives, spades, hoes, and adzes, as a form of money. When one considers how an early metalworker might fashion a hoe blade from bronze, it is easy to appreciate that the basic shape would be cast, with a suitable hole included for later attaching to a handle. If the hole were made square, the handle would not as easily twist during use. After casting, the blade would be pounded flat and filed into the proper shape.

Apparently, a very similar process was used to fashion early Chinese money. Early Chinese coins (and even some Chinese coins today) have a square or round hole in the center. It is believed that, during manufacture, a rod was inserted through the hole of the rough cast coin. The rod served as a handle that the worker could grip while filing the edges of the coin into a round shape. As many as 50 or more coins could be filed at the same time, by stringing them on a common rod.[7]

Whether knowledge of these early Chinese coins spread to the Lydians, no one can be sure. However, there was a significant difference between the coins of the Lydians and the coins of the Chinese. Whereas Lydian coins were made of precious metal, Chinese coins were made of cheap base metal, the same metal used to fashion knife blades, hoes, and adzes. This difference caused the Lydian and Chinese monetary systems to diverge in a significant way. Lydian coins of precious metal fully supplanted the need to weigh gold and silver when making even large purchases. In contrast, Chinese coins were individually of little value — pocket change by today's standards — and did not supplant the practice of using weighed amounts of gold and silver for large purchases. The Lydian economy took a giant leap forward, creating for the first time in recorded history a nation of retail merchants. The Chinese economy failed to fully exploit the possibilities of coinage and remained rooted in the slower-paced barter system.[8]

§ 4.05   The Greeks Spread the Lydian Invention of Money Throughout the Western World

Greek city-states in Asia Minor were under the rule of the Lydian king Croesus until his empire fell to the Persians. This gave the Greeks first hand access to the Lydian coinage system. As the concept of coinage spread both east and west from Lydia, an interesting variation arose. The Persians to the east developed a preference for coins of gold, although they also used coins of silver; whereas the Greeks to the west developed almost an exclusive preference for

---

[7] *Supra*, note 1 at 55.
[8] *Supra*, note 1 at 55.

silver. Thus, the bimetal monetary standard of Croesus seemed to grow stronger as one traveled east into the Persian empire, but diminished rapidly as one traveled west through the Greek islands toward the powerful city-states of Greece.

The spread of coinage to the Greek city-states acted as a rapid growth fertilizer, causing the Greek civilization to leap ahead of the remaining world. To most city-states the minting of coins carried great civic pride Each city-state minted its own, thus a traveler from city to city might need to exchange coin denominations several times. Not every city-state was so quick to adopt coinage, however. The Trojans, in particular, seemed to reject the concept of money, or at least, they did everything they could to discourage its use. Trojan money took the form of heavy iron bars that could barely be lifted.

One of the strongest coin-based economies grew up in Athens. Athens had rich silver deposits nearby and the Athenians dug deep mines in the nearby hillsides to retrieve it. Over 2,000 ancient mineshafts have been discovered in an area called Laurion, 25 miles south of ancient Athens. The deepest mine is 386 feet, having main shafts of up to six feet in diameter, with tributary shafts of about two feet square.

Because each city-state jealously guarded its own sovereignty and thus minted its own coins, trade between cities dictated the service of money exchangers. That need lead to another innovation: banking. Although banking became widespread in ancient Greece, it was not the Greeks themselves who invented banking, but rather the foreigners or Metoikos, or "metics" living among the Greeks. Davies reports that the Greek aristocracy looked down upon manual labor and the everyday boredom of business life, placing more importance on being able to enjoy the leisure necessary for a cultured life. Thus the Greek aristocracy was perfectly happy to leave the field of banking wide open to the metics.[9]

One had to be quite astute to be a banker in ancient Greece. Counterfeit coins were commonplace. The authentic Athenian "owl" coin, so named because it bore the picture of an owl on one side, was made of silver. Unscrupulous, get-rich-quick counterfeiters made knockoffs of these popular coins by casting imitation owl coins of copper with a thinly applied outer coating of silver. The counterfeit coins were quite good and easily passed in commerce. The accepted test for detecting counterfeit coins was to use a knife and cut a slash across the face or into one edge of the coin, to reveal if it had a fraudulent copper core. Although this defaced the coin, the coin could then pass in commerce more freely, because everyone could see that it was genuine.

Unfortunately for honest Greeks, this counterfeit spotting tactic was met by the cheater's counter tactic. As if engaged in a game of point-counterpoint, clever counterfeiters began casting their knockoff copper coins with the slash mark already cast or cut into the face of the coin. When the copper slug was then

---

[9] *Supra*, note 1 at 68-71.

coated with silver, the precast slash mark also took on a silver coating, making it look as if the coin had already been tested and its core was solid silver.

The spread of counterfeiting, while regrettable and certainly not limited to the ancient Greek civilization, demonstrates the importance of the Lydian invention, coinage. It was evident that the idea of money had taken root when the criminals decided to adopt it.

Of course, we are still struggling with the criminal element over counterfeiting today. Counterfeit coins are less of a problem today, no doubt because coins have deflated in value so much that it is no longer "cost effective" for the criminal to devote his or her time to this practice. However, the game of point-counterpoint continues. Money has evolved to a new frontier and with it has evolved money's dark side. Stolen credit card numbers, counterfeit identities, and fraudulent authenticity certificates have become the counterfeit coin of the e-commerce marketplace.

When the ancient metic bankers were not occupied detecting counterfeit coins, they were busy making money for themselves. Part of the early banker's money-changing function involved converting foreign currencies — converting currency from one city-state to that of another, much like the Cambio de Change found in today's international airports. But, in ancient times, all different coins of local denominations were not always readily available. Thus, the early bankers also served the important function of "making change" in the local currency by converting one coin denomination into another. Metic bankers charged a fee for these money-changing operations. When one considers the skill needed to filter out counterfeit coins and the large variety of different coins these bankers had to deal with, the fee was probably well earned.

Unlike modern day coinage systems, which are typically based on a simple decimal system — the U.S. system being a modified decimal system that includes nickels and quarters — ancient coinage systems could be quite complex. Davies gives an account of this complexity as he describes the popular Athenian monetary system of ancient Greece.

> Taking the silver drachma as the main, central, standard monetary unit, one moved down to the less valuable and proportionally lighter sub-unit, the obol, six of which made one drachma. . . . Below the obol came the chalkous, in normal times the smallest monetary unit, and made, as its name implied, of copper, just as our use of 'coppers' for small change. Eight chalkoi — usually — made one obol. Moving above the central unit of the drachma, and ignoring for the moment the stater and other multiples of the drachma, we come first to the mina, roughly a pound in weight, equivalent to one hundred drachmae, and finally the talent, equivalent to sixty minae . . .
>
> . . . In the west and mainland Greece it was initially the two-drachma coin, the didrachm which became the standard, while a number of eastern city-states preferred their own three-drachma staters. It was however the Athenian double stater, the four-drachma or tetradrachm, with the owl on one side and head of the goddess Athena on the other, which eventually became the ancient world's most popular coin by far and therefore in practice the

most common standard or stater by which other coins were weighed and judged.[10]

Every era has its financial success story. In modern times every investor knows the story of Bill Gates and how he rose from teenage computer hacker to billionaire. In ancient Greek times it was a man named Pasion. Pasion was born a slave in the 4th Century B.C. As a young man, Pasion had the fortune to work for the prominent Greek bankers, Antisthenes and Archestratus. Antisthenes and Archestratus, two of Athens's earliest bankers, had built up an enormously successful banking business in Athens during the latter part of the 5th Century. Although Pasion was technically a slave, Antisthenes and Archestratus treated Pasion more as a son and later as a business partner. Pasion had a talent for business. Antisthenes and Archestratus found they could trust him to handle even the most intricate banking transactions, and Pasion never disappointed.

When his masters passed on, Pasion continued to operate the banking business, taking it to heights never achieved by his masters. Pasion became the most wealthy of all the Greek bankers. With his wealth he bought his freedom and gained Athenian citizenship. Pasion retired in 371 B.C., one of the wealthiest men in ancient Greece. Following in his former masters' footsteps, Pasion passed his business to his son, Apollodorus, and to his former salve Phormion. Unfortunately, after Pasion's death, Apollodorus and Phormion squabbled over the huge estate, eventually taking their dispute to court.

Pasion's rise from slavery to riches demonstrates how invention of money could powerfully influence the social advancement of an individual in ancient times. The cumulative effect was even more far-reaching. Money liberated the individual and this transformed society. In his book, *The History of Money*, Jack Weatherford tells the story of how ancient society evolved through stages, from kinship-based communities, to tributary-based states, to financially-liberated democracies. Kinship-based communities were small, 60 to 100 people, tied through kinship and marriage. Barter was quite adequate to support the economy of the kinship-based community. Tributary-based states were larger, often as many as a million people. These states were held together through military strength, bureaucracy, and recordkeeping. In a tributary system, wealth flowed through taxation from the outer reaches of the realm to the capital city and markets functioned primarily as subservient to the political structure.

The introduction of money drove an invisible wedge into both kinship and tributary systems, by giving the individual the power to change his or her destiny. Weatherford explains:

> The use of money does not require face-to-face interaction and the intense relationships of a kinship-based system. Nor does it require such extensive administrative, police, and military systems. Money became the social nexus connecting humans in many more social relationships, no matter how distant

---

[10] *Supra*, note 1 at 74.

or how transitory, than had previously been possible. Money connected humans in a more extensive and more efficient way than any other known medium. It created more social ties, but in making them faster and more transitory, it weakened the traditional ties based on kinship and political power.[11]

There are, of course, many reasons why the ancient Greek civilization advanced so quickly beyond neighboring civilizations. However, money played a key role by shaping the political system and freeing the Greeks to follow other scholarly pursuits. In 594-593 B.C., the Athenian ruler, Solon, known as the great lawgiver, abolished the traditional practice of limiting public office to only those of noble birth — a vestige of kinship-based society. Instead, Solon advocated that anyone who owned land could be elected to public office. This move untied one of the strings that had tightly bound earlier government. Now, anyone with money could buy land and thereby become eligible for public office. No longer was political power tied up by the aristocracy. Solon had given the Athenian populous the power to unleash democracy.

Not all Greek city-states followed Solon's reasoning. Sparta, which had always been strongly tied to traditional military values, resisted the democratic movement. Even Sparta's money illustrates this resistance. Sparta allowed only heavy iron bars and spear tips to be used as money. Spartan money was not convenient, or easy to carry, and thus it did not lubricate the private Spartan economy the way silver coins did in neighboring Athens.

Spartan reluctance notwithstanding, the Greek money-based economy spread throughout the civilized world. It is this economy that some say was directly responsible for the ancient Greek intellectual wealth.

After thousands of years of empires throughout the world, the marketplace emerged during the Greek era and changed history. Every great civilization prior to Greece had been based on political union and force backed by military might. Greece, which by then was unified, arose from the marketplace and commerce. Greece had created a whole new kind of civilization.

The wealth generated by this commerce expanded the leisure time of the Greek elite, thus allowing the opportunity to create a rich civic life and to pursue social luxuries including politics, philosophy, sports, and the arts as well as good food and festive celebrations. Never before in history had so many people had so much wealth; yet in a world with only a few luxury goods, they spent that wealth on leisure consumption. Scholars still today mine the rich intellectual deposits of words and ideas laid down by these Greeks, and their era marks the beginning of the academic disciplines of history, science, philosophy, and mathematics.[12]

---

[11] *Supra*, note 9 at 35.
[12] *Supra*, note 9 at 38.

§ 4.06   Money Marches Throughout the Roman Empire

The Roman republic and later the Roman empire spanned some 22 centuries, from about 750 B.C. to 1450 A.D. In duration, our present civilization pales in comparison. During the first third of this long, 22 century history, Rome expanded voraciously into virtually all of the western world. Roman emperors seized all of the land worth taking, while ignoring the poverty stricken desert land that was home to small, profitless tribes of barbarian people. From the lands they conquered, Roman emperors extracted great wealth: precious metals, slaves to work the mines, and taxes.

During this expansionary period, Rome took the idea of coins, which it had learned from the Greeks, and spread it throughout the Mediterranean and across southern and western Europe. In 269 B.C., the Romans began issuing a new silver coin, the *denarius*, which they minted in the temple of Juno Moneta. The coin bore the name and image of the goddess Moneta. Like many other artifacts of the Roman empire, the coin itself faded into antiquity, but the name, Moneta, became the root of the modern English words, money and mint.

The Roman empire lies so far in our distant past, that it is difficult to imagine its greatness today. If time were miles, and one could arrange the Roman empire side-by-side with our present U.S. civilization, both beginning at, say, Washington, D.C., the Roman empire would extend all the way to California, while the United States would barely reach Cleveland.

When it came to money management, the Roman emperors didn't quite get it. Budgets were unheard of. Roman emperors simply spent the money they had and then marched off to conquer more wealthy land when the money ran out. Thus Roman armies conquered and looted Syria, Egypt, Judea, Gaul, Spain, Assyria, Mesopotamia, and all of the other nations along the Mediterranean. After each conquest, Roman coffers would bulge with bounty; but soon the wealth would be squandered.

Rome's conquer and loot strategy would ultimately prove unsustainable. It cost each emperor a sizeable fraction of his loot to fund the armies to make future conquests. As the boundaries of the Roman empire pressed outward, Roman armies became more widely dispersed. To concentrate sufficient military force to continue the conquest, it took increasingly larger and larger armies, and every year they had to travel longer and longer distances.

What Rome was not spending on its armies it spent on goods from other lands. Rome produced little. It therefore had to import most of its goods from other lands, notably from Asia. Thus Rome experienced a phenomenon that perhaps no prior civilization ever had: a trade imbalance. Rome had no modern day economic theory to understand and deal with this — not that the condition went unnoticed. The emperor Tiberius was heard to complain that Rome's wealth was being transferred to foreign and even hostile nations. In 77 A.D., Pliny the Elder complained that as much as 550 million sesterces (138 million denarii) a year were squandered in India on luxury goods. Rome simply lacked the sophistication to address this trade imbalance, and that ultimately contributed greatly to its downfall.

Eventually, the Roman empire stopped growing in size. It was simply impossible to fund further expansion, given the amount it cost versus the amount it returned. Yet the Roman emperors, seeking to hold on to their conquests, continued to pour increasingly large sums of money into the Roman armies, which were now equipped with elaborate and expensive equipment. Meanwhile, at home the government bureaucracy grew at roughly the same rate as the army. In an effort to govern the massive Roman empire, emperors subdivided the realm into increasingly smaller parts, each with its own layers of local government administration. This placed an even greater burden on Rome's money supply.

In an effort to combat the dwindling money supply, emperor Nero in 64 A.D. conceived a plan to would ultimately lead to financial disaster. Nero recalled all of the coins in circulation. He ordered his mints to produce new gold and silver coins, naturally stamped with his portrait, but slightly smaller than before. He also ordered his mints to reduce the amount of silver by approximately 15 percent and the amount of gold by approximately 11 percent. Later emperors copied Nero's strategy. In a span of about 200 years, Roman emperors eventually leached out all of the silver content from the "silver" coins.

While this practice did increase the number of coins in circulation, it did not increase the money supply. As coins fell in silver and gold content, prices rose. Wheat that sold for one-half a denarius in the 2nd century sold for 100 denarii in the 3rd century. Naïve when it came to economic theory, the Roman emperors were wholly unprepared to address the inflation their monetary policy had caused.

Not satisfied with simply manipulating the money supply, Roman emperors relied increasingly on heavy taxation to support the empire's military and bureaucratic weight. Initially tax revenue came from two sources, a head tax, or *tributum capitis*, levied annually against every adult between the ages of 12 and 65, and a property tax, or *tributum soli*, levied annually against all land, ships, slaves, animals, and other personal property. The famous biblical story of the birth of Jesus in a stable in Nazareth took place during a Roman census, ordered by Caesar to collect the *tributum capitis*.

As the reign was handed down from emperor to emperor, new forms of taxation were invented to supplement the two main staple taxes, *tributum capitis* and *tributum soli*. One notable tax was a sales tax, levied on many traded commodities. It began as a special, temporary tax called an *indicto*, intended to support the Roman armies. It quickly became a permanent fixture of the Roman tax structure. The sales tax did little to help as the tax collectors were often quite corrupt. Anyone deemed a "traitor" could have all his property seized. Thus as the need for money grew, Roman tax collectors became quite adept at discovering the many "traitors" in the realm, so that they could be brought before the eagle of justice to be picked clean of their property.

As the Roman economy slowly deteriorated, associated problems grew. Unable to support themselves, many citizens had to rely on the Roman welfare system, which doled out free wheat so that the poor might eat. At the time of Julius Caesar, nearly one-third of the population was on the public dole.

Although it took centuries, the Roman empire finally crumbled into financial ruin. Rome's economic apogee probably occurred sometime during the reign of Marcus Aurelius, who ruled Rome from 161 A.D. to 1800 A.D. Unlike some of the later rulers who exploited all for their personal gain, Marcus Aurelius was a benevolent philosopher-king who befriended the poor, freed slaves at every opportunity, ordered gladiators to fight with blunted points, and even resisted the urge to tax. Aurelius kept a journal, in which he wrote down his thoughts on life. "If thou canst see sharp, look and judge wisely, says the philosopher."[13] Aurelius wrote ostensibly to guide himself in the ruling of Rome. "That which is not good for the swarm, neither is it good for the bee."[14]

Marcus Aurelius, as Stoic philosopher, saw man as a part of nature:

> The universal nature out of the universal substance, as if it were wax, now molds a horse, and when it has broken this up, it uses the material for a tree, then for a man, then for something else . . . Nature which governs the whole will soon change all things which you see, and out of their substance will make other things, and again other things . . . in order that the world may be ever new.[15]

Although the Roman empire was clearly in decline after Aurelius, the empire would continue to spread its influence for centuries to come. One lasting influence was a fusion of money and societal mores that remain fused to this day. Perhaps it was Aurelius's open-minded approach to government, or perhaps it was simply in the nature of the Roman citizens, but Rome evolved in a way that other ancient civilizations had not. Rome organized its society around money. Other civilizations of that day were far more traditional and organized themselves around a strong central government. That is not to say that Rome did not provide a strong central government, but rather that Rome allowed money to be used by the citizens to buy and sell land. This gave rise to an equestrian class that began to rival the patrician class of the traditional aristocracy.

We use terms today like "new money" and "old money" to distinguish between classes, in much the same way as the Romans would have distinguished the equestrian class from the patrician class. We take for granted such axioms as "money is power." Little to we realize today that *how* money becomes power is a complex set of societal values that we learned from the successes and failures of the ancient Romans.

---

[13] Marcus Aurelius, The Meditations, Book 8.
[14] *Id.*, Book 6.
[15] *Id.*, Book 7.

§ 4.07 The Chinese Invent Paper Money and Discover Inflation

Compared to the Lydians, the Chinese were slow to exploit the full power of coinage. This led to a different Chinese innovation that put the Chinese about 500 years ahead of their European counterparts: paper money. Because Chinese coins were individually of such little value, long strings of coins were needed to make even medium-sized purchases. Facing severe copper shortages that limited the coin supply, Chinese emperors began printing paper money. The first emperor to print paper money was Hien Tsung, who reigned from 806 to 821 A.D. The early Chinese treated this new paper money as a temporary substitute for copper money, which was then in short supply. What befell this early paper money economy is uncertain. However, later paper money issues in the 1020-1030 A.D. timeframe lead to another Chinese discovery: hyperinflation.

At that time China was exporting huge amounts of cash to maintain China's considerable appetite for imported goods and to buy off potential invaders from the North. To cover the cash flow out of China, the emperor simply printed more paper money. This produced an inflationary spiral, rendering the Chinese paper money worthless. Later emperors simply replaced the old paper money issue with new, but these new issue also eventually fell into the inflationary spiral.

§ 4.08 The Knights Templar Invent Banking

After the decline of the Roman empire, the European world fell into an economic malaise, known as the Dark Ages. Although money had been in use for a thousand years by this stage, very little of it was seen changing hands during the Dark Ages. For the most part, all of western Europe reverted to barter. Only the eastern Mediterranean region, where money had begun, continued to use it.

In about 1120, well over six hundred years after the fall of the Roman empire, a religious order known as the Order of the Knights of the Temple of Jerusalem — The Templars — formed in Jerusalem. The Order was sponsored by King Henry II of England, who donated to it a large sum of money, as atonement for having ordered the murder of the Archbishop of Canterbury, Thomas à Becket. Although the order was formed primarily to rid the Holy Land of the Infidels, they also managed to become the world's first bankers. The Templars were fierce soldiers, spartan in lifestyle and strict in their religious practices.

In medieval warfare, at which the Templars were superbly adept, the building and maintaining of castles at strategic locations was key. The Templars maintained a number of heavily protected castles at strategic cities and ports throughout Europe. They transported wealth seized during the Crusades back to England, using these well positioned castles to store the wealth while in transit.

With their reputation as fierce solders, trading merchants soon discovered that these Templars made excellent couriers, and that their castles made excellent strongholds. Trade with people of foreign lands was very risky in the Middle Ages. Bands of robbers might jump out from any forest or outcropping of rock. Trading merchants saw the Templars as the ideal solution. Traveling with them was safe, and their castles made excellent banks. The Templars even had their own Mediterranean fleet, which facilitated trade with that region.

Eventually, the Templars recognized the lucrative nature of their protection service and began acting as bankers. They charged a fee for each transaction handled, and often took an additional cut from the exchange rate applied when converting one currency to another. They made loans to kings and even administered wealthy estates. Knights departing for the Holy Wars would deposit their valuables with the Templars and would file with the Templars their last will and testament, in the event they did not return from the wars.

For 175 years the Templars prospered. They had heavily guarded fortresses in major cities, including a particularly wealthy one in Paris. At their peak they were said to have 870 castles strewn from England to Jerusalem, and a clientele ranging from kings to the Catholic church. The Templars became the treasury for all of western Europe and occasionally they found themselves funding wars between neighboring kingdoms. Loaning money to kings, the Templars learned, was not always wise business, for not all kings shared the Templars code of honesty.

It happened in 1295. King Philip IV of France had managed to spend his country into serious debt. Philip had tried Roman emperor Nero's trick of minting new money with lower precious metal content, but that got him nowhere because his subjects simply began paying their taxes in this new, devalued money. Philip had more coins in his coffer, but no more wealth. The answer to Philip's financial woes, he figured, was just inside the walls of the Templar's Paris castle.

Rather than make a frontal attack upon the castle, which would probably have failed, Philip devised a more clever plan. He waited until the Templars held a scheduled business meeting in France and then ordered his soldiers to arrest the Templar leaders. Concurrently, Philip spread rumors that the Templars, who had all taken vows of celibacy, were child molesters, devil worshipers, and heretics. Riding on a wave of public outcry, which Philip's rumors had started, he began prosecuting the Templar members one by one. The Templars initially fought back, but public sentiment was running strongly against them. Under pressure from King Philip, Pope Clement V issued an order on March 22, 1312, that the Templar Order was abolished. The Templars were then rounded up and imprisoned. Philip then seized the castle and had all of the imprisoned Templars put to death in a mass execution.

So ended the first world banking system. The extermination of the Templar's left a void that neither government nor the Church had sufficient power to fill. This left the field of banking open to private enterprise, which several astute Italian families discovered and cultivated to amass a fabulous wealth.

§ 4.09   Italian Banking Families Exploit the Bill of Exchange

With the Templars gone, merchants once again faced the old familiar problem, how to do business in foreign lands without being robbed en route. This time, several wealthy families in northern Italy set up innovative private banking businesses through which merchants could do business securely. Operating without fortress walls as the Templars had done, the Italian innovation was the *bill of exchange*. Italy was predominately Catholic. The Catholic church strictly prohibited the lending of money. Any person caught lending money for interest was excommunicated.

The bill of exchange created a clever loophole in the usury laws. The bill of exchange was a contractual order requiring the payment of a certain amount of money at a certain time and place. Usually the exchange involved a change from one currency to another. It was considered lawful to charge a fee for the service of providing the monetary conversion. Thus a merchant in Venice could go to the Italian banker and receive cash in the form of, say ducats of Venice. Both merchant and banker would execute a bill of exchange whereby the merchant agreed to pay a slightly higher amount at a later date in, say, Florence, making payment in florins, the currency of Florence. The differential between the ducats received and the florins repaid was identified as the fee for making the currency exchange. This maneuver sidestepped the usury laws.

The beauty of the bill of exchange was that it allowed money to flow without the need to cart boxes of coins from place to place. Robbers seeking to make an easy plunder found merchants carrying nothing but paper documents, contracts for which the robbers could receive nothing of value. The bill of exchange proved to be an economic lever of tremendous power. No longer requiring transportation of heavy and easily stolen coins, commerce accelerated. Everybody was protected. Everybody got rich.

The bill of exchange gave western civilization the power of a single currency, without treading on the monetary domains of individual governments. Eventually, bills of exchange themselves were being traded between merchants, as money. Thus the bills of exchange put more money into circulation. It worked like this. Duke Leto has 100 florins in a strongbox in his castle keep, doing nothing. He takes the 100 florins and deposits it with an Italian banker in Pisa. The banker lends the 100 florins to an English merchant, circulating a bill of exchange as money. The result: the spending power of Duke Leto's original 100 florins is multiplied many fold. The Duke has 100 florins on deposit with the bank; the bank has 100 florins on its books; the borrowing merchant has 100 florins to spend, and the person holding the bill of exchange has 100 florins as well. Each party may be in a different city. The spending power of Duke Leto's 100 florins is now spread throughout Europe.

Italian bills of exchange fueled the European economy and pulled Europe toward the Renaissance. By 1422, seventy years before Columbus discovered America, the city of Florence boasted 72 banks. Among the most powerful and influential was the bank of the Medici family. The Medici were supremely successful at banking and poured vast amounts of their wealth into art.

[A]   Arabic Numerals and Arithmetic Fuel the Italian Renaissance

Today we take for granted such basic innovations as numbers and arithmetic. These innovations were key to the success of the Italian banks, and were probably far more responsible for the Renaissance than the upsurge in art and literature. The Arabic numerals that we use today were actually invented in India. Arab mathematicians undoubtedly perfected their use. Leonardo Fibonacci introduced the Arabic numbering system to Europe in 1202, as a replacement for the cumbersome Roman numeral system that was being used at the time.

It seems like a simple thing today, but one extremely significant advantage Arabic numerals had over Roman numerals is size. Arabic numerals take far less space to write. Compare Arabic numeral ''8'' with Roman numeral ''VIII.'' The compact size makes Arabic numerals better suited to performing arithmetic. Adding and subtracting Roman numerals is difficult enough, but imagine doing multiplication and division. It is next to impossible.

The use of the numeral zero as a place holder, also an Indian invention, is also key to performing arithmetic. As every grade school child knows, the numeral zero makes all the difference between 10 and 100. Without it, simple arithmetic would be impossible.

Arabic numerals, with zero as place holder, and arithmetic made it possible for Italian bankers to keep books of accounts. Armed with arithmetic, bankers and merchants were able to dispense with the abacus, and could easily jot down prices and keep records on slips of paper.

[B]   The Bank of Florence Fails When King Edward Defaults

Just as in Templar times, the banks of the Italian Renaissance got into the habit of loaning large sums of money to kings and queens. That would prove to be their undoing.

At the outset of what would later be called the Hundred Years War between England and France, King Edward III borrowed heavily from several of the major Florentine banks. As the name suggests, the war did not go well for either side and caused both nations to run up enormous debt. In 1343, when it was time for Edward III to pay back his loans, he was unable to do so and simply defaulted. The Florentine bankers were trapped. With no way to cover the deposits of their many good customers, they failed and the economy built upon bills of exchange collapsed like a house of cards. As if that were not enough, as the economic downturn was at its worst, a deadly epidemic, the Black Death, swept through northern Italy throwing the economy into utter disaster.

Fortunately history gave bankers another chance. After Europe recovered from the Black Death, banking again took root, much under the same model as developed by the Italians.

[C] Banking Changes the Social Structure — Feudalism Is Destroyed

The system of banking that took hold after the Italian Renaissance, and the use of bills of exchange and paper money, changed the world economy. That, in turn, changed society. The plentiful supply of money, which could be exchanged for a day's labor, made it possible for peasants to buy land. Under the feudal system land passed by inheritance. The nobility owned the land and the peasants were relegated to the lower class position of having to work it. Money gave the peasant class a choice. They became skilled tradesmen, merchants, and ultimately land owners themselves. The feudal system was no longer needed and it disappeared.

While the ability to own land was a big advance for the working class, the changes brought about by money did not stop there. Gradually, even land ownership became less important. As businesses formed and grew in sophistication, and the focus shifted from agriculture to industry, owning stocks, bonds, and corporations became more important than owning land. Money had several times nearly choked civilization, but each time civilization recovered and grew stronger and more sophisticated. The next challenge would be gold.

§ 4.10 Spanish Gold Floods Europe and Bankrupts Spain

Gold had been used in coins since the very invention of money by the Lydians. It had always been in short supply and thus it was highly valued. When Columbus discovered America, the Spanish almost immediately began scouring the coasts of the Americas looking for gold. They found it. The Spanish discovered the Aztec civilization in Central America had amassed great quantities of gold and silver. The Aztecs used gold and silver for artwork and jewelry. Later the Spanish discovered the Inca civilization in Peru. They too had rich hordes of gold and silver.

For the next 50 years the Spanish looted all the gold and silver they could find from the native population. Oblivious to the value of the gold and silver as art objects, the Spanish kings melted down the gold and silver artifacts to make coins, which they promptly spent to finance wars with the rest of Europe. When the supply of gold and silver artifacts finally ran out, the Spanish established mines in the Americas. Most of the mines were owned and operated by the Spanish Crown. The Crown staffed its mining operations with officials from Spain, who in turn inducted the native American people into the mining operation. Spanish galleons transported tons of gold and silver back to Spain, and returned with provisions from Europe, which were heavily taxed.

From 1,500 to about 1,800 mines in the Americas produced approximately 70 percent of the world's gold and 85 percent of the world's silver. Much of this flowed through the treasuries of Spain and Portugal. This flow of gold and silver may have temporarily advanced the political causes of Spain and Portugal, but it brought on a devastating side effect: inflation. In the first century in which mining operations began, prices in Spain rose 400 percent.

Even worse, at around the time Columbus sailed for America, the Spanish monarchy had expelled the Jews and Muslims, which constituted nearly the entire merchant class within Spain. Thus, now with the influx of gold and silver at unprecedented high levels, there was no one in Spain to support a mercantile economy. Merchants in neighboring Italy, Germany, and Holland stepped in to fill the void, but that meant that all the wealth extracted from the Americas simply slipped through Spanish and Portuguese fingers, as they were forced to purchase everything from other countries.

It is difficult to imagine how Spain, with every ship in its fleet bringing gold and silver by the ton, could possibly go into financial ruin, but that is what happened. King Philip II had launched the Spanish Armada against England in 1588 at considerable expense. He also waged war against the Netherlands, the Germans, and the Turks. He found he was spending his Galleon fleet's cargo before it even arrived in port, and often borrowed from bankers in Italy, Germany, and the Netherlands. In 1575, King Philip defaulted on a loan from the Italian bankers, and they cut off funds. Philip was unable to pay his soldiers. They took matters into their own hands and sacked the city of Antwerp, where Philip had been conducting a focused campaign against the Protestants. Sacking Antwerp disrupted trade, causing considerable damage to the Spanish economy which had relied on the merchant class there to supply its needs.

While Spain and Portugal were struggling with the inflation-induced poverty the influx of gold and silver had produced, other nations were busy building financial institutions that would serve them to this day. In 1656, the Riksban bank of Sweden was founded. In 1694, the Bank of England was founded. These were significant steps forward because they represented recognition that government involvement in the once privately handled banking function was warranted. However, even government involvement in the banking system would not manage to tame this Lydian invention called money. How to use gold and silver as the basis for the international monetary system would prove to be a continuing challenge for at least the next 400 years.

§ 4.11   The United States Government Is Founded and Enters the Banking Business

Money was scarce in colonial America. Although most of the colonial settlers had come from England, there was very little English money to be had. Spanish and Portuguese coins were far more prevalent, and these too were in short supply. As a result, many colonists used tobacco leaves, furs, and indigenous wampum for their day-to-day transactions. When the colonists revolted in 1776, the Continental government began printing paper money, called appropriately, *continentals*. Unfortunately, the Continental Congress had little of value to back up this paper money, and inflation ran rampant. During 1775 to 1780, the Continental government printed some $241 million continental dollars. Individual states issued another $210 million. Only $11.5 million of this issued sum was backed by real money, mostly Spanish dollars. As inflation took its toll, the value of the continental dropped to near worthlessness. By 1781 the value of 100 continental dollars had fallen to 10 cents. When the revolutionary war ended, the government ceased issuing the worthless currency. Several of the States did so as well.

After the war, the new nation's economy was in shambles. Paper money had become the political hot potato of the day, the issue even inciting several riots. We may look back upon the Constitutional Convention of 1787 with nostalgic appreciation for the wisdom of the founding fathers, but the Convention was by no means a picnic for the participants at the time. At the very top of the list of topics under debate was money. Who should be in control of the money supply? The Constitution tells us who won the debate:

> Article I, Section 8.
>
> Clause 1: The Congress shall have Power To lay and collect Taxes, Duties, Imposts and Excises, to pay the Debts and provide for the common Defence and general Welfare of the United States; but all Duties, Imposts and Excises shall be uniform throughout the United States;
>
> Clause 2: To borrow Money on the credit of the United States;
>
> Clause 3: To regulate Commerce with foreign Nations, and among the several States, and with the Indian Tribes;
>
> Clause 4: To establish an uniform Rule of Naturalization, and uniform Laws on the subject of Bankruptcies throughout the United States;
>
> Clause 5: To coin Money, regulate the Value thereof, and of foreign Coin, and fix the Standard of Weights and Measures;
>
> Clause 6: To provide for the Punishment of counterfeiting the Securities and current Coin of the United States;

The United States of America began operating under its current Constitution in 1789. In 1791, Congress founded the first nationally chartered bank, aptly named the First United States Bank. The bank had a Congressional charter to operate from 1791 until 1811. There were also a number of state chartered banks in existence at the time, and these were permitted to remain in operation.

Pursuant to its Constitutional responsibility, Congress established, through the Coinage Act of 1792, that the United States unit of legal tender would be the *dollar* and that it would be tied to the value of gold and silver. The official exchange rate set one U.S. dollar at 371.25 grains of silver or 24.75 grains of gold. Thus a person wishing to purchase an ounce of gold could pay for it with $19.75 U.S. dollars. The first mint was built in Philadelphia. It began minting gold, silver, and copper coins in 1794.

It was Thomas Jefferson who suggested that a gold and silver, bimetal standard should be used. Copper coins were also permitted, although these were used, as today, to make small change. Jefferson believed that reliance upon a gold-only standard might jeopardize the U.S. economy because gold was in short supply. Jefferson understood that if sufficient coinage were not available, the citizens might revert to tobacco leaves and wampum.

[A] The First United States Bank Opens and then Closes Its Doors

The founding of the First United States Bank was a politically charged struggle. Even though the Constitution had placed the power to control the money supply with the federal government, there were many who argued for State control at every opportunity and considered a National Bank to be unconstitutional. Alexander Hamilton, who was then serving as Secretary of the Treasury, introduced the bill to charter a national bank to Congress, where it passed by slim margins. President George Washington is said to have been leaning toward veto of the bill, when Hamilton prevailed upon him to see the merits of what a national bank could provide.

Throughout its 20 year charter, there were many who were opposed to the federal government being engaged in the banking business. The national bank was an economic success, which undoubtedly irritated the private banking community. Cries of ''monopoly'' were heard within the private banking circles. This may have been at least partially justified, because the U.S. Treasury placed up to 90 percent of its treasury deposits with the First United States Bank.

When the Bank's 20 year charter was about to expire, a bill was drawn up to renew it. The House voted to renew the Bank's charter by a vote of 65 to 64. The Senate deadlocked in a tie vote, 17 yea, 17 nay. The vice-president cast the tie breaking vote against renewal, and the First United States Bank ceased to be.

[B] The Second United States Bank Is Chartered

Even as the First Bank was being dismantled, details for a Second United States Bank were in the works. As proponents of the First Bank had predicted, as soon as the First Bank ceased operation, the nation began to experience inflation. Indeed, the First Bank had closed its doors in 1811 and in 1812 the United States found itself pitted against England in the War of 1812. Without the discipline and stability of a national bank to regulate the money supply, paper money from small, weak banks began flooding financial markets. Serious inflation set in.

This produced a political climate that allowed the Second United States Bank to be chartered in April 1816. It began operations in Philadelphia in January 1817 and financially operated much as the First Bank had. Unlike the First Bank, however, the Second Bank pursued a branching policy, establishing at least one branch bank in every State. The Second Bank restored the nation to a financially sound footing; however, not without making enemies. As with the First Bank, the Second Bank eventually amassed enough political detractors to defeat its charter renewal. The political demise of the Second Bank began in 1828 with the election of Andrew Jackson to the Presidency.

Jackson didn't trust bankers, having seen how other economies had failed in a bubble burst of defaulted loans. In Jackson's first address to Congress after election, he questioned the necessity for a national bank. Several States had earlier tried to ban the national bank from within their borders. The State of Maryland had even gone so far as to impose a tax on the national Bank. The Supreme Court, in its famous *McCulloch v. Maryland*[16] decision, ruled that Maryland's action violated the supremacy clause of the Constitution and struck down the law as unconstitutional. However, even the Supreme Court could not stop Andrew Jackson. When renewal of the Second Bank's charter passed Congress and arrived on his desk for signature, he vetoed the bill.

When Jackson ran for reelection in 1832, his opponent, Henry Clay made Jackson's veto of the Second Bank's charter his battle cry. Clay lost the election, 49 electoral votes to Jackson's 219. With Clay's overwhelming loss, the chances for the Second Bank's renewal came to an end.

§4.12   Wildcat Bankers Scatter Coherent Banking Strategy to the Wind

After the close of the Second United States Bank, a large number of independent banks, called "wildcat banks" by their detractors, rushed in to fill the void. The total number of banks doubled between 1830 and 1836, and more than doubled again by 1861. Largely unregulated, these banks freely issued bank notes of their own origin. A business person in the 1850s would encounter so many different types of bank notes that he or she needed a book to determine which were genuine. *Hodges Genuine Bank Notes of America* was such a book. In 1859, it listed 9,916 notes issued by 1,365 banks. Actually, a business person of that day probably needed two books. The *Nicholas Bank Note Reporter* listed some 5,400 different known counterfeit notes in circulation. Perhaps Samuel Morse, who invented the telegraph in 1837, might have carried a copy of *Hodges* with him as he negotiated with businesses to popularize his newfangled mode of communication. Little did he know that one day his telegraphic form of communication would be used to transfer money between America and Europe, with no gold or paper money changing hands.

Amid the economic turmoil of the wildcat banking era, America dove headlong into civil war. The war lasted from 1861 to 1865. Finally realizing that something had to be done to contain the complexity to which unregulated banking had evolved, Congress passed the National Banking Act of 1863. The Act established a national banking system which ended the wildcat era by taxing all notes issued by state banks. The act also established a national currency controlled by the federal government. Although there was no national bank, there were federally chartered banks. Congress also passed a series of acts whereby it issued legal tender in the form of paper money that were actually non-interest-bearing bonds. Green in color, the money was called "greenbacks," a term we still use colloquially today. In theory, the greenbacks would be redeemable in gold after the war.

---

[16] 17 U.S. 316, 4 L. Ed. 579 (1819).

Like many governments before, the U.S. government issued more greenbacks than they had gold and silver to back them, and the value of each greenback dollar plummeted. The Confederate government issued paper notes as well, and these suffered even greater devaluation as the Confederacy failed.

After the Civil War, the South's economy was wrecked and the North's economy was in serious trouble as well, thanks to the inflationary effect of the $450 million greenback dollars in circulation. The Civil War did accomplish one thing, however. It destroyed the State's power to issue currency and put the federal government finally in control of the national money supply.

[A] The Wizard of Oz

After the Civil War, people's focus turned sharply on the national monetary policy. Poor farmers of the West and South wanted the government to issue more paper money and silver coins. In 1874, they formed the Greenback Party and the Populist Party to promote their positions. They reasoned that with more money in circulation it would be easier for them to sell their crops and purchase the things they needed. These farmers distrusted the wealthy bankers, and their gold-oriented monetary policy. Even those who did not support a proliferation of greenback dollars still wanted a bimetal, gold and silver standard. The bankers, in contrast, wanted the nation to move to a singular gold standard. Even the bimetal, gold and silver, standard initiated by Thomas Jefferson seemed outdated and too difficult to control.

Few recognize this today, but L. Frank Baum's novel, *The Wonderful Wizard of Oz*, published in 1900, was actually a political satire on the gold standard debate. This may spoil your childhood memories of the movie but here is the analysis in a nutshell:

- Populist orator, Leslie Kelsey, spoke out against the gold standard. His nickname was "the Kansas Tornado."

- Dorothy follows the yellow brick (gold) road to a land called Oz, where wicked witches and wizards of banking live.

- Oz is an abbreviation for ounce, the standard by which gold is measured.

- In route to Oz, Dorothy first meets the Scarecrow. He represents the American farmer. She next meets the Tin Woodsman. He represents the factory worker. Then she meets the Cowardly Lion, who represents William Jennings Bryan.

- The Wizard of Oz is Marcus Hanna, leader of the Republican Party and power behind the McKinley administration that wanted to adopt the gold standard.

- The Wizard lives in Emerald City. The Populists believed that the gold standard was designed to favor wealthy city dwellers at the expense of the populace. Behind curtains of secrecy, the Wizard pulls the levers and cords that manipulate the economy.

- Munchkins were simpleminded people of the East who did not understand how the Wizard was manipulating them.

- In the book, the inhabitants of Oz were required to wear green-colored sunglasses, attached by a gold buckle.

- In the end Dorothy triumphs. Her friend the farmer, Scarecrow finds he has a brain after all. The factory worker Tin Man (in the book) receives a golden ax with a silver blade that will never rust again. The Cowardly Lion discovers his courage and speaks out for what he knows is right.[17]

§ 4.13   The Panic of 1907 Spawns the Federal Reserve System

The crisis began when five New York banks could not cover their accounts on October 14, 1907. A week later several of the nation's largest trust companies failed. Bankers panicked and the nation went into economic shock. In the two years that followed, some 246 banks failed. Why had this happened? Economists at the time attributed it to the fact that because the stock market was centered in New York, most of the nation's bank reserves were also there, nearly 75 percent in 1900. Thus the nation's money supply was in the hands of a very few New York bankers. When the New York banks got into trouble, it put the entire nation's economy at risk.

The proposed solution, which was signed into law December 23, 1913, by President Woodrow Wilson, was the Federal Reserve System. The law established federal reserve banks upon which the member banks could draw in time of need to cover short-term demand for gold. The Federal Reserve System was a major step toward an elastic currency system because the federal reserve banks had the power to control the discount rate at which it loaned money to the member banks. Under the plan, each member bank was required to deposit a minimum reserve amount in its Reserve Bank and the Reserve Bank, in turn, was required to hold a minimum of 35 percent of the balance in legal tender. The law required each Reserve Bank to back its reserve notes with 40 percent in gold.

Considerable attention was given to how the Reserve System would govern itself. The Reserve Banks were organized into geographic regions, called Reserve Districts, to which the member banks belonged. The Reserve Districts were collectively governed by a Central Board, based not in the financial capital, New York, but in the political capital, Washington, D.C. No two Central Board members could come from the same Reserve District, thus ensuring the entire country would be well represented.

At the outset, the Federal Reserve System had been designed to allow discount rates to be established on a regional basis. By the 1920s however, the New York Federal Reserve District had assumed the dominant position, largely because New York was the nation's dominant financial center. This had the

---

[17] For a more in-depth analysis of the Wizard of Oz, see J. Weatherford, *supra*, note 6 at 175-177.

effect of centralizing control over the nation's money supply, even though the Federal Reserve System had been designed to be regionally controlled.

At the heart of the Federal Reserve System's control over the money system is the discount rate. Each Reserve District could set the rate at which it would re-discount commercial paper, based on the perceived needs of this region. These discount rates were subject to the approval of the Federal Reserve Board. The Board also had the power to select which forms of commercial paper were eligible to be re-discounted. Thus short-term notes needed to cover production of goods already ordered could be given preferential treatment over long-term notes covering more speculative ventures.

§ 4.14   The Stock Market Crash of 1929 Brings Major Changes in U.S. Monetary Controls

The Federal Reserve System was not insurance against all economic woe, as the Crash of 1929 demonstrated. Indeed, some say "the Fed" may have been partially at fault by affording the easy credit that promoted speculation and then failing to react when things got out of hand. Why the Crash of 1929 happened is a complex issue, upon which noted experts certainly do not agree. According to Professor Galbraith,[18] the mid-1920s had boasted a get-rich-quick mentality. Speculators first poured money into Florida land speculation and then moved from land into the stock market when a series of hurricanes wiped out the beachfront. As demand for stocks increased, the Federal Reserve Bank relaxed credit by cutting its discount rate from four percent to three-and-a-half percent. Galbraith does not believe the reduction in discount rate can be blamed for stock speculation. Milton Friedman, on the other hand, seems to place at least some blame on the Fed for stimulating the stock market boom prior to the crash.[19]

Regardless of the cause, the Crash of 1929 precipitated a series of banking reforms, many of which were spearheaded by newly elected President Franklin D. Roosevelt. Through the Banking Act of 1933, Congress created the Federal Deposit Insurance Corporation, or FDIC. This required all banks to pay a small insurance premium which generated a substantial sum to be used to guarantee the repayment of bank customers' deposits up to a certain amount. Nearly every bank joined, such that by January 1934, 97 percent of all bank deposits were guaranteed.

The FDIC was designed to prevent future runs on the bank. However, President Roosevelt had a present financial crisis to deal with. On March 9, 1933, Congress enacted a law that gave Roosevelt the power to prevent the hording of gold, and in an executive order one month later, Roosevelt nationalized the gold supply. Citizens who voluntarily surrendered their gold were paid $20.67 per ounce in paper money. One year later, the government devalued its paper money from $20.67 to $35 for each ounce of gold. This meant that everyone who had

---

[18] As reported by G. Davies, *supra*, note 1 at 508.
[19] *Id.*

complied with Roosevelt's order lost 41 percent of their gold's value. In 1934, Roosevelt issued a similar order, nationalizing silver. Both the gold and silver were melted down into ingots, much like the Spanish had done with the Aztec and Inca gold and silver art objects four centuries before. To house the horde of gold and silver, the U.S. Treasury built Fort Knox. It was completed in December 1936.

President Roosevelt's action in nationalizing the United States gold and silver reserves was a significant first step towards detaching the U.S. money supply from bullion and any other commodity. As we shall see, the second and final step was taken by President Nixon in 1971, after which time the U.S. money supply became truly an abstract measure of worth, tied to nothing but trust.

§ 4.15   The Bretton Woods Agreement

According to British Economist, Glyn Davies, the American dollar became the most important currency in the world during the 20th century, replacing the British pound sterling for that honor.[20] It is therefore quite understandable that the rest of the world considers itself to have a vested interest in the well being of the American dollar. Davies puts it this way:

> The dollar climbed to its international eminence on the back of its factories and farms. The strength of the dollar was not derived from the strength of the American financial system; rather the reverse, for time and time again throughout its history the dollar, weakened by endemic failures, has been restored to strength through the robust power of American agriculture, forestry, mining, manufacturing industry and managerial expertise. The fact that major international institutions like the World Bank and IMF set up their headquarters in Washington simply endorsed and reinforced a choice that had already been made by the world's markets and underlined universal acceptance of the belief that the value of the dollar was too important a matter to be left to the decisions of American politicians alone.[21]

International involvement in the well being of the American dollar took on a formal tone in 1944. In July 1944, one month after the World War II invasion of Normandy, 700 Allied delegates from 44 nations met at Bretton Woods hotel in New Hampshire to discuss how they could stabilize the World economy. The delegates laid the groundwork for the World Bank and the International Monetary Fund, or IMF, and they agreed to peg their currencies to the American dollar. The United States, in turn, agreed to peg the dollar at $35 per ounce of gold. The Bretton Woods agreement required constant cooperation by all participating nations to adjust their exchange rates so the buying power of their currencies would remain pegged to the $35 per ounce U.S. dollar.

---

[20] *Id.* at 455.
[21] *Id.* at 455.

§ 4.16   Nixon Pulls Out of Bretton Woods and Takes the United States Off Gold Standard

Throughout history the need for money to fight wars has lured rulers into tinkering with the monetary system. Such was the case in America during the Vietnam War. President Johnson had spent heavily on that war, and when President Nixon took office he continued. The Vietnam War was quite unpopular in America, and both Johnson and Nixon had difficulty getting Congress to levy higher taxes to support it. Thus both presidents borrowed heavily, increasing the national debt to astronomical levels. Government spending on the war effort placed billions of additional dollars into circulation, and this caused inflation. The combination of massive borrowing and massive spending drove the U.S. economy into a state where the United States could no longer redeem dollars for gold in the world market. Something had to be done.

President Nixon and Treasury Secretary, John Connally, devised a plan. Under the Nixon-Connally plan, Nixon, by executive order, froze all prices, wages, and rents in an effort to put inflation in check. He also levied a ten percent surcharge on most imports to slow the growing foreign trade imbalance. Finally, on August 15, 1971, in a surprisingly bold move, Nixon told world financial leaders that the United States would no longer agree to redeem dollars at $35 per ounce of gold. This move repudiated the Bretton Woods agreement, and it demonstrated that the U.S. dollar peg, against which all other currencies were measured, had slipped. At first the United States struggled to reset the peg at a new, slightly weaker set point: 1972, $38 per ounce; 1973, $42.77 per ounce. However by November of 1973, it was apparent that the peg could not be artificially maintained. Nixon responded by taking the United States off the gold standard altogether. Now the U.S. dollar was truly free to float in value relative to all other currencies.

What was the effect of Nixon's bold move? The dollar dropped precipitously in value. In November of 1973, when Nixon took this final bold move, the dollar was valued at $42.77 per ounce. By 1995, the dollar had fallen to $400 per ounce.

§ 4.17   Alternate Forms of Money Are Invented

While the United States was struggling to maintain the gold standard, which it ultimately abandoned, there were other more modern forms of money being devised. Largely a product of the electronic computer age, these new forms of money caused commerce to once again speed up. The new forms of money began with a 1950 invention that today we take for granted: the credit card.

[A]   The First General Purpose Credit Card — Diners Club 1950

In and of themselves, credit cards were nothing new in 1950. Many department stores and gas stations had issued credit cards for some time. Gasoline credit cards may be traced well back to the 1930s. However, in 1950 a banking operation known as Diners Club began issuing a general purpose credit card that could be used to make purchases on credit at a variety of different participating establishments. At first the establishments were restaurants. However, eventually Diners Club persuaded other merchants to accept their card in lieu of cash payment. The advantage to the merchant was that it increased business because customers could purchase goods and services with money they had not yet earned. The credit card thus accelerated the economy by giving consumers the ability to put *future* money into current circulation.

As we all know, the credit card innovation caught on. In 1958, BankAmericard issued its general purpose credit card. The same year American Express did the same. MasterCard came later. It evolved from the Everything Card issued by New York's Citibank in 1967.

The credit card issuing companies were quick to employ the latest technology to make their products quicker and easier to use. American Express already had experience with electronic banking innovations. In 1960 it had began issuing travelers checks with electronic ink, allowing the checks to be sorted by machine. As soon as electronic communication was feasible, the credit card companies quickly adopted it as a means to facilitate faster, more secure commerce.

[B] Green Stamps and Frequent Flyer Miles

Credit cards allowed consumers to spend dollars they didn't have. Thus credit cards may be said to have created credit dollars. There were, however, several other forms of money that did not directly correlate to dollars. In the 1960s, for example, it was quite common to receive S&H Green Stamps at the grocery checkout counter. These stamps were awarded in measure based upon the amount spent at checkout and could be redeemed for prizes detailed in a catalog. The Green Stamps represented, in effect, a small refund of the grocery purchase price. The stamps were bearer instruments, which could be redeemed by anyone in possession of them.

Today, airline frequent flyer miles represent a similar form of money that is not directly tied to a dollar equivalent. Frequent flyer miles are awarded by airlines, hotels, grocery stores and even credit card companies. Frequent flyer miles are redeemable for airline tickets or airline seat upgrades and have no direct dollar equivalent. Frequent flyer miles are thus quite a bit like the S&H Green Stamps of the 1960s. They are "earned" by making purchases and are "redeemed" in goods or services other than cash dollars.

[C] Electronic Commerce 1970s Style

The first automated teller machine (ATM machine) appeared in Burbank, California in 1971. That same year the NASDAQ stock exchange went electronic. In 1972, the Federal Reserve Bank of San Francisco began using electronic funds transfer, or EFT. It took six years, but by 1978, all Federal Reserve Banks were linked electronically. What made EFT important is that it considerably lowered the cost and increased the speed of moving money. Prior to electronic funds transfer, money moved by written instruments, such as checks. It took a great deal of time to process these paper documents. Amounts had to be entered into ledgers, signatures had to be verified, and the paper documents had to be physically stored where they could be conveniently retrieved if needed. EFT eliminated all of that. By 1977, electronic commerce had evolved to the point where funds could be sent electronically across international borders. The international exchange of funds electronically was established by an organization called the Society for Worldwide Interbank Financial Telecommunication, or SWIFT. Based in Brussels, it grew to enormous size, connecting some 100 nations in an electronic cash flow network.

[1] Debit Cards

In 1974, a new kind of money card appeared on the scene: the debit card. The card was issued jointly by the First Federal Savings and Loan of Lincoln, Nebraska, and the Hinky Dinky grocery store. The card made it possible for shoppers to purchase groceries, using their own banked funds, but without taking time to write a check or stop at an ATM machine. Anyone who has experienced how slow cashing a check at a grocery store can be will appreciate why the debit card was an improvement. From a monetary standpoint, the debit card was significant because it was the first time a card actually became a substitute for money.

The actual debit card implementation of that day was physically not much different from a charge card. The card required access to a bank's computer system to obtain the account number, look up the account balance and effect the appropriate funds transfer. Future debit cards may not have this limitation. Indeed, if you happen to live in proximity to the Westminster Bank of England, you may be able to acquire a card issued under their Mondex system, which allows the card to disburse and receive funds without the need for an intermediary computer operated by a bank.

Introduced in 1995, the Mondex system uses a card with an embedded processor chip that stores the account balance and has the necessary software to know how to disburse and receive funds. When party *A* wishes to pay party *B* a certain sum of money, both parties insert their cards into a special reader device and the cards exchange electronic data — transferring funds from card *A* to card *B* — without any intervention by a bank. The Mondex cards thus act as portable electronic wallets.

[D] The Monetary Control Act of 1980

In 1980, Congress passed the Monetary Control Act. Its purpose was to enlarge the electronic network used by Federal Reserve Banks, so that other banks could use them to exchange funds in the Federal Reserve System. The initial implementation was hardly what one today would call a network. Banks wishing to transact business with the Federal Reserve system recorded electronic funds transfer data onto magnetic tapes. The tapes were then transported from one bank to another to allow the data to be exchanged. This was hardly convenient or particularly secure, but it did eliminate a great deal of paperwork that would otherwise have to be done.

The monetary network became truly networked, eliminating magnetic tape exchange, in the early 1990s. By 1992 the network was moving $200 trillion annually. By the mid-1990s the Federal Reserve System alone was moving upwards of $20 billion a day in electronic funds. Paper checks still outnumbered electronic transfers, at $47 billion a day, but electronic transfer is gaining on paper checks by the day.

[E] The Electronic Funds Transfer Act of 1996

On April 26, 1996, President Clinton signed into law an act that signaled the federal government's intention to fully support electronic funds transfer. Officially called the Debt Collection Improvement Act,[22] the act provided, subject to regulations to be developed, that all $1 billion in annual payments made by federal agencies, except for tax refunds, would be by electronic funds transfer. The regulations were passed two years later, on September 25, 1998, and are codified at 31 C.F.R. § 208.

The act is significant because it represents a clear signal, at least as far as the U.S. government is concerned, that money is shifting from paper form to electronic. With the concurrent advances in Internet technology, it seems likely that money and the Internet are going to merge. Money, after all, is a form of information, information about value, information about what a buyer will pay, and what a seller will accept. How this merger will take place remains to be seen. One thing is certain, whatever mechanisms are built and whatever processes are implemented, they will be the subject of many business model and e-commerce patents.

---

[22] 31 U.S.C. § 3332.

CHAPTER 5

THE NATURE OF COMMERCE TODAY—ELECTRONIC COMMERCE

[1] Open-architecture System for Real-time Consolidation of Information from Multiple Financial Systems — U.S. Patent No. 6,128,602

[2] System for Automatically Determining Net Capital Deductions for Securities Held, and Process for Implementing Same — U.S. Patent No. 6,144,947

[3] Facility-based Financing System — U.S. Patent No. 6,154,730

[4] System and Method for Creating and Managing a Synthetic Currency — U.S. Patent No. 6,188,993

[5] Global Financial Services Integration System and Process — U.S. Patent No. 6,226,623

[H] Inventory Management Patents

[1] Integrated System Monitoring Use of Materials, Controlling and Monitoring Delivery of Materials, and Providing Automated Billing of Delivered Materials — U.S. Patent No. 5,983,198

[2] System and Method for Dynamic Assembly of Packages in Retail Environments — U.S. Patent No. 6,138,105

[3] Supply Chain Financing System and Method — U.S. Patent No. 6,167,385

[4] Inventory Control System — U.S. Patent No. 6,188,991

[I] Marketing Patents

[1] Method and Apparatus for Selling an Aging Food Product as a Substitute for an Ordered Product — U.S. Patent No. 6,052,667

[2] Method and Apparatus for Issuing and Managing Gift Certificates — U.S. Patent No. 6,193,155

[3] Computer Implemented Marketing System — U.S. Patent No. 6,236,977

[J] New Money Patents

[1] Trusted Agents for Open Electronic Commerce — U.S. Patent No. 5,557,518

[2] Currency and Barter Exchange Debit Card and System — U.S. Patent No. 5,592,376

[3] Electronic Funds Transfer Instruments — U.S. Patent No. 5,677,955

[4] Untraceable Electronic Cash — U.S. Patent No. 5,768,385

[5] System and Method for Detecting Fraudulent Expenditure of Electronic Assets — U.S. Patent No. 5,878,138

139

[6] Electronic-monetary System — U.S. Patent No. 6,047,067

[7] Executable Digital Cash for Electronic Commerce — U.S. Patent No. 6,157,920

[8] Self-contained Payment System with Circulating Digital Vouchers — U.S. Patent No. 6,205,435

[K] Resource Allocation Patents

[1] Reorder System for Use with an Electronic Printing Press — U.S. Patent No. 6,246,993

[2] Integrated Business-to-business Web Commerce and Business Automation System — U.S. Patent No. 6,115,690

[3] Method for Production Planning in an Uncertain Demand Environment — U.S. Patent No. 6,138,103

[4] Method and Computer System for Controlling an Industrial Process by Analysis of Bottlenecks — U.S. Patent No. 6,144,893

[5] Method of Managing Contract Housekeeping Services — U.S. Patent No. 6,144,943

[6] Method and Apparatus to Connect Consumer to Expert — U.S. Patent No. 6,223,165

[L] Electronic Shopping Patents

[1] Method and System for Placing a Purchase Order Via a Communications Network — U.S. Patent No. 5,960,411

[2] Universal Web Shopping Cart and Method of On-line Transaction Processing — U.S. Patent No. 6,101,482

[3] Method and Apparatus for Executing Electronic Commercial Transactions with Minors — U.S. Patent No. 6,173,269

[4] System and Method for Home Grocery Shopping Including Item Categorization for Efficient Delivery and Pick-up — U.S. Patent No. 6,246,998

[M] Supply-chain Management Patents

[1] Method and Apparatus for Collaboratively Managing Supply Chains — U.S. Patent No. 6,157,915

[2] Method for Part Procurement in a Production System with Constrained Resources — U.S. Patent No. 5,970,465

[3] System and Process for Inter-domain Planning Analysis and Optimization Using Model Agents as Partial Replicas of Remote Domains — U.S. Patent No. 5,995,945

        [4]   **Projected Supply Planning Matching Assets with Demand in Microelectronics Manufacturing — U.S. Patent No. 6,049,742**

        [5]   **System and Method for Allocating Manufactured Products to Sellers — U.S. Patent No. 6,167,380**

[N]   **Other Business Model Patents**

        [1]   **Methods and Apparatus for Determining or Inferring Influential Rumormongers from Resource Usage Data — U.S. Patent No. 6,151,585**

        [2]   **Method and Apparatus for Surveying Music Listener Opinion About Songs — U.S. Patent No. 5,913,204**

        [3]   **Process to Convert Cost and Location of a Number of Actual Contingent Events Within a Region into a Three Dimensional Surface over a Map That Provides for Every Location Within the Region Its Own Estimate of Expected Cost for Future Contingent Events — U.S. Patent No. 6,186,793**

        [4]   **Service Business Management System — U.S. Patent No. 6,216,108**

        [5]   **Method and System for Accommodating Electronic Commerce in the Semiconductor Manufacturing Industry — U.S. Patent No. 6,240,400**

§ 5.01   Understanding Commerce

Like it or not, the U.S. patent system and commerce are intimately tied. The U.S. Patent Office is part of the U.S. Department of Commerce. Yet this observation hardly begins to explain the scope of the term "commerce," nor the intricate way in which commerce completely permeates and defines modern civilization.

Lexicographers define commerce on a large scale. The Microsoft Office 2000 on-line dictionary defines commerce quite succinctly: Commerce — the large-scale buying and selling of goods and services. The *Merriam Webster New Collegiate Dictionary* also adopts this definition, although it reveals the term commerce has a far broader meaning:

> com•merce \"ka-(')mers\ noun [MF, fr. L commercium, fr. com- + merc-, merx merchandise] (1537)
>   1: social intercourse: interchange of ideas, opinions, or sentiments
>   2: the exchange or buying and selling of commodities on a large scale involving transportation from place to place

For our purposes, the Microsoft Office 2000 definition will suffice. However, we should make it clear that implicit in the concept of buying and selling is the exchange of information and money. Advertising and marketing functions rely heavily on the exchange of information. Buyers will not buy unless they know something about what it is they are buying. Moreover, in today's information economy, information itself may be the product. In this sense, information can be treated as goods; or the supplying of information can be treated as services. Either viewpoint is valid.

The topic of money was discussed in the previous chapter. The use of money is the quintessential business method. Unquestionably, money is the glue that binds commerce together. Without money, modern commerce would cease. Thus it may be helpful here to repeat that money serves at least six basic functions. Money is:

1.   a medium of exchange
2.   a means of payment
3.   a common measure of value
4.   a store of value
5.   a unit of account
6.   a standard for deferred payments

At a fundamental level, money and information are intimately tied. As a common measure of value, money is a form of information. Money is the information exchanged between buyer and seller that determines the value of goods and services. It is not surprising then, that the electronic means to communicate information have been directly used to effect the flow of money. At the binary digit level there is no difference between one million words and one million dollars.

§ 5.02   A Generalized Commerce Model

In developing an understanding of any business concept or commerce-related invention, it is helpful to have a model. As Chapter 10 on claim drafting explains, inventions are abstractions given substance through practical application. Models can be very helpful in analyzing and describing these abstractions.

In the business model or e-commerce context, the invention will often reside in some specific commerce domain, such as marketing, sales, payment, shipment, product support or disposal. Thus your model will likely focus on the specific aspects that make the invention work in that domain. However, do not overlook how the invention fits into the bigger picture of commerce. To assist you in developing models for business-related inventions, this section will present a generalized model that can be used for all commerce. Constructing such a model is a tall order, because commerce is a complex and multifaceted topic. Use the commerce model developed here to help you think about your invention, in terms of these "big picture" concepts. Doing that should help you define what is important about your invention and to help you ensure that no important details are left out.

To develop the generalized commerce model, we first consider the basic process by which commerce flows. Raw materials and labor are input into the commerce system, which in turn produces products and services that are then marketed and sold to consumers: see Figure 5-1. If you were to examine any stage in this basic process under a magnifying glass, you would discover an intricate detail built upon many centuries of innovation.

Thus in preparing to draft a business model or e-commerce patent, it helps to understand where your magnifying glass is positioned. Is the invention an improvement in the way raw materials are obtained? Is the invention an improvement in the way labor is expended? Perhaps the innovation lies in how the resulting goods and services are marketed, sold or delivered to consumers. Many business-to-business (B2B) e-commerce systems are designed to facilitate the logistics of obtaining and utilizing raw materials and labor. On the other hand, customer relationship management (CRM) systems are typically directed to the marketing and sales end of the process.

It is usually easy to identify where along the basic commerce process your magnifying glass needs to be positioned. Inventors of business model and e-commerce innovations understand this instinctively, and may wonder why you would even ask such a basic question. The reason for asking is to make clear how the invention fits within the larger commerce framework before delving into the details. An invention focused on one stage in the commerce process may offer improvements downstream that could go undeveloped in the patent application if you do not start with the big picture first.

Having identified the basic commerce process flow, we need to develop a language that can be used to describe the component parts. We can do so by identifying the basics: what is being exchanged? by whom? by what methods? using what medium? Using these basic questions, we can define the component parts of our commerce model in terms of these four categories:

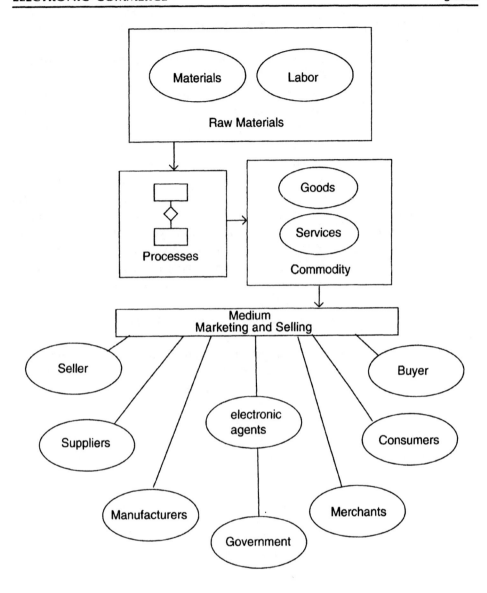

FIGURE 5-1: GENERALIZED COMMERCE MODEL

1. the Commodity — these are the goods and services being exchanged
2. the Parties — these are the many players that influence how goods and services are exchanged
3. the Processes — these are the fundamental operations performed by the Parties to effect commerce
4. the Medium — this represents the various media of exchange by which commerce is consummated and the communication channels through which commercial information and money flow.

[A] The Commodity Component of the Commerce Model

Taking these categories one by one, the Commodity component is probably the most straightforward. Commerce can be built around any goods and services. Often an invention will focus on a particular class of goods or services, and the inventor will likely understand his or her invention in that context. You should challenge the inventor to examine whether the invention may be applicable to other goods or services. For example, an improved assembly line to make and sell hamburgers could be extended to fried chicken, or pizza. It is doubtful that the method could be extended to manufacturing and selling books, however.

If you express the invention broadly enough to encompass hamburgers, fried chicken, pizza, and books, you could be setting yourself up for problems. When you generalize an invention sufficiently to cover a wide range of goods and services you may find the claims of this broadly stated invention will lead directly to the prior art. Therefore, you should be circumspect in how far you push the scope of the invention beyond what the inventor originally conceived.

[B] The Parties Component of the Commerce Model

The Parties component of our commerce model gets a bit more interesting. Often there will be influential players that you initially did not think of. Here is a top level list. Under each listed party you can expect to find numerous other players.

- Government
- Suppliers
- Manufacturers
- Merchants
- Consumers
- Electronic agents for any of the above

[1] Government as a Player in the Commerce Model

Suppliers, manufacturers, merchants, and consumers are players you would expect to find in a commerce model. Government might not have occurred to you. Government regulates commerce in numerous ways. Government regulation will often influence how a business model or e-commerce invention is carried out, or provide the impetus for such invention, even though government may not be a direct player in the commercial transaction. A case in point is the invention that was the subject of litigation in *State Street Bank v. Signature Financial*. The Hub and Spoke system of the Boes patent, U.S. Patent No. 5,193,056 (the '056 patent), was specifically designed to give the participating mutual funds a favorable position under the U.S. tax laws. Indeed, on appeal to the Federal Circuit Court of Appeals, State Street Bank, which was trying to have the patent declared invalid, argued that the Boes patent was an overreaching attempt to preclude others from taking advantage of certain tax laws.

> The federal tax laws and regulations describe a certain safe harbor (the "substantial economic effect" regulations) that a partnership can use to ensure its tax allocation scheme is respected. See 26 U.S.C. §§ 701-706; 26 C.F.R. §§ 1.701-1 to 1.771-1. In the case of a Hub-and-Spoke configuration, a pro rata allocation scheme is used to reflect the economic interest of the partners. See '056 patent col. 3. The critical aspects of the accounting and record keeping necessary to ensure that the tax allocations pertaining to a pro rata partnership economic scheme will be respected are defined by the federal tax regulations largely finalized in 1985 and incorporating long-known business concepts. Yet the '056 patent claims a system for meeting them. The effect is to preempt calculations practically necessary to ensure that a pro rata partnership's tax allocation scheme will be respected.[1]

As an example of how government influences business model innovation, the Boes '056 patent does not stand alone. Many e-commerce inventions employ security mechanisms such as encryption and digital signatures. These are regulated by government: see Figure 5-1. Indeed, when it comes to innovations touching upon commerce, the United States Constitution practically guarantees the government will be involved:

> Clause 1: The Congress shall have Power To lay and collect Taxes, Duties, Imposts and Excises, to pay the Debts and provide for the common Defence and general Welfare of the United States; but all Duties, Imposts and Excises shall be uniform throughout the United States;

> Clause 2: To borrow Money on the credit of the United States;

> Clause 3: To regulate Commerce with foreign Nations, and among the several States, and with the Indian Tribes;

> Clause 4: To establish an uniform Rule of Naturalization, and uniform Laws on the subject of Bankruptcies throughout the United States;

> Clause 5: To coin Money, regulate the Value thereof, and of foreign Coin, and fix the Standard of Weights and Measures;

> Clause 6: To provide for the Punishment of counterfeiting the Securities and current Coin of the United States;

> Clause 7: To establish Post Offices and post Roads;[2]

---

[1] State Street Bank v. Signature Financial Group, Brief of Appellant, State Street Bank, pp. 28-29.

[2] U.S. Const., art. I, § 8, cls. 1-7.

[2] Electronic Agents as Players in the Commerce Model

Any of the previously mentioned players can be represented by electronic agents. With today's Internet technology, electronic agents can be devised to do the bidding for a human player. How does agent technology work?

Imagine a free marketplace where buyers and sellers negotiate to trade goods and services. Each buyer has a limited available budget to spend and knows what a particular good or service is worth to that buyer. Similarly, each seller knows what it cost to develop or generate the offered good or service and may have a measure of its "opportunity cost" of parting with that good or service. Turn these buyers and sellers loose in the marketplace and let them negotiate. Assuming each buyer and seller has equal access to information, the free marketplace system "solves" how goods and services should be allocated.

The preceding describes one way in which software developers now exploit a popular technology called software agent technology. The programmer constructs a virtual "free marketplace" filled with software agents, and then turns those agents loose to work out solutions to problems through negotiation. Software agent technology has many uses. You may find it in Internet information retrieval systems, for example.

The software agent, sometimes called the intelligent agent, is a construct that performs tasks and makes decisions based on an internal representation of its world. The construct operates autonomously, or nearly so, with little or no control from a central intelligence. The construct can communicate with its world (such as with other entities that make up its world) and can make changes to its world by executing actions. Often that world is a multi-agent one, where individual agents take on specific roles and interact with other agents taking on different roles.

The term "software agent" has yet to acquire a well-accepted meaning. One way to define the term is to list its key attributes. Again, there is no generally accepted list of attributes. The most commonly encountered attributes include autonomy, cooperation, learning, and mobility.

Autonomy, the ability to make independent decisions, and *cooperation*, the ability to communicate and interact with other entities in its world, are key attributes of the software agent. Independent decision-making can range from deliberative to reactive. Deliberative agents possess an internal, symbolic reasoning mechanism that can engage in planning and negotiation with other agents. Reactive agents behave in a simple stimulus-response fashion, typically lacking higher order reasoning capability.

Learning is another key attribute of the software agent. The agent needs to learn as it reacts with or interacts with its world. Learning may range from simple storage of information to altering the agent's behavior.

In addition, sometimes the software agent will also have *mobility*, the ability to transport itself into different memory spaces, such as into different computers distributed across a computer network. Mobility allows the software agent to enlarge its world view and to interact with systems beyond the confines of its original residence. Mobility is not a required attribute, as multi-agent systems can rely entirely on communication to gather information and to perform operations.

Do not confuse software agents with computer viruses, despite what the property of mobility may imply. A computer virus has mobility to seek out new memory space in which to replicate. Ordinarily the software agent does not employ replication. Some software agent systems could employ replication, however.

Software agent technology itself is not new. Computer scientists in the field of distributed artificial intelligence have been working with software agent techniques since the 1970s. More recently, during the 1990s, software agent technology has begun to find its way into mainstream software development.

As previously noted, the term "software agent" or "intelligent agent" is widely used to describe a variety of different constructs, often without regard to any precise or formal definition. In the Internet information retrieval world alone, you will find software agents called Internet agents, spiders, webcrawlers, wanderers, search agents, brokers, and bots. In the desktop computing world, you will also find software agents called system agents, interface agents, application agents, wizards, and desktop agents. There are undoubtedly countless other examples.

[C] The Process Component of the Commerce Model

As with the players component, there is a plethora of processes employed in commerce today. In broad terms, one can define a handful of processes that could each become chapter headings in a primer on commerce. Some primary commerce process categories are:

- research
- product development
- manufacturing
- marketing
- sales
- payment
- fulfillment
- support
- disposal

The following subsections will explore these primary commerce process categories and will suggest where you may want to look to find innovative aspects to patent.

[1] Research and Product Development

Research and product development certainly can involve business model innovation. Finding unique, new ways to innovate should be a key goal of companies seeking growth. Why are some companies better able to develop innovative new products and services? The answer may well lie in how they are structured, how their innovators are rewarded and what tools their workers are provided. Perhaps these structures will warrant patent protection.

For example, there is currently a race to decode the human genome, that is, to map out the structure and function of each protein in the human DNA molecule. The commercial benefits are clear. With an understanding of the human genetic structure, scientists hope to cure disease and the improve the human condition. However, the secret of our DNA structure has been locked inside our bodies for ten thousand years. The tools for picking that lock have only recently been discovered and are improving every year. Companies who develop these tools may well want to consider obtaining patent protection on their novel aspects. Where the commercial benefits of the underlying technology are clear, it often makes good sense to patent the tools used to develop that underlying technology.

[2] Manufacturing

Clearly manufacturing is also a major area that is ripe for business model innovation. The McDonald brothers' assembly line food preparation innovation is a classic example. Before the McDonalds' innovation, fast food was certainly not fast by today's standards. Short order cooks prepared each order as it came in. There was no preplanning, no preparation in advance. The McDonald brothers' seemingly simple assembly line preparation technique cranked up the speed, lowered the cost, and significantly boosted profits.

Henry Ford used the same assembly line technique to outperform the other automobile manufacturers of his day. His superior assembly line manufacturing methods eventually spread across the globe. During the 1960s and 70s, Toyota took Henry Ford a step further with its just-in-time (JIT) manufacturing process. Instead of stockpiling axles, engines, chassis, and bumpers, as the American and European automobile manufacturers did, Toyota employed a manufacturing innovation developed by Taiichi Ohno in which the necessary auto parts were built just in time to be assembled onto the vehicle. Ohno's just-in-time technique required accurate recordkeeping and a high degree of process control. However, it allowed Toyota to eliminate costly stockpiles of warehoused parts waiting to be assembled. This gave Toyota a tactical advantage that allowed it to gain a foothold in the American automotive market in the 1970s.

[3] Marketing, Sales, and Fulfillment

Marketing is the business activity of presenting products or services to potential customers. It can encompass such additional activities as advertising, catalog distribution, and the like. Sales represents the logical next step, that of negotiating the offer and acceptance leading up to the exchange of goods or services for an agreed consideration. Sales can include such additional activities as order processing. Payment and fulfillment follow naturally from the sales process. Payment is the act of paying money and fulfillment is the act of supplying the goods or services in response to payment. In an e-commerce environment, payment will likely involve some form of credit card processing or electronic funds transfer. Fulfillment runs the gambit from warehouse and inventory management to shipping and logistics.

[4] Support and Disposal

Support and disposal processes typically occur after the sale. In the software context, support has come to encompass both help desk functions and software maintenance functions. In the leased vehicle context, support encompasses periodic service and maintenance functions as well as warranty claims processing. Disposal is far less frequently considered, but it is an ever-present component of virtually all commerce involving goods. In the software context, disposal may be effected by the simple act of hitting the delete key, or it may involve specialized software programs designed to remove an old version of an application program before a new version is loaded. In the automotive context, disposal involves considerable physical effort. The automotive disposal process includes business processes ranging from used vehicle trade-in and resale programs to auto wrecking and landfill programs.

[D] The Medium Component of the Commerce Model

Money is by far the prevalent medium of exchange for today's commerce. Invented before the ancient Greek civilization, money is perhaps the quintessential business model innovation. The Internet pales in comparison. However, money is not the only thing that flows in the stream of commerce. Goods and services flow in that stream. Customer loyalty and goodwill flow in that stream as well. Thus in developing a generalized commerce model, it is helpful to identify all things of value that flow in the stream of commerce. Sometimes the term *value chain* is used to describe this flow. The term value chain carries the connotation that things of value are passed from entity to entity, from buyer to seller.

It can often be useful to conceptually separate the physical things of value (goods and services) from the information *about* those goods and services. Many electronic commerce systems are designed to do this. To illustrate, money can be treated as a form of information, or at least a component of information is intimately tied to the concept of money. When you deposit 500 dollars in your child's college savings account, that money becomes a journal entry — information — in a bank ledger. When you purchase goods with a credit card, money flows from the credit bank to the payee and a journal entry — information again — is made against your account so that the expenditure will show up on your next credit card statement. In today's information economy, money flows through commercial channels almost entirely as information. Few commercial transactions involve the actual passing of greenbacks.

Thus in commerce systems today, it is critical to have information flow. Without information flow, money flow would be reduced to the hand-to-hand exchange of our great grandfathers. Therefore, when considering the *medium* of commerce, our commerce model should describe the different media or channels by which information flows to effect today's commerce. These media or channels include, of course, the Internet, as well as corporate intranets, commercial networks, electronic funds transfer networks, telecommunications networks, wireless networks, and the like.

Because electronic information flow virtually dominates today's commerce, the following sections will explore electronic commerce, or e-commerce. In the examples that follow, several popular information media and channels will be featured. Often a new media will give rise to new business models. As proof of this, consider the numerous new business models (and numerous issued patents) that arose from the construction of the Internet.

The generalized commerce model presented here can help you organize your thoughts when trying to assess a new business model. The model can be used with all forms of business inventions, including those that employ high-tech electronic commerce components and those that do not. If yours is an e-commerce invention, the next section may provide further help.

§ 5.03   Fundamental E-Commerce Building Blocks

In their book, *Place to Space*,[3] Peter Weill and Michael Vitale describe how e-commerce can be modeled in terms of fundamental components or, to use their terminology, "atomic e-business models." All e-commerce, they maintain, can be expressed in terms of these fundamental components, which comprise the following:

- Content Provider — provides information, digital goods, or services through intermediaries;

- Direct to Customer — provides goods, services, or information directly to the customer;

- Full-Service Provider — provides a full range of services under one domain either directly or through allies in an attempt to control the primary consumer relationship;

- Intermediary — provides a concentration of information to assist in bringing buyers and sellers together;

- Shared Infrastructure — provides a common information technology (IT) infrastructure to allow multiple competitors to cooperate by sharing those IT resources;

- Value Net Integrator — collects, synthesizes and distributes information to coordinate activities across a value net;

---

[3] P. Weill and M. Vitale, *Place to Space* (Harvard Business School Publ., 2001).

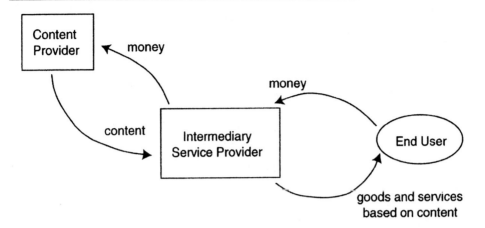

FIGURE 5-2: CONTENT PROVIDER MODEL

- Virtual Community — creates and facilitates an online community of people with common interests;

- Whole-of-Enterprise/Government — provides a firm-wide, single point of contact through which all services provided by a company or governmental agency are disseminated.

Some e-commerce businesses may fit one of these atomic business models exactly. Others may be a combination of several. The following subsections will give brief examples of each of these business models. You may find these models helpful in understanding your client's e-commerce invention.

[A] Content Provider

In the entertainment industry, content is king. Consumers understand this instinctively. Everybody wants a glimpse of the famous movie actress, sports superstar, rock idol, or even infamous villain. Nobody will give even a second look to yet another law firm Web site featuring a rambling primer explaining patent, trademark, and copyright basics to the masses. That is why fans will flock to a Web site featuring the famous movie actress, the sports superstar, the rock idol, and even the villain, and are probably avoiding your law firm primer like the plague.

In a content provider business model, the content provider produces something people want to see, or hear, or download, and then distributes this content through an intermediary. Some content provider models extract payment directly from users of the content; other content provider models extract payment indirectly through advertising sponsorship. Many Web sites that provide daily news updates, sports scores, stock market quotes, and weather reports operate on this basic content provider model.

Figure 5-2 shows a simple content provider model. Content is provided through an intermediary service provider to the end user. The content can be provided, as-is, or it may be bundled into other goods and services. Note the flow

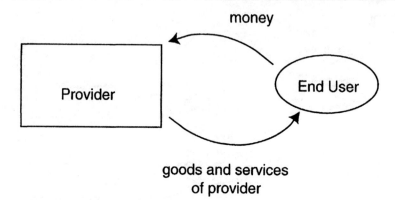

FIGURE 5-3:   DIRECT-TO-CUSTOMER MODEL

of money from the end user to the intermediary service provider and from the intermediary service provider to the content provider.

[B]   Direct-to-customer

If you log onto Dell Computer's Web site, or Hewlett-Packard's Web site, or Apple Computer's Web site you can see good examples of direct-to-customer e-commerce. The customer logs onto the site, makes a product selection, enters the necessary shipping and payment information, and within a few mouse clicks, the order is registered. The direct-to-customer business model eliminates the need for both the distributor and the dealer. Many traditional manufacturing businesses are using this direct-to-customer model. The model gives manufacturers better control over how its products are represented to consumers. What's more, the model allows the manufacturer to move towards just-in-time production. The product ordered by the customer does not need to be built ahead of time, but rather it can be built to the customer's specifications after the order is placed. This eliminates the need to warehouse large quantities of finished product.

Figure 5-3 shows a direct-to-customer model. The model is quite simple. Goods and services flow from the provider to the end user, while money flows from the end user to the provider.

[C]   Full-service Provider

General Electric, one of America's powerhouse brick and mortar corporations, is also a leader in exploiting e-commerce business models. If you visit its GE Supply Company <gesupply.com> you can get an idea what being a full-service provider is all about. GE Supply offers over one thousand different electrical, voice and data communication, lighting and power distribution products, including those made by GE as well as products of GE's competitors. The site advertises local delivery, low transaction cost, and world class product support. The site is designed to be so useful — in providing full-service products and services — that the consumer will conclude, "Why shop anywhere else." At least that is GE's hope.

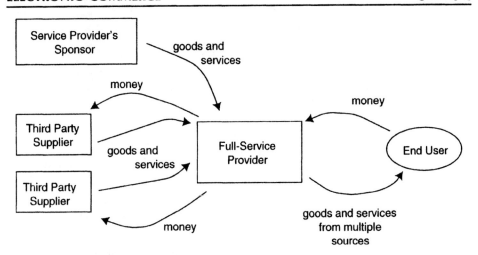

FIGURE 5-4: FULL-SERVICE PROVIDER MODEL

The full-service provider model depends on a thorough understanding of the consumer's needs. The site provides the consumer a portal through which the products of many different suppliers are offered for sale. Figure 5-4 shows a simple full-service provider model. The full-service provider receives goods and services from its own sponsor, and also from other third party suppliers. It passes these goods and services onto the end user. Money flows from the end user to the full-service provider and from the full-service provider to the third party suppliers. In the figure it is assumed that the service provider's sponsor owns the full-service provider. Hence a flow of money from provider to sponsor has not been explicitly shown. Of course, if desired, the business model could be constructed so the sponsor also receives money from the provider.

[D] Intermediary

The intermediary typically serves a vertical market. It pulls together information, tools, and products that would be useful to that vertical market. As an example, <Quicken.com> acts as an intermediary for the vertical market of personal financial management. If you visit the Quicken.com site, you will find financial information, stock quotes, financial advice, and an assortment of financial tools, such as retirement planners, car loan evaluators, and college tuition expense calculators. The site offers links to several on-line brokerage firms, for which Quicken.com undoubtedly receives compensation for the business it directs to those firms. Additional revenue comes from advertising displayed on the site.

Figure 5-5 shows one example of a simple intermediary model. Advertisers and third party sellers pay the intermediary a fee or commission to be represented on the intermediary's site. End users access the intermediary either directly, or through an ally site with which the end user may have a preexisting relationship. The intermediary is positioned to collect usage information, which it can sell to

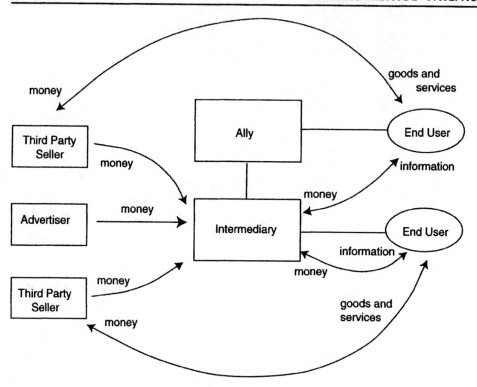

FIGURE 5-5: INTERMEDIARY MODEL

advertisers or third party sellers. The primary source of revenue for the intermediary derives from its concentration of information which it passes to the end users.

[E] Shared Infrastructure

A shared infrastructure model involves strategic cooperation among entities that traditionally view each other as competitors. Why would competitors want to do this? The answer is simple. The shared infrastructure provides a benefit that is not otherwise available in the marketplace.

A case in point is COVISINT, a shared infrastructure of the ''Big Three'' U.S. automobile manufacturers, General Motors, Ford, and DaimlerChrysler. The COVISINT infrastructure is designed to provide supply-chain logistics, a place in cyberspace where manufacturers and their suppliers can exchange information and transact business. What is the value proposition behind COVISINT? The automakers determined that, in this case, it made sense to cooperate with one another because having a unified supply-chain interface would help each increase the pace of vehicle implementation. Each manufacturer had previously experimented with its own supply-side electronic exchange, but that placed a costly burden on suppliers: suppliers had to implement all three systems if they wanted to do business with all three manufacturers. In the end, the manufacturers concluded that it would be more efficient to have one supply-side exchange. Not

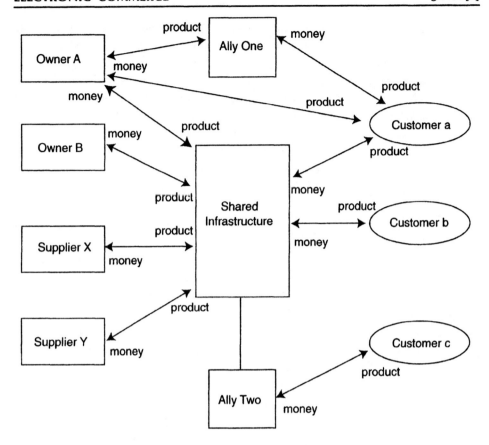

FIGURE 5-6: SHARED INFRASTRUCTURE MODEL

only would that make it easier to train suppliers in the system's use, but it would give the manufacturers a unified front in the effort to improve product quality, shorten delivery times, and lower cost.

Figure 5-6 shows how a shared infrastructure model might be configured. *Owners A* and *B* develop the shared infrastructure, which they and other third party suppliers may access. The owners and suppliers offer products to the customers through the shared infrastructure. Money flows through the shared infrastructure, from customers to the owners and suppliers. As the depicted model illustrates, it is possible to have associated ally sites through which additional transactions can take place. In the figure, *Ally One* transacts business between *Owner A* and *Customer a*, wholly outside the shared infrastructure system. *Ally Two* transacts business with *Customer c* and then funnels the transaction information through the shared infrastructure.

[F] Value Net Integrator

The value net integrator model is perhaps the most sophisticated of the business models presented here. At its core is a fairly simple, yet profound, concept that *information* about the sale of goods and services can be managed separately from the physical sale of those goods and services. The model identifies two separate value chains: the physical value chain along which goods and services travel from manufacturer to distributor to merchant to customer; and the virtual value chain through which *information* about the members of the physical chain is gathered, analyzed, and distributed. The value net integrator model is quite powerful because it permits a highly profitable business to be developed with minimal capital investment in physical assets.

Case in point is Seven-Eleven Japan. Most Americans are familiar with the ubiquitous Seven-Eleven convenience store. However, most Americans do not know that the U.S. parent company is now largely owned by its highly successful subsidiary, Seven-Eleven Japan. How the subsidiary grew to become the market leader and how it ultimately acquired a controlling interest in its parent company makes an interesting case study.

One factor in Seven-Eleven Japan's success is Japan's crowded living conditions. Real estate is costly, homes are small, and storage space is at a premium. Thus most Japanese people shop daily for provisions needed for the night's supper. Often such shopping is done on the way home from work. Since the vast majority use public commuter transportation, the daily shopping trip is likely made by walking to a local convenience store, such as Seven-Eleven.

That explains, perhaps, why convenience stores are so popular in Japan, but that does not explain why Seven-Eleven rose up as the market leader, allowing it to become wealthy enough to buy out its U.S. parent company. One reason for Seven-Eleven Japan's success is its value net integrator strategy. As we shall see, Seven-Eleven Japan is no longer in the convenience store business today. It is in the convenience store *information* business.

Seven-Eleven franchisees own their own stores. Typically, behind each Seven-Eleven storefront sign you will find a family run, mom and pop operation. The franchisees pay Seven-Eleven 45 percent of their profits (other franchises charge only 35 percent) in exchange for the right to use the Seven-Eleven trademarks, the right to purchase from the Seven-Eleven sponsored delivery center, and the right to receive up to the minute marketing information on what products are selling well at the moment.

Seven-Eleven's value net gathers sales information from all its franchisees and it makes this information available to suppliers of goods in exchange for favorable pricing, the benefit of which Seven-Eleven passes to its franchisees. The value net gives all franchisees collective bargaining power over suppliers, and the suppliers get valuable, real-time marketing data about how its products are selling. By carefully monitoring the supply chain, the Seven-Eleven value net ensures that hot selling products can be delivered overnight, allowing the franchisees to operate leanly.

Seven-Eleven does not need to own physical distribution centers. It outsources that function, using the information about what is selling to know which products to stock and at what frequency. By continually improving its operation, Seven-Eleven stays ahead of the competition. For instance, whereas it

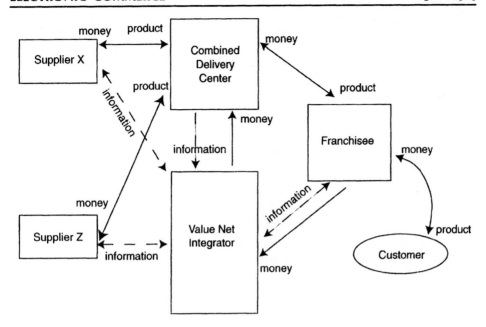

FIGURE 5-7: VALUE NET INTEGRATOR MODEL

once made as many as 40 deliveries per day, by analyzing its sales data, Seven-Eleven was able to reduce this to less than seven.

By separating the physical product from the business information about the product, the value net integrator adds value by using the business information better than a conventional physical business can. Unfettered by physical assets and the need to deliver products in a physical world, the value net integrator is able to concentrate on gathering, synthesizing, and distributing information that drives the physical business. The value net integrator succeeds because it owns the information that gives the physical world its operating data.

Figure 5-7 shows one example of a value net integrator model. The value net integrator is positioned to gather and control information from its franchisees and from the suppliers of products being offered to the franchisees. Note the combined delivery center through which the physical products flow is funded by the value net integrator, although it does not necessarily need to be owned by the value net integrator.

[G] Virtual Community

You can think of the virtual community as a meeting place where members of the pubic can exchange their views. Under the virtual community model, a sponsor builds the site, hoping that members of the public will frequent the site to exchange their views. Often the site will include bulletin boards, on-line chat services, or Web conferencing services though which the public can communicate. The benefits to the sponsor are two-fold. The site can promote and support the sponsor's products, through user groups. The site can also generate advertising revenue by selling advertising space or by charging a fee for forwarding "click-through" contacts to another site. Many Web sites today have at least one component built on the virtual community model.

Of all the e-commerce business models in existence today, the virtual community model is perhaps the oldest. Computer users of the 1980s will remember electronic bulletin boards, such as CompuServe and later Prodigy and America Online (AOL), which both hosted many user group bulletin boards. CompuServe, Prodigy, and AOL each predate the Internet. They extracted payment from usage fees — users paid by the minute, or by the month, to use the service. After the Internet came into existence, other public access bulletin boards began to spring up, ultimately challenging services like CompuServe, Prodigy, and AOL. The WELL (Whole Earth 'Lectronic Link), developed by Stewart Brand, editor of the Whole Earth Catalog in 1985, is an early example. The WELL was also, and still is, a subscription service. Today there are millions of non-subscription virtual communities that are supported by individual corporate sponsors or advertising revenue.

Figure 5-8 shows a simple virtual community model. Members connect to the virtual community site to obtain information, including information about products offered by the suppliers. The virtual community may be funded by usage fees from its members or from fees or commissions extracted from suppliers whose products are represented in the virtual community.

[H] Whole-of-enterprise/Government

This model seeks to bring together all different business units within a company, or departments within a governmental agency, for access through a single point of customer contact. Many companies have a single Web page that serves as the access point by which the customer can obtain information about a specific business unit. A more sophisticated whole-of-enterprise model integrates information that may be found under different business unit headings. The integrated model allows customers to search for information based on their needs, without knowing in advance which business units or departments may have the answers.

To illustrate an integrated whole-of-enterprise model, say that you have just moved to a new community and would like to get all utilities connected. If your new community has an Internet site designed around a whole-of-enterprise model, you might simply select "I just moved to the community," and the Web site would lead you through the process of getting the electricity turned on, the water bill transferred, connecting to your selected phone service, notifying the police regarding the particulars of your security system, and the like. Some of these services are probably not within the domain of the local government, however, the system would still provide the necessary contact information to third party suppliers. For example the system might notify the user that the cable TV service provider in that area is XYZCom, and provide phone numbers and Internet addresses for that service.

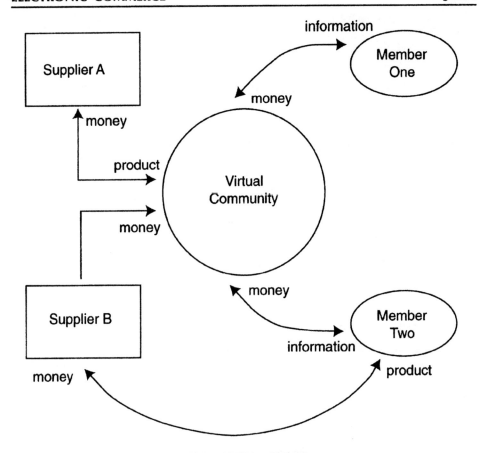

FIGURE 5-8: VIRTUAL COMMUNITY MODEL

Figure 5-9 depicts a whole-of-enterprise model. Although users are free to contact the business units directly, the model is designed to allow users to interact with all business units through the whole-of-enterprise server. The server thus serves as a conduit through which products flow from business unit to users, and through which money flows from users to the business units.

§ 5.04 Interviewing the Inventor — Different Vantage Points for Assessing E-Commerce Business Models

E-commerce business models can be viewed from several different vantage points. Each is helpful in understanding how the model works. In discussing an e-commerce invention with your client, here are some vantage points you may want to consider. Chances are your client has thought a lot about these and will be glad you asked:

- Business Strategy — How is the business going to make money? Who are the target customers? What products or services will the company be providing? What are the business strengths and weaknesses? Why would the customer choose this product or service over that of a competitor?

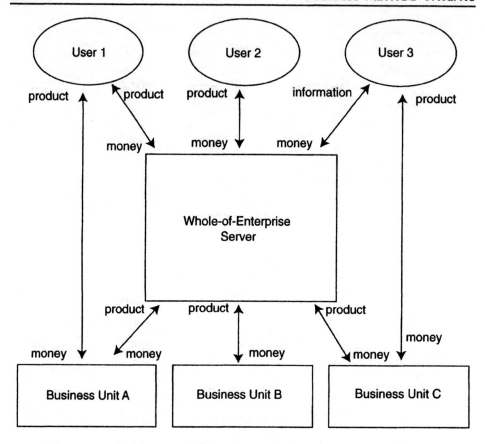

FIGURE 5-9: WHOLE-OF-ENTERPRISE MODEL

- Organizational Structure — How is the enterprise organizationally structured? Who reports to whom? What are the respective duties of each person or department involved in the business?

- Business Processes — What are the core processes that the business will perform in delivering products and services to the customer? How are the processes allocated to the persons or departments within the company? What measurement techniques will be used to ensure the processes are being performed optimally?

- Value Chain — Look at the business in terms of the value added by the business to the raw materials it receives. Why is the value added of interest to the target customer? How does the value added compare with other segments of the industry and with other competitors in the field?

- Core Competencies — If you had to strip the business down to the things it does best, what are they? How were these skills acquired and how are they maintained? How are these skills exploited to add value to the end product?

§ 5.05   Tracking the Value Flow

In analyzing a business model or e-commerce invention it is always a good idea to track the flow of value. Which ways do the goods, services and information flow? Which way does the money flow? We saw examples of these in the atomic models presented in § 5.03. These are important considerations and they relate primarily to the customer transaction.

In addition to the customer transaction, you should also identify where and how any data is collected and to what use that data is thereafter put. Often e-commerce transactions involve collecting data that can be quite valuable. For example, an e-commerce Web site that sells books may also collect data on the preferences of its customers. That preference information could be used to cross-sell items other than books.

Finally, always look to see who owns the customer relationship, and how that relationship is maintained. The customer relationship is a key component of many e-commerce Web sites. E-commerce technologies today permit the measurement of customer relationships by monitoring how customers access commercial Web sites. An example of such a Microsoft system for this purpose, is described in § 5.07[N][1] below. In low-tech terms, the customer relationship forms part of a company's goodwill. Trademark attorneys have long understood and worked with this concept in attributing value to the intangible assets of a company.

§ 5.06   Examples of Electronic Commerce Innovations

Electronic commerce has taken root in many important domains. This section will provide a snapshot of some of these domains by featuring example patents from various e-commerce and business model disciplines. A sample claim from each featured patent is provided.

[A]   Accounting Patents

[1]   Time and Expense Logging System — U.S. Patent No. 5,991,742

Tran   November 23, 1999

Abstract

A portable computer system manages data conveniently for a user. The system has a processor, a program storage device coupled to the processor, an input recognizer adapted to receive data from the user, and a computer readable code embodied in the program storage device for storing and processing the data.

Inventors: Tran; Bao Q. (10206 Grove Glen, Houston, TX 77099)
Appl. No.:   650293
Filed: May 20, 1996

Current U.S. Class: 705/32; 700/14; 702/176; 702/178; 702/187; 705/11
Intern'l Class: G06F 013/00
Field of Search: 705/41,11,28,30,32,34,40,404,418 704/875,251 382/187,186
                      702/176,178,187

Sample Claim:

1. A program storage device having a computer readable code embodied therein for recording time and expense data pertaining to the performance of a task by a user, said task having a start time and a completion time, said program storage device comprising:

an input recognizer adapted to receive non-cursive handwritings from said user, said input recognizer converting said non-cursive handwritings into text data;

a timer adapted to receive instructions from said user to measure a duration count between said task start time and said completion time, said timer having an active state in which said timer increments said duration count and a suspended state in which said timer maintains said duration count; and

a database coupled to said input recognizer and said timer, said database having at least one record, each of said records having data fields adapted to store said duration count and said text data.

[2] Method for Processing Business Activity Financial Data — U.S. Patent No. 6,085,173

Suh   July 4, 2000

Abstract

A method for processing business activity data including providing a selected one of business activity fields, where each field represents at least one business activity, and information data associated with the selected business activity field. The method further includes providing a selected one of second business activity items associated with the selected business activity field, finding a corresponding account name in a lookup table based on the selected second business activity item and the information data, and processing the information data in accordance with the corresponding account name.

Inventors: Suh; Won-kyo (Kwangju-si, KR)
Assignee: Eastern Consulting Co., Ltd. (Seoul, KR)
Appl. No.:   714497
Filed: September 16, 1996
Foreign Application Priority Data

Jul 27, 1993[KR]   93-14324

Current U.S. Class: 705/30; 705/11
Intern'l Class: G06F 157/00
Field of Search: 705/30,33,34,11

Sample Claim:

1. A computer implemented method for processing business activity data in a data processing system containing a plurality of secondary business activity items and corresponding account names, comprising the steps of:

displaying a plurality of primary business activity categories selected from a group comprising a purchasing activity, a selling activity, a financial activity, a production activity, and a management activity;

receiving a user selection of one of the displayed primary business activity categories;

displaying a plurality of secondary business activity items associated with said selected primary business activity category;

receiving a user selection of one of the displayed secondary business activity items;

displaying at least one possible settlement category in response to the selected secondary business activity item;

receiving a user selection of one of the displayed settlement categories;

receiving from the user an amount of account to settle the selected primary business activity category;

retrieving a corresponding account name from the data processing system in response to the selected secondary business activity item and the selected settlement category; and

processing the received amount in accordance with the retrieved account name to be assigned to a respective field of said account name.

[3] Method and System For Tracking Employee Productivity Via Electronic Mail — U.S. Patent No. 6,141,649

Bull   October 31, 2000

Abstract

A method and system in a computer system for tracking the productivity of technical support specialists in a call center environment. The system displays on a display device a form for entry of normal activity and exceptions to normal activity. The system receives from the technical support specialist indications of normal activity and exceptions. Each indication of an exception includes its type and its length. The system stores indications of the normal activity and exceptions. Upon receiving a request to transmit an exception report, the system retrieves the stored indications and formats the stored indications into an electronic mail message. The system then presents the formatted electronic mail message to the technical support specialist so that modifications to formatted electronic mail message can be made. Finally, the system sends the electronic mail message to a central location so that electronic mail messages from multiple technical support specialists can be collected and processed.

Inventors: Bull; Jeffrey A. (Nampa, ID)
Assignee: Micron Electronics, Inc. (Nampa, ID)
Appl. No.:   956180
Filed: October 22, 1997

Current U.S. Class: 705/11; 235/377; 700/111; 705/32; 708/110
Intern'l Class: G06F 017/50
Field of Search: 705/32,9,11,10 364/468.18,705.06 235/377

Sample Claim:

1. A method in a computer system for contemporaneously tracking the productivity of technical support specialists in a call center environment, the method comprising:

displaying on a display device a form for entry of exceptions to normal activity;

using a timer to time the length of the exception to normal activity; receiving from the technical support specialist indications of at least one exception, each indication including a type of the exception and the time length of the exception;

storing each of the indications of the plurality of exceptions;

receiving a request to transmit an exception report; and

in response to receiving the request to transmit an exception report,

retrieving the stored indications;

formatting the stored indications into an electronic mail message;

presenting the formatted electronic mail message to the technical support specialist so that modifications to formatted electronic mail message can be made;

sending the electronic mail message to a central location so that electronic mail messages from multiple technical support specialists can be collected and processed; and

analyzing the indications to determine productivity.

[4] **Accounting and Billing Based on Network Use — U.S. Patent No. 6,208,977**

Hernandez, et al.   March 27, 2001

Abstract

A method and a system determines billing information for use of a network. For a data transmission on a path between a pair of node locations, the price for the transmission depends, *inter alia*, on the amount of data transported and the price of using the path. The method collects traffic data for data transmissions sent over links of the network, selects a set of paths connecting each pair of node locations, and determines a price for data transported between the pairs. The method bills an allocated portion of the total price for each pair to each node location of the pair.

Inventors: Hernandez; Milton B. (River Edge, NJ); Kusnetsov; Dmitry (Brooklyn, NY); Tapia; Pablo E. (Maywood, NJ)
Assignee: Apogee Networks, Inc. (Rochelle Park, NJ)
Appl. No.:   394004
Filed: September 10, 1999

Current U.S. Class: 705/34; 705/30; 705/400
Intern'l Class: G06F 15//60
Field of Search: 377/13,14,15,16 379/100.04,111,114 705/30,34,400,418

Sample Claim:

1. A method of billing users according to their usage of a network having a plurality of links interconnecting a plurality of node locations, comprising:

> determining a usage price per unit bandwidth for a virtual path between a first and second node locations, the usage price being proportional to a sum of usage prices of links of a set of paths between the first and second node locations;

> collecting traffic date for an amount of data transported between the first and second node locations;

> calculating a total price equal to the amount of data transported times the usage price per unit bandwidth; and

> transmitting a bill including the total price to each of the users.

[5]   **Financial Services Account Manager System — U.S. Patent
No. 6,246,999**

Riley, et al.   June 12, 2001

Abstract

A method of reporting accounting information to acquiring banks includes compiling accounting information from different processing systems to a mainframe. The accounting information is then routed to a database residing on a server operable with the World Wide Web (WWW). Different reports of the accounting information are then accessed with a client station operable with the server.

Inventors: Riley; James F. (Frederick, MD); Sudec; John A. (Hagerstown, MD); Abram; Timothy James (Coral Springs, FL)
Assignee: First Data Corporation (Hackensack, NJ)
Appl. No.:   100434
Filed: June 19, 1998

Current U.S. Class: 705/30; 705/35; 705/38; 705/39; 705/40; 705/42
Intern'l Class: G06F 017/60
Field of Search: 705/30,35,38,39,40,42 707/10,1,4,201

Sample Claim:

1. A method of reporting accounting information to acquiring banks, the method comprising:

compiling accounting information from different processing systems to a mainframe, the different processing systems including a debit processing system, a funding processing system, an interchange processing system, a chargeback processing system, a master processing system, a submissions processing system, and a summaries processing system;

routing the accounting information from the mainframe to a database residing on a server operable with the World Wide Web (WWW);

receiving a request for a report of the accounting information at the server from a client station operable with the server via the World Wide Web, wherein reports available for request include a cash position report, a funding report, an interchange report, a chargeback report, and a submissions report;

executing queries associated with the requested report on the accounting information stored in the database to generate the requested report; and accessing the requested report of the accounting information with the client station operable with the server.

[B]   Agent Technology Patents

[1]   Apparatus and Method for Communicating Between an Intelligent Agent and Client Computer Process Using Disguised Messages — U.S. Patent No. 6,085,178

Bigus, et al.   July 4, 2000

Abstract

An intelligent agent and its client communicate using a selector known by both parties to generate and interpret messages and thereby effectively disguise confidential information transmitted in the messages from third parties. Moreover, a neural network is used to implement the decision logic and/or the message disguising functions of an agent such that the logic employed in such functions is not readily reverse compiled or scanned by third parties.

Inventors:	Bigus; Joseph Phillip (Rochester, MN); Cragun; Brian John (Rochester, MN); Delp; Helen Roxlo (Rochester, MN)
Assignee:	International Business Machines Corporation (Armonk, NY)
Appl. No.:	822119
Filed:	March 21, 1997

Current U.S. Class:	705/80; 283/17; 283/56; 283/73; 283/901; 380/56; 380/59; 705/50
Intern'l Class:	G06F 013/00
Field of Search:	395/200.32 706/25 380/25,3,21,56,59 705/50,80 283/17,56,73,901

Sample Claim:

1. A method of communicating between an intelligent agent computer program and a client computer process, comprising the steps of:

(a) selecting a message from a plurality of messages based upon a selector and information to be transmitted between the intelligent agent and the client computer process to disguise the information from discovery by a third party; and

(b) transmitting the selected message between the intelligent agent and the client computer process.

[C]   Auction Patents

[1]   Method of Conducting an On-line Auction with Bid Pooling — U.S. Patent No. 5,794,219

Brown   August 11, 1998

Abstract

A method of conducting an on-line auction that permits individual bidders to pool bids during a bidding session. The auction is conducted over a computer network that includes a central computer, a number of remote computers, and communication lines connecting the remote computers to the central computer. A number of bidding groups are registered in the central computer, each bidding group having a total bid for the item being auctioned. Bids entered from the remote computers are received in the central computer, each bid including a bid amount and a bid designation. Each bid amount is contributed to the total bid of the bidding group indicated by the bid designation. The bidding group having the largest total bid at the end of the bidding session wins the item being auctioned.

Inventors: Brown; Stephen J. (Mountain View, CA)
Assignee: Health Hero Network, Inc. (Mountain View, CA)
Appl. No.:   603131
Filed: February 20, 1996

Current U.S. Class: 705/37; 705/26; 705/27; 705/39
Intern'l Class: G06F 017/60
Field of Search: 705/1,26,27,35,37,38,39,44   395/200.47,200.49,200.57
                    340/825.26,825.27,825.3,825.31,825.33

Sample Claim:

1. A method of conducting an auction using a computer network, said computer network comprising a central computer, a plurality of remote computers, and a plurality of communication lines connecting said remote computers to said central computer, said method comprising the following steps:

a) registering a plurality of bidding groups in said central computer, each of said bidding groups having a total bid;

b) receiving in said central computer bids entered from said remote computers, each of said bids comprising a bid amount and a bid designation, said bid designation indicating a chosen group comprising one of said bidding groups for which said bid is intended;

c) contributing said bid amount to said total bid of said chosen group; and

d) declaring a winning group, said winning group being the one bidding group having the largest total bid at the end of said bidding session.

[2] Facilitating Internet Commerce Through Internet Worked Auctions — U.S. Patent No. 6,202,051

Woolston   March 13, 2001

Abstract

Auctioning a uniquely identified item (e.g., used goods or collectibles) with a computerized electronic database of data records on the Internet includes creating a data record containing a description of an item, generating an identification code to uniquely identify the item, and scheduling an auction for the item at the computerized database of records. The item is presented for auction to an audience of participants through a worldwide web mapping module executing in conjunction with the computerized database. The data record connotes an ownership interest in the item to a seller participant on the computerized electronic database of data records. The worldwide web mapping module translates information from the data record on the computerized database of records to a hypertext markup language (HTML) format for presentation through the Internet. Bids are received on the item from participants on the Internet through an auction process that executes in conjunction with the computerized database of data records. Auctioning of the item is terminated when the auction process reaches predetermined criteria. The auction participant is notified of the high bid in the auction process. The unique identification code is provided to the auction participant with the high bid to uniquely identify the item.

Inventors: Woolston; Thomas G. (8408 Washington Ave., Alexandria, VA 22309)
Assignee: Merc Exchange LLC (Alexandria, VA)
Appl. No.:   253021
Filed: February 19, 1999

Current U.S. Class: 705/27; 705/26; 705/37
Intern'l Class: G06F 017/60
Field of Search: 705/27,37,26,35,36,39 235/383,381

Sample Claim:

1. An automated method, performed by a computer-based auction system, for enabling a seller to auction a uniquely identified item via the Internet to one or more potential buyers, the method comprising:

requiring the seller to establish a seller's account, the seller's account being based at least on the seller's identity and a financial instrument associated with the seller;

receiving information from the seller including a description of an item offered for auction by the seller;

creating a data record containing a description of the item based on the information received from the seller, the data record connoting an ownership interest by the seller in the item, the data record being stored in a computerized electronic database maintained by the computer-based auction system;

generating an identification code to uniquely identify the item; scheduling an auction for the item, the auction to be hosted by the computer-based auction system;

presenting the item for auction to an audience of participants through a worldwide web mapping module executing in conjunction with the computerized database, the worldwide web mapping module translating information from the data record to a hypertext mark up language format for presentation through the Internet;

receiving bids on the item from participants via the Internet through an auction process that executes in conjunction with the computerized database;

terminating the auction for the item when the auction process encounters predetermined criteria;

notifying a winning auction participant that the winning auction participant has entered a high bid in the auction process;

providing the unique identification code to the winning auction participant to uniquely identify the item; and

charging a fee to the seller's account based on an amount of the high bid.

[3] Method and System for Processing and Transmitting Electronic Auction Information — U.S. Patent No. 6,243,691

Fisher, et al.   June 5, 2001

Abstract

A system and method for conducting a multi-person, interactive auction, in a variety of formats, without using a human auctioneer to conduct the auction. The system is preferably implemented in software. The system allows a group of bidders to interactively place bids over a computer or communications network. Those bids are recorded by the system and the bidders are updated with the current auction status information. When appropriate, the system closes the auction from further bidding and notifies the winning bidders and losers as to the auction outcome.

Inventors: Fisher; Alan S. (Fremont, CA); Kaplan; Samuel Jerrold (Hillsborough, CA)
Assignee: Onsale, Inc. (Mountain View, CA)
Appl. No.:   624259
Filed: March 29, 1996

Current U.S. Class: 705/37; 705/26
Intern'l Class: G06F 017/60
Field of Search: 705/37,26,27,39 395/500 364/578 703/22

Sample Claim:

1. A computer system for conducting auctions over a computer network, comprising:

a posting means for posting on a computer merchandise information describing each lot of a plurality of lots that are available for bidding by a plurality of bidders, each lot including at least one item;

an auction selection means for associating each lot of the plurality of lots with an auction format selected from a plurality of available auction formats;

a bid receiving means for receiving a bid related to at least a portion of a lot of the plurality of lots; and

a bid categorization means for automatically categorizing the received bid as successful or unsuccessful in accordance with the selected auction format for the lot.

[D]   Business-to-business Patents

[1]   Internet-based Customer Referral System — U.S. Patent No. 6,029,141

Bezos, et al.   February 22, 2000

Abstract

Disclosed is an Internet-based referral system that enables individuals and other business entities ("associates") to market products, in return for a commission, that are sold from a merchant's Web site. The system includes automated registration software that runs on the merchant's Web site to allow entities to register as associates. Following registration, the associate sets up a Web site (or other information dissemination system) to distribute hypertextual catalog documents that includes marketing information (product reviews, recommendations, etc.) about selected products of the merchant. In association with each such product, the catalog document includes a hypertextual "referral link" that allows a user ("customer") to link to the merchant's site and purchase the product. When a customer selects a referral link, the customer's computer transmits unique IDs of the selected product and of the associate to the merchant's site, allowing the merchant to identify the product and the referring associate. If the customer subsequently purchases the product from the merchant's site, a commission is automatically credited to an account of the referring associate. The merchant site also implements an electronic shopping cart that allows the customer to select products from multiple different Web sites, and then perform a single "check out" from the merchant's site.

Inventors: Bezos; Jeffrey P. (Seattle, WA); Kaphan; Sheldon J. (Seattle, WA); Ratajak; Ellen L. (Seattle, WA); Schonhoff; Thomas K. (Seattle, WA)
Assignee: Amazon.com, Inc. (Seattle, WA)
Appl. No.:   883770
Filed: June 27, 1997

Current U.S. Class: 705/27; 705/10; 705/26
Intern'l Class: G06F 017/60

Field of Search: 705/27,10,14,26 707/513
                        395/200.3,200.33,200.53,200.54,200.57

Sample Claim:

1. A method of selling items with the assistance of associates, the method comprising:

    providing a Web site system that includes a browsable catalog of items and provides services for allowing customers to electronically purchase the items;

    providing an associate enrollment system which allows users to electronically apply to operate as associates that select and recommend items from the catalog and refer customers to the Web site system in exchange for compensation;

    in response to a submission to the enrollment system by a user, assigning an associate identifier to the user and recording the associate identifier within a computer memory;

    electronically providing to the user instructions for generating hypertextual documents with item-specific links that, when selected by a customer, cause the user's associate identifier and an identifier of a recommended item to be transmitted to the Web site system in a request message;

    receiving a request message which contains an associate identifier and an item identifier and extracting the associate and item identifiers from the message, the request message generated by a computer of a customer in response to selection by the customer of an item-specific link provided by an associate in conjunction with a recommendation of the item;

    transmitting to the customer's computer a Web page which corresponds to the item identifier extracted from the request message;

    transacting a sale of the item and/or other items of the catalog with the customer through the Web site system;

    using the associate identifier extracted from the request message to identify the associate; and

    determining and recording within a computer memory compensation for the associate for the sale.

[E] Cryptography Patents

[1] System for Identifying, Authenticating, and Tracking Manufactured Articles — U.S. Patent No. 5,768,384

Berson   June 16, 1998

Abstract

This invention relates to a system for identifying, authenticating and tracking articles of manufacture throughout their manufacturing and distribution channels. The foregoing system utilizes: manufacturing meters that are located at authorized manufacturing locations and produce encrypted data that is uniquely associated with each manufactured article; a printer located at the authorized manufacturing locations so that the printer will print the information encrypted by the meter, which encrypted information is affixed to the manufactured article; a data center coupled to the manufacturing meters and located at a site remote from the manufacturing meters; means for producing information that identifies the manufactured articles; and a plurality of means located where the authenticity of the manufactured articles are checked by comparing the encrypted information on the article with the information produced that identifies the article.

Inventors: Berson; William (Weston, CT)
Assignee: Pitney Bowes Inc. (Stamford, CT)
Appl. No.:   623078
Filed: March 28, 1996

Current U.S. Class: 705/50; 235/385; 380/51; 705/11; 705/28; 713/178
Intern'l Class: H04L 009/00
Field of Search: 380/51,23 283/74 705/11,28,29

Sample Claim:

1. A system for identifying, authenticating and tracking articles of manufacture, said system comprising:

one or more manufacturing meters that are located at authorized manufacturing location, said meters produce encrypted information that is uniquely associated with each manufactured article and the operator of the equipment that produced the manufactured article;

one or more printers located at the authorized manufacturing locations wherein each of said printers is coupled to one of said manufacturing meters, that is located at the same location as said printer, so that said printers print the information encrypted by said meters, which encrypted information is affixed to the manufactured article;

means for producing information that is used to identify the manufactured article; and means for identifying the authenticity of the manufactured articles by comparing the encrypted information printed on the article with the information produced by said producing means.

[2] Method and Apparatus for Computer-based Educational Testing—U.S. Patent No. 5,947,747

Walker, et al.   September 7, 1999

Abstract

Methods and apparatuses are disclosed for computer-based evaluation of a test-taker's performance with respect to selected comparative norms. The system includes a home testing computer for transmitting the test-taker's test results to a central computer which derives a performance assessment of the test-taker. The performance assessment can be standardized or customized, as well as relative or absolute. Further, the transmitted test results are configured to reliably associate the student with his test results, using encoding, user identification, or corroborative techniques to deter fraud. Thus, for example, the system allows a parentally-controlled reward system such that children who reach specified objectives can claim an award that parents are confident was fairly and honestly earned without the parent being required to proctor the testing. Fraud, and the need for proctoring, is also deterred during multiple student testing via an option for simultaneous testing of geographically dispersed test-takers.

Inventors: Walker; Jay S. (Ridgefield, CT); Schneier; Bruce (Oak Park, IL); Jorasch; James A. (Stamford, CT)
Assignee: Walker Asset Management Limited Partnership (Stamford, CT)
Appl. No.:   647301
Filed: May 9, 1996

Current U.S. Class: 434/354; 380/251; 705/50
Intern'l Class: G09B 005/14; G09B 007/00
Field of Search: 380/2,9 434/118,350,354

Sample Claim:

1. A computer system, comprising:

means for storing instructions and data comprising a test-taker database having a record containing a test-taker field representing a test-taker identifying datum and a reward threshold field representing a reward threshold provided by an end user;

means for receiving a score code having said test-taker identifying datum and a test score incorporated therein encoded by an electronic testing device associated with a test-taker;

means for decoding said score code and distinguishing said test-taker identifying datum from said test score;

means for retrieving said record of said test-taker database corresponding to said test-taker identifying datum and locating said corresponding reward threshold;

means for comparing said test score relative to said reward threshold; and

means for providing a reward attainment message to said end user if said test score exceeds said reward threshold.

[3] Method of Debiting an Electronic Payment Means — U.S. Patent No. 6,076,073

Pieterse, et al.   June 13, 2000

Abstract

The invention provides a method for protectedly debiting an electronic payment mechanism, such as a telephone card. In the communication protocol between the payment mechanism and a payment station, an authentication is used to identify the payment mechanism during various steps of the protocol. According to the invention, the authentications are mutually related by states of a cryptographic process in order to be capable of detecting an interference in the protocol. The method may be applied to existing payment cards having a dynamic memory, it being guaranteed that the contents of the dynamic memory, in which there is located information related to the authentication, is not lost during the protocol.

Inventors: Pieterse; Rob (Aerdenhout, NL); Rombaut; Willem (The Hague, NL)
Assignee: Koninklijke KPN N.V. (Groningen, NL)
Appl. No.:   259584
Filed: March 1, 1999 Foreign Application Priority Data

Nov 15, 1995[NL]   1001659

Current U.S. Class: 705/39; 705/50
Intern'l Class: G06F 017/60
Field of Search: 705/39,50 380/21,24 235/380,383

Sample Claim:

1. A method of performing a transaction between an electronic payment mechanism associated with a first payment institution, and a payment station associated with a second payment institution, comprising:

requesting, by said payment station, payment data from said payment mechanism;

generating a plurality of authentication codes in which each authentication code of said plurality of authentication codes is linked to a preceding authentication code in a same transaction between said payment mechanism and said payment station;

transmitting, by said payment mechanism, said requested payment data including a respective authentication code of said plurality of authentication codes, to said payment station; and

exchanging payment institution data, which is derived from payment data transacted between the payment mechanism and the payment station, between the first payment institution and the second payment institution.

[4] Method and Apparatus for Enforcing Software Licenses — U.S. Patent No. 6,188,995

Garst, et al.   February 13, 2001

Abstract

The present invention comprises a method and apparatus for enforcing software licenses for resource libraries such as an application program interface (API), a toolkit, a framework, a runtime library, a dynamic link library (DLL), an applet (e.g. a Java or ActiveX applet), or any other reusable resource. The present invention allows the resource library to be selectively used only by authorized end user software programs. The present invention can be used to enforce a "per-program" licensing scheme for a resource library whereby the resource library is licensed only for use with particular software programs. In one embodiment, a license text string and a corresponding license key are embedded in a program that has been licensed to use a resource library. The license text string and the license key are supplied, for example, by a resource library vendor to a program developer who wants to use the resource library with an end user program being developed. The license text string includes information about the terms of the license under which the end user program is allowed to use the resource library. The license key is used to authenticate the license text string. The resource library in turn is provided with means for reading the license text string and the license key, and for determining, using the license key, whether the license text string is authentic and whether the license text string has been altered. Resource library functions are made available only to a program having an authentic and unaltered license text string.

Inventors: Garst; Blaine (Belmont, CA); Serlet; Bertrand (Palo Alto, CA)
Assignee: Apple Computer, Inc. (Cupertino, CA)
Appl. No.:   901776
Filed: July 28, 1997

Current U.S. Class: 705/59; 380/201; 380/202; 705/50; 705/51; 705/57; 705/58
Intern'l Class: G06F 017/60
Field of Search: 705/1,50-52,57-59 380/4,201,202

Sample Claim:

1. In a computer operating environment comprising a software program and a software resource, an apparatus for limiting use of said software resource by said software program, comprising:

an access authorization indicator associated with said software program, said access authorization indicator comprising one or more license terms for use of said software resource;

a digital signature of said access authorization indicator; means in said software resource for reading said access authorization indicator;

means in said software resource for determining whether said access authorization indicator is valid;

means for allowing said software program to use said software resource only if said access authorization indicator is determined to be valid.

[F] Electronic Negotiation Patents

[1] Ideal Electronic Negotiations — U.S. Patent No. 5,615,269

Micali   March 25, 1997

Abstract

There is described an electronic communications method between a first party and a second party, with assistance from at least a plurality of trustees, enabling an electronic transaction in which the first party having a selling reservation price (SRP) and the second party having a buying reservation price (BRP) may be committed to a transaction if a predetermined relationship between SRP and BRP is established, but not otherwise. The method begins by having each of the parties transmit shares of their respective reserve prices to the trustees. These shares are such that less than a given number of them does not provide enough useful information for reconstructing the reserve prices while a sufficiently high number of them allows such reconstruction. The trustees then take some action to determine whether the predetermined relationship exists without reconstructing SRP and BRP. If the predetermined relationship exists, then the trustees provide information that allows a determination of the sale price according to a given formula. Otherwise, the trustees determine that no deal is possible. As used herein, ''sale'' is merely respresentative as the transaction may be of any type including, without limitation, a sale, lease, license, financing transaction, or other known or hereinafter created financial commercial or legal instrument.

Inventors: Micali; Silvio (459 Chestnut Hill Rd., Brookline, MA 02146)
Appl. No.:   604870
Filed: February 22, 1996

Current U.S. Class: 705/80; 705/7; 705/75
Intern'l Class: H04L 009/00; H04K 001/00; G06F 017/60
Field of Search: 380/23,49,24 364/402,227.3,918.9

Sample Claim:

1. An electronic process executed by a first party and a second party, with assistance from at least a plurality of trustees, wherein the first party has a selling reservation price (SRP) and the second party has a buying reservation price (BRP) and the parties are committed to a transaction if a predetermined relationship between the reservation prices is established, but not otherwise, comprising the steps of:

> initiating the electronic process by having the first and second parties compute data strings encoding their respective reservation prices, wherein at least one of said parties uses an electronic device for such computation;

> having each of the first and second parties transmit to the trustees the data strings that encode their respective reservation prices, wherein at least one of these transmissions is carried out electronically, and wherein a subset of trustees containing less than a given number of trustees does not possess any useful information sufficient for reconstructing the reservation prices; and

> having the plurality of trustees participate in the electronic process by taking action to thereby determine whether the predetermined relationship exists, wherein the determination is made without reconstructing the reservation prices.

[2] Method for Electronic Merchandise Dispute Resolution — U.S. Patent No. 5,878,139

Rosen   March 2, 1999

Abstract

A system for open electronic commerce having a customer trusted agent securely communicating with a first money module, and a merchant trusted agent securely communicating with a second money module. Both trusted agents are capable of establishing a first cryptographically secure session, and both money modules are capable of establishing a second cryptographically secure session. The merchant trusted agent transfers electronic merchandise to the customer trusted agent, and the first money module transfers electronic money to the second money module. The money modules inform their trusted agents of the successful completion of payment, and the customer may use the purchased electronic merchandise.

Inventors: Rosen; Sholom S. (New York, NY)
Assignee: Citibank, N.A. (New York, NY)
Appl. No.:   774248

Filed: October 16, 1996

Current U.S. Class: 705/75; 705/24; 705/26; 705/65; 705/80; 902/2
Intern'l Class: H04L 009/32; G07F 019/00
Field of Search: 380/4,24,49 705/24,26 902/2

Sample Claim:

1. A method for resolving a dispute over electronic merchandise utilizing a customer trusted agent, a first host processor, a merchant trusted agent, and a second host processor, comprising the steps of:

(a) establishing a cryptographically secure session between said customer trusted agent and said merchant trusted agent;

(b) said customer trusted agent sending transaction log data to said first host processor for choice of a dispute corresponding to an electronic ticket stored in said customer trusted agent;

(c) said first host processor sending dispute information to said customer trusted agent;

(d) said customer trusted agent sending a copy of said electronic ticket and said dispute information to said merchant trusted agent, via said cryptographically secure session;

(e) said merchant trusted agent validating said electronic ticket;

(f) said merchant trusted agent sending said electronic ticket and said dispute information to said second host processor;

(g) deciding to deny said dispute relating to said electronic ticket and said dispute information;

(h) said second host processor sending a dispute denied message to said merchant trusted agent;

(i) said merchant trusted agent reporting said dispute denied message to said customer trusted agent;

(j) said customer trusted agent committing; and

(k) said merchant trusted agent committing.

[3] Method and Apparatus for the Sale of Airline-specified Flight Tickets — U.S. Patent No. 5,897,620

Walker, et al.   April 27, 1999

Abstract

An unspecified-time airline ticket representing a purchased seat on a flight to be selected later, by the airlines, for a traveler-specified itinerary (e.g., NY to LA on March 3rd) is disclosed. Various methods and systems for matching an unspecified-time ticket with a flight are also disclosed. An exemplary method includes: (1) making available an unspecified-time ticket; (2) examining a plurality of flights which would fulfill the terms of the unspecified-time ticket to determine which flight to select; and (3) providing notification of the selected flight prior to departure. The disclosed embodiments provide travelers with reduced airfare in return for flight-time flexibility and, in turn, permits airlines to fill seats that would have otherwise gone unbooked. Because of the flexibilities required of the unspecified-time traveler, unspecified-time tickets are likely to attract leisure travelers unwilling to purchase tickets at the available published fares and, at the same time, are likely to "fence out" business travelers unwilling to risk losing a full day at either end of their trip. Moreover, the flexibilities required of the unspecified-time traveler need not be limited to a departure time; the flexibilities may also include the airline, the departing airport, the destination airport, or any other restriction that increases the flexibility afforded the airline in placing the traveler aboard a flight. The disclosed embodiments thus permit airlines to fill otherwise empty seats in a manner that stimulates latent and unfulfilled leisure travel demand while leaving their underlying fare structures intact.

Inventors: Walker; Jay S. (Ridgefield, CT); Sparico; Thomas M. (Riverside, CT); Case; T. Scott (Darien, CT)
Assignee: priceline.com Inc. (Stamford, CT)
Appl. No.:   889304
Filed: July 8, 1997

Current U.S. Class: 705/5; 705/6
Intern'l Class: G06F 017/60
Field of Search: 705/5,6,7,9,28 707/1,2,3,102,104

Sample Claim:

1. A method comprising the steps of:

viewing, using a computer, special fare listing information for air travel to a specified destination location from a specified departure location within a specified time range, said special fare listing information excluding a specified departure time;

transmitting, using a computer, a request to purchase a commitment for carriage corresponding to said special fare listing information;

receiving a commitment for carriage, including an obligation by an airline to provide a seat on a flight, that satisfies said request but does not specify a departure time;

accepting said commitment for carriage; and

receiving at a time subsequent to said accepting an identification of said departure time.

**[4]   Framework for Negotiation and Tracking of Sale of Goods —
U.S. Patent No. 6,055,519**

Kennedy, et al.   April 25, 2000

Abstract

A computer implemented system and process are provided for negotiation and
tracking of sale of goods. In this system and process, a negotiation engine (16)
operates to store data representing a current state (18) of a negotiation between a
seller and buyer. The negotiation engine (16) stores the data within a framework
for representing aspects of the negotiation between the seller and buyer. The
framework includes a request object, a promise object and an acceptance object
that can store a current description of a contract. The framework also includes a
set of one or more delivery deals determined by the contract. Each delivery deal
can have a delivery request object, a delivery promise object, and a delivery
acceptance object that can store associated item deals and time periods for
delivery of item deals. Each item deal can have an item request object, an item
promise object and an item acceptance object that can store individual sales-order
line-items. The negotiation engine (16) thereby allows a user to monitor the
current state of the negotiation over a range of prices, a range of dates, ranges of
quantities of a set of goods, and a range of configurations of the goods in the set.

Inventors: Kennedy; Brian M. (Coppell, TX); Burchett; Christopher D.
                (Carrollton, TX)
Assignee: i2 Technologies, Inc. (Dallas, TX)
Appl. No.:   947544
Filed: October 11, 1997

Current U.S. Class: 705/80
Intern'l Class: G06F 015/00
Field of Search: 705/26,4,1,6,31,34,35,37,23,80 706/10

Sample Claim:

1. A computer implemented system for negotiation and tracking of sale of goods,
comprising:

        a computer system having a processor and memory, the computer
system executing a software application that provides a negotiation engine;

        the negotiation engine operating to store data representing a current
state of a negotiation between a seller and a buyer;

        the negotiation engine storing the data within a framework for
representing aspects of the negotiation between the seller and a buyer, the
framework including:

        a request object, a promise object and an acceptance object that can
store a current description of a contract;

wherein the request object represents a request from a buyer that initiates negotiation, the promise object represents a promise to sell from seller in response to the request, and the acceptance object represents the buyer's acceptance of the promise;

wherein the current state of negotiations is determined by means of state transitions between the request object, promise object, and acceptance object, said state transitions having at least three tiers;

a set of one or more delivery deals determined by the contract, each delivery deal including a delivery request object, a delivery promise object, and a delivery acceptance object that can store associated item deals and time periods for delivery of item deals; and

the item deals, each item deal including an item request object, an item promise object and an item acceptance object that can store individual sales-order line-items;

such that the negotiation engine allows a user to monitor the current state of the negotiation over a range of prices, a range of dates, ranges of quantities of a set of goods, and a range of configurations of the goods in the set.

[5] Method and Apparatus for Facilitating Transactions on a Commercial Network System — U.S. Patent No. 6,236,972

Shkedy   May 22, 2001

Abstract

A method and device for using a computer to facilitate a transaction of secondary market shares of an investment company such as a mutual fund between a buyer and a seller, having the steps of: a customer determining the mutual fund to be traded receiving a schedule of fees from the central controller, the customer selecting the class of shares and inputting the quantity to be traded, the customer selecting the order type and adding any special instructions. The customer then submits the order to the central controller. The central controller will match buyers and sellers and determine which orders were executed. For all executions, the central controller will provide the seller with payments and the buyer with shares in the selected mutual fund.

Inventors: Shkedy; Gary (455 E. 86th St., Apt. 22A, New York, NY 10028)
Appl. No.:   217663
Filed: December 21, 1998

Current U.S. Class: 705/1; 705/26; 705/27; 705/37; 705/75; 705/77; 705/80; 380/25
Intern'l Class: G06F 017/60

Field of Search: 705/1,26,37,75,77,80   713/173,186   340/825.31,825.34
                    380/25

Sample Claim:

1. A computer based method for facilitating a secondary market transaction of shares of investment companies between at least one buyer, an intermediate, and a seller, comprising:

>   receiving at the intermediate a sell order to sell in the secondary market shares of an investment company from the seller, binding the seller to said intermediate, said intermediate having a central controller including database storage;

>   assigning a sell tracking identification, corresponding to said sell order and said seller;

>   storing said sell order and sell tracking identification in said database;

>   receiving a buy order from the buyer, assigning a buy tracking identification company to said buy order and said buyer;

>   storing said buy order and buy tracking identification to said database; scanning the database for matching sell and buy orders; binding the buyer to the intermediate;

>   executing trade upon a match;

>   forwarding trade confirmation to a transfer agent; and

>   forwarding confirmations of transaction to the buyer and the seller.

[G]   Finance Patents

[1]   Open-architecture System for Real-time Consolidation of Information from Multiple Financial Systems — U.S. Patent No. 6,128,602

Northington, et al.   October 3, 2000

Abstract

An open-architecture system automatically consolidates information from a plurality of financial systems into a single accounting system without the need for expensive and time-consuming backroom procedures. The system enables an entity to use multiple independent and potentially incompatible financial systems to facilitate, control and monitor its spending, purchasing and other financial activities, while also enabling the entity to monitor and control all of these activities in real time. The system receives, processes and stores information obtained from a plurality of financial and/or other external computerized systems, and provides one or more authorized users with the ability to monitor financial transactions on-line and manipulate and control all financial transactions of the entity in real time using, for example, Web-browser software technology. Different users may have different levels of access to the financial transaction data obtained, processed and stored by the system. The system may also be readily integrated with the entity's existing computer systems.

Inventors: Northington; Cathy C. (Virginia Beach, VA); Goodson; Louis J. (Virginia Beach, VA)
Assignee: Bank of America Corporation (Charlotte, NC)
Appl. No.:   166069
Filed: October 5, 1998

Current U.S. Class: 705/35; 705/40; 705/42; 705/44; 713/187
Intern'l Class: G06F 017/60
Field of Search: 705/35,40,42,44,1 380/24

Sample Claim:

1. A computerized system for consolidating, monitoring and controlling financial transactions of an entity, comprising:

a network and gateway services element for accessing a plurality of financial systems;

a data repository services element for storing data generated, received and processed by said system;

an application services element, comprising

a processor for formatting data received and transmitted by said system,

a navigator for identifying one or more sources of information required by said system to consolidate, monitor, and control the financial transactions of the entity in response to control commands received from a remote access terminal or from one or more of said plurality of financial systems, and for generating corresponding source access command signals, and

a session manager coupled to said navigator for generating session commands in response to said source access command signals generated by said navigator,

wherein said network and gateway services element accesses said financial systems in response to session commands received from said session manager;

a web services element for providing access to said system from said remote access terminal; and

a communications network for enabling communication among said network and gateway services element, said data repository services element, said application services element, and said web services element.

[2] System for Automatically Determining Net Capital Deductions for Securities Held, and Process for Implementing Same — U.S. Patent No. 6,144,947

Schwartz   November 7, 2000

Abstract

A computer system is provided for automatically calculating broker-dealer disclosure data requirements for a publicly traded security that the broker-dealer holds. The computer system includes a MODEM in electrical/electronic communication with a source of encoded publicly traded security market data and a digital computer comprising a first I/O port for electronic communication with broker-dealer data processing apparatus, a second I/O port for electronic connection to the MODEM for processing the encoded data, and a memory device within which is stored a set of computer instructions defining computer system operation. The digital computer communicates with the broker-dealer apparatus to identify a broker-dealer position in the publicly traded security and processes MODEM-received data relating thereto to calculate a haircut coefficient for adjusting the position to comply with Rule 15-c 3-1 of the Securities Exchange Act, as amended.

Inventors: Schwartz; Peter A. (1011 Park Ave. #5, Hoboken, NJ 07030)
Appl. No.:   092114
Filed: June 5, 1998

Current U.S. Class: 705/37; 705/35; 705/36
Intern'l Class: G06F 017/60
Field of Search: 705/37,36,35,30 235/379

Sample Claim:

1. A computer system for automatically calculating broker-dealer disclosure data for a publicly traded security that the broker-dealer holds to comply with current SEC rules for regulating said broker-dealer data, comprising:

a MODEM in electrical/electronic communication with a source of encoded publicly traded security market data; and

a digital computer comprising a first I/O port for electronic communication with broker-dealer data processing apparatus, a second I/O port for electronic connection to the MODEM for processing the encoded data, and a memory device within which is stored a set of computer instructions defining computer system operation;

wherein the digital computer communicates with the broker-dealer apparatus to identify a broker-dealer position in the publicly traded security and processes MODEM-received data relating thereto to calculate a haircut coefficient for adjusting the position, adjusting the position based on the haircut and communicating data reflecting the adjusted position.

[3]   **Facility-based Financing System — U.S. Patent No. 6,154,730**

Adams, et al.   November 28, 2000

Abstract

A system for employing the projected receipts of a public facility to finance the construction of the facility itself, or the acquisition of a team to play in the facility. A preferred system includes: a method for projecting future cash flows (e.g., gate receipts) associated with the operation of the facility; pooling rights to receive those cash flows; transferring the pooled rights to a special purpose vehicle; and issuing securities on behalf of the special purpose vehicle in order to generate revenues for the construction and/or operating costs of the facility, or for the purchase of the team itself. The system also includes a computerized method for the ongoing implementation of such a financing system, including the steps of: inputting estimated cash flows and actual cash receipts; comparing the estimated and actual values in order to determine adjusted amounts to allocate between investors in the special purpose vehicle and ongoing operations. The invention further provides a facility (such as a stadium) and/or a team funded or acquired by such a system.

Inventors: Adams; Edward S. (2010 W. 49th St., Minneapolis, MN 55409);
               Goldman; Philip M. (1926 South La., Mendota Heights, MN
               55118)
Appl. No.:   175822
Filed: October 20, 1998

Current U.S. Class: 705/35; 705/36; 705/40; 705/45
Intern'l Class: G06F 017/60
Field of Search: 705/35,36,40,45 52/9,8,10

Sample Claim:

1. A system for financing a stadium facility, the system comprising a computerized system for:

   a) projecting one or more sources of future cash flow expected to be generated by the originating entity in the operation of the facility, said sources being selected from the group consisting of gate receipts, season ticket revenue, individual game ticket revenue, and vending revenue;

   b) collecting and pooling rights to receive said cash flow(s) over a period of years;

   c) transferring said rights to one or more special purpose vehicles, each in the form of a separate legal entity, in a manner that effectively removes the respective cash flow(s) from the originating entity's bankruptcy estate;

   d) issuing securities on behalf of said special purpose vehicle(s), using said pooled rights as collateral; and

   e) employing the revenue generated from the sale of said securities to finance at least a portion of the construction and/or operating costs of the facility.

[4]   **System and Method for Creating and Managing a Synthetic Currency — U.S. Patent No. 6,188,993**

Eng, et al.   February 13, 2001

Abstract

A synthetic currency transaction network which performs transactions with near real time finality of transaction between potential borrowers and potential lenders. Synthetic currency is created by pooling and dividing into shares a portfolio of highly liquid assets and frequent evaluation and disbursements of dividends on those assets so as to hold the value of the synthetic currency share at unity with the underlying currency. The synthetic currency network provides for interfacing users to the synthetic currency transaction network. A database is used for storing and maintaining records and other information of the network. A transaction manager manages network users' accounts and all network transactions. A fund accountant manages network information regarding the synthetic currency. A deposit bank acts as custodian for the portfolio of highly liquid assets which underlie the synthetic currency. An investment manager manages the measure of synthetic currency and directs investment decisions. A loan accountant manages all lending and borrowing activities in the synthetic currency transaction network.

Inventors: Eng; Alvin (New York, NY); Brill; Curtis (London, GB)
Assignee: Citibank, N.A. (New York, NY)
Appl. No.:   840133
Filed: April 11, 1997

Current U.S. Class: 705/37; 705/35
Intern'l Class: G06F 017/60
Field of Search: 705/35,36,37

Sample Claim:

1. A synthetic currency transaction network performing transactions with near real time finality of transaction between potential borrowers and potential lenders, said synthetic currency created by pooling and dividing into shares a portfolio of highly liquid assets and by frequent evaluation and disbursements of dividends on those assets so as to hold the value of the synthetic currency share at unity with the underlying currency, said synthetic currency network comprising:

> means for interfacing synthetic currency network users to said synthetic currency transaction network;

> database means for storing and maintaining records and other information used by said synthetic currency transaction network;

> transaction management means operatively connected to said means for interfacing and said database means for managing network users' accounts and all network transactions;

fund accounting means operatively connected to said transaction management means for managing network information regarding said synthetic currency;

deposit bank means which acts as custodian for said portfolio of highly liquid assets which underlie said synthetic currency;

investment management means operatively connected to said transaction management means and said deposit bank means for managing the measure of synthetic currency and directing investment decisions; and

loan accountant means operatively connected to said transaction management means for managing all lending and borrowing activities in said synthetic currency transaction network.

[5] Global Financial Services Integration System and Process — U.S. Patent No. 6,226,623

Schein, et al.   May 1, 2001

Abstract

A global standard messaging system and process for allowing customers to access a full range of global financial services using a variety of access points. The system includes a global communications network that integrates customer information and makes the information accessible from remote locations. The system includes a comprehensive database assembled from diverse sources and systems and processes for retrieving the information from the central database in a meaningful and practical way. The system includes several levels of access communications as well as built-in flexibility so that it can be accessed by a variety of remote systems of varying degrees of complexity and language. System and process permit rapid communication among worldwide users of the service as may be desired by industries relating to the transfer of finances.

Inventors: Schein; Arthur A. (Rockville Center, NY); Aron; Paul (Scotch Plains, NJ); Demeter; Dan A. (Woodmere, NY); Ataie; Faraz (New York, NY); Bamberger; Frank (Brooklyn, NY); McGlynn; John (Stamford, CT); Musalo; Florence (Manhasset, NY); Paul; Margot (Harrington Park, NJ); Poplizio; John (Milford, CT); Rico; Lucila (Uchie) (New York, NY); Tsien; Michael (Tuckahoe, NY); Yorke; Michael (Port Washington, NY)

Assignee: Citibank, N.A. (New York, NY)
Appl. No.:   077458
Filed: May 29, 1998
PCT Filed: May 23, 1997
PCT NO: PCT/US97/08413
371 Date: May 29, 1998
102(e) Date: May 29, 1998

PCT PUB. NO.:   WO97/43893
PCT PUB. Date: November 27, 1997

Current U.S. Class: 705/35; 709/217; 709/218; 709/230; 709/238
Intern'l Class: G06F 017/60
Field of Search: 705/35,36 709/217,218,219,230,238

Sample Claim:

1. A global communications network for use by a financial institution, the global communications network including a plurality of local area networks; a plurality of distribution points for allowing access to the global communications network; a plurality of service providers for providing information in response to data level commands; and an integration facility for decomposing high level business language requests into data level commands that are understandable by the service providers so as to allow end users located at distribution points to relay information to and receive information from the network, the integration facility comprising:

   means for determining whether an information request is simple or complex;

   means for receiving information requests from a distribution point and relaying-information the requests received from a distribution point to the logical router for determination of whether the request is simple or complex;

   a router for routing simple requests to a service provider that can satisfy the distribution point's request;

   means for generating a message relating to the satisfaction of the request;

   a router for routing messages relating to the satisfaction of the request back through the network to the originating distribution point;

   a plurality of messaging service agents, each messaging service agent including means for consulting script and workflow data model rules, and sending messages to logical servers which determine which service provider is appropriate to receive the complex request;

   means for routing complex requests to messaging service agents, which consult script and workflow data model rules, send one or more messages to logical servers which determine which service provider is appropriate to receive the complex request;

   a router for routing complex requests to the appropriate service provider which performs the request;

   means for allowing information relating to the satisfaction of the request to be exchanged between the distribution point and the service provider until the request is satisfied;

means for generating a message relating to the satisfaction of the request;

a router for routing messages relating to the satisfaction of the request back through the network to the originating distribution point.

[H] Inventory Management Patents

[1] Integrated System Monitoring Use of Materials, Controlling and Monitoring Delivery of Materials, and Providing Automated Billing of Delivered Materials — U.S. Patent No. 5,983,198

Mowery, et al.   November 9, 1999

Abstract

The system and method include a fleet of vehicles to provide material to a plurality of tanks at various customer locations. An inventory indicator associated with each of the tanks provides a quantity signal to the central station indicating the quantity and temperature of each of the tanks. A processor at the central station monitors the quantity signals of each of the tanks to determine past usage rates of the contents of each of the tanks. The processor projects future tank quantities based on the past usage pattern and determines possible routes for each of the vehicles to each of the tanks. The processor optimizes the routes, delivery amounts, and delivery schedule to minimize total delivered cost for the products based on the projected future tank levels and the possible routes to dispatch each of the vehicles.

Inventors: Mowery; Kevin M. (St. Louis, MO); Bartley; John P. (St. Charles, MO); Hantak; Robert J. (St. Louis, MO); Etling; Richard E. (St. Louis, MO); Read; Joseph R. (Chesterfield, MO)
Assignee: Novus International, Inc. (St. Louis, MO)
Appl. No.:   636289
Filed: April 23, 1996

Current U.S. Class: 705/22; 340/989; 340/990; 701/202; 701/300; 705/16
Intern'l Class: G06F 017/60
Field of Search: 701/300,202 340/989,990 705/22,16

Sample Claim:

1. A system for using a fleet of vehicles to provide material to a plurality of tanks at various customer locations comprising:

a central station;

an inventory indicator associated with each of the tanks and providing a quantity signal to the central station indicating the quantity in each of the tanks; and a processor at the central station for:

collecting and storing information from the inventory indicators;

monitoring the quantity signals of each of the tanks to determine past usage rates of the contents of each of the tanks;

projecting future tank quantities of each of the tanks based on the determined past usage rates;

determining possible delivery routes for each of the vehicles to each of the tanks;

optimizing the delivery routes for each of the vehicles based on the projected future tank quantities and based on the possible routes to minimize delivered cost;

optimizing the delivery amounts for the routes for each of the vehicles based on the projected future tank quantities and based on the possible routes to minimize delivered cost;

optimizing the delivery schedule for the routes for each of the vehicles based on the projected future tank quantities and based on the possible routes to minimize delivered cost; and

dispatching each of the vehicles in accordance with its optimized route and delivery schedule.

[2] System and Method for Dynamic Assembly of Packages in Retail Environments — U.S. Patent No. 6,138,105

Walker, et al.   October 24, 2000

Abstract

A system and method for managing the sale of a group of products at a single price based on sales performance data of the products is presented. The method and apparatus include offering a plurality of products by identifying products that are complementary, verifying acceptable sales performance for the complementary products, identifying a package including the complementary products having acceptable sales performance, determining a package price for the products included in the package, and offering the products included in the package at the package price. The status of the package is set to invalid when a time interval in which the package is available has expired. The method and apparatus further include package offer redemption by identifying a package including the products identified, determining a package price for the products included in the package, processing a sale of the products included in the package, adjusting sales performance data based on the sale of the products, and setting a status of the package to invalid when the sales performance data for the products included in the package fail to meet limits.

Inventors: Walker; Jay S. (Ridgefield, CT); Tedesco; Daniel E. (New
                  Canaan, CT); Van Luchene; Andrew S. (Norwalk, CT)
Assignee: Walker Digital, LLC (Stamford, CT)
Appl. No.:   085424
Filed: May 27, 1998

Current U.S. Class: 705/10; 235/378; 235/385; 705/16; 705/22
Intern'l Class: G06F 017/60
Field of Search: 705/26,27,16,22,28,10 235/378,385,380 345/962

Sample Claim:

1. A method for package offer redemption, the method comprising the steps of:

        receiving a plurality of product identifiers indicative of a plurality of
products;

        identifying a package including the received product identifiers;

        determining a package price for the products identified by the product
identifiers included in the package;

        processing a sale of the products identified by the product identifiers
included in the package;

        adjusting sales performance data based on the sale of the products; and

        setting a status of the package to invalid when the sales performance
data for the products included in the package fail to meet limits.

[3] Supply Chain Financing System and Method — U.S. Patent No. 6,167,385

Hartley-Urquhart   December 26, 2000

Abstract

A method for financing a supply of goods (a supply chain) from a supplier to a
buyer in which the buyer has a lower cost of funds than the supplier. According
to the method, the buyer generates a purchase order for the goods which is
forwarded to the supplier who in turn ships the goods to the buyer. The supplier
sends an invoice to the buyer which stores the invoice data in a database. The
financing institution electronically accesses the database to retrieve the daily
invoices. The financial institution then calculates the financing applicable to the
shipped good and forwards a payment to the supplier. Upon maturity of the
financing, the buyer settles with the financial institution by remitting the gross
proceeds.

Inventors: Hartley-Urquhart; William Roland (New York, NY)
Assignee: The Chase Manhattan Bank (New York, NY)
Appl. No.:   203208
Filed: November 30, 1998

Current U.S. Class: 705/35; 705/7; 705/8; 705/28; 705/37; 705/39
Intern'l Class: G06F 017/60
Field of Search: 705/7,8,35,39,28,37

Sample Claim:

1. A method for financing, through a third party, a supply of goods from a supply chain to a buyer, the supply chain consisting of a number of participants, wherein the buyer and each participant in the supply chain have a cost of financing, the method comprising the steps of:

> identifying a first participant that has a cost of financing that is the greatest above the buyer's cost of financing;

> establishing rules between the buyer and the third party including establishing, in a first rule, a manner in which the third party can identify acceptance data representing acceptance of goods by the buyer;

> the third party identifying acceptance data related to goods accepted by the buyer;

> calculating the financing for the goods in response to the identified acceptance data and in response to the established rules;

> forwarding payment for the goods from the third party to the first participant prior to a time the payment for the goods would normally be payable; and

> at maturity of the financing, settling the financing between the buyer and the third party.

[4] Inventory Control System — U.S. Patent No. 6,188,991

Rosenweig, et al.   February 13, 2001

Abstract

An inventory control system includes a computer system which may be coupled to a bar code scanner to allow items added to the inventory or deleted from the inventory to be scanned. The scanner may be mounted on an inventory storage container to facilitate inventory control. When the inventory of a particular item falls below a given level, an indication can be provided to assist in inventory control. A given list of items can be compared to the available inventory to determine whether the list of items is available. In addition, the inventory can be compared to a plurality of such lists to determine which lists are available in inventory.

Inventors:	Rosenweig; Michael (Hillsboro, OR); Rahman; Rezaur (Beaverton, OR)
Assignee:	Intel Corporation (Santa Clara, CA)
Appl. No.:	071559

Filed: May 1, 1998

Current U.S. Class: 705/29; 340/568.1; 700/214; 705/28
Intern'l Class: G06F 017/60
Field of Search: 235/383,385 340/568 377/6,13 700/213,214 705/1,22,28,29

Sample Claim:

1. A method of providing a set of producible products comprising:

preparing an inventory of items on a computer;

obtaining a set of lists of items, each list comprising a set of items needed to produce a product;

determining by computer those lists for which all the items on a list are available in inventory; and

reporting products that have an associated list which can be filled from available inventory.

[I] Marketing Patents

[1] Method and Apparatus for Selling an Aging Food Product as a Substitute for an Ordered Product—U.S. Patent No. 6,052,667

Walker, et al.   April 18, 2000

Abstract

A POS terminal receives an order for a food product. The POS terminal in turn selects a complementary product based on the food product. The selected complementary product has an age within a predetermined age range, such as between seven and ten minutes since the product was assembled. The POS terminal outputs an offer to substitute the complementary product for the food product. In one embodiment, the substitution is performed with no additional charge, so the customer will pay the price of the food product for the complementary product if he accepts the offer. The customer's response to the offer is received. If the response indicates acceptance of the offer, then the complementary product is sold in place of the food product.

Inventors: Walker; Jay S. (Ridgefield, CT); Van Luchene; Andrew S. (Norwalk, CT); Rogers; Joshua D. (New York, NY)
Assignee: Walker Digital, LLC (Stamford, CT)
Appl. No.:   157837
Filed: September 21, 1998

Current U.S. Class: 705/15; 705/20; 705/22
Intern'l Class: G06F 017/60
Field of Search: 705/1,7,14,15,16,20,21,22,24

196

Sample Claim:

1. A computer based method for selling an aging food product, comprising:

receiving an order for a food product;

selecting a complementary product based on the food product, the complementary product having an age within a predetermined age range;

outputting an offer to substitute the complementary product for the food product;

receiving a response to the offer; and

selling the complementary product in place of the food product if the response indicates acceptance of the offer.

[2] Method and Apparatus for Issuing and Managing Gift Certificates — U.S. Patent No. 6,193,155

Walker, et al.   February 27, 2001

Abstract

The present invention relates to a method and apparatus for issuing and redeeming a gift certificate drawn on a credit card or other financial account. The present invention includes a first aspect directed to a merchant card authorization terminal and a second aspect directed to a credit card issuer central controller. According to the first aspect, a method for redeeming a gift certificate drawn on a financial account is disclosed including the steps of receiving a gift certificate for payment of an identified value, transmitting a request for authorization to a central server, receiving an authorization signal, representing an indication that redemption of the gift certificate is authorized, from said central server and receiving a payment from the account issuer based on said identified value. A system is also disclosed for implementing the methods in all aspects of the present invention.

Inventors: Walker; Jay S. (Ridgefield, CT); Tedesco; Daniel E. (Stamford, CT); Jorasch; James A. (Stamford, CT); Lech; Robert R. (Norwalk, CT)
Assignee: Walker Digital, LLC (Stamford, CT)
Appl. No.:   997680
Filed: December 23, 1997

Current U.S. Class: 235/381; 705/44; 902/22
Intern'l Class: G06F 007/00; G06F 017/00
Field of Search: 235/380,381,382,382.5,495,379 902/22 705/47,39

Sample Claim:

1. A method for issuing a gift certificate corresponding to a financial account, comprising the steps of:

identifying and accessing stored account data associated with a financial account, said stored data including an account identifier;

determining a certificate identifier corresponding to said account identifier;

producing a gift certificate including thereon said certificate identifier; and

distributing said gift certificate to an owner of said financial account.

[3] Computer Implemented Marketing System — U.S. Patent No. 6,236,977

Verba, et al.   May 22, 2001

Abstract

A state-event engine manages behavior over time in response to events in deterministic fashion from user-supplied rules and policies for a set of dynamic objects. The campaign engine selectively generates and stores a campaign population, representing different types of marketing campaigns. A collection of dynamic object data stores maintains data on the corresponding marketing agent population, resource population, and listing population. A matching process orders members of the listing populations based on the target of at least two members of the campaign population such that a set of offers to buy and offers to sell the same resources is created. A prediction engine processes historical data to predict how campaigns can best match buyer to seller. Software agents negotiate on behalf of buyer and seller identify potential deals. The system is capable of being implemented using computer network and web-based technology.

Inventors: Verba; Stephen M. (Kirtland, OH); Ciepiel; Anthony M. (Hudon, OH)
Assignee: Realty One, Inc. (Cleveland, OH)
Appl. No.:   225283
Filed: January 4, 1999

Current U.S. Class: 705/10; 705/14; 705/26; 705/27; 705/37
Intern'l Class: G06F 017/60
Field of Search: 705/37,10,14,26,27,44 706/925

Sample Claim:

1. A computer-implemented self-optimizing marketing system comprising:

a campaign engine for selectively generating and storing a campaign population, said campaign population having members comprising stored first data representing a plurality of marketing campaigns each campaign characterized by a plurality of campaign attributes including a plurality of campaign activities;

said campaign engine having processing functionality to assemble campaign population members from said campaign activities;

a customer population data store for storing a customer population, said customer population having members comprising stored second data representing a plurality of customers and potential customers, characterized by a plurality of customer attributes; and

an optimization engine for accessing said first and second data to optimize at least one of said campaign population and said customer population, said optimization engine including a scoring system for ordering the members of at least one of said campaign population and said customer population, said scoring system employing adaptive scoring process that alters said scoring process based upon relations among at least some of said campaign attributes and said customer attributes.

[J] New Money Patents

[1] Trusted Agents for Open Electronic Commerce — U.S. Patent No. 5,557,518

Rosen   September 17, 1996

Abstract

A system for open electronic commerce having a customer trusted agent securely communicating with a first money module, and a merchant trusted agent securely communicating with a second money module. Both trusted agents are capable of establishing a first cryptographically secure session, and both money modules are capable of establishing a second cryptographically secure session. The merchant trusted agent transfers electronic merchandise to the customer trusted agent, and the first money module transfers electronic money to the second money module. The money modules inform their trusted agents of the successful completion of payment, and the customer may use the purchased electronic merchandise.

Inventors: Rosen; Sholom S. (New York, NY)
Assignee: Citibank, N.A. (New York, NY)
Appl. No.:   234461
Filed: April 28, 1994

Current U.S. Class: 705/69; 235/375; 235/379; 235/380; 235/381; 705/17; 705/44; 705/68; 705/74; 705/75; 705/76; 705/77; 902/1; 902/2; 902/7; 902/26

Intern'l Class:　　G06F 017/60; G06G 007/52; G07D 007/00; G06K 019/02
Field of Search:　　364/401,405,406,407,408,409 235/375,379,380,381,382,384
　　　　　　　　　　340/541,550 380/1,2,3,4,21,23,24,25,28,29,30

Sample Claim:

1. A system for open electronic commerce where both customers and merchants can securely transact comprising:

　　a customer trusted agent;

　　a first money module associated with said customer trusted agent, and capable of securely communicating with said customer trusted agent;

　　a merchant trusted agent capable of establishing a first cryptographically secure session with said customer trusted agent;

　　a second money module associated with said merchant trusted agent and capable of securely communicating with said merchant trusted agent, and capable of establishing a second cryptographically secure session with said first money module;

　　where said merchant trusted agent transfers electronic merchandise, via said first cryptographically secure session, to said customer trusted agent which provisionally retains said electronic merchandise;

　　where said customer trusted agent provides first payment information to said first money module and said merchant trusted agent provides second payment information to said second money module;

　　where said first money module transfers electronic money, in an amount consistent with said first and second payment information, to said second money module via said second cryptographically secure session;

　　where said first money module informs said customer trusted agent upon successful transfer of said electronic money, whereupon said retention of electronic merchandise is no longer provisional, and where said second money module informs said merchant trusted agent upon successful receipt of said electronic money.

[2] Currency and Barter Exchange Debit Card and System — U.S. Patent No. 5,592,376

Hodroff　　　　January 7, 1997

Abstract

The Currency Exchange Network transaction management and accounting system assists businesses, employees, and consumers to engage in productive economic activity that is not supported by traditional cash-and credit-based transaction systems. The system functions as a currency exchange between the non-cash, volunteer and barter economies and the mainstream cash economy.

This dual-currency system handles transactions for goods and services using a combination of cash and Community Economic Development Scrip, a new currency based on non-cash service credits.

Inventors: Hodroff; Joel (Minneapolis, MN)
Assignee: Commonweal Incorporated (St. Louis Park, MN)
Appl. No.:   261459
Filed: June 17, 1994

Current U.S. Class: 705/14; 705/32; 705/40
Intern'l Class: G06F 157/00
Field of Search: 235/379,380

Sample Claim:

1. A record management system for use in accounting for sales by a Vendor of a quantum of goods or services in return for a payment in a dual-currency combination of cash and service credits, said system comprising:

a) input means for receiving data indicative of the cash and service credit balances credited those respective accounts of each participant Member and on each Member's Cash Discount Category (CDC);

b) register means connected for receiving data from said input means and for storing that data for each Member and each account; and

c) control means for implementing and accounting for a sale of a particular quantum of goods and services to be sold through the dual-currency system, said control means carrying out the operations for each transaction of:

1. obtaining information indicative of the TRANSACTION PRICE for the selected quantity of the goods or services;

2. calculating the minimum cash percentage of the CASH BASE price required for the selected quantity of the goods or services;

3. accessing cumulative records of cash discounts made available to Members over a specified time interval and cumulative records of all service credits earned by all Members in the same selected time interval to enable the calculation of a current ratio of goods and services to service credits and storing that current ratio in the register means;

4. accessing CDC and current ratio information in the register means and calculating the amounts of cash and service credits required to be paid by a participant to complete the transaction, with the minimum cash portion equaling the CASH BASE; and

5. debiting the Member's dollar and service credit accounts in the register means by an amount equal to the dollars and service credits necessary to complete the transaction and crediting designated accounts with those dollars and service credits in accordance with a distribution plan implemented by the control means.

**[3] Electronic Funds Transfer Instruments — U.S. Patent No.
5,677,955**

Doggett, et al.   October 14, 1997

Abstract

An electronic instrument is created in a computer-based method for effecting a
transfer of funds from an account of a payer in a funds-holding institution to a
payee. The electronic instrument includes an electronic signature of the payer,
digital representations of payment instructions, the identity of the payer, the
identity of the payee, and the identity of the funds-holding institution. A digital
representation of a verifiable certificate by the institution of authenticity of the
instrument is appended to the instrument. This enables a party receiving the
instrument, e.g., the payee or a bank, to verify the authenticity of the account or
account holder. The invention may be generally applied to any financial
electronic document.

Inventors:	Doggett; John (Brookline, MA); Jaffe; Frank A. (Sharon, MA); Anderson; Milton M. (Fair Haven, NJ)
Assignee:	Financial Services Technology Consortium (New York, NY); The First National Bank of Boston (Boston, MA); Bell Communications Research, Inc. (Livingston, NJ)
Appl. No.:	418190
Filed:	April 7, 1995

Current U.S. Class:	705/76; 235/379; 380/30; 705/1; 705/44; 705/65; 705/75; 713/173; 902/2
Intern'l Class:	H04K 001/00; G06F 017/60; G06G 007/52
Field of Search:	380/30,24 235/379 902/2 395/201

Sample Claim:

1. A computer-based method comprising

creating an electronic instrument for effecting a transfer of funds from
an account of a payer in a funds-holding institution to a payee, the
instrument including an electronic signature of the payer, and

appending, to the electronic instrument, digital representations of a
verifiable certificate by the institution of the authenticity of the account or
the account holder.

[4] Untraceable Electronic Cash — U.S. Patent No. 5,768,385

Simon   June 16, 1998

Abstract

An electronic cash protocol including the steps of using a one-way function
f.sub.1 (x) to generate an image f.sub.1 (x.sub.1) from a preimage x.sub.1;

sending the image f.sub.1 (x.sub.1) in an unblinded form to a second party; and receiving from the second party a note including a digital signature, wherein the note represents a commitment by the second party to credit a predetermined amount of money to a first presenter of the preimage x.sub.1 to the second party.

Inventors: Simon; Daniel R. (Redmond, WA)
Assignee: Microsoft Corporation (Redmond, WA)
Appl. No.: 521124
Filed: August 29, 1995

Current U.S. Class: 705/69; 380/30; 705/74
Intern'l Class: H04L 009/00; H04L 009/30
Field of Search: 380/23,24,25,29,30,49,59,4,9,50

Sample Claim:

1. A method of implementing an electronic cash protocol comprising the steps of:

using a one-way function f.sub.1 (x) to generate an image f.sub.1 (x.sub.1) from a preimage x.sub.1;

sending the image f.sub.1 (x.sub.1) in an unblinded form to a second party; and

receiving from the second party a note including a digital signature, said note representing a commitment by the second party to credit a predetermined amount of money to a first presenter of said preimage x.sub.1 to the second party.

[5] System and Method for Detecting Fraudulent Expenditure of Electronic Assets — U.S. Patent No. 5,878,138

Yacobi   March 2, 1999

Abstract

An electronic asset system includes tamper-resistant electronic wallets that store non-transferable electronic assets. To break such tamper-resistant wallets, the criminal is expected to spend an initial investment to defeat the tamper-resistant protection. The electronic assets are uniquely issued by an institution to a wallet (anonymously or non-anonymously). During expenditure, the electronic assets are transferred from the wallet to a recipient. Since the assets are non-transferable, they are marked as exhausted assets upon expenditure. The recipient then batch deposits the received electronic assets with a collecting institution (which may or may not be the same as the issuing institution). A fraud detection system samples a subset of the exhausted assets received by the recipient to detect "bad" assets which have been used in a fraudulent manner. Upon detection, the fraud detection system identifies the electronic wallet that used the bad asset and marks it as a "bad wallet." The fraud detection system then compiles a list of bad electronic wallets and distributes the list to warn potential recipients of the bad electronic wallets. When a bad wallet next attempts to spend assets (whether fraudulently or not), the intended recipient will check the local hot list of bad wallets and refuse to transact business with the bad wallet.

Inventors: Yacobi; Yacov (Mercer Island, WA)
Assignee: Microsoft Corporation (Redmond, WA)
Appl. No.: 600409
Filed: February 12, 1996

Current U.S. Class: 705/69; 340/5.41; 705/41; 705/44; 705/76
Intern'l Class: H04K 001/00; H04L 009/00; G06F 007/04; G07D 007/00
Field of Search: 380/23,24,25 705/41,44 340/825.3,825.33,825.34,825.35

Sample Claim:

1. An electronic asset system comprising:

a plurality of electronic wallets;

a plurality of non-transferable electronic assets stored on the electronic wallets, the electronic assets being removed from the wallets when used and marked as exhausted assets; and

a probabilistic fraud detection system to sample a subset of less than all of the exhausted assets to detect bad assets that have been used in a fraudulent manner, the fraud detection system further identifying the electronic wallets that used the bad assets.

[6] Electronic-monetary System — U.S. Patent No. 6,047,067

Rosen April 4, 2000

Abstract

An electronic-monetary system having (1) banks or financial institutions that are coupled to a money generator device for generating and issuing to subscribing customers electronic money including electronic currency backed by demand deposits and electronic credit authorizations; (2) correspondent banks that accept and distribute the electronic money; (3) a plurality of transaction devices that are used by subscribers for storing electronic money, for performing money transactions with the on-line systems of the participating banks or for exchanging electronic money with other like transaction devices in off-line transactions; (4) teller devices, associated with the issuing and correspondent banks, for process handling and interfacing the transaction devices to the issuing and correspondent banks, and for interfacing between the issuing and correspondent banks themselves; (5) a clearing bank for balancing the electronic money accounts of the different issuing banks; (6) a data communications network for providing communications services to all components of the system; and (7) a security arrangement for maintaining the integrity of the system, and for detecting counterfeiting and tampering within the system. An embodiment of the invention includes a customer service module which handles lost money claims and links accounts to money modules for providing bank access.

Inventors: Rosen; Sholom S. (New York, NY)
Assignee: Citibank, N.A. (New York, NY)
Appl. No.: 994088
Filed: December 19, 1997

Current U.S. Class: 705/68; 235/379; 705/41; 705/69; 902/26
Intern'l Class: G07F 019/00
Field of Search: 902/2,26 235/379 705/41 380/24

Sample Claim:

1. An electronic monetary system comprising:

an issuing bank having an on-line accounting system;

a money issued reconciliation system;

electronic representations of money that are accounted for in said on-line accounting system;

a money generator module associated with said issuing bank, for generating said electronic representations of money;

a teller module associated with said issuing bank, capable of storing said electronic representations of money;

a transaction money module capable of transferring said electronic representations of money;

where a transfer record is appended to said electronic representations of money upon each transfer between any two of said modules, and where said electronic representations of money are periodically passed to said money issued reconciliation system;

said money issued reconciliation system having a processor for analyzing said transfer records for each electronic representation of money to identify electronic representations of money that are lost or duplicated; and

where a subscriber of a given transaction money module submits to said issuing bank a lost money claim that identifies one or more lost electronic representations of money associated with the given transaction money module, and said issuing bank remunerates said subscriber based on said lost money claim and on said money issued reconciliation system confining the validity of said lost money claim for each said one or more lost electronic representations of money.

[7]   **Executable Digital Cash for Electronic Commerce — U.S. Patent No. 6,157,920**

Jakobsson, et al.   December 5, 2000

Abstract

The invention provides techniques for implementing secure transactions using an instrument referred to as "executable digital cash." In an illustrative embodiment, a first user generates a piece of digital cash representing an offer made by that user. The piece of digital cash includes a digital certificate authorizing the first user to make specified transfers, and an offer program characterizing the offer. The piece of digital cash is broadcast or otherwise transmitted to one or more additional users, utilizing a mobile agent or other suitable mechanism, such that a given one of these users can evaluate the offer using the offer program. For example, a second user could execute the offer program with a specific bid as an input to determine what that user would receive upon acceptance of his bid. If the result is acceptable to the second user, that user generates a bid capsule including the bid, the corresponding output of the offer program, and another certificate authorizing the second user to make the transfer specified in the bid. The bid capsule is submitted to an institution for processing in accordance with a policy which may be specified in the piece of digital cash. The institution selects one or more winning bids and implements the corresponding transactions. Digital signatures generated using secret keys associated with the certificates of the first and second users are utilized to ensure adequate security for the transmitted offer and bid information.

Inventors: Jakobsson; Bjorn Markus (Hoboken, NJ); Juels; Ari (Cambridge, MA)
Assignee: Lucent Technologies Inc. (Murray Hill, NJ); RSA Security Inc. (Bedford, MA)
Appl. No.:   134012
Filed: August 14, 1998

Current U.S. Class: 705/69; 705/64; 705/65; 705/67; 705/74; 705/76; 705/80
Intern'l Class: G06F 017/60
Field of Search: 705/37,69 235/380

Sample Claim:

1. A method comprising the steps of:

   generating a piece of digital cash representing an offer made by a first user, the piece of digital cash including: (i) a digital certificate, and (ii) an offer program characterizing the offer, the offer program specifying for a given input a particular transfer to be made by the first user and authorized by the digital certificate; and transmitting the piece of digital cash, whereby a second user can evaluate the offer by supplying a bid as input to the offer program, and if the bid is indicated as acceptable by the offer program the second user can bind the first user to make the corresponding transfer authorized by the digital certificate.

[8] **Self-contained Payment System with Circulating Digital Vouchers — U.S. Patent No. 6,205,435**

Biffar   March 20, 2001

Abstract

A self-contained payment system uses circulating digital vouchers for the transfer of value. The system creates and transfers digital vouchers. A digital voucher has an identifying element and a dynamic log. The identifying element includes information such as the transferable value, a serial number and a digital signature. The dynamic log records the movement of the voucher through the system and accordingly grows over time. This allows the system operator to not only reconcile the vouchers before redeeming them, but also to recreate the history of movement of a voucher should an irregularity like a duplicate voucher be detected. These vouchers are used within a self-contained system including a large number of remote devices which are linked to a central system. The central system can be linked to an external system. The external system, as well as the remote devices, are connected to the central system by any one or a combination of networks. The networks must be able to transport digital information, for example the internet, cellular networks, telecommunication networks, cable networks or proprietary networks. Vouchers can also be transferred from one remote device to another remote device. These remote devices can communicate through a number of methods with each other. For example, for a non-face-to-face transaction the internet is a choice, for a face-to-face or close proximity transactions tone signals or light signals are likely methods. In addition, at the time of a transaction a digital receipt can be created which will facilitate a fast replacement of vouchers stored in a lost remote device.

Inventors: Biffar; Peter (1060 High St., Palo Alto, CA 94301)
Assignee: Biffar; Peter (Palo Alto, CA)
Appl. No.:   458117
Filed: December 8, 1999

Current U.S. Class: 705/41; 235/380; 705/39; 705/65; 705/69
Intern'l Class: G06F 017/00
Field of Search: 705/39,65,69,41 235/380

Sample Claim:

1. In a system for making payments using vouchers having an identifying element and a dynamic log to which additional data representing the transfer of

the voucher is added whenever there is a transaction involving the voucher, wherein a transaction includes a creation of, use of, or movement of the voucher, a remote device comprising:

an input signal receiver for receiving said vouchers sent to the said remote device;

a memory for storing said vouchers and storing an identification number of said remote device and storing an account number;

an output signal generator for sending said vouchers;

a processor storing said received vouchers in said memory, wherein said processor generates and attaches a remote device set of log digits to said voucher, and wherein said processor retrieves vouchers from said memory and sends them to said output signal generator linked to said processor; and

user controls for inputting operating instructions to said remote device.

[K] Resource Allocation Patents

[1] Reorder System for Use with an Electronic Printing Press — U.S. Patent No. 6,246,993

Dreyer, et al.   June 12, 2001

Abstract

A method of, and system for, selectively reordering the reprinting of books on one or more electronic presses are disclosed. In one embodiment, sensors and a processor are utilized to reorder books fouled by an auxiliary device. In another embodiment, a sensor, and a global processor in communication with first and second local processors respectively associated with first and second printing presses are utilized to reorder books on one of the first and second printing presses selected to minimize processing time and/or to maximize postal rate discounts. In another embodiment, a sensor and a press processor are utilized to reorder errored books by inserting a book back into a stream of books being printed if re-printing the book with the stream entitles the errored book to a predefined postal discount.

Inventors:	Dreyer; Mark G. (Aurora, IL); Warmus; James L. (LaGrange, IL); Gill; Robert W. (Plainfield, IL); Kitzmiller; Roger (Charlotte, NC)
Assignee:	R. R. Donnelley & Sons Company (Downers Grove, IL)
Appl. No.:	959994
Filed:	October 29, 1997

Current U.S. Class: 705/9; 700/100; 700/110; 705/8; 705/410
Intern'l Class: G06F 017/60

Field of Search: 53/52,147,443,447,540 270/1.01
                      364/148.01,468.01,468.03,468.05,468.06,468.15,468.16,468.17
                      705/8,9,401,402,410 700/28,95,99,100,108,110

Sample Claim:

1. For use with an auxiliary device located downstream from a first printing press printing a stream of books with fixed and variable data, the auxiliary device being adapted to perform a process on books printed by the first printing press, a reorder system comprising:

> a first sensor located downstream of the auxiliary device to monitor books output by the auxiliary device;

> a local processor associated with the auxiliary device, the local processor being in communication with the first sensor and being adapted to recognize an errored book based at least in part on an output of the first sensor; and,

> a global processor in communication with the local processor, the global processor being responsive to a communication from the local processor identifying the errored book to cause the errored book to be reprinted with fixed and variable data after at least one other book in the stream of books has been printed by the first printing press.

[2] Integrated Business-to-business Web Commerce and Business Automation System — U.S. Patent No. 6,115,690

Wong   September 5, 2000

Abstract

A software system business-to-business Web commerce (Web business, or e-business) and automates to the greatest degree possible, in a unified and synergistic fashion and using best proven business practices, the various aspects of running a successful and profitable business. Web business and business automation are both greatly facilitated using a computing model based on a single integrated database management system (DBMS) that is either Web-enabled or provided with a Web front-end. The Web provides a window into a ''seamless'' end-to-end internal business process. The effect of such integration on the business cycle is profound, allowing the sale of virtually anything in a transactional context (goods, services, insurance, subscriptions, etc.) to be drastically streamlined.

Inventors: Wong; Charles (14250 Miranda Rd., Los Altos Hills, CA 94022)
Appl. No.:   995591
Filed: December 22, 1997

Current U.S. Class: 705/7; 700/237; 705/1; 705/8; 705/30; 705/34; 708/136
Intern'l Class: G06F 017/60

Field of Search: 235/380
364/468.02,468.14,468.21,479.06,479.07,479.08,705.06,709.06 705/34,1,30,7,8

Sample Claim:

1. An automated end-to-end business process for product sales that uses a relational database management system, the process comprising the steps of:

a first user inputting a sales record to the database for an order of a customer;

automatically generating a customer invoice;

a second user inputting a customer payment record to the database, wherein system privileges of the first user and the second user are at least partially mutually exclusive;

automatically determining a status of the customer payment as reconciled or not reconciled; and

during each of the foregoing inputting steps, qualifying user inputs using experiential constraints, based on the then-current state of the database as a whole.

[3] Method for Production Planning in an Uncertain Demand Environment — U.S. Patent No. 6,138,103

Cheng, et al.   October 24, 2000

Abstract

A decision-making method suitable for production planning in an uncertain demand environment. To this end, the method comprises combining an implosion technology with a scenario-based analysis, thus manifesting, a *sui generis* capability which preserves the advantages and benefits of each of its subsumed aspects.

Inventors: Cheng; Feng (Elmsford, NY); Connors; Daniel Patrick (Wappingers Falls, NY); Ervolina; Thomas Robert (Hopewell Junction, NY); Srinivasan; Ramesh (San Jose, CA)
Assignee: International Business Machines Corporation (Armonk, NY)
Appl. No.:   362010
Filed: July 27, 1999

Current U.S. Class: 705/7; 700/103; 705/8; 705/19
Intern'l Class: G06F 017/60; G06F 015/21; G06F 015/60
Field of Search: 705/7,8,9,10,29,28,20,35,14 700/106,103

Sample Claim:

1. A program storage readable by a machine, tangibly embodying a program of instructions executable by a machine to perform method steps for production planning in an uncertain demand environment, said method steps comprising:

a) inputting a plurality of demand scenarios over a timing horizon and a probability associated with each of said demand scenario which represent uncertainties in a demand environment;

b) employing a scenario-based analysis including the steps of performing multiple optimization runs against different demand scenarios;

c) combining an implosion technology with the scenario-based analysis for generating for any one individual demand scenario, a deterministic solution which is optimal for the particular demand scenario; and,

d) generating an output comprising a payoff table that includes said solution for each demand scenario, each said solution being optimized in accordance with a selected performance measure.

[4] Method and Computer System for Controlling an Industrial Process by Analysis of Bottlenecks — U.S. Patent No. 6,144,893

Van Der Vegt, et al.   November 7, 2000

Abstract

A method and computer system for controlling an industrial process are disclosed. The industrial process has problems which adversely effect its performance. A personal computer is programmed with a database and a custom application. The database contains data describing the attributes and performance of the process to be controlled. The custom application calculates the financial value of each of the problems. The calculation of the financial value of each problem takes into account the impact of each problem on the process bottleneck. The financial values of the problems are used to prioritize the problems, allowing them to be remedied in accordance with their priority.

Inventors: Van Der Vegt; Anton Hans (Bondi Beach, AU); Thompson; Ian Chetwynd (Clareville, AU)
Assignee: Hagen Method (Pty) Ltd. (Newport, AU)
Appl. No.:   027098
Filed: February 20, 1998

Current U.S. Class: 700/108; 700/95; 700/100; 700/109; 700/117; 705/7; 705/8; 705/9; 705/400
Intern'l Class: G06F 019/00
Field of Search: 700/95,108,97,99,100,101,109,115,116,117 705/7,8,9,400

Sample Claim:

1. A method of controlling a process for producing product, the process being designed to run at an optimum performance, the method comprising the steps of:

identifying a problem in the process, which problem causes the process to perform at less than the optimum performance, thereby causing the process to lose processing time;

identifying a bottleneck in the process, such that the bottleneck dictates the maximum speed at which the process runs;

determining how much processing time the process loses due to the problem by the steps of:

determining a slow running time for the problem indicative of how much processing time the process loses due to the problem causing the bottleneck to run at a speed slower than an expected speed;

determining a downtime for the problem indicative of how much processing time the process loses due to the problem causing the bottleneck to stop running;

determining a bottleneck waste time for the problem indicative of how much processing time the process loses due to the problem causing product to be scrapped at or after the bottleneck;

calculating a financial value of the problem based on how much processing time the process loses due to the problem;

prioritizing the problem based on the financial value of the problem.

[5] Method of Managing Contract Housekeeping Services — U.S. Patent No. 6,144,943

Minder   November 7, 2000

Abstract

The invention provides a method and apparatus including a computer system for managing contract housekeeping services so as to improve the quality and value of the housekeeping services received. The method invention comprises a series of actions in order' to generate a grade representative of the quality of housekeeping services. This grade is then used to make at least one decision regarding the management of housekeeping services. The management having an impact upon the physical appearance and maintenance of a given facility.

Inventors: Minder; Diane L. (Richmond, VA)
Assignee: Virginia Commonwealth University (Richmond, VA)
Appl. No.:   955103

Filed: October 21, 1997

Current U.S. Class: 705/11; 235/375; 235/376; 705/8; 705/9
Intern'l Class: G06F 017/60
Field of Search: 705/1,7-9,11 235/375,376

Sample Claim:

1. A method of using a computer system to manage contract housekeeping services comprising:

> storing in said computer system a plurality of variables describing specific housekeeping service tasks;

> defining a plurality of possible performance criteria scores probative of the plurality of variables;

> receiving in said computer system user input operative to select a subset of said plurality of variables;

> receiving in said computer system actual performance criteria scores for the selected subset of variables;

> using said computer system to process the actual performance criteria scores and to output a grade representative of the quality of housekeeping services;

> using the grade in making at least one decision regarding management of housekeeping services, the management having an impact upon the physical appearance and maintenance of the given facility.

[6] Method and Apparatus to Connect Consumer to Expert — U.S. Patent No. 6,223,165

Lauffer   April 24, 2001

Abstract

This invention provides for a method of (or apparatus for) facilitating the delivery of advice to consumers using a server unit which can store and display the names and characteristics of experts and then rapidly assist in connecting the expert and consumer for real-time communication. The server can also have the ability to receive keywords from the consumer, match those keywords to one or more experts, and tell the consumer how to contact an expert.

Inventors: Lauffer; Randall B. (Brookline, MA)
Assignee: Keen.Com, Incorporated (San Francisco, CA)
Appl. No.: 488130
Filed: January 20, 2000

Current U.S. Class: 705/8; 705/1; 705/26
Intern'l Class: G06F 017/60

Field of Search: 705/26,3,1,7,8

Sample Claim:

1. A method of connecting two parties in real time, the method comprising:

    displaying a list of experts to a consumer via an Internet connection with said consumer prior to the consumer submitting a question;

    the list indicating individually whether each expert is currently available to telephonically communicate with said consumer at a time when said consumer is viewing the list, said list includes a compensation rate for each expert;

    in response to the consumer selecting a displayed icon corresponding to an expert from the list, automatically establishing a telephone connection between the expert and the consumer prior to the consumer submitting a question to the expert; and

    said automatically establishing the connection includes a central controller placing a telephone call to said consumer via a connection separate from said Internet connection, and said central controller placing a telephone call to said expert.

[L]   Electronic Shopping Patents

[1]   Method and System for Placing a Purchase Order Via a Communications Network — U.S. Patent No. 5,960,411

Hartman, et al.   September 28, 1999

Abstract

A method and system for placing an order to purchase an item via the Internet. The order is placed by a purchaser at a client system and received by a server system. The server system receives purchaser information including identification of the purchaser, payment information, and shipment information from the client system. The server system then assigns a client identifier to the client system and associates the assigned client identifier with the received purchaser information. The server system sends to the client system the assigned client identifier and an HTML document identifying the item and including an order button. The client system receives and stores the assigned client identifier and receives and displays the HTML document. In response to the selection of the order button, the client system sends to the server system a request to purchase the identified item. The server system receives the request and combines the purchaser information associated with the client identifier of the client system to generate an order to purchase the item in accordance with the billing and shipment information whereby the purchaser effects the ordering of the product by selection of the order button.

Inventors: Hartman; Peri (Seattle, WA); Bezos; Jeffrey P. (Seattle, WA);
               Kaphan; Shel (Seattle, WA); Spiegel; Joel (Seattle, WA)
Assignee: Amazon.com, Inc. (Seattle, WA)
Appl. No.:   928951
Filed: September 12, 1997

Current U.S. Class: 705/26; 345/962; 705/27
Intern'l Class: G06F 017/60
Field of Search: 705/26,27   380/24,25   235/2,375,378,381   395/188.01
               345/962

Sample Claim:

1. A method of placing an order for an item comprising:

> under control of a client system,

> displaying information identifying the item;

> in response to only a single action being performed, sending a request to order the item along with an identifier of a purchaser of the item to a server system;

> under control of a single-action ordering component of the server system, receiving the request;

> retrieving additional information previously stored for the purchaser identified by the identifier in the received request;

> generating an order to purchase the requested item for the purchaser identified by the identifier in the received request using the retrieved additional information; and

> fulfilling the generated order to complete purchase of the item whereby the item is ordered without using a shopping cart ordering model.

[2] Universal Web Shopping Cart and Method of On-line Transaction Processing — U.S. Patent No. 6,101,482

DiAngelo, et al.   August 8, 2000

Abstract

A method of purchasing products and services on-line using a client connectable to a plurality of servers via a computer network. The method begins by initiating from the client two or more independent transaction sessions, each of the independent transaction sessions established as a connection between the client and one of the plurality of servers is active. During each independent transaction session, transaction information is collected at the client to facilitate a purchase of products and services after the connection between the client and the server is closed and the transaction session is completed. According to the invention, the transaction information is maintained persistent across multiple independent transaction sessions. At a given time, for example, after all Web sites have been visited and the information gathered, the transaction information (as originally collected and/or as filtered, updated or enhanced) is then used to effect a purchase of given products and services.

Inventors:	DiAngelo; Michael F. (Round Rock, TX); Fox; Valerie J. (Etobicoke, Canada)
Assignee:	International Business Machines Corporation (Armonk, NY)
Appl. No.:	929044
Filed:	September 15, 1997

Current U.S. Class: 705/26; 705/27; 709/227
Intern'l Class: G06F 017/60
Field of Search: 705/26,27,17   395/200.57,200.31,200.47,200.48,200.49,
                    200.54,200.58 380/24,25713/200,201

Sample Claim:

1. A method of purchasing products and services on-line using a client connectable to a plurality of servers via a computer network, comprising the steps of:

> initiating from the client two or more independent transaction sessions, each of the independent transaction sessions established as a connection between the client and one of the plurality of servers is active;

> during each independent transaction session, collecting transaction information at the client to facilitate a purchase of products and services after the connection between the client and the server is closed and the transaction session is completed;

> maintaining the transaction information persistent across multiple independent transaction sessions;

> processing the transaction information maintained across multiple independent transaction sessions;

> filtering information from the independent transaction sessions to generate a selection; and at a given time, using the information to effect a purchase of given products and services.

[3] Method and Apparatus for Executing Electronic Commercial Transactions with Minors — U.S. Patent No. 6,173,269

Solokl, et al.   January 9, 2001

Abstract

A method and apparatus is provided for executing electronic transactions with teens, especially where such transactions are limited only to those vendors that have been approved by the teen's parents. In one embodiment, a virtual automatic teller machine (VATM) is provided in which funds are transferred from an existing account, such as a saving account, checking account, or credit card account, to an Internet passport account. The VATM account mimics a bank account, i.e. it gives the user the appearance of an ATM machine. Functionally, the VATM allows the user to transfer funds from an existing account into the Internet passport account. The VATM does this by emulating an ATM machine as it appears to the Automated Clearing House (ACH) system. The ACH system is a separate network from the Internet. Rather than acting as a trustee for a teen account, the invention provides a method and apparatus that allows a merchant to withdraw funds directly from the teen's account automatically at the time of purchase. In this way, the invention provides a system in which funds are not held, thereby eliminating cash advance fees and liabilities associated with trusteeship. A second embodiment of the invention, a global gift certificate, is provided. The preferred second embodiment of the invention is configured to appear as a debit card to the ACH system. In this regard, the gift certificate thus generated is truly global in that it is accepted anywhere it is presented.

Inventors: Solokl; Daniel David (San Jose, CA); Knight; Kirk Hoyt (Alameda, CA); Corsini; Frank Anton (Tiburon, CA)
Assignee: Zowi.com, Inc (San Francisco, CA)
Appl. No.:   288046
Filed: April 7, 1999

Current U.S. Class: 705/35; 705/39; 705/42
Intern'l Class: G06F 017/60; G06F 015/00
Field of Search: 705/35,39,42,14

Sample Claim:

1. A method for executing electronic transactions with an individual lacking contractual capacity, comprising the steps of:

> funding a separate account held by a financial institution with funds from a fund source account of a parent or guardian of said individual, wherein said separate account is independent and unsupervised relative to said fund source account; and

> providing a service for supervising access by said individual who is enrolled with said service to funds in said separate account, said service executing binding transactions with third parties on behalf of said individual.

[4] **System and Method for Home Grocery Shopping Including Item Categorization for Efficient Delivery and Pick-up — U.S. Patent No. 6,246,998**

Matsumori   June 12, 2001

Abstract

A system for organizing and categorizing purchases made through an Internet based home shopping system is disclosed. Products offered for sale over the Internet are contained in a server's PLU table and are each associated with weights and measures metrics, environmental storage metrics and nutritional content indicia. As items are selected for purchase, and placed in a virtual shopping basket, a user may evaluate the contents of a virtual shopping basket in accordance with any of the physical property, environmental storage, or nutritional content indicia. Goods designated for either pickup or delivery are packaged in accordance with their size and weight criteria and are stored in an appropriate environmental storage facility in accordance with their environmental storage indicia.

Inventors: Matsumori; Kunihiko (San Diego, CA)
Assignee: Fujitsu Limited (Kanagawa, JP)
Appl. No.:   257851
Filed: February 25, 1999

Current U.S. Class: 705/27; 705/26; 345/349
Intern'l Class: G06F 017/60
Field of Search: 705/26,27,29,28 345/349

Sample Claim:

1. An Internet based home grocery shopping system including a personal computer system configured for Internet access, the system comprising:

a store server configured to host an Internet access application program and including an Internet communication interface and a control processor;

a mass storage device coupled to the store server;

a master PLU database stored in the mass storage unit, the PLU database including data representing item identification, price and weights and measures metrics, each said metric associated with an item of merchandise, the PLU database further including an environmental storage metric identifying an environment suitable for storing said item of merchandise; and

means for summing the merchandise item information metrics, including the weights and measures metrics, the means for summing further including means for displaying the summed metrics on a display screen of a personal computer system, thereby allowing a consumer to visually inspect a size and weight characteristic of an electronic order.

[M] Supply-chain Management Patents

[1] Method and Apparatus for Collaboratively Managing Supply Chains — U.S. Patent No. 6,157,915

Bhaskaran, et al.   December 5, 2000

Abstract

An active collaboration technology in an open architectural framework that delivers information and decision support tools in a timely, contextual and role sensitive manner to present a collaborative dynamic decision making capability to a community of role players within a supply chain process. Such a comprehensive collaborative dynamic decision making capability is made possible through the integration of the business process, the organization of role players and relevant business applications.

Inventors: Bhaskaran; Kumar (Kearny, NJ); Connors; Daniel P. (Wapping-
                ers Falls, NY); Heath, III; Fenno F. (New Haven, CT); Nayak;
                Nitin (Ossining, NY)
Assignee: International Business Machines Corporation (Armonk, NY)
Appl. No.:   131114
Filed: August 7, 1998

Current U.S. Class: 705/7; 705/1; 705/8; 705/9; 705/10; 705/11; 705/14;
                        705/26; 705/27
Intern'l Class: G06F 017/60; G06F 015/00
Field of Search: 705/7,1,8,9,10,11,28,29,26,27,14

Sample Claim:

1. A communication and decision support method of collaboratively devising, disseminating, deciding and communicating a business plan in a community of role players within a supply chain, comprising:

  collaboratively creating a business scenario by role players having permission to access interactive role sensitive active documents for supply, production and distribution of a product, wherein the interactive role sensitive active documents comprise relevant business information related to the supply chain and business support tools enabling role players to make appropriate decisions for maintaining congruity and efficiency of the supply chain, and wherein the interactive role sensitive active documents are dynamic documents rendering data to a role player in the community of role players based on a predefined role of the role player and within a context of business processes within the business scenario;

collaboratively submitting the business scenario, via the interactive role sensitive active documents on a distributed network, to the several role players having permission to access the interactive role sensitive active documents in response to collaboratively creating the business scenario;

collaboratively dynamically formulating the business plan based on the collaboratively created and submitted business scenario; and

nesting and maintaining the business scenario and the business plan in the interactive role sensitive active documents including updates of the business scenario and the business plan in response to the collaboratively dynamically formulating the business plan.

[2] Method for Part Procurement in a Production System with Constrained Resources — U.S. Patent No. 5,970,465

Dietrich, et al.   October 19, 1999

Abstract

A method for determining procurement for parts (P) in a production system having constraints comprising at least one of constrained resources ($r.sub.i$) and known maximum demands ($d.sub.j$). The method comprises two steps. Step 1 includes constructing a production planning decision space comprising independent sets of hyperplanes defined by decision variables ($q.sub.j$) corresponding to product quantities for products (j). The constructing step subsumes steps of expressing a potential usage of part (p) as a linear combination of production quantities ($q.sub.j$) based on bill of material usage rules; limiting the production quantities ($q.sub.j$) so that each is less than or equal to the maximum demand quantity ($d.sub.j$); and limiting the production quantities ($q.sub.i$) so that the usage of each resource (r) is based on bill of material and bill of capacity usage rates less than or equal to the availability of that resource. For each part p, the second step includes locating a region in the decision space corresponding to a high level of usage of part (p).

Inventors: Dietrich; Brenda Lynn (Yorktown Heights, NY); Lin; Grace Yuh-Jiun (Goldens Bridge, NY); Srinivasan; Ramesh (Yorktown Heights, NY)
Assignee: International Business Machines Corporation (Armonk, NY)
Appl. No.:   871567
Filed: June 4, 1997

Current U.S. Class: 705/7; 700/103; 705/10
Intern'l Class: G06F 017/60
Field of Search: 705/7,8,10,28,29 364/468.05,468.09,468.13,468.16

Sample Claim:

1. A computer implemented process for determining procurement level for parts (P) in a production system having constraints comprising at least one of constrained resources (r) and known maximum demands d.sub.j, comprising:

1) inputting to the computer, data including decision variables, production quantity, demand quantity, bill of material, bill of capacity and usage rate;

2) constructing in computer memory a production planning decision space consisting of independent sets of hyperplanes defined by decision variables q.sub.j,t corresponding to production quantities for products (j), said constructing including the steps of:

a) expressing a potential usage of part (p) as a linear combination of production quantities (q.sub.j);

b) limiting the production quantities (q.sub.j) so that it is less than or equal to the maximum demand quantity (d.sub.j); and

c) limiting the production quantities (q.sub.j) so that the usage of each other resource (r) based on multiple level bill of material and bill of capacity usage rates is less than or equal to the availability of that resource;

3) locating a region in the decision space corresponding to a high level of usage of part (p); and

4) interactively displaying to a user candidate values in said decision space corresponding to the highest level of usage of part (p) so assist the user in deciding quantities of part (p) to procure necessary to enable achievement of the highest level of service, corresponding to the highest level of usage of part (p), while maintaining the lowest possible WIP and inventory on the basis of linear demand, resource, and supply constraining information.

[3] System and Process for Inter-domain Planning Analysis and Optimization Using Model Agents as Partial Replicas of Remote Domains — U.S. Patent No. 5,995,945

Notani, et al.   November 30, 1999

Abstract

A computer implemented system for inter-domain analysis is disclosed. The system includes a domain engine (234), associated with a first supply chain entity, having a local model representing a supply chain activity of the first supply chain entity. The domain engine (234) operable to generate a model agent (240) that represents a partial replica of the local model. The system also includes another domain engine (236), associated with a second supply chain entity, having a local model representing a supply chain activity of the second supply chain entity. The other domain engine is operable to generate a model agent representing a partial replica of the local model. The domain engines (234 and 236) further operate to expand the respective local models using the remote model agents (240) to accomplish local inter-domain analysis. In one embodiment, the model agents (240) can be data model agents, object model agents and behavior model agents.

Inventors: Notani; Ranjit N. (Irving, TX); Mayer; John E. (Dallas, TX)
Assignee: i2 Technologies, Inc. (Irving, TX)
Appl. No.:   917151
Filed: August 25, 1997

Current U.S. Class: 705/28; 705/1; 705/7; 705/8; 705/10
Intern'l Class: G06F 017/60
Field of Search: 705/7,8,28 395/500 364/468,578

Sample Claim:

1. A computer implemented system for inter-domain analysis related to a supply chain activity, comprising:

> a first domain engine associated with a first supply chain entity, the first domain engine having a first domain model representing a supply chain activity of the first supply chain entity; and

> the first domain engine operable to generate a first model agent that represents a partial replica of the first domain model; and

> a second domain engine associated with a second supply chain entity, the second domain engine having a second domain model representing a supply chain activity of the second supply chain entity; and

> the second domain engine operable to generate a second model agent representing a partial replica of the second domain model;

> the first domain engine further operable to expand the first domain model using the second model agent to accomplish inter-domain analysis related to the supply chain activity at the first domain engine; and

> the second domain engine further operable to expand the second domain model using the first model agent to accomplish inter-domain analysis related to the supply chain activity at the second domain engine.

[4] Projected Supply Planning Matching Assets with Demand in Microelectronics Manufacturing — U.S. Patent No. 6,049,742

Milne, et al.   April 11, 2000

Abstract

A computer-implemented decision-support tool serves as a solver to generate a projected supply planning (PSP) or estimated supply planning (ESP) match between existing assets and demands across multiple manufacturing facilities within the boundaries established by the manufacturing specifications and process flows and business policies to determine what supply can be provided over what time-frame by manufacturing and establishes a set of actions or guidelines for manufacturing to incorporate into their manufacturing execution system to ensure that the delivery commitments are met in a timely fashion. The PSP or ESP tool resides within a data provider tool that pulls the required production and distribution information. PSP matching is driven directly by user-supplied guidelines on how to flow or flush assets "forward" to some inventory or holding point. After the supply plan is created, the analyst compares this plan against an expected demand profile.

Inventors: Milne; Robert J. (Jericho, VT); O'Neil; John P. (Essex Junction, VT); Orzell; Robert A. (Essex Junction, VT); Tang; Xueqing (Naperville, IL); Wong; Yuchung (Silver Spring, MD)
Assignee: International Business Machines Corporation (Armonk, NY)
Appl. No.: 938764
Filed: September 26, 1997

Current U.S. Class: 700/99; 700/90; 700/97; 705/7; 705/8; 705/28
Intern'l Class: G06F 019/00
Field of Search: 700/95,97,90,99,100-213 705/7,8-28

Sample Claim:

1. A computer-implemented decision-support method to generate a projected supply planning (PSP) or estimated supply planning (ESP) match between existing assets and demands within boundaries established by manufacturing specifications and process flows and business policies using manufacturing specification and calendar information accessed by material requirements planning (MRP) matching tools to ensure synchronization between these matching tools to determine what supply can be provided over what time-frame, the method comprising the steps of:

selecting user provided part numbers of interest and modifying a user provided production specification information to reflect this selection; specifying substitution rules, wherein the substitution rules allow one part to be substituted for another part;

specifying a start date and shutdown dates;

optionally allowing a user to modify any due dates for projected receipts;

specifying a "from/to" split of allocation fractions to reflect business plans, and to allow the projection of assets across multiple bills of material levels;

selectively inputting new starts for an individual date and dates propagated across a span of dates;

executing a forward-flush solver to generate a projected or estimated supply plan;

aggregating the supply plan and required starts into user-defined time buckets;

creating and analyzing explanation reports of the solution generated by the forward flush solver; and

comparing the supply plan with the required demands to assess a "fitness" of the plan relative to meeting demand.

[5]   System and Method for Allocating Manufactured Products to Sellers — U.S. Patent No. 6,167,380

Kennedy, et al.   December 26, 2000

Abstract

A software system is provided for managing available to promise (ATP) and making promises to fulfill customer requests. The software system includes a plurality of generic product models (600) each representing a generic product. Each generic product model (600) specifies one component of a plurality of possible components. A plurality of specific product models (602) each represent a specific product and each specify all components of the specific product. Each component specified by each specific product model (602) is specified by one of the generic product models (600) such that each specific product model (602) is related to a subset of the generic product models (600). A customer request matching a specific product then can be fulfilled by available-to-promise of the specific product or by available-to-promise of all related generic products. In addition, an organization in a seller hierarchy can retain product and designate first-come-first-served product. The organization also can define an ATP horizon to specify when forecasted product is actually available. Further, the organization can use an automatic allocation policy to allocate to members, and some forecast entries can be designated zero-ATP entries.

Inventors: Kennedy; Brian M. (Coppell, TX); Burchett; Christopher D. (Carrollton, TX)
Assignee: i2 Technologies, Inc. (Irving, TX)
Appl. No.:   802434
Filed: February 18, 1997

Current U.S. Class: 705/10; 705/28
Intern'l Class: G06F 153/00
Field of Search: 705/10,22,28,29 364/468.05,468.13

Sample Claim:

1. A system operating on a computer for allowing an organization to allocate manufactured products among sellers, comprising:

a seller model stored in a memory of the computer and representing sellers that sell at least one product, the seller model forecasting for at least one product and defining commitment levels for creating at least one forecast request value;

at least one supplier model stored in the memory, the supplier model receiving the forecast request value and providing at least one supplier promise value to the seller model, the supplier promise value being based on the forecast request value; and

at least one customer model stored in the memory, the customer model providing a customer order value to the seller model and receiving an actual customer promise value;

a forecast model for the product that is stored in the memory and maintains an allocation value, a forecast request value, and a supplier promise value, the allocation value representing the sum of promises from different suppliers plus any promises to customers;

wherein the system displays allocation data to make product allocation available to the sellers; and

wherein the seller model comprises a hierarchy of sellers, the forecast data representing an aggregate of forecasts from members of the hierarchy, the allocation data being distributed as sub-allocation data to the members.

[N]   Other Business Model Patents

[1]   Methods and Apparatus for Determining or Inferring Influential Rumormongers from Resource Usage Data — U.S. Patent No. 6,151,585

Altschuler, et al.   November 21, 2000

Abstract

Resource usage data is used to infer degrees of influence between users. Once such inferences are made, a directed graph representation of the users and the inferred "influence" between the users can be generated. "Influential rumormongers" can then be determined from the directed graph, for example, by using a greedy graph covering algorithm. In this way, marketing information can be targeted to "influential rumormongers" to optimize its dissemination and impact. If actual (explicit) data regarding the influence between users is known, such data may be used to refine or replace at least some edge values.

Inventors: Altschuler; Steven J. (Redmond, WA); Wu; Lani F. (Redmond, WA); Ingerman; David (New York, NY)
Assignee: Microsoft Corporation (Redmond, WA)
Appl. No.:   065970
Filed: April 24, 1998

Current U.S. Class: 705/10; 705/1; 705/7; 705/8; 705/9; 705/500

Intern'l Class: G06F 017/60
Field of Search: 705/10,1,500,8,9,7

Sample Claim:

1. A computer-implemented method for ascertaining an influential rumormonger from amongst a plurality of users through a directed graph from resource usage log data including records having (i) user information, (ii) resource identification information, and (iii) resource request time information, the method comprising steps of:

a) inferring a corresponding measure of influence between each of a plurality of pairs of said users from the usage log data based on (i) the user information, (ii) the resource identification information, (iii) the resource request time information, and (iv) a memory length parameter so as to define a plurality of influence measures, wherein, for each of said pairs of users, one of the users in said each pair appears to exhibit influence, as reflected in said corresponding influence measure, over action taken by another one of the users in said one pair;

b) generating the directed graph having, for said each one of the pairs of users, first and second vertices corresponding to the one and the other one, respectively, of the users and having an edge connecting the first and second vertices and associated with the corresponding influence measure; and

c) determining the influential rumormonger from amongst the users and the influence measures provided in the directed graph.

[2] Method and Apparatus for Surveying Music Listener Opinion About Songs — U.S Patent No. 5,913,204

Kelly   June 15, 1999

Abstract

A method and apparatus for surveying and reporting listener opinion of a list of songs such as the songs comprising a radio station music library. A group of music listeners is selected from whom individual listener opinions are recorded. A home music preference test kit is fielded to the select listeners for use in each listener's residence. Data from each music preference test kit is then collected, compiled and tabulated to reflect a mean or average listener opinion for each song in the list. The home music preference test kit includes instructions for completing the test and returning the survey form, a music medium having a preselected number of song hooks representing each song in the library of the radio station. The kit also includes a survey form for recording the listener's opinion of each song based on the song hooks and an honorarium for agreeing to properly complete a survey form.

Inventors:	Kelly; Thomas L. (139 E. Hathaway, Havertown, PA 19083-1517)
Appl. No.:	693355

Filed: August 6, 1996

Current U.S. Class: 705/500; 434/363; 455/2.01; 705/10; 705/30
Intern'l Class: G06F 017/60; G06F 017/30; G06F 017/00; G06F 019/00
Field of Search: 705/10,30 348/1,2,3 434/363

Sample Claim:

1. A method of surveying and reporting listener opinion of a list of songs from the music library of a radio station including the steps of:

a) selecting a group of music listeners from which individual listener opinions are to be recorded;

b) fielding from a survey taker a home music preference test kit including a music audio medium containing a number of song hooks to the select listeners at each listener's residence;

each listener listening to a list of music of music song hooks, entering his preferences as to the song hooks on a test sheet without the presence of a survey interviewer;

returning the completed test sheets to the survey taker,

c) collecting and compiling data from each music preference test kit test sheet answered by each listener;

d) tabulating the test sheet data in a computer to reflect listener opinion for each song in the list, reporting the tabulated data to the radio station, and modifying the music programming of the radio station based on the results of the survey.

[3] Process to Convert Cost and Location of a Number of Actual Contingent Events Within a Region into a Three Dimensional Surface over a Map that Provides for Every Location Within the Region Its Own Estimate of Expected Cost for Future Contingent Events — U.S. Patent No. 6,186,793

Brubaker   February 13, 2001

Abstract

A method of establishing the insurance rate at a desired location comprising the steps of selecting a plurality of predetermined points, calculating the insurance rate at each predetermined point, and interpolating among the values of insurance rates at predetermined points adjacent to the desired location to calculate the insurance rate at the desired location.

Inventors: Brubaker; Randall E. (230 Maple Rd., Barrington, IL 60010)
Assignee: Brubaker; Randall E. (Barrington, IL)
Appl. No.:   970964
Filed: November 14, 1997

Current U.S. Class: 434/107; 705/4; 705/500
Intern'l Class: G09B 019/18; G06F 017/60
Field of Search: 434/107 705/500,4 345/440

Sample Claim:

1. A method that utilizes and transforms historic insurance data such as losses, policyholder location, coverage provided, and other factors affecting likelihood of loss into a three dimensional surface over a map that provides a uniquely estimated rate value for every location desired, comprising the steps of:

> selecting a plurality of predetermined points such that every location for which rate values are desired is located at or among the predetermined points;

> determining a rate value for each predetermined point by using data uniquely selected and weighted from other locations;

> interpolating among the rate values at the predetermined points to determine the rate values at desired locations which are located among the predetermined points; and

> specifying the rate values calculated at the predetermined points and, by formula, the rate values at desired locations among the predetermined points to create a three dimensional surface over a map that provides a uniquely estimated rate value for every location desired.

[4]   Service Business Management System — U.S. Patent No. 6,216,108

LeVander   April 10, 2001

Abstract

The invention provides a micro processor, an input device for entering job costing data and job parameter information, an output device such as a printer to generate contract proposals and/or management reports. A display device is preferably also provided to display the minimum labor rate calculated by a program executed on the microprocessor from the job costing data. Memory is preferably also provided to store at least the minimum labor rate for use in generating management reports.

Inventors: LeVander; Mark R. (273 Thunder Lake Rd., Wilton, CT 06897)
Appl. No.:   909341
Filed: August 11, 1997

Current U.S. Class: 705/7; 705/400; 705/500
Intern'l Class: G06F 017/60
Field of Search: 705/400,500,7

Sample Claim:

1. A service business management system comprising:

a microprocessor;

an input device for entering job costing data, job parameter information, and a proposal price to said microprocessor;

a program executing on said microprocessor which calculates a minimum labor rate from the job costing data, and calculates a minimum job price from the minimum labor rate and the job parameter information;

a display device for displaying the minimum job price;

an output device for generating a contract proposal having a proposal price entered after review of the displayed minimum job price; and

wherein the minimum labor rate calculation includes as factors, the equity investment in the business (EI) and a desired rate of return on the equity investment (ROE).

[5] Method and System for Accommodating Electronic Commerce in the Semiconductor Manufacturing Industry — U.S. Patent No. 6,240,400

Chou, et al.   May 29, 2001

Abstract

A method for accommodating electronic commerce in a semiconductor manufacturing capacity market. The method comprises the steps of identifying a plurality of players in the semiconductor manufacturing capacity market, each of which players can solicit capacity in semiconductor manufacturing capacity market; providing a neutral third-party, the neutral third party and the plurality of players configured in a hub arrangement for communicating with each of the plurality of players in semiconductor manufacturing capacity trades; and realizing an open market conditionality between each of the plurality of players and the neutral third party so that the semiconductor manufacturing capacity supplied by one or more of the players can be bought and sold among the players; and, the neutral third party can preserve anonymity of each of the plurality of players soliciting semiconductor manufacturing capacity.

Inventors: Chou; Yu-Li (White Plains, NY); Garg; Amit (White Plains, NY)
Assignee: International Business Machines Corporation (Armonk, NY)
Appl. No.:   024526
Filed: February 17, 1998

Current U.S. Class: 705/37; 705/1; 705/500
Intern'l Class: G06F 017/60
Field of Search: 705/37,1,500

Sample Claim:

1. A computer implemented method for accommodating electronic commerce in a semiconductor manufacturing capacity market, the method comprising:

1) identifying a plurality of players in the semiconductor manufacturing capacity market, each of which players can solicit capacity in semiconductor manufacturing capacity market,

2) providing a neutral third-party, the neutral third party and the plurality of players configured in a hub arrangement for communicating with each of the plurality of players in semiconductor manufacturing capacity trades; and

3) realizing an open market conditionality between each of the plurality of players and the neutral third party so that:

i) the semiconductor manufacturing capacity supplied by one or more of the players can be bought and sold among the players; and

ii) each of the players may individually select at any time during said market conditionality, one of the following regardless of selection of the other players: identifying themselves to the other players and preserving their anonymity among each of the plurality of players soliciting semiconductor manufacturing capacity.

CHAPTER 6

JUDICIAL DECISIONS—BEFORE
STATE STREET BANK

§ 6.01   Why Study the Early Business Method Cases

The Federal Circuit Court of Appeals ruled in *State Street Bank v. Signature Financial Group*[1] that business methods can constitute statutory subject matter. That being said, why bother reading the old cases in which business method patents were rejected? There are at least three reasons. First, there is indeed something curiously unique about business methods. Business methods can fall well outside the technological arts, and often do. Was our patent law designed to reach beyond the technological arts? Reading the old cases sheds light on this nagging question. Although *State Street Bank* has opened the patentability door for most business methods, did *State Street Bank* remove that door from its technological hinges? We may find in the old cases that history can answer that question.

Second, Judge Rich, author of the *State Street Bank* decision, was right. The early cases really do not support the proposition that everyone had been attributing to them. Quite simply, the pronouncements against business method patents in the early cases was *dicta*. In virtually every early case where patentability of a business method was involved, the issue was decided on other grounds — most notably on novelty or obviousness grounds. One can simply take Judge Rich's word for it; or one can explore the issue for herself or himself. The old cases make fascinating reading; and they give considerable force to Judge Rich's decision.

Third, the early cases help us understand what influenced Judge Rich in writing the *State Street Bank* decision. Judge Rich sat on judicial panels that decided some of the seminal software patent decisions, as well as a number of business method patent decisions. Reading these cases, and those that preceded them, helps put Judge Rich's views on business method patents into proper context. For instance, Judge Rich wrote an opinion as a member of the Court of Customs and Patent Appeals, striking down a business method patent application in the 1960s. The application involved a method of converting bank checks into promissory notes with installment payment features. Did Judge Rich hold anti-business method patent views in the 1960s? Did something subsequently change his mind? Judge Rich also wrote a seminal pro-software patent appellate court decision that was reversed by the Supreme Court in *Gottschalk v. Benson*. The invention involved a mathematical algorithm for converting numbers from one form to another (binary to binary coded decimal). What impact did that reversal have on Judge Rich's thinking regarding patentable subject matter?

The following sections briefly visit several dozen business method patent cases, spanning roughly the last 150 years. When perusing these cases, take note that the subject matter of many could arise today in the context of e-commerce. If history repeats itself, expect to read *Munson.com v. Mayor of New York* in next week's advance sheets.

---

[1] 149 F.3d 1368 (Fed. Cir. 1998).

§ 6.02   Coupon Bond Tracking System — *Munson v. The Mayor, Etc., of New York*

The year was 1867. Munson was awarded U.S. Patent No. 63,419 for a bond and coupon register. In Munson's day, the coupon bond was a popular investment instrument that entitled the bearer to periodic interest payments. To receive the interest payment when it fell due, the bearer would clip the corresponding coupon from the bond and present it to the bond issuer for payment. Bonds were issued in great numbers. Thus, the bond issuer had the practical problem of keeping track of the bonds and their coupons when presented.

Munson's invention was a book, not unlike a common scrapbook, with pages spaced to receive each bond and its coupons. The spaces were numbered to individually identify each coupon of a given bond. As coupons were presented, they were placed in the appropriate numbered space. A space was also provided on the same page in which to place the bond itself after it had matured and was presented for payment. Using Munson's register system, the bond issuer could tell at a glance which coupons were paid and which were outstanding. The system conveniently correlated each bond with its coupons on the same page.

The City of New York adopted Munson's register system without compensating Munson for its use. Munson sued for patent infringement. The City of New York defended, arguing that Munson's register was not proper subject matter for patent. The City argued that Munson's book was properly the subject of copyright, not patent. The City also argued that Munson's invention lacked novelty and patentability in view of a prior system taught by Warren and O'Brien.

Judge Wheeler, District Judge for the Circuit Court of the Southern District of New York, rejected the City of New York's arguments and ruled that the City was liable in its corporate capacity for infringement of Munson's patent.

Judge Wheeler expended surprisingly little effort in rejecting the City's statutory subject matter argument. Although books were commonly the subject of copyright, Judge Wheeler saw this invention as an apparatus:

> Here the principle of the invention is embodied in each register, as the principle of a machine is embodied in each machine. There is no difference because the contrivance is in the form of a book, although books are commonly copyrighted.[2]

In reaching his decision, Judge Wheeler had the benefit of the Supreme Court's analysis in *Baker v. Selden*,[3] in which that court had explained that copyright law and patent law were intended to cover different things. *Baker v. Selden* involved a claim of copyright in a book with a brief introductory chapter

---

[2] Munson v. The Mayor, Etc., of New York, 3 F. 338, 339 (S.D.N.Y. 1880).
[3] 101 U.S. 99 (1879).

describing a bookkeeping method with the remainder of the book being ruled lines designed for practicing the method. In holding that the copyright did not give its owner the exclusive right to the bookkeeping method, the Court suggested that patent law was the appropriate protection for the underlying method.

> To give the author of the book an exclusive property in the art described therein, when no examination of its novelty has ever been officially made, would be a surprise and a fraud upon the public. That is the province of letters patent, not of copyright. The claim to an invention or discovery of an art or manufacture must be subjected to the examination of the Patent Office before an exclusive right therein can be obtained; and it can only be secured by a patent from the government.[4]

Judge Wheeler was ahead of his time in seeing the Munson invention in terms of its practical utility. In a single sentence, this prescient Judge placed this invention in the domain of the patent law: "The proof is abundant that the invention is useful, and that the defendant's officers make use of it for the bonds of the city."[5] Years later, after much legal haggling, the courts would ultimately determine that a useful application of an abstract idea is proper subject matter for a patent, even if the abstract idea involves the art of business.

However, Judge Wheeler's views would ultimately not dictate the outcome of this case. The City of New York appealed to the Second Circuit and then ultimately to the U.S. Supreme Court. The Supreme Court ruled that Munson's invention was obvious in view of the Warren prior art. The Court did not elect to comment on Judge Wheeler's view that Munson's book was statutory subject matter, although it could have. On this point, Justice Gray wrote:

> If upon the face of the specification this could be considered as an "art, machine, manufacture, or composition of matter," within the meaning of the patent laws (upon which we express no opinion), it is quite clear that, in the state of previous knowledge upon the subject, there was no patentable novelty in the plaintiff's scheme; inasmuch as the only difference between it and the earlier schemes of Warren was that in Warren's books there was no place for the bonds, and the coupons were grouped according to their dates of payment, instead of being grouped together with the bonds to which they respectively belonged. The providing of spaces for the bonds, and the change in the order of arrangement of the coupons, cannot, upon the most liberal construction of the patent laws, be held to involve any invention.[6]

---

[4] 101 U.S. 99 (1879).

[5] 3 F. at 339.

[6] Mayor, Alderman and Commonalty of the City of New York and the Comptroller Thereof v. Munson, 124 U.S. 586, 588 (1888).

§ 6.03   Business Forms — *Waring v. Johnson*

One year after the decision in *Munson v. the Mayor*, discussed above, the Circuit Court for the Southern District of New York was asked to decide another patent case involving business forms. That case involved a reissue of U.S. Patent No. 8,199 for an "improvement in pocket check-books." The improvement was the brainchild of James R. Osgood. Osgood was awarded his original patent on the idea on October 17, 1876. He then assigned his patent to Waring, a reissue patent was sought, and a suit for patent infringement was brought against Johnson.

The problem with checkbooks, which Osgood's improvement sought to correct, was they did not fit easily in one's pocket. One conventional design placed the check stub at the side edge of each check, connected by perforations; when the check was torn out, the stub remained bound in the checkbook. This design was too long to fit within the average pocket. Another conventional design placed the check stub at the top edge of each check, connected by perforations; again, when the check was torn out, the stub remained bound in the checkbook. This design had a similar problem. It was too wide to fit in the average pocket.

Osgood's improvement employed one check-sized stub for every two checks, which he interleaved between every two checks. Each stub was divided into two regions to provide space to record the particulars of the two checks immediately in front of and behind that stub. Thus Osgood's checkbook had substantially the same dimensions as the check and would easily fit into the average pocket.

Osgood's claim read quite simply:

> The combination in a check-book of checks and stub pieces of substantially the same size, so united that two checks lie between every two stub pieces, substantially as specified and set forth.[7]

The infringement lawsuit centered around whether the patent claimed a novel invention and whether Johnson's checkbook infringed. Johnson also raised an argument that the reissue application improperly enlarged the scope of the patent. The Court found the patent valid and infringed, and rejected Johnson's argument that the reissue was improper.

Unlike the defendants in *Munson*, Johnson raised no argument that the checkbook was not statutory subject matter. *Waring v. Johnson* was brought in the same district court (Southern District of New York) as *Munson*. Perhaps Johnson was aware that the *Munson* court had ruled that a coupon book was proper subject matter for patent, and thus elected not to raise that issue again. In any event, the decision in *Waring v. Johnson* demonstrates at least one court's acceptance that business forms constitute statutory subject matter. However, as several later cases discussed in this chapter demonstrate, the uncertainty

---

[7] 6 F. 500, 502 (S.D.N.Y. 1881).

surrounding business forms was not laid to rest with *Munson*. Numerous later litigants raised the issue that patents on business forms, in effect, preempted the underlying business methods. That issue, they argued, placed the inventions outside the proper domain of statutory subject matter.

§ 6.04   Book Publishing Format — *Dugan v. Gregg*

Nine months after its decision in *Waring v. Johnson*, discussed in the preceding section, the District Court for the Southern District of New York upheld the validity of another improvement relating to books. The invention was an improved book and index combination that allowed the reader to consult the index without losing the place in the book which the reader was consulting. The combination was conceived by Robert M. Rigby, who obtained U.S. Patent No. 383,543 for it on May 29, 1888, and assigned the patent to George Dugan, who then brought suit against Thornton F. Gregg for infringement.

Unlike a conventional book in which the index is bound in the back, in the same manner as the other pages of text, Rigby's idea was to bind the index as a separate collection of pages, placed in the back of the book, but attached to the exposed edge of the end leaf of the main text instead of to the spine. Being attached in this fashion allowed the reader to pull the index out to the side of the main text and flip through its pages, while keeping the main text open to the page the reader had been consulting. When finished, the reader simply slipped the index portion back in place adjacent to the rear cover of the book. Thus Rigby's combination allowed the index to be confined between the covers of the book, or withdrawn from them to be consulted, with a simple movement of the hand and without turning any of the pages of the book.

Rigby's claim read as follows:

> In a combined book and index, the combination, with a book provided with a leaf, C, free of the book-cover, to its rear edge, of an index provided with a leaf or cover, F, the free edge of the latter being flexibly united to the free edge of the leaf, C, whereby the index is independent of the book-cover and may be inserted and confined between the book-cover and the leaf, C, with the front edges of its leaves outermost, substantially as described.[8]

In the patent infringement suit, Gregg denied infringement and argued that the patent lacked novelty and invention. The district court disagreed and found the patent valid and infringed. As in *Waring v. Johnson*, discussed in the preceding section, no question was raised whether this invention was statutory subject matter. Indeed, the patent claims were so clearly directed to the construction of a book, that it would have been quite a stretch to argue that this invention raised questions about the patentability of business methods facilitated by use of the book.

---

[8] Dugan v. Gregg, 48 F. 227, 227 (S.D.N.Y. 1881).

§ 6.05   Bank Account Ledgers — *Thomson v. Citizens' Nat. Bank of Fargo*

Proof that the patentability of business forms was far from having wide acceptance came about one year after the Southern District Court decisions in *Dugan* and *Waring*. The patent involved an improvement in the manufacture of bank account books.

In the late 1800s, banks kept ledgers of accounts in horizontally ruled books having both long leaves and short leaves. On the left-hand page of the open book each long leaf contained a place in the left-hand margin to write the name of each account holder. The short leaves, ruled with columns to record account balances, could be successively turned without covering up the names of the account holders written on the long leaves. The long and short leaves had horizontal rules that lined up, so that each horizontal line would define a row containing the account data for one account holder. Account balances were filled into these rows, proceeding from left to right.

The problem was what to do when the right-most column of a row was reached. Conventionally, bank clerks had to copy the account balance from the right-most column on one page to the left-most column on the succeeding page. This took time and invited error. Walter Thomson devised a solution. He creased the short leaves along the right-most column and pre-folded the creased portions towards the reader, so that the clerk writing in the right-most column was actually writing on the reverse side of the leaf. Then when the short leaf was turned, it could be unfolded to reveal the current account balance, already written from the page before. Thomson's invention eliminated the need to recopy account balances from one page to another, saving time and reducing clerical error.

Thomson applied for and was awarded U.S. Patent No. 385,648 on July 3, 1888. He reissued the patent as no. 10,977 on December 25, 1888.

The Citizens National Bank of Fargo used Thomson's invention for several years, obtaining its books from a licensed source. Then, it elected to purchase books from an unlicensed source and Thomson sued for patent infringement. The Bank defended by asserting obviousness, that it did not constitute patentable subject matter, and that the Bank had an implied license to use the subject matter. Its implied license argument arose because one bookkeeper of the Banks' predecessor had learned the idea from Thomson and had folded an existing ledger book to exploit Thomson's idea prior to Thomson's filing for his patent.

The district court ruled the patent valid, but agreed that the Bank had an implied license. Both parties appealed.

[A]   The Claims at Issue in *Thomson v. Citizens' Nat. Bank of Fargo*

Thomson's reissue patent had two claims:

> 1. The bank account book, A, having a suitable number of full leaves, C, and alternate series of short leaves, B, each of said short leaves having margin, b, creased or perforated at b' substantially as and for the purposes set forth.

2. The bank account book herein described, composed of alternate long and series of short leaves, the several long leaves prepared to receive the depositors' names on the left-hand side, and both also prepared to receive the accounts for several successive days, the right-hand end of each of said short leaves having a margin adapted by folding to receive the last day's balance, so that when the leaf is turned, and said margin straightened out, said balance forms the beginning of the next day's account on the next page.[9]

[B]   The Bank Account Ledger Patent Upheld

The parties took their appeals to the Eighth Circuit Court of Appeals. Judge Sanborn wrote the opinion for the court, which reversed the district court on its implied license ruling and found the patent was valid and infringed.

Judge Sanborn began his opinion with an insightful statement about patent law that bears repeating:

Letters patent grant a public franchise, giving to the inventor some compensation for the exercise of his inventive genius in the discovery of, and his labor and ingenuity in reducing to practice and describing, novel and useful inventions, by which the public may attain beneficial results with less expenditure of time and labor. From every patented invention of value the compensation derived by the inventor is small in proportion to the benefit conferred upon the public. The inventor's reward is limited to a few years, at most, while the benefit to the public continues forever. No patented invention can, in the nature of things, be valuable to its owner unless it is of greater value to the public even during the term of his franchise, since the latter will not purchase the right to vend or use it unless it is more profitable to do so than to do without it. Letters patent issued under our constitution and laws thus offer the necessary pecuniary inducement to those gifted with inventive genius, without which they would not be able to bestow the thought, time, and toil required to find, perfect by experiment, reduce to practice, and give to the public many of those useful inventions which have enabled us to excel in the manufacture and use of machinery, and in progress in all the useful arts, that promote the efficiency and comfort of our citizens. The time, thought, labor, and expense that produce a valuable invention are the inventor's, the completed invention which is their product is his, and when, in consideration of his describing it, and making it useful to the public forever, he is granted the exclusive right to use and vend it for a limited time, this franchise should not be regarded as a monopoly conferred on him at the expense and to the prejudice of the public, but as a just and fair consideration, granted for valuable services rendered, which both equitably and legally entitle him to the same protection for this property that the owner of any other species of property enjoys.[10]

---

[9] Thomson v. Citizens' Nat. Bank of Fargo, 53 F. 250, 251 (8th Cir. 1892).
[10] *Id.* at 252.

Judge Sanborn then rejected the Bank's obviousness argument. He considered it significant that the bookkeepers had

> [G]one on, from generation to generation, copying off this last column, turning the leaf, and then recopying it on the succeeding page, and no one of them, during all the time since such account books have been in use, by any display of the expected skill of his calling, or by the exercise of his faculties of reasoning, had ever arrived at the conclusion that [Thomson's] improvement would accomplish.[11]

On the Bank's statutory subject matter argument, the Judge rejected it as a technical argument with these words:

> The technical claim of the defendant that this invention is not patentable because it is not embraced within the terms of the patent laws is without merit. The statute provides that "any person who has invented and discovered any new and useful art, machine, manufacture, or composition of matter, or any new and useful improvement thereof, may obtain a patent therefor," when his invention or discovery is made under the circumstances proved in this case. Section 4886, Rev. St. It is not perceived why one who invents a new and useful improvement in the manufacture of bank account books is not equally entitled to a patent with one who invents an improvement in the manufacture of barbed wire for fencing, of packing for the stuffing boxes of pistons, of safety valves for engines, of looms for weaving cloth, or of any other article upon improvements in the manufacture of which the Supreme Court has sustained patents.[12]

The implied license argument is probably the most interesting aspect about this case, although it has little relevance to the business method topic of this book. Section 4899 of the Revised Statutes that were in force at the time stated that

> [E]very person who purchases of the inventor or discoverer, or with his knowledge and consent constructs, any newly invented or discovered machine, or other patentable articles, prior to the application by the inventor or discoverer for a patent, or who sells or uses one so constructed, shall have the right to use, and vend to others to be used, the specific thing so made or purchased without liability therefor.

The court ruled that the only right the Banks' predecessor had, and hence the only right that predecessor could have conveyed to the Bank, was the right to use the specific book that the predecessor's bookkeeper had made with Thomson's knowledge and consent. This right did not convey any rights under the patent.

---

[11] *Id.* at 254.
[12] *Id.* at 255.

[C] Understanding *Thomson v. Citizens' Nat. Bank of Fargo*

Thomson v. Citizens' Nat. Bank is not so much about business method patents as it is about business forms. Unlike some of the cases from the Southern District Court of New York which preceded it, this case addressed whether bank account books were patentable subject matter, simply because the defendant raised that issue. In *Waring v. Johnson*,[13] involving checkbooks, and *Dugan v. Gregg*,[14] involving books and their indexes, the statutory subject matter issue had not even been raised.

The principal expressed in this case extends naturally to modern day computer systems for maintaining banking records. Suitably programmed, a computer can comprise a new and useful improvement in bank account recordkeeping, entitling its inventor to a patent, just as one who invents an improvement in the manufacture of barbed wire for fencing, of packing for the stuffing boxes of pistons, of safety valves for engines, of looms for weaving cloth, or of any other article upon improvements in the manufacture of which the Supreme Court has sustained patents.

§ 6.06 Insurance — *United States Credit System Co. v. American Credit Indemnity Co.*

Levy Maybaum applied for a patent on a "means for securing against excessive losses for bad debts." The Patent Office granted him U.S. Patent No. 465,485 on December 22, 1891. Maybaum assigned the patent to United States Credit System.

Maybaum's patent described his invention as a plan of insurance against excessive losses for bad debts in excess of the usual percentage of such losses in a given line of business. The patent provided forms for ruling paper with spaces for entering various details of the insurance transactions.

Maybaum's assignee, United States Credit System, brought a patent infringement suit against American Credit Indemnity Co. in the northern district of Illinois. The Illinois District Court ruled the patent invalid. In an oral opinion, (transcribed), the District Judge pronounced:

> I cannot see any ground on which to sustain this patent. It is void for want of invention. The guarantying [sic] of men's financial ability to pay is not an invention of the complainant. Nearly all forms of guarantying [sic] or insuring have been in existence for many years, notably fidelity, casualty, fire, lightning, and other forms of insurance, all of which are based upon averages obtained from practical experience. It required no inventive genius to form and plan the insurance on this basis. One is not entitled to a patent for

---

[13] 6 F. 500 (S.D.N.Y. 1881).
[14] 48 F. 227 (S.D.N.Y. 1881).

a plan or method of business which only requires good judgment and foresight.[15]

The Illinois District Judge saw both business methods and abstract ideas lurking behind the curtain of Maybaum's claims; but he saw it unnecessary to pull the curtain back and directly confront these issues:

> I do not intend to decide that a man may not have a patent for a mode of keeping accounts, or for a form of tabulating accounts or statistics; but am clearly of opinion that this patent cannot be construed to cover a business principle such as a law of averages, which seems to have been the purpose of the specification in this patent.

Not to be easily dissuaded, United States Credit System filed another action against American Credit in the Southern District of New York. American Credit again defended on the grounds that the patent was invalid.

How United States Credit System was able to have a second district court consider the validity of its patent raises historical procedural questions that the reported decisions do not answer. Being afforded this second bite at the apple seems dubious by today's procedural standards. *Why* United States Credit System sought out the Southern District of New York is clear. Judge Wheeler, District Judge of that court, had ruled valid the *Munson* patent on a bookkeeping system for coupon bonds. From the plaintiff's viewpoint, the *Maybaum* patent was likewise a bookkeeping system.

Judge Wheeler refused to go along with United States Credit System. He distinguished this case from *Munson* and ruled the patent was invalid for lack of novelty. Looking first at the business forms described in the patent, Judge Wheeler found these to lack novelty.

> The arrangement of such spaces, headings, and margins to show the parties to, and terms and details of, any particular contract, would not be new to persons skilled in that art, and could not amount to any patentable invention or discovery.[16]

The Judge cited *Baker v. Selden*[17] for that proposition. In that case, the Supreme Court addressed whether copyright protection for a book describing a system of bookkeeping would extend protection to the bookkeeping system itself. The Court said no. The specific subject of that suit involved a book containing a brief introductory chapter describing the Selden bookkeeping system. The remainder of the book was a blank ledger with ruled lines and headings into which bookkeeping entries could be recorded.

---

[15] United States Credit System Co. v. American Indemnity Co., 51 F. 751, 754 (N.D. Ill. 1892).
[16] United States Credit System Co. v. American Credit Indemnity Co., 53 F. 818, 819 (1893).
[17] 101 U.S. 99 (1879).

Next District Judge Wheeler turned from the pure physical manifestation of the ruled book to the underlying business concepts. He addressed Maybaum's plan of indemnity first:

> The novelty, if any, must consist in the terms of the contract of indemnity arising out of the plan of insuring only against excess above the average of losses from bad debts in similar lines of business, and in the embodiment of these terms upon the sheets or pages.[18]

He then rejected this subject matter as lacking novelty. Thus he wrote:

> But plans of indemnity against losses or parts of losses from casualty or misfortune by contracts of insurance or indemnity in various forms were in common use before, and not, in any sense, novel.[19]

Judge Wheeler did not stop there. He further considered whether Maybaum's patent might be considered a novel method of making *proposals* for contracts of indemnity — the asserted utility being in making proposals of such terms as would be desirable to those wanting indemnity. Judge Wheeler concluded that the art of making proposals for contracts was no more patentable than the making of contracts themselves.

Finally, Judge Wheeler turned to the Munson patent, which he had found to be patentable subject matter 13 years before. He saw a difference between the subject matter of the Munson patent and the subject matter of the Maybaum patent-in-suit:

> This patent is different in this respect from that in *Munson v. Mayor*. That was for a contrivance to preserve paid coupons and bonds, and might be patentable as a machine or manufacture; this is for a method of transacting common business, which does not seem to be patentable as an art.[20]

The preceding quote can be read that methods of transacting common business are not patentable subject matter. However, the statement is also *dicta*, inasmuch as Judge Wheeler found that the Maybaum patent lacked novelty and did not reach the patentable subject matter question. Nevertheless, there is a grain of truth in what Judge Wheeler is saying. Depending on how the invention is claimed, some methods of business are patentable subject matter and some are not. A method of business claimed so broadly as to constitute an abstract idea — without substantive ties to the physical world — would fall outside the domain of the patent law; whereas a method of business that produces a useful, concrete, and tangible result would fall inside the domain of the patent law.

---

[18] 53 F. at 819.
[19] *Id.* at 819.
[20] *Id.* at 819.

[A] The Claims at Issue in *United States Credit System Co. v. American Credit Indemnity*

The Maybaum patent had three claims. Each claim represented a different variation on Maybaum's insurance ledger sheet, used in practicing the insurance method he described in his patent specification. Claim 1, reproduced below, is exemplary.

> The means for securing merchants and others from excessive losses by bad debts, which consist of a sheet provided with separate spaces and suitable headings, substantially as described, for the name of the assurer, the name of the assured, the percentage or amount beyond which assurance is given, the class or classes of persons, as to rating, capital, or otherwise, in respect to whom said losses are guaranteed against, and the percentage of said capital or the amount which said losses must not exceed.[21]

[B] The Insurance Method Was Not New — Patent Invalid

Sometimes a case must be disposed of three times before it is finally put to rest. Such was the case here. After losing its patent battle in two separate federal district courts, United States Credit System Co. took an appeal to the Second Circuit Court of Appeals.

Before the court of appeals, both sides argued at length whether this was or was not a patentable method of doing business. The court of appeals sidestepped this issue altogether. It did so by focusing, not on Maybaum's patent specification, but on his claims. The Maybaum patent specification began with a statement that he had invented a "new and useful improvement in means for securing merchants and others from excessive losses by bad debts." He then went on to describe an exact description of the method by which he proposed to insure against only those risks that were above the statistical or actuarial average risk.

However, as can be seen from Maybaum's claim reproduced in the preceding section, his patent claim was directed, not to the business method, but to the sheet of paper designed to record the information used in practicing the method. The court of appeals seized upon this to reject the Maybaum patent as lacking novelty.

> [W]hat is claimed is not what is stated in the title and declaration of invention, viz. "Means for securing against excessive losses by bad debts." The sheets described in the claims may be printed by the ream, and may even be filled in interminably with details appropriate to each heading, "the several details of the transaction hereinbefore described," as the patentee expresses it, and yet not a single dollar of loss by bad debt will be secured

---

[21] United States Credit System Co. v. American Credit Indemnity Co., 59 F. 139, 142 (2d Cir. 1893).

against. Nor are the "sheets," the "forms of contract," or "guaranty" referred to in the specifications. The three claims of the patent are concerned solely with the providing of sheets with appropriate headings, adapted to be used in preparing historical records of certain business transactions. There is nothing peculiar or novel in preparing a sheet of paper with headings generally appropriate to classes of facts to be recorded . . . Given a series of transactions, there is no patentable novelty in recording them, where, as in this case, such record consists simply in setting down some of their details in an order or sequence common to each record.[22]

[C] Understanding *United States Credit System Co. v. American Credit Indemnity*

United States Credit System ultimately lost its case against American Credit Indemnity because its claim was to the paper form designed to capture the key business information needed to practice the disclosed business method. Whether the patent would have fared better had the claim been tied more tightly to the business method, no one will ever know.

The case does contain a lesson for practitioners today. The paper forms of Maybaum's day would today be embodied as data structures within a computer memory. Although there was a time when computers were viewed by the judiciary as technically complex data processing instruments — and their data structures of inherent patentably novel — that time has come and gone. Computers have become so commonplace as to now function as the familiar paper sheets of Maybaum's day. Whereas courts initially upheld the validity of computer-implemented data structures, that trend may be near the end. Certainly it is easy to envision a modern court using these words — adapted from the *United States Credit System* case — in declaring a patent invalid.

There is nothing peculiar or novel in preparing a computer data structure with headings generally appropriate to classes of facts to be recorded . . . Given a series of transactions, there is no patentable novelty in recording them in a computer, where, as in this case, such record consists simply in setting down some of their details in an order or sequence common to each record.

What can be done to avoid losing a patent on the above grounds? While there are never any guarantees, one course of action is to include claims that focus more directly on the inventive business method or business model. While some aspects of the method or model may involve computer data structures, consider ways of strengthening the ties between the business method or model and the data structures. The goal is to avoid sole reliance on a claim that is nothing more than the mere recording of information.

---

[22] *Id.* at 143.

§ 6.07   Recording Deeds — *Johnson et al. v. Johnston*

Anyone who has visited a county recorders office, particularly in a small, rural county, has probably seen rows and rows of record books in which deeds are recorded and records of births, deaths, and marriages are kept. Although the computer has, today, supplanted most of these paper tomes, many of the records kept in the old record books will never be transferred to computer.

In Montgomery H. Watson's day, that is, in the late 1800s the record books used in recorders offices had an index table in the front or back of the book to allow the user to find the exact page where records for a particular person could be found. Typically, the recorders office would use a separate bound volume for each letter of the alphabet, corresponding to the person's last name. The index table of that volume then designated what page to consult, based on the person's first name.

The conventional book system, known as the Campbell system, required much time flipping back to the index, with great wear and tear on the book. Watson conceived a better way. He proposed printing the index table on every page of the book. Because the pages for particular letters of the alphabet were preassigned, each index table was the same. Watson introduced his better indexing scheme and it quickly replaced the Campbell system.

Watson was awarded U.S. Patent No. 461,878 on October 20, 1891. He assigned the patent to his company Johnson & Watson. When William G. Johnston, clerk for the county of Allegheny, Pennsylvania, began using Watson's indexing system without permission, Johnson & Watson sued.

[A]   Claims at Issue in *Johnson et al. v. Johnston*

Watson's patent contained the following two claims:

> 1. As a new article of manufacture, an index book or volume consisting of numbered pages suitably ruled, headed, and numbered, and of a table composed of the letters of the alphabet appearing on said pages, such letters representing the initials of Christian names, and a figure or figures associated with each of said initial letters, and corresponding with a page or pages in said book, the book being designated by a letter of the alphabet.
>
> 2. As a new article of manufacture, the herein described index book or volume, the same being designated by a letter of the alphabet, and consisting of a suitable number of pages consecutively numbered, one or more pages of such book being devoted to Christian names commencing with a certain letter of the alphabet, and each page being suitably headed and ruled, and a table on each page, consisting of the letters of the alphabet progressively arranged, and a figure or figures corresponding with a page or pages of said book.[23]

---

[23] Johnson et al. v. Johnston, 60 F. 618, 619 (W.D. Penn. 1894).

[B] The Deed Recording Index Book Declared a Patentable Article of Manufacture

Judge Acheson, District Judge for the Western District of Pennsylvania, concluded that infringement was clearly shown. The only issues before him were the defenses that the Watson patent was invalid for failing to recite patentable subject matter as being obvious in view of the prior art.

Judge Acheson looked at some of the items that had been awarded patents and upheld in the past and had little difficulty concluding that Watson's index book was patentable subject matter. On this point, Judge Acheson wrote:

> The term "manufacture," as used in the patent law, has a very comprehensive sense, embracing whatever is made by the art or industry of man, not being a machine, a composition of matter, or a design.[24]

Judge Acheson then went on to list some of the prior cases that he found particularly persuasive on the patentable subject matter issue:

> In *Waring v. Johnson, 6 Fed. 500*, letters patent for an improvement in pocket check books were sustained by Judge Blatchford; and in *Norrington v. Bank, 25 Fed. 199*, Judge Colt sustained a patent whose subject-matter was of a like nature. In *Dugan v. Gregg, 48 Fed. 227*, a combined book and index, so connected as to facilitate the more ready and convenient handling thereof, was held to be a patentable improvement by Judge Coxe, who, also, in *Carter & Co. v. Wollschlaeger, 53 Fed. 573*, upheld a patent for an improvement in duplicate memorandum sales slips, following a decision of Judge Colt in *Carter & Co. v. Houghton, 53 Fed 573*, sustaining the same patent. In *Thomson v. Bank, 3 C.C.A. 518, 53 Fed. 250*, the United States circuit court of appeals for the eighth circuit sustained a patent for a bank account book, the improvement consisting in a suitable number of full leaves and alternate series of short leaves, each of the latter being creased or perforated for folding in such a manner as to transfer the column of balances on the right-hand page to the succeeding left-hand page. I have no difficulty in holding that the subject-matter of the patent in suit is patentable.[25]

On the question of obviousness, Judge Acheson found particularly persuasive the fact that the county of Allegheny had paid $4,000 for the right to use the prior art Campbell index. Yet, once the county saw the Watson index, it switched to Watson. That, in the judge's view, greatly strengthened the validity of the patent.

---

[24] *Id.* at 620, citing Curt. Pat. § 27; 1 Rob. Pat. § 183.

[25] *Supra*, note 23 at 620.

[C] What *Johnson et al. v. Johnston* Teaches Us Today

This case, like others before it, which Judge Acheson cited in his opinion, demonstrates that courts were willing to uphold patents on the manufactured articles used in commerce, bookkeeping, banking, and business. It also demonstrates that patent attorneys of the late 1800s had determined that, for their day, the best way to protect business-related inventions was to patent the devices used in carrying out that business.

§ 6.08 Anti-fraud Restaurant Menus — *Benjamin Menu Card Co. v. Rand, McNally & Co.*

Frank C. Gellenbeck worked as a dining car waiter for the railroad in the late 1800s. When he took a customer's order he filled out a numbered check drawn from a book issued by his employer. Each check was perforated into three parts, designated "cook's check," "guest's check" and "waiter's check." Customers would place their order after consulting a separate bill of fare and then the waiter would record their order on the check. The respective parts of the check were then delivered to the cook, as a voucher for the meal, to the guest who would present it when paying the bill, and to the conductor or steward. At the end of the journey all three portions of the check were collected and returned to the railway company as a record of the meals served. The intent of this system was to prevent fraud.

In his position as waiter, Gellenbeck learned the weakness of this system. In collusion, the waiter, cook, and conductor could easily cheat the railway company. Although the checks were individually numbered, they could simply re-use the checks for several customers, taking the revenue from second and subsequent sales for themselves.

Gellenbeck's idea was simple, yet ingenious. He reasoned that while there was little to prevent collusion among the railroad employee staff, it was far less likely that railroad customers, at large, would actively partcipate in a fraudulent scheme. Gellenbeck proposed dispensing with the separate bill of fare. Instead, he proposed that the bill of fare or menu card be printed on the back of each numbered check so that it would be automatically destroyed when the perforations in the check were split. Once split up, no one part of the check contained the entire bill of fare. This made it more difficult to reuse one check with multiple customers.

Gellenbeck applied for and was awarded U.S. Patent No. 482,899 on September 20, 1892, for his menu card-meal check combination. He assigned his patent to Benjamin Menu Card Company, which marketed the printed menu card-meal check to railway companies. The Gellenbeck system was an immediate success and rapidly spread throughout the industry. Rand, McNally & Co. then began printing and selling menu cards of identical design. Benjamin Menu Card sued for patent infringement.

[A] The Claims at Issue in *Benjamin Menu Card Co. v. Rand, McNally & Co.*

The claims of Gellenbeck's patent were as follows:

1. The combination, with a menu card, of two or more checks detachably secured thereto, two of said checks being designated respectively as ''guest's check'' and ''cook's check.'' So as to make the remainder incomplete as a bill of fare, and hence useless for another guest.
   2. The combination of a menu card having a bill of fare on its face, with a check or checks upon the reverse side thereof, and placed so that when removed the bill of fare will be mutilated, so as to make the remainder incomplete as a bill of fare, and hence useless for another guest.

[B] The Menu Card Is Proper Subject Matter for Patent

Rand McNally admitted infringement. It defended solely on grounds that the patent was invalid. It argued that the menu card was not proper subject matter for a patent, essentially because it was merely a piece of perforated cardboard bearing printed matter for use in a system of doing business.

The District Judge rejected this argument, stating:

I find no ground for objection that the elements or ingredients here employed are not themselves of patentable nature. The fact that the structure may be of cardboard with printed matter upon it does not exclude the device from patentability according to the practice of the Patent Office, as shown by the numerous patents introduced for the defense of anticipation.[26]

Rand, McNally raised additional grounds of objection, one being that the claimed invention was anticipated by the prior art. The prior art included such devices as a railway ticket with map printed on the reverse side and was designed to prevent counterfeiting and fraud. Taking the position that the patent examiner had originally taken with respect to the Gellenbeck invention, Rand McNally urged that the patent was invalid.

The patent examiner had initially determined that Gellenbeck's menu card was simply a printed ticket, put to double use. In the initial rejection the examiner said:

The railroad tickets in common use all have checks or vouchers attached thereto, and many of them have a series of vouchers or checks, one for each officer of the road to whom such voucher or check should be properly sent. To apply this same scheme or idea to a bill of fare is merely a double use, and does not involve invention; that is to say, a construction of a ticket with a voucher or check attached thereto being old, it is, of course, immaterial whether the matter printed upon the ticket be such as to adapt it to one use or

---

[26] Benjamin Menu Card Co. v. Rand, McNally & Co. et al., 210 F. 285, 286 (N.D. Ill. 1894).

another, and it is also equally immaterial whether the printed matter be on either or both sides of the ticket or check. It is, of course, obvious that the removal of a coupon or check from any ticket mutilates the ticket to a certain extent. It is believed therefore that applicant's claim is fully answered by the state of the art above referred to.[27]

Upon reconsideration, the Patent Office rejected the examiner's view and allowed the patent.

In essence, Rand, McNally's argument asked the court to reconsider and adopt the patent examiner's view. The District Judge declined. He wrote:

> Cardboard or paper have long been used for coupon tickets of various kinds, for various uses, perforated upon the desired lines for separation, with printed matter arranged upon opposite sides as required. Printed bills of fare and printed meal checks are old. Nothing new can be claimed in either of these simple elements, but there may be invention in the thought to bring together the bill of fare, with its three courses, and the three required meal checks, so that they shall co-operate for common object and enlist the passenger (involuntarily) in the work of detection or avoidance of fraud. This device seems simple enough now, but the record shows that although some such simple means was long sought and with all the light of these various devices of coupon tickets, it had not occurred to any of the seekers to utilize the bill of fare until the thought came to this patentee.[28]

[C] Understanding *Benjamin Menu Card Co. v. Rand, McNally & Co.*

This case is one of several where the patented device was of paper or cardboard. What is interesting is how Gellenbeck's patent attorney integrated the business purpose into the patent claim. Without this business purpose, the perforated cardboard device, printed on both sides was otherwise known in the art.

In some sense, this menu card could be characterized as a new use for an old device. However, the patent claims do better than this. In surprisingly few words, the claims link different elements of the underlying business method with different elements of the physical cardboard device. This claim drafting technique is perhaps easiest to envision in claim 2. The elements of the business method may be expressed as follows:

1. Provide customer with a bill of fare;
2. Provide waiter with a check upon which to record the customer's order;
3. Require the waiter to separate the check into parts and to distribute those parts to the cook and to the customer, respectively;

---

[27] *Id.* at 288.
[28] *Id.* at 288.

4. Require the waiter to mutilate the bill of fare after taking the customer's order, so that the bill of fare may not be used with a second customer;
5. Later collect the parts of the check and use them to account for the number of meals sold.

The inventive aspect of the Gellenbeck menu card is how steps 3 and 4 are enforced and interrelated. Step 3 is enforced by a business method rule. Simply, the waiters are instructed to separate the check into parts and distribute them as a means of communicating the customer's order to the cook. Step 4 is enforced — thanks to the double-side printed card — as an automatic consequence of performing step 3.

Now reconsider Gellenbeck claim 2, linking up each element of the claim with the elements of the business method:

*   The combination of a menu card having a bill of fare on its face (this enables step 1 of the method),
*   With a check or checks upon the reverse side thereof (this enables step 2 of the method),
*   And placed so that when removed the bill of fare will be mutilated, so as to make the remainder incomplete as a bill of fare, and hence useless for another guest (this links steps 3 and 4 of the method).

The combination, when considered as a whole, enables step 5 of the method, although the claim does not necessarily require the portions of the check to be collected at a later time. The fifth step of the method has been included here to express the business method utility in preventing fraud. The fifth step can be readily omitted if the utility is simply in taking customer orders.

§ 6.09 Perforated Bookkeeping Forms — *Safeguard Account Co. v. Wellington*

In June of 1887 and November of 1988, John W. Horne filed applications for a patent on a blank book used for bookkeeping. In Horne's day, bookkeepers commonly used ledger books that had columnar-ruled pages of two different widths. Some pages extended the full width of the book's cover and other pages extended less than full width. When the book was laid open, and the user began writing on the left-hand page, the left margin of the full width page presented a column that was not covered up by the shorter pages. This column was used to record the name of the account holder, eliminating the need to recopy the name as successive short pages were used to record financial data.

It was more difficult to manufacture books with both short pages and long pages, because the bookbinder could not cut and finish the edge of the book in the usual fashion. More care was required to ensure that the correct long and short pages were interleaved, and the edges had to be cut before the final book was bound.

Horne's idea was to use one size for all pages, but to perforate some of the pages before the book was bound. Perforated and unperforated pages were interleaved in a predetermined fashion. The book was then bound, edges cut and finished in the usual fashion. In use, the bookkeeper could tear selected pages at the perforation, to convert those pages into short pages, as needed.

Horne assigned his patents to Safeguard Account Co., which sold many thousands of books employing the perforated pages as taught by Horne's patents. Safeguard brought suit for patent infringement against Edward F. Wellington in the Circuit Court for the District of Massachusetts. In a decision that is difficult to rationalize today, the Court found the earlier of Horne's patents valid and infringed.

[A] The Claims at Issue in *Safeguard Account Co. v. Wellington*

Horne's earlier patent, no. 393,506, made the following claim. The court did not discuss, or report on the claim of the later patent, no. 393,507.

> A blank book having full leaves of the same, or substantially the same, width, a part of which are provided near their outer edges with longitudinal lines of perforations to form removable margins, and the rest of which are unperforated, the perforated and unperforated leaves being interposed between each other throughout the book with one or more of the perforated leaves between the unperforated ones, substantially as set forth.[29]

[B] Using Perforations in Bookkeeping Forms Deemed Obvious — But Patent Surprisingly Upheld

Judge Putnam, Circuit Judge for the District of Massachusetts, rendered the decision that the earlier patent was valid and infringed. He ruled that the later patent was not patentably different from the earlier patent and ruled it void.

Judge Putnam first noted the advantages the bookbinder gained from using Horne's invention, and the better quality that resulted from having all pages the same width during manufacture. However, Judge Putnam noted, "none of these alleged advantages are suggested in the patent."[30] This, in the Judge's view, diminished the importance of those advantages in assessing the strength of the patent.

> None of these alleged advantages are suggested in the patent; and while, of course, this fact does not deprive the holder of the patent of the benefit of them, if they in fact exist, it weighs in favor of the proposition that they do not exist, or at least are not of importance.[31]

---

[29] Safeguard Account Co. v. Wellington, 86 F. 146, 147 (D. Mass. 1898).
[30] *Id.* at 148.
[31] *Id.* at 148.

Judge Putnam then opined that the idea of perforations was lacking the spirit of invention. In today's patent vernacular, Judge Putnam found the use of perforations was obvious.

> Sitting as triors of the facts, as well as of the law, we must take notice that the use of perforated lines in paper for the purpose of permitting easy separation has for many years been common to all the arts where paper is used. Under these circumstances, its application to any particular art, or any subdivision thereof, cannot be regarded as involving invention, unless under special circumstances. We have great doubts whether the advantages claimed in this case exhibit utility of the actual presence of the spirit of invention.[32]

However, having stated his first impression, that adding perforations to the pages of the book seemed to involve no invention, Judge Putnam then proceeded to find the patent was not obvious by drawing upon analogy to the Eighth Circuit Court of Appeals decision *Thomson v. Citizens Nat. Bank*,[33] discussed in preceding § 6.05. Judge Putnam wrote:

> Nevertheless, in *Thomson v. Bank*, . . . the court of appeals for the Eighth circuit found invention in a patent in this same art issued for a device of substantially the same nature as that at bar, though not the same. We feel constrained to follow the analogy of that decision, especially as the case at bar is one of doubt.[34]

In the *Thomson* case, the Eighth Circuit found nonobvious the concept of perforating and folding pages at the margins, so that once written upon as part of one page, the folded part could be flipped over for reading as part of the next page. If the perforated pages in Thomson were patentably inventive, Judge Putnam reasoned, then he should reach the same result in this case. He therefore found the patent was valid and infringed.

[C] Understanding *Safeguard Account Co. v. Wellington*

This case is peculiar in that Judge Putnam seemed inclined to find the invention obvious; yet he ultimately found the patent valid by drawing an analogy to a different case where similar subject matter was held to be inventive. The *Thomson v. Citizens Nat. Bank* case that Judge Putnam found so persuasive was an 1892 reported decision, involving a patent originally issued in 1888. Horne had applied for the first of the two patents-in-suit in this case in June of 1887. Thus, the patent involved in the *Thomson* case was not published at the time Horne applied for patent on his invention.

In essence, Judge Putnam did not find the invention in *Thomson* to be any more significant than the invention of Horne. Both were for account books, both

---

[32] *Id.* at 148.
[33] 53 F. 250 (8th Cir. 1892).
[34] 86 F. at 148.

contemplated using both wide and narrow pages, both used perforations to define special columns. The basic difference was the Thomson invention involved folding the page at the perforation, so that writing on one side could be viewed when looking at the reverse side of the page; whereas the Horne invention involved tearing the page at the perforation, so it would not cover up information written in a column on another page.

Given Judge Putnam's initial conclusions, that he found the use of perforations to be obvious, it seems likely that he could have ruled the patent invalid for lacking invention, if the *Thomson v. Citizens Nat. Bank* had not already been decided. Although Judge Putnam was not bound to follow this decision from another circuit, he followed it out of deference and because he saw the obviousness question as a close case.

Judge Putnam could have distinguished the invention in *Thomson* from the Horne invention before him. Both involved a perforation (however, the pages in Thomson could also be creased without first being perforated). The invention in *Thomson* was different in that the perforated and folded pages were designed to be flipped over so that writing on one side would be useful when viewing the page from the opposite side. Thus the invention in *Thomson* involved somewhat more than just a defined perforation. The objectives of the two inventions were different, and Judge Putnam could have used that fact in distinguishing the two cases.

Horne's objective was to find a way to have short pages — as the prior art ledger books did — without complicating the binding process. His solution was to make the pages short after binding, by tearing away part of the page. By comparison, the invention in *Thomson* was something more. The objective of the invention in *Thomson* was to find a way to eliminate the need to recopy account balances from one page when it was time to turn the page and start a new one. The solution was to fold the page in advance, so that writing in the folded right margin could be flipped out and made visible when the page was turned to begin writing on the reverse side.

Other factors may have swayed Judge Putnam not to distinguish Horne's invention from the invention in Thomson. Unfortunately, such other factors are not reported in his opinion. Thus today we can only speculate how Judge Putnam might have ruled if the distinctions between Horne's invention and the invention in *Thomson* had been more fully pointed out.

§ 6.10   Numbering Scheme for Insuring Against Shipping Loss — *Hocke v. New York Cent. & H. R. R. Co.*

If shipping by rail was how America spread its goods across the continent at the turn of the 20th Century, it was also how a certain percentage of those goods got lost. Today, shipping companies such as Federal Express have highly computerized systems for tracking goods in shipment. Likewise, today manufacturing companies have cleverly thought out, systematized techniques for pulling commodities from the supply chain, just in time for their use in the manufacturing process. However, in 1893 when Joseph Babbitt Mockridge was awarded his patent, the state of the shipping art was far less advanced.

Mockridge had been granted a patent, U.S. Patent No. 493,595, on a means for securing railroads and shippers against loss of freight. The system consisted of applying numbers to all freight cars and all packages, and duplicating those numbers on all checks and shipping receipts. Mockridge employed a temporary receptacle on each car for holding the checks during loading. After the car was loaded a checker would compare checks in the receptacle with articles in the car to determine if anything was misplaced before the train left the station.

The process worked like this:

1. Number each train car to designate the car's particular destination;
2. Prepare numbered shipping receipts for each package to be shipped; include on the shipping receipt the number of the car on which the package is to be loaded as well as the number assigned to the package;
3. Prepare checks for each package; these checks should be prepared simultaneously with the shipping receipts and shall include both the number of the car and the number of the package;
4. Place a removable box on each train car; when loading the car with packages, place the check for each loaded package in the removable box;
5. The clerk in charge of the station checks opens the removable box and compares the checks with the shipping receipts for that car, noting any discrepancies before the train leaves the station.

[A] The Claims at Issue in *Hocke v. New York Cent. & H. R. R. Co.*

The Mockridge patent issued with three claims, two of which include the removable box as an element. The third claim recites only the documents employed. Claims 1 and 3 are presented below.

> 1. The means for securing against loss of freight consisting of a receipt or other like document containing characters indicating the car designed to receive the merchandise, a separate independent check or ticket containing duplicate characters of the one on the said receipt so that the check and receipt control each other, and a receptacle held temporarily on the car, and adapted to receive the ticket being deposited in the receptacle at the time the merchandise indicated on the ticket is loaded in the car, substantially as shown and described.
>
> 3. The means for securing against loss of freight, consisting of a receipt or other like document containing characters indicating the car designed to receive the merchandise, and characters to designate the merchandise to be shipped, and a separate independent ticket or check containing sets of

characters which are a duplicate of the characters on the said receipt or document, substantially as shown and described.[35]

[B] The Numbering Scheme to Insure Against Loss Deemed Obvious

The Second Circuit looked at the Mockridge claims and concluded that the invention was obvious. To reach its decision it admitted having difficulty deciding whether the subject was better classified as a process or as an apparatus. Without deciding which, the Court concluded that Mockridge's contribution did not rise to the level of invention.

> It is as difficult to classify the subject of the patent as it was in the case of *Jacobs v. Baker*, 7 Wall. 295 and *Munson v. Mayor, etc.*, 124 U.S. 601; but whether the patent is to be regarded as one for an improvement in an art, or as one for a machine, we are of the opinion that it discloses nothing of patentable novelty.[36]

§ 6.11   Anti-fraud Cash Registering — *Hotel Security Checking Co. v. Lorraine Co.*

On June 20, 1893, John Tyler Hicks received a patent on a simple bookkeeping method that would prove to have a profound effect upon the patent laws of the United States for much of the next century. Struck down as unpatentable by the courts in 1908, the Hicks's patent's demise touched off an avalanche of later court decisions that snowballed into the widely accepted premise that business methods are not patentable.

Hicks was an observant man, with a great experience in the restaurant business. No doubt he had watched the occasional waiter succumb to temptation and pocket a few coins that rightfully belonged in the cash register of his employer. Hicks apparently knew enough about the subject to author a treatise on restaurant account keeping. In it he proffered these words of advice: "employ a competent and observant head waiter, have at least one honest man in charge, give each waiter a number and slips with a corresponding number, stamp the price of the articles ordered by him on the slip, and charge the amounts to him on a sheet of paper under his number, printed or written at the top of the sheet."[37]

Hicks obtained his U.S. Patent No. 500,071 for "a Method and Means for Cash Registering and Account Checking," for a system that bore resemblance to the method described in his earlier published treatise. In his patent, Hicks recommended the use of a vertically ruled sheet of paper having a numbered column for each waiter. Each waiter was assigned a different number, which was written on his badge and also on each of a set of 3 x 5 slips upon which the waiter would write the customer's order.

---

[35] Hocke et al. v. New York Cent. & H. R. R. Co., 122 F. 467, 468 (2d Cir. 1903).
[36] *Id.* at 468.
[37] Hotel Security Checking Co. v. Lorraine Co., 160 F. 467, 470 (2d Cir. 1908).

When the order was filled, a person in the kitchen or other department wrote the amount of the order on the slip and also on the horizontally ruled sheet. The waiter would then deliver the order to the customer along with the slip, which the customer would then take to the cashier when checking out.

At the end of the day, the slips for each waiter were totaled up and compared with the totals of the entries recorded on the ruled sheet. If there had been no carelessness or dishonesty, the amounts would agree, and if not, the a comparison between the ruled sheet and the slips would readily show who was at fault.

Hicks's company, Hotel Security Checking, sued upon his patent in the Southern District of New York and the Court dismissed the suit. The company appealed to the Second Circuit Court of Appeals to no avail. The appeals court ruled Hicks's patent was invalid. The court ruled the invention lacked novelty and patentability. In expressing the basis for its ruling, the court had these words regarding business methods:

> A system of transacting business disconnected from the means for carrying out the system is not within the most liberal interpretation of the term, art. Advice is not patentable.[38]

The term "art" was used in those days to refer to a process. The patent statute at that time read that "any person who has invented or discovered any new and useful art, machine, manufacture or composition of matter" may obtain a patent therefor.

[A] The Claims at Issue in *Hotel Security*

The claim drafting practice when the Hicks patent issued in 1893 permitted far more reliance on the specification than is permitted today. Thus, the following Hicks patent claims make explicit reference to certain features as being "herein described" or "as substantially described" in the specification. While that claim drafting style is not used today, it is clear that Hicks was claiming sheets of paper in combination with the waiters' slips.

> 1. The herein-described improved means for securing hotel or restaurant proprietors or others from losses by the peculations of waiters, cashiers or other employees, which consists of a sheet provided with separate spaces, having suitable headings, substantially as described, said headings being designatory of the several waiters to whom the several spaces on the sheet are individually appropriated, in conjunction with separate slips, each so marked as to indicate the waiter using it, whereby the selling price of all the articles sold may be entered in duplicate, once upon the slip of the waiter making the sale, and once upon his allotted space upon the main sheet, substantially and for the purpose specified.

---

[38] *Id.* at 469.

2. The herein-described improvement in the act of securing hotel or restaurant proprietors and others from losses by the peculations of waiters, cashiers or other employees, which consists in providing separate slips for the waiters, each so marked as to indicate the waiter using it, and in entering upon the slip belonging to each waiter the amount of each sale that he makes, and also in providing a main sheet having separate spaces for the different waiters and of their slips, and in entering upon said main sheet all the amounts marked upon the waiters slips so that there may thus be a duplication of the entries, substantially in the manner and for the purpose specified.

[B] Transacting Business Deemed Outside the Patentable Arts — But Case Decided on Other Grounds

In assessing the validity of Hicks's patent, the Second Circuit Court of Appeals began by extracting the business method "essence" from Hicks's patent claims. It started with the patent statute, which provided that "any person who has invented or discovered any new and useful art, machine, manufacture or composition of matter" may obtain a patent therefor.[39] Of the Hicks invention, the Court said:

> It is manifest that the subject-matter of the claims is not a machine, manufacture or composition of matter. If within the language of the statute at all, it must be as a "new and useful art."[40]

Relying on Webster, the term "art" was construed to mean, "the employment of means to accomplish some desired end; the adaptation of things in the natural world to the uses of life; the application of knowledge or power to practical purposes." In this sense, the court said, "art is not a mere abstraction."[41]

Having narrowed the issue to one of whether the claimed invention was art, as that term was used in the patent law of the day, the court ruled that it was not:

> A system of transacting business disconnected from the means for carrying out the system is not, within the most liberal interpretation of the term, an art. Advice is not patentable.[42]

The effect of this reasoning was to remove from Hicks's claims all patentable weight in the business system he proposed. The court thus reduced Hicks's claims to a bookkeeping system comprised of mere sheets of paper, which it then had little difficulty in dismissing as lacking novelty:

---

[39] Revised Statutes 4886 (U.S. Comp. St. 1901, p. 3382).
[40] 106 F. at 469.
[41] *Id.* at 469.
[42] *Id.* at 469.

It cannot be maintained that the physical means described by Hicks, — the sheet and the slips, — apart from the manner of their use, present a new and useful feature. A blank sheet of paper ruled vertically and numbered at the top cannot be the subject of a patent, and, if used in carrying out a method, it can impart no more novelty thereto, than the pen and ink which are also used. In other words, if the "art" described in the specification be old, the claims cannot be upheld because of novelty in the appliances used in carrying it out, — for the reason that there is no novelty.[43]

While the court viewed a system of transacting business as nonstatutory subject matter, it did suggest that a method of bookkeeping (e.g., the use of books and ledgers to keep records of business transactions) was a statutory "art." The problem for Hicks was that bookkeeping was already a well known art, and Hicks's adaptation — the replacement of ponderous books and ledgers with simple sheets of paper — was obvious. Indeed, it did not help Hicks's case that he had published a treatise on the topic of restaurant bookkeeping, describing the idea of keeping duplicate sets of records to prevent mistake and fraud.

Importantly, the court found it did not need to reach the question of whether bookkeeping was a statutory art, because bookkeeping was already in the prior art, and Hicks's invention was not a significant enough advance over that art to leap the patentability (obviousness) threshold. In the words of the court:

> If at the time of Hicks's application, there had been no system of bookkeeping of any kind in restaurants, we would be confronted with the question whether a new and useful system of cash-registering and account-checking is such an art as is patentable under the statute. This question seems never to have been decided by a controlling authority and its decision is not necessary now unless we find that Hicks has made a contribution to the art which is new and useful. We are decidedly of the opinion that he has not, the overwhelming weight of authority being that claims granted for such improvements as he has made are invalid for lack of patentability.[44]

[C] Understanding *Hotel Security Checking Co. v. Lorraine Co.*

The *Hotel Security Checking* case forms the basis for numerous later decisions that reject the patentability of business methods. The foundation stone of these later decisions is the Court's statement: "A system of transacting business, disconnected from the means for carrying out the system is not . . . an art." Many have read this as saying that a system of transacting business is not statutory subject matter.

However, the foundation stone of *Hotel Security Checking* is *dicta*. The Court ruled that the Hicks's claim lacked novelty and patentability. It openly conceded that it did not reach the underlying issue of whether the Hicks business system, itself, was statutory subject matter.

---

[43] *Id.* at 469.
[44] *Id.* at 472.

If at the time of Hicks's application, there had been no system of
bookkeeping of any kind in restaurants, we would be confronted with the
question whether a new and useful system of cash-registering and account-
checking is such an art as is patentable under the statute. This question seems
never to have been decided by a controlling authority and its decision is not
necessary now unless we find that Hicks has made a contribution to the art
which is new and useful. We are decidedly of the opinion that he has not.[45]

As will be seen, the Hicks claim legacy persisted for nearly 100 years,
resulting in repeated rejection of business methods as nonstatutory subject matter.
Indeed, the *Hotel Security Checking* decision was for many years enshrined in the
Patent Office in support of an express prohibition against business method
patents.[46]

If the Second Circuit Court of Appeals ultimately struck down the Hicks
patent for lack of novelty and patentability — and admittedly did not reach the
underlying statutory subject matter question — why did that court begin its
analysis by stating that a business system was not an art?

Following the court's analysis closely, it began from the premise that
"advice is not art." In other words, advice alone is not statutory subject matter.
At the highest level, the Hicks patent contained advice to restaurant owners on
how to structure the business to prevent theft. It was in this context that the court
made the oft quoted statement that a system of transacting business, apart from
the means for carrying out the system, is not art. Perhaps the court started its
analysis at this point to avoid giving patentable weight to that which was mere
business advice in the Hicks patent.

While it avoided giving patentable weight to Hicks's business advice, the
court did appear to treat the physical manifestation of bookkeeping as an art.
Indeed, throughout the opinion the court referred to bookkeeping as an art. "The
fundamental system [Hicks] is as old as the *art of bookkeeping*."[47] The court
continued, "[T]he ponderous system of *bookkeeping* is unnecessary; the
substitution of a blank sheet laid on the desk for a blank sheet bound in a book
may require ingenuity and be more convenient, but it adds nothing of substance
to the *art*."[48]

What may have troubled the court was not bookkeeping as a physical
manifestation of books and records, but rather the more abstract concept of a
business *system* as a series of human acts. Perhaps that is why the court made
clear that it did not wish to address this issue. "If . . . there had been no *system
of bookkeeping* of any kind in restaurants, we would be confronted with the

---

[45] 160 F. at 472.
[46] *U.S. Patent Office, Manual of Patent Examining Procedure*, § 706.03(a) (August 1993).
[47] 160 F. at 469.
[48] *Id.* at 470.

question whether a new and useful *system* of cash-registering and account-checking is such an art as is patentable under the statute.''[49]

Whether human acts alone may be the subject of a patent does seem to be the fundamental question. That question appears time and time again throughout many of the cases involving business method patents.

§ 6.12 Electronic Commerce in 1910 — *Berardini v. Tocci*

Michael Berardini, a New York banker, had many Italian customers who, having found their fortunes in America, wanted to send money back home to Italy. Transatlantic cable communication was at the height of popularity by the early 1900s and Berardini thought that a coded message sent by wire would be an excellent way to transmit funds, if it could be made more cost effective. In those days the transatlantic cable cost a fortune to use. Users paid by the word. Each number sent counted as one word, provided that number did not exceed five digits. Thus the number 12345 was charged at the one word rate; the number 123456 was charged as two words, because it exceeded the five digit word limit.

Berardini devised a code that would allow him to pack all pertinent wire transfer information — amount, remitter's name, payee's name and address — into a single number that did not exceed the five digit limit. Berardini used the left-most digit of the five digit word to represent amount, in multiples of 100 lire. The next two digits represented the remitter's name; and the last two the payee's name and address. Berardini kept duplicate sets of books in New York and in Italy in which he kept lookup tables to convert the coded names and addresses of remitters and payees. Berardini's first patent covered this coded message.

Sending important financial information by wire was risky, particularly when all information is packed so tightly into only five digits. An error in any one of the digits was fatal. Either the wrong amount was sent, or the wrong account was charged, or the wrong payee got the money. Berardini devised what would today be called a checksum to handle this problem. He arranged his coded messages in blocks of five and then added the digits both horizontally and vertically to produce a checksum which he sent as an additional coded message. At the receiving end the numbers were similarly added up and compared with the checksum to ensure all numbers were received without error.

Berardini applied for patents on both his basic coded message scheme and also his checksum technique. He was awarded patents for both concepts. Berardini's idea caught on and it was not long before other bankers were using his patented technique. Berardini brought suit against one such banker, a Felice Tocci. Tocci defended on the basis that the patent was not valid. The District Court for the Southern District of New York heard the case and rendered a decision in July 1911 that the patents were not valid.

---

[49] 160 F. at 472.

[A] Claims at Issue in *Berardini v. Tocci*

Berardini's first patent made the following claim:

A code message comprising a series of elements, the number of which is a multiple of the number of elements constituting a code unit, each code unit consisting of two portions, one of which indicates the value or amount of the order, while the other is a record mark identifying the parties to the transaction, the message also including means indicating the value of the elements representing the amounts of the orders.[50]

His second patent made the following claim:

A code message comprising a number of code units each consisting of a number of elements having a numerical significance and test totals indicating the proper results of the addition of said elements upon the arrangement of said units in columns and rows.[51]

[B] Use of Codes to Make Electronic Commerce Secure Deemed Obvious

The District Judge, Judge Hough, ruled both Berardini patents invalid for lack of invention. The prior art admitted in evidence showed that the idea of assigning arbitrary code meanings to words and numbers was very old in cable code systems. Indeed, Samuel Morse had in the late 1800s experimented with a number of different coding schemes before settling on his now famous Morse Code.

Having found both inventions obvious, Judge Hough proceeded to analyze the patent claims to explore whether they represented statutory subject matter. The patent statute at that time allowed patents on a new and useful machine, art, manufacture or composition of matter. Looking at the language of Berardini's claims, Judge Hough quickly concluded that these claims were not for any composition of matter, manufacture, or machine. If anything, he concluded, the claims would have to be for an ''art,'' as that term was used in the patent statute.

Judge Hough defined ''art'' as earlier courts had defined it:

The statute term ''art,'' used as it is in the sense of the employment of means to a desired end or the adaptation of powers in the natural world to the uses of life, . . . an art may require one or more processes or machines in order to produce a certain result or manufacture. It is for the discovery or invention of some practical method or means of producing an essential result or effect that a patent is granted, not for the result or effect itself.[52]

---

[50] Berardini v. Tocci, 190 F. 329, 330 (S.D.N.Y. 1911).

[51] *Id.* at 331.

[52] *Id.* at 332.

Using this distinction between the result and the means of producing the result, Judge Hough ruled that this patent was not seeking to claim a "coded message" but rather a system or method for devising a coded message. Admittedly the language of the claim recited the invention as a "code message." However, Judge Hough refused to consider this code message as the true invention. It troubled Judge Hough that there was no single message claimed, but a potentially infinite number of messages, all coded according to the teachings of the patent. Thus, instead of assessing whether the code message was statutory, Judge Hough concentrated on the process for producing the message and found it to be nothing but unpatentable "advice."

> No particular code message can be produced which in every exemplar thereof is the single subject of this patent. Indeed, the claims are misnomers. The patent is not intended to be for a code message, in the sense that patents have been granted for books of a peculiar kind. The patent is really for a system of devising code messages, and as such (upon a most liberal reading of the claims) it is in my judgment obnoxious to the remarks above quoted from *Hotel Security Co. v. Lorraine Co.* The patent is really for advice. It is for an art only in the sense that one speaks of the art of painting, or the art of curving the thrown baseball. Such arts, however ingenious, difficult, or amusing, are not patentable within any statute of the United States.[53]

The language Judge Hough had referred to was this quote from *Hotel Security Checking Co. v. Lorraine Co.* That case is discussed in preceding § 6.11.

> In the sense of the patent law an art is not a mere abstraction. A system of transacting business, disconnected from the means for carrying out the system, is not, within the most liberal interpretation of the term, an art. Advice is not patentable.[54]

[C] Understanding *Berardini v. Tocci*

Judge Hough's finding that the claimed inventions were obvious presents little difficulty. More interesting, however, is his analysis of whether the claims were for a statutory art. Judge Hough dismissed, out of hand, that the claimed invention was for a machine, manufacture or composition of matter. This left him to assess whether the claimed invention was for an art.

Today, a coded message embodied in a computer memory or embodied in a carrier wave is considered statutory subject matter. The Patent Office treats these as "manufactures" within the meaning of the patent statute (35 U.S.C. § 101). If Berardini were claiming his "code message" invention today, he might have presented claims expressing the code message as embodied in a carrier wave. It is that carrier wave (of electromagnetic energy) that carried Berardini's message,

---

[53] *Id.* at 333.
[54] Hotel Security Checking Co. v. Lorraine Co., 160 F. 469 (2nd Cir. 1908).

guided by cable, across the Atlantic. Consider the following hypothetical claims, based on the Berardini invention.

1. A code message embodied in a carrier wave, comprising:
a series of elements, the number of which is a multiple of the number of elements constituting a code unit;
each code unit consisting of two portions: one portion indicating the value or amount of an order; the other portion indicating the parties to the transaction; and
means indicating the value of the elements representing the amounts of the orders.
2. The coded message of claim 1 wherein said means indicating the value is a checksum.

Aside from a few stylistic changes, and the express recitation of a "carrier wave," the above hypothetical claims differ little from the subject matter Berardini had sought to patent. Today, the Patent Office grants claims of this form, in apparent recognition that such carrier waves are made by the hands of man and hence patentable subject matter. Of course, few courts in 1911 would have understood the carrier wave concept, and certainly few would have seen such waves as statutory "manufactures." The carrier wave claiming innovation did not come into bloom until the late 1990s. Nevertheless, it is interesting to consider how Berardini might have exploited modern claiming techniques if he were alive today. By characterizing his invention as a manufacture, Berardini parries Judge Hough's main thrust that "advice is not patentable."

Whether advice is unpatentable remains an interesting question to consider. Judge Hough drives home his point with the example that the art of throwing a curve ball is not patentable subject matter. While Judge Hough may be correct, the Patent Office is currently issuing patents for advice on similar sporting activities, such as the following patent on a golf swing, specifically an improved method of putting.

1. Method for putting, comprising the sequential acts of:
(a) providing a putter, the putter having handle and a body having a face, the body being configured to allow the putter handle to be inclined towards the ground with the face in a generally upward direction, such that the handle may be smoothly pivoted about the head in an arc to an upright position such that the face is generally perpendicular to the ground, the body being further provided with an alignment mark to allow a golfer to sight along the aligning mark and a golf ball disposed between the putter and a golf cup;
(b) positioning the putter proximate a golf ball which is on the ground, such that the golf ball is disposed between the putter and a golf cup disposed within the ground;
(c) moving the handle of the putter to a position inclined towards the ground such that the handle is generally aligned in a direction away from the golf cup and the face of the putter is facing in a generally upward direction;

(d) sighting along the alignment mark on the body of the putter and adjusting the position of the handle of the putter to align the alignment mark with the golf ball in the direction of the golf cup while the handle of the putter is in the position inclined towards the ground;

(e) pivoting the putter handle about the head to move the handle of the putter from the position inclined towards the ground to a generally upright position such that the face of the putter is essentially perpendicular to the ground and is proximate the golf ball; and

(f) hitting the golf ball with the face of the putter to urge the golf ball in the general direction of the golf cup.[55]

§ 6.13 Script for the Traveler, A New Form of Money — *Rand, McNally & Co. v. Exchange Scrip-Book Co.*

Richardson and Langston received their patent, U.S. Patent No. 669,489, for a scrip-book of coupons on March 5, 1901. In its broadest aspect, the patent described a new form of money.

At the turn of the 20th century, it was the custom for different railway companies to issue booklets of tear-out coupons, each coupon was numbered from 1 to 1000, or from 1 to 2000, depending on the size of the booklet. Each coupon represented a certain distance traveled, for example, one mile. By agreement of the various railway companies, the booklets were useable over different rail lines. Thus a traveling salesman could travel from Milwaukee to Chicago over a first railway and then from Chicago to Detroit over a second railway, paying for each leg of the journey with coupons torn from the book.

Richardson and Langston's innovation was conceived to overcome the problem that different lines sometimes charged different rates. For example, travel by rail was more expensive than travel by boat. If the traveling salesman were not in too much of a hurry, he might save money by taking a steamship over Lake Michigan from Milwaukee to Chicago and then complete his trip to Detroit by rail. However, it was cumbersome to use the conventional mileage book for such a journey, because the passenger and ticket agent had to calculate a rate conversion to know how many tickets to tear from the book.

Richardson and Langston's solution to the problem was quite simple. Treat the numbered coupons as money instead of miles. If each numbered coupon is worth one penny, it becomes simple to determine how may coupons to exchange, regardless of what mode of transportation is taken. The idea instantly spread and soon supplanted the prior mileage books.

Richardson and Langston assigned their patent to Exchange Scrip-Book Company. When Rand, McNally began issuing coupon books expressed in money instead of miles, their company sued in the Northern District of Illinois. The District Court ruled in favor of Exchange Scrip-Book Company and Rand, McNally appealed.

---

[55] U.S. Patent No. 6,019,689, ''Method of Putting,'' issued to Hogan on January 1, 2000.

[A] The Claims at Issue in *Rand, McNally & Co. v. Exchange Scrip-Book Co.*

Exemplary of Richardson and Langston's patent position is claim 4 from their patent:

> A scrip-book having as integral parts thereof a strip or series of coupons to be detached and exchanged at ticket-offices for passage-tickets, certificates or stubs having appropriately-designated spaces to receive descriptions of such tickets respectively and the signatures of successive conductors honoring each ticket, and ownership certificates having appropriately-designated spaces for successive signatures of the passenger in the presence of the respective conductors, the several parts being mutually identified by identical marks such as numbers.

[B] Ticket Valued Using Money Instead of Miles Deemed Patentable

The Court of Appeals for the Seventh Circuit ruled that the patent was valid and infringed. How it reached this conclusion is interesting and demonstrates that courts of that era did not necessarily feel as strongly bound to the claim language as are courts today. The court began, stating,

> Apart from the main idea of the patentees, that the unit in their patented ticket should be expressed in money instead of miles, we do not see anything in the patent that the defendants have infringed; for whether the physical differences, introduced by the patentees, are patentable invention or not, they are so narrow, and make the patent so limited, that the alleged infringing device (differing also in form) does not seem to us to be included.[56]

To readers today it is curious that the "main idea of the patentees" as expressed by the court does not appear to be explicitly recited in the claim. Nevertheless, the court expressed the issue thus,

> Does the adaptation of this new ticket to a unit, based on money instead of miles, make it patentable invention?

Answering the question in the affirmative, the court dismissed Rand, McNally's argument that the invention was obvious. It found that the money-based scrip had not been used before Richardson and Langston's patent; and that its use spread quickly after their introduction. The court seemed to give Richardson and Langston the benefit of the doubt because it viewed this rapid acceptance as evidence that the invention was not obvious.

> Unquestionably, the idea embodied in this ticket is a happy one. Possibly it has followed naturally, and without the employment of inventive thought, the

---

[56] Rand, McNally & Co. v. Exchange Scrip-Book Co., 187 F. 984, 985 (7th Cir. 1911).

wider use of interchangeable tickets. Possibly the wider use of interchangeable tickets has followed the embodiment of this idea. Which is cause and which is effect, and to what extent one is cause and the other is effect, we have no means of determining. The record does not help us on that inquiry. All that we know is, that it was not used before appellee's patent, and that it has been followed by the widest use. This, it seems to us, implies that the conception of this ticket, at least, helped to bring about the idea of a wide interchangeable use. There is no ground, under such circumstances, for saying that the concept is necessarily obvious.[57]

Rand, McNally had also argued that the invention was nothing more than a business method. The court rejected this argument as well. It did so by characterizing the invention as an apparatus to facilitate business, and not a method at all.

Nor do we think that this patented concept is nothing more than a business method. Its use is a part of a business method. The ticket patented is not a method at all, but a physical tangible facility, without which the method would have been impracticable, and with which it is practicable. And this is the status of thousands of like facilities that, once designed and put into use, have become the first of a new business method; and patents on such facilities have been sustained.[58]

[C] The Dissent Questions Whether Business Advice Is Patentable Subject Matter

Of the three judges who decided this case, one was a district judge — a common practice. The district judge, Judge Carpenter, disagreed with his circuit judge brethren and wrote a strong dissenting opinion. It bothered Judge Carpenter that this patent purportedly covered the practice of using coupon books that were physically unchanged from the coupon books of the prior art. The only difference was that the numbers printed on the coupons now represented money instead of mileage. To Judge Carpenter, this was an obvious change, given that it was well-known in the commercial world to express value in dollars on bank checks, letters of credit and express company's checks and orders.

Judge Carpenter also questioned whether this invention was statutory subject matter. He characterized this invention amounting to mere business advice, which he felt could not form the basis of patentable invention.

The inventors in this case have not devised ''any new and useful art, machine, manufacture or composition of matter, or any new and useful improvement thereof. . .

* * *

---

[57] *Id.* at 986.
[58] *Id.* at 986.

They have, at the most, advised the railroads of a desirable change in their method of doing business. . .

* * *

I admit the usefulness of the new system, just as I would admit the usefulness of various short-cut methods of computing interest or of multiplication or subtraction; just as I would admit the usefulness of a business house keeping a complete set of books. I cannot admit that a new and shorter method of computing interest, or a new and less complicated system of bookkeeping, involving fewer books, and less labor, while highly useful and desirable, can form (there being no new physical device) the basis of a patentable invention. The idea of the inventors in the present case involves merely a change of method of doing business, following simply and obviously from a change in conditions.

* * *

The form of the patent in this case has been anticipated. The new system of its use, in my opinion, is not patentable.[59]

[D] Understanding *Rand, McNally & Co. v. Exchange Scrip-Book Co.*

Judge Carpenter's observation that this invention was nothing more than the mere switch from miles to dollars is appealing. In this sense, Richardson and Langston's coupon book was simply a new use for an old apparatus. Troubling is the fact that the patent claim reads on the prior coupon books as well as it does Richardson and Langston's invention. None of the judges seemed to focus on this point. Instead, all seemed willing to import the new business system (dollars for miles) into their construction of the claim. That is, of course, the judge's prerogative.

Was Judge Carpenter on solid ground in characterizing this invention as a business method? His remarks in this regard are *dicta*. He ultimately concluded that the invention was obvious, and thus did not address the statutory subject matter question. Nevertheless, Judge Carpenter did seem to lump methods of doing business with methods of performing mathematical calculations and methods of bookkeeping and conclude that none of these could form the basis of a patent.

In Judge Carpenter's day, methods of performing mathematical calculations were regarded as abstract ideas — involving mental steps and perhaps pencil and paper or abacus. Today, the language of mathematics is understood as being capable of expressing attributes of both abstract ideas and practical inventions alike. However, even in Judge Carpenter's day, the art of bookkeeping (to the extent it involved physical objects, such as books and ledgers) was viewed as patentable subject matter.[60]

Its similarity to the prior art coupon book aside, the Richardson and Langston invention was, after all, a manmade article. Thus it certainly fell into

---

[59] *Id.* at 991.
[60] *See, for example*, Munson v. The Mayor, Etc., of New York, 3 F. 338 (S.D.N.Y. 1880).

the statutory category of a manufacture or an apparatus. What Judge Carpenter did when he looked at the Richardson and Langston invention was to identify what was *new* about their creation — the idea of switching from mileage to money. Viewed in that light, their creation was an abstract idea, and abstract ideas are not patentable subject matter. However, most courts today would reject Judge Carpenter's analysis, for he focused on only what was new, at the expense of ignoring the claim as a whole.

Whether Judge Carpenter's view will someday be vindicated remains to be seen. Although the current Federal Circuit Court of Appeals rejects Judge Carpenter's approach of extracting from the patent the essence of what was new, the statutory language that defines what is patentable subject matter, 35 U.S.C. § 101, does suggest that a focus on that which is new may not be unwarranted:

> Whoever invents or discovers any *new* and useful process, machine, manufacture, or composition of matter, or any new and useful improvement thereof, may obtain a patent therefor, subject to the conditions and requirements of this title.[61]

§ 6.14 Commuter Transfer Tickets — *Cincinnati Traction v. Pope*

The Pope patent, U.S. Patent No. 805,153, was issued November 21, 1905, for a "time limit" transfer ticket for streetcars. The ticket had a body portion and an attached stub that could be removed by tearing along a perforation. When the stub was attached, the ticket was good for afternoon use; when removed, the ticket was good for morning use. The ticket also bore preprinted numbers corresponding to hours of the day, which could be punched to denote the hour during which the transfer ticket was useable.

Pope sued Cincinnati Traction for infringement in the District Court for the Southern District of Ohio. The district court found the patent valid and infringed. Cincinnati Traction appealed to the Sixth Circuit Court of Appeals.

[A] Claims at Issue in *Cincinnati Traction v. Pope*

Exemplary of the Pope patent are claims 1 and 8, which read:

> 1. A transfer ticket comprising a body portion and a coupon, said body portion and coupon bearing conventional indications to constitute an ante meridian transfer ticket when said body portion is used separately and a post meridian transfer ticket when used together.
> 8. A transfer ticket comprising a body portion and a coupon and further provided with conventional indications to constitute a complete transfer ticket for one part of the day when said body portion is used separately and a complete transfer ticket for another part of the day when said body portion and coupon are used together.

---

[61] 35 U.S.C. § 101.

[B] Business Method Embodied in Transfer Ticket Provided No Grounds for Invalidity

The court of appeals separately addressed the questions of (a) whether the patent recited statutory subject matter; (b) whether the claims were to a novel and nonobvious invention; and (c) whether the claims were infringed.

Cincinnati Traction had argued that the subject matter was merely "a method of transacting business, a form of contract, a mode of procedure, a rule of conduct, a principle or idea, or a permissive function, predicated upon a thing involving no structural law." Cincinnati Traction's counsel had argued that the ticket in question had "no physical characteristics which enable it to be distinguished from any other transfer ticket or from any other printed slip of paper." For these reasons, it urged that the patent was not drawn to statutory subject matter.

The court declined to accept Cincinnati's argument, ruling:

> But while the case is perhaps near the border line, we think the device should be classed as an article to be used in a method of doing business, and thus a "manufacture" within the statute . . .

> * * *

> The device of the patent clearly involves physical structure. The claims themselves are, in a proper sense, limited to such structure . . . But the alleged patentable novelty does not reside in the arrangement of the printed text, nor does such text constitute merely a printed agreement. The argument to that effect overlooks the important consideration that the body alone is good at one time, and that the body and coupon are required for the other portion of the day; and that the ticket bears on its face, whether the body is used alone or with the coupon, the distinguishing indications.[62]

On the issue of novelty and invention (nonobviousness), the court distinguished several prior art references including one transfer ticket product, the Ham ticket, that was sold in competition with Pope. The court weighed evidence of the Pope ticket's imitation and immediate acceptance in the marketplace as proof of the invention's novelty and inventiveness.

> The "tribute of its imitation" is cogent evidence of the utility of the Pope transfer. Indeed, this utility is readily apparent as respects both facility and rapidity of use and effectiveness in preventing misunderstandings and misuse. It is novel and distinctive in form. It has been favorably received. Although its average price was slightly higher than that of the Ham, it has largely superseded that device.[63]

---

[62] Cincinnati Traction v. Pope, 210 F. 443, 446 (6th Cir. 1913).
[63] *Id.* at 449.

[C] Understanding *Cincinnati Traction v. Pope*

It is perhaps difficult to understand how a simple ticket with tear off stub could be awarded a patent, even in 1905. Tickets comprising body portion and tear away stub or coupon were then known in the prior art. What was new about Pope's ticket was the use to which he ascribed to the body portion and stub. With attached coupon meant PM; without attached coupon meant AM.

Thus, the Pope ticket carried with it an integral means for conveying a message: AM or PM. Proprietors of streetcar businesses were already issuing transfer tickets, valid for either AM or PM when Pope invented his ticket. Pope's ticket was designed to provide a better way to communicate AM or PM information. Thus Pope simply provided a better way for streetcar companies to communicate information they were already using.

In this sense, Pope's invention may clearly be characterized as an apparatus or article of manufacture. The court chose to call it a manufacture. The term manufacture corresponds to one of the four statutory categories of patentable subject matter. It is tempting to conclude that anything altered by the hand of man may be classified a manufacture. As the Supreme Court advised us in *American Fruit Growers, Inc. v. Brogdex Company*, even the term manufacture has its limits.

§ 6.15 Anti-mold Fruit Treatment — *American Fruit Growers, Inc. v. Brogdex Company*

In 1923, Brogden and Trowbridge applied for a patent on a technique for treating fresh fruit so that it would not be inflicted with blue mold and spoil so quickly. The technique involved treating the skin of the fruit with borax. The borax could be quite simply applied to the wash water into which the fruit was dipped prior to packaging and shipping.

The Patent Office awarded Brogden and Trowbridge U.S. Patent No. 1,529,461 on March 10, 1925. One year later their company, Brogdex, marched into court to seek an injunction against the American Fruit Growers, Inc. The district court found the patent valid and infringed and enjoined American Fruit Growers from using the process. American Fruit Growers appealed and the Circuit Court for the Third Circuit affirmed. American Fruit Growers then petitioned the Supreme Court to review the case. The Supreme Court granted *certiorari*.

[A] The Claims at Issue in *American Fruit Growers, Inc. v. Brogdex Company*

Brogden and Trowbridge claimed their invention both as a process and as a manufacture. Claim 3 is representative of the process and claim 26 the manufacture.

3. In the preparation of fresh fruit for market, the process which comprises subjecting fruit to the action of an aqueous solution of borax, the fluidity, strength and temperature of the treating solution, and the duration of the treatment, being such that exposed rind or skin tissues of the fruit are effectively impregnated with borax and rendered resistant to blue mold decay, while at the same time the fruit is not scalded nor is its freshness or edibility otherwise substantially impaired.

26. Fresh citrus fruit of which the rind or skin carries borax in amount that is very small but sufficient to render the fruit resistant to blue mold decay.[64]

[B] An Orange Coated with Borax Is Still an Orange — Patent Invalid

American Fruit Growers had argued that the process, as exemplified by claim 3, was not sufficiently different from the prior art to warrant a patent. It argued that oranges, as defined by claim 26, are not an article of manufacture. It admitted that it dipped its fruit in a borax solution and thus admitted infringement, if the patent were valid. Brogdex responded that the orange, once coated with borax, was no longer a natural product, but a manufacture, because the complete product is not found in nature. It distinguished the prior art as dipping fruit into boric acid, whereas their patent called for the use of borax, an alkaline.

Justice McReynolds delivered the opinion of the Court. He rejected Brogdex's argument that an orange coated with borax is an article of manufacture, citing the Century Dictionary to define "manufacture."

Manufacture . . . is "the production of articles for use from raw or prepared materials by giving to these materials new forms, qualities, properties, or combinations, whether by hand-labor or by machinery." Also, "anything made for use from raw or prepared materials."[65]

At first blush, that definition could cover oranges coated with borax, it would seem. However, Justice McReynolds saw things differently. He wrote:

Addition of borax to the rind of natural fruit does not produce from the raw material an article for use which possesses a new or distinctive form, quality, or property. The added substance only protects the natural article against deterioration by inhibiting development of extraneous spores upon the rind.[66]

How did Justice McReynolds propose to distinguish raw materials that were given new form from raw materials that were merely given a new package? He explained it this way:

---

[64] American Fruit Growers, Inc. v. Brogdex Co., 283 U.S. 1, 6 (1930).
[65] *Id.* at 11.
[66] *Id.* at 11-12.

> There is no change in the name, appearance, or general character of the fruit.
> It remains a fresh orange fit only for the same beneficial uses as theretofore.
>
> * * *
>
> There must be a transformation; a new and different article must emerge
> having a distinctive name, character or use.[67]

Thus, in Justice McReynolds's view, a wolf in sheep's clothing is still a wolf. An orange in borax coating is still an orange. An orange coated with borax is not a manufacture for which patent protection may be had under the patent laws.

Turning next to the prior art, Justice McReynolds noted that boric acid (used in the prior art) and borax (an alkaline) were both known to inhibit the rapid development of blue mold. Both, he noted, are compounds of boron and contain the boric acid radical. Thus, he concluded, for purposes of this patent dispute, boric acid and borax are equivalents, such that the mere substitution of one for the other would not involve invention or avoid infringement.

Thus Justice McReynolds concluded that Brogden and Trowbridge's claims to a new article of manufacture did not recite statutory subject matter and their claims to the process of applying borax to fruit was obvious and lacked patentability.

[C] Understanding *American Fruit Growers, Inc. v. Brogdex Company*

The *American Fruit Growers* case provides important guidance of what a statutory manufacture consists. There must be a transformative effect or state change in the raw material to which the claimed result is attributed. In the case of the orange, the claimed result was increased immunity to blue mold. The borax coating, however, did not transform the orange or its rind into anything new to achieve heightened mold resistance. The borax merely took up residence within the textured surface of the orange rind. Neither borax nor orange rind were changed. The borax continued to do what it always did — suppress blue mold — and the orange rind continued to do what it always did — protect and contain the inner moisture of the fruit.

Within the limits set down by *American Fruit Growers*, the scope of what may be termed a manufacture has been greatly extended since that case was decided in 1930. Today, computer programs are classified as manufactures if they are embodied in a computer readable medium. The computer readable medium is viewed as having been changed or transformed into a new article by the addition of particular computer code.

Going even a step further, the Patent Office even suggested that computer programs may be classified as manufactures if they are embodied in a propagating carrier wave. This proclamation occurred shortly after the Patent Office released its Examination Guidelines for Computer-Related Inventions in

---

[67] *Id.* at 12.

1996. The Patent Office determined that "additional training materials were needed to address how to apply the Guidelines in the areas of business, artificial intelligence and mathematical processing applications."[68] It issued a set of training materials that include specific examples of "manufacture" claims comprising computer memories and propagating carrier waves.

Are these modern-day claims truly within the limits set down by the Supreme Court in *American Fruit Growers*? If so, then there must be some state change, some transformation; a new and different article must emerge having a distinctive name, character or use. What is it? In the case of the computer memory, the new article is a computer-readable "book," if you will, containing instructions that the computer physically executes. By placing new computer code into the memory, a new use emerges.

Would the same be true if the memory were not computer-readable, but human-readable? Is a book containing instructions for performing a new business process a "manufacture"? Instinct says no, but why? Would not the instructions contained in that book effect a change in human behavior, thereby causing them to practice the new business method, for example?

The answer is no. While humans may elect to follow the instructions in a book, thereby practicing a new business method, they cannot by the words of the book be forced to do so. The instructions in that book do not change human beings into something they are not. Unless someday it is scientifically proven otherwise, human thought is not something that can be controlled.

§ 6.16   Buying and Selling Stock — *In re Wait*

If you happen to think that these pre-computer era cases have little relevance in today's Internet-charged business markets, it is time to consider the *Wait* application that was prosecuted during the throes of the Great Depression. Wait applied for a patent for "Improvements in Process of Vending," in which he described a method of buying and selling stock. Wait's patent specification suggested that electrical systems could be used to remotely post prices of stocks and commodities offered for sale as well as offers for those commodities. The Court of Customs and Patent Appeals that ultimately decided the fate of Wait's patent application described his system this way:

> From the recitals in the specification we conclude that appellant expects his process to be used chiefly in connection with the business of buying and selling stocks and other commodities usually dealt upon stock and commodity exchanges, and that through it there may be an elimination of brokers and like agents as well a quickly made and accurately kept records of the transactions which take place, together with other advantages.
>
> It also appears with reasonable clearness that appellant expects his process to be carried out by means of an electrical system, through the

---

[68] U.S. Patent Office, *Examination Guidelines For Computer-Related Inventions Training Materials Directed To Business, Artificial Intelligence, And Mathematical Processing Applications.*

medium of which when prices at which stocks, commodities and the like are offered for sale, or at which offers of purchase are made, are posted upon suitable devices at given points, such prices will be transmitted speedily to remote points and there posted upon suitable devices where they may be visualized by those interested. Also through the medium of the system, acceptances by those wishing to buy or sell may be speedily transmitted to the proper point, and the system is to have the requisite means for accurately recording all necessary elements of the transactions.[69]

The patent examiner rejected Wait's application on dual grounds: first, that the application did not disclose an operative system and second, that the claims were drawn to a nonstatutory business method.

Wait appealed to the Board of Appeals, which rejected the examiner's first contention but affirmed the second. The Board ruled that Wait's invention was a business method and that business methods were not statutory subject matter. Wait then appealed to the Court of Customs and Patent Appeals for one last bite at the apple.

[A] Claims at Issue in *In re Wait*

Wait submitted 17 process claims. All of them were variations with minor limitations. Claim 1 reads as follows:

The process which comprises posting an offer figure for a commodity causing the figure to be visible at a remote point, contacting the stations of a buyer and a seller through a central point, causing such contact to be indicated at the point of posting and consummating a sale through such connection and removing said posted figure.

[B] Buying and Selling Stock Using Remote Terminal Deemed Obvious

The Court of Customs and Patent Appeals affirmed the rejection. However, it did so on lack of novelty and expressly without endorsing the Patent Office's view that business methods did not constitute patentable subject matter. The court held:

The process, when analyzed carefully, appears to comprise, in its essence, nothing more than the advertising of, or giving publicity to, offers of purchase or sale by one party, the acceptance thereof by another and the making of a record of the transaction followed by a withdrawal of the offer. Surely these are, and always have been, essential steps in all dealings with this nature, and even conceding, without holding, that some methods of doing

---

[69] *In re* Wait, 24 U.S.P.Q. (BNA) 88, 89 (C.C.P.A. 1934).

business might present patentable novelty, we think such novelty is lacking here.[70]

[C] Understanding *In re Wait*

In re Wait represented an opportunity for the court to expel the business method from the body of law governing patentable inventions. Had it chosen to do so, the court could have reversed the Board's ruling that business methods are not patentable subject matter. That would have ended the business method controversy in 1934 and this book would never have been written.

Wait was represented by George E. Tew and John C. Wait (the relationship with applicant Justin F. Wait is not known). Tew and Wait saw this case as an opportunity to finally put the business method issue to rest. They filed an extensive brief, urging that the time had come for the court to decide this issue once and for all. Unfortunately for Tew and Wait, the judicial system was not yet ready to take this bold step. In the words of the court,

> Before this court there has been filed an elaborate and, we may add, a quite interesting brief in which a large number of authorities are cited and analyzed. There was also an oral argument on appellant's behalf during the course of which it was suggested that an opportunity is here afforded this court to render a decision which might possibly clarify questions growing out of applications for patents relating to what is called "methods of doing business."
>
> However inviting this field may be, the court does not deem it proper to deviate from its usual practice of determining only the relevant questions presented by the application actually before it, avoiding *dicta* insofar as possible.[71]

Three months before the decision in *In re Wait* was handed down, George Tew published a law review article in the Journal of the Patent Office Society. In his article, he described the generally accepted belief that methods of doing business were not patentable, and he noted that he could find no cases actually holding this. He wrote,

> It is probably settled by long practice and many precedents that "methods of doing business," as these words are generally understood, are unpatentable, notwithstanding the absence in decided cases of any logical or statutory reason or rule why they are unpatentable. It is accepted as sufficient that it is so held.[72]

---

[70] *Id.* at 89.

[71] *Id.* at 89.

[72] Tew, Geo. E., "Method of Doing Business," 16 J. Pat. [& Trademark] Off. Soc'y 607.

George Tew explored the nature of business methods, observing that "method of doing business" had become a label used — rightly or wrongly — when courts declared certain subject matter unpatentable:

> We are not informed why in theory a patentable process must change the state of material objects or reduce something to a different state or thing. The expression "methods of doing business are unpatentable" has come to be an accepted slogan, perhaps because it is fairly well understood in the profession as imparting unpatentability to only a limited kind or class of methods, usually amounting to mere paper work, or involving the mental reaction or sense of an individual.[73]

Tew hit the nail on the head. If a change in state is indeed required, can that change in state be a change in the mental state of an individual? This is a statutory subject matter question that is explored more deeply in Chapter 3. The short answer is no, for this reason. At the foundation of our patent system is an exchange between the inventor and the public. The inventor discloses knowledge that he or she alone possesses, in exchange for a public grant (patent) of the right to exclude others from using that knowledge for a limited time. Possession is the key. The public understandably is not willing to give this public grant to an individual who is not in possession of the knowledge he or she purports to convey.

Possession implies control. An inventor must exert control over the idea before it is his or hers to convey. If an inventor claims to know how to extract salt from seawater, the public can observe the process, taste the results and know that it will get a fair benefit of the bargain if it agrees to grant a patent in exchange for this knowledge.

On the other hand, if an inventor claims to know how to change someone's mind on a particular issue — say on how to vote in an upcoming election — the public can observe the process, see statistics on the purported results and still not know whether it was the inventor's method, or some other factor, that caused the election to turn out one way or the other. Simply stated, the public has no proof that the inventor has actually controlled the thinking of the human mind, hence it should understandably refuse to grant a patent for the technique.

George Tew probably did not use the above analysis. However, he did arrive at the insightful observation that there is indeed a need to separate patentable business methods from nonpatentable business methods:

> A distinction may probably be drawn between a method *of* doing business and a method *used* in doing business, because many patentable processes, those found in telegraphy and telephony for example, are used in doing business, and in a large sense substantial portions of the whole field of patentable processes are used in doing business of some sort. As to methods of doing business we find such expressions as that in *Ex parte Abraham*,

---

[73] *Id.* at 607-608.

1869 C.C. 59 — "it is contrary to the spirit of the patent law as construed by the Office for many years, to grant patents for methods or analogous systems of bookkeeping" etc.[74]

George Tew identified the problem that sooner or later, courts would have to face. The mere fact that an inventive method impacts business, as Tew noted nearly every inventive method does, should not per se render the method unpatentable. On the other hand, there is a point at which the method so involves natural human functions that the public is unwilling to grant to one person the sole right to exploit it. Tew did not see a court resolve this question during his lifetime. However, the ideas that he and others expressed that day spread from courtroom to courtroom until Judge Rich, on July 23, 1998, finally stated it openly that business methods are patentable subject matter.[75]

§ 6.17   National System for Fire Fighting — In re Patton

Wilbur and Orville Wright delivered the first military aircraft to the U.S. military in 1909. A reconnaissance tool at first, the airplane gradually developed into an offensive weapon during the first World War. Although the first bombs were dropped by hand over the side of the cockpit, by the end of World War I, military engineers had outfitted special bombers with standardized bomb fittings and bombsights to enable systematic bombing with far greater precision. Wilbur and Orville's amazing flying machine had become a threat in the skies that must have greatly troubled the populace.

One such person was Ross Cummings Patton. Living in those troubled times between the two World Wars, Patton worried about his country's national defense and what would happen if an airborne enemy might attack American cities with incendiary bombs. Patton believed that our nation should have a nationally operated system of fire fighting, employing self-propelled fire trucks with electric motor driven pumps. The components should be standardized and interchangeable, Patton reasoned, to allow the equipment to work together no matter where a fire might break out.

Patton applied for a patent on his fire-fighting system, but met with resistance from the U.S. Patent Office. The Patent Examiner rejected Patton's claims because, in the examiner's view, the claimed invention was a mere aggregation of already well-known technologies. Patton appealed to the Board of Patent Appeals, where again his claims were rejected. Determined to prevail, Patton appealed the Board's decision to the Court of Customs and Patent Appeals (C.C.P.A.). Patton represented himself in that final appeal.

---

[74] Id. at 608.

[75] State Street Bank v. Signature Financial Corp., 149 F.3d 1368 (Fed. Cir. 1998).

[A] The Claims at Issue in *In re Patton*

Patton had a variety of claims in his application. Each was directed to a different combination of basic fire-fighting components. Claim 1 is exemplary of the scope of Patton's claims on appeal.

> 1. The combination, in a unified mobile, intracity, intrastate, interstate and national, fire protective system against aircraft attack, but suitable for general use; utilizing a conventional type self propelled chassis, electric motor driven pressure pump, combination chemical and service hose lines; also, an electric cable and reel with suitable end plug, for making connection to independent or interlocking, intracity, intrastate, interstate and national, electric power transmission supply lines.[76]

[B] System of Fire Trucks, Electric Pumps, and Hose Lines Deemed Obvious

Patton did not dispute that fire trucks, electric pumps, hose lines, and electric cables were known in the art. Patton's position was that his invention encompassed the combination of these components to create a national fire-fighting system to combat mass aircraft attack. The court gave Patton's claims little weight in this regard, stating:

> Appellant contended before the tribunals of the Patent Office and contends here that the appealed claims are patentable because they provide for a novel "interstate and national fire-fighting system to combat mass aircraft, incendiary-explosive bombing attack"; that such a system was "the essential aim" of appellant's alleged invention; that his system has been utilized by the United States Government, the city of New York, and manufacturers unknown to appellant; and that it has practical application.
>
> In this connection it is sufficient to say that a system of transacting business, apart from the means for carrying out such system, is not within the purview of section 4886,[77] . . . nor is an abstract idea or theory, regardless of its importance or the ingenuity with which it was conceived, apart from the means for carrying such idea or theory into effect, patentable subject matter.[78]

In support of its statement regarding what is patentable subject matter, the court cited *In re Moeser*,[79] *Hotel Security Checking Co. v. Lorraine Co.*,[80]

---

[76] *In re* Patton, 127 F.2d 324, 325 (C.C.P.A. 1942).

[77] Revised Statutes 4886 (35 U.S.C. § 31).

[78] 127 F.2d at 327.

[79] 27 App. D.C. 307 (1906 U.S. App. LEXIS 5169).

[80] 160 F. 467 (2d Cir. 1908).

Berardini v. Tocci,[81] *In re Dixon*,[82] *In re McKee*,[83] and *In re Lockert*.[84] Some of these cases are discussed in preceding sections of this chapter.

Having stated its intention to focus on the structure of Patton's system and not the underlying business purpose, the Court framed the issue on appeal as follows:

> [T]he question here presented is whether the structure defined in appealed claims 1, 7, 8 and 12, for carrying out appellant's fire-protective system against aircraft attack, involves invention in view of the [prior art patent] disclosures.[85]

The Court assessed the prior art and concluded that Patton's invention was obvious.

[C]   Understanding *In re Patton*

Although the court cited the proposition that a business method is not patentable, apart from the means to carry it out, it actually decided this case on obviousness grounds. Thus this decision lies with the majority of decisions reviewed in this chapter that raise the statutory subject matter issue in *dicta*, but then decide the outcome on other grounds.

When Patton's claims are carefully analyzed it seems Patton was simply attempting to do what many patent applicants attempt to do: patent an old combination for a new purpose. This often leads to an obviousness rejection. In Patton's case, the fire-fighting equipment was well-known. His innovation was simply to deploy such equipment on a national scale. Unfortunately for Patton, this new purpose — national scale deployment — bore no operative nexus to the equipment he selected. Scaling up from township to city, from city to state, and from state to nation required no inventive act. The basic fire-fighting components, fire trucks, pumps, hoses, etc., remained essentially unchanged.

Patton's case helps shed light on an often encountered problem where the applicant has identified a "new use" for an existing technology, but that "new use" is technically no different from the existing use. To illustrate, consider this hypothetical example. *Applicant John Doe* observes the success of prepaid telephone cards. After purchasing the card, the customer inserts it into a public telephone and may then use the phone to place a telephone call. *Doe* adapts the prepaid phone card system to encompass a prepaid fax business. The customer inserts card into public telephone to which a fax machine is attached. The customer may then use the system to send a fax over the public phone lines. *Doe* applies for a patent.

---

[81] 190 F. 329 (S.D.N.Y. 1911).
[82] 44 F.2d 881 (C.C.P.A. 1930).
[83] 64 F.2d 379 (C.C.P.A. 1933).
[84] 65 F.2d 159.
[85] 127 F.2d at 328.

Because no one has ever proposed the prepaid fax card before, he argues that it is patentable. The Patent Office rejects *Doe's* application as being obvious. *Doe* appeals. Does *Doe* win in the appellate court? Probably not. Aside from the content of the message — fax data versus voice data — *Doe's* system is no different from the prior art.

Having failed in his first patent attempt, *Doe* tries again with a slightly different embodiment: the prepaid Internet access card. The customer inserts card into public telephone, attaches a computer to the phone's data port and uses it to gain access to the Internet. So far, this is not much different from the fax card example. However, the card does more than just permit a phone connection to be made. It also contains an embedded user ID and password that allow the computer to log onto a specific Internet service provider's site that hosts the prepaid Internet access account. After a telephone connection is established, the system interrogates the card, obtains the prepaid user ID and password, and authorizes the user's computer to log onto the Internet. Does *Doe* win this patent? More likely yes. In this case the card is not simply designated for a particular use, it contains specific elements to facilitate that use. The embedded user ID and password make the card uniquely suited to the Internet access application.

§ 6.18 Drive-in Movies — *Loew's Drive-in Theatres, Inc. v. Park-in Theatres, Inc.*

Shortly before the 1939 World's Fair, one Richard M. Hollingshead, Jr. hit upon a novel way of cashing in on the motion picture craze. Popularized by Thomas Edison in the late 1800s, the motion picture industry had evolved from silent films of the 1910s and early 1920s to "talkies" of the late 1920s to the technicolor films of the 1930s. Indeed, the motion picture industry captured on film an image of its very evolution. We can still relive the moment — as the first audiences did — when black and white film evolved into color before the viewers very eyes in the 1939 classic, *The Wizard of Oz.*

By the early 1930s, motion picture theaters had sprouted like daisies across the American landscape. Hollingshead conceived of a new type of theater in which patrons could drive in and watch the latest movie from the comfort of their cars. Hollingshead devised a parking layout that would permit persons seated in either the front or back seat to see the screen without obstruction. The parking lot was graded slightly downward toward the screen so that patrons seated in cars in the back row could see over the cars in the front rows.

Unlike automobiles of today, automobiles of the 1930s typically had comparatively small windshields, making it hard to see the screen when parked in the front rows. To address this problem, Hollingshead constructed his preferred parking area as a series of elevated terraces with individual ramps at the front of each parking space. By driving the front wheels partially or fully up these ramps, patrons could adjust the viewing angle of their car with respect to the screen. Thus the ramps enabled each driver to adjust the viewing angle of his or her vehicle, to selectively accommodate individual vehicle occupants, without interfering with any other automobiles.

Hollingshead applied for a patent on his drive-in theater concept and was awarded U.S. Patent No. 1,909,537 on May 16, 1933. He assigned the patent to Park-in Theatres, Inc., which, in turn, licensed Loew's Drive-in Theatres, Inc. to commence operation of a drive-in theater in Providence, Rhode Island, in June 1937.

Loew's Drive-in paid Park-in a percentage of its gross receipts as royalties until the close of the 1937 season in November. The next spring, Loew's again opened its theater, but refused to pay further royalties. Loew's Drive-in did so on advice of its legal counsel that the license was illegal and the Hollingshead patent was invalid for absence of patentable invention.

What was behind Loew's Drive-in's counsel's criticism that the invention was invalid for absence of "patentable" invention? In today's patent law terminology, this requirement is characterized as the non-obviousness requirement.

During the 1930s, and indeed up until 1952 when the patent statute was revised, the patent statute contained only the express requirement that the invention must be novel to be eligible for a patent grant.[86] The statute did not expressly require a "patentable invention." Nevertheless, court interpretation of the patent laws as they existed at that time did require the invention to be "patentable," as a way of screening out inventions that were mere obvious applications of the state of the art. This judicially created "patentable invention" requirement was in 1952 codified into the patent statute as the "non-obviousness" requirement:

> A patent may not be obtained though the invention is not identically disclosed or described as set forth in section 102 of this title, if the differences between the subject matter sought to be patented and the prior art are such that the subject matter as a whole would have been obvious at the time the invention was made to a person having ordinary skill in the art to which said subject matter pertains.[87]

The District Court judge ruled in favor of patentee, Park-in, finding that the Hollingshead patent was valid and infringed. The District Judge found Hollingshead's drive-in theater to be:

> "ingenious," "practical," "a good business investment," "a completely new idea of giving entertainment to people who could not ordinarily go to the theater," and "is a beneficial contribution to the art of exhibition of motion pictures."[88]

---

[86] Revised Statutes 4886, 4887, 4923 (35 U.S.C. § 102; 1946 Ed, 31, 32, 72).

[87] 35 U.S.C. § 103.

[88] Loew's Drive-in Theatres, Inc. v. Park-in Theatres, Inc., 174 F.2d 547, 552 (1st Cir. 1949).

Loew's Drive-in appealed to the First Circuit Court of Appeals. Today all patent appeals are heard by the Court of Appeals for the Federal Circuit. However, in 1949 when this appeal was heard, patent appeals were heard by the appellate court for the circuit in which the district court was located.

[A] The Claims at Issue in *Loew's Drive-in*

In his patent, Hollingshead had claimed an outdoor theater. His patent contained 20 claims. The following claim is exemplary.

An outdoor theater comprising a stage, alternate rows of automobile stall-ways arranged in front of the stage, said stall-ways being adapted to receive automobiles disposed adjacent to each other and facing the stage; — said automobile stall-ways being at a vertical angle with respect to the stage such as will produce a clear angle of vision from the seat of the automobile, through the windshield thereof to the stage, free of obstruction from the automobiles ahead of it.[89]

[B] Overlooking the Underlying Business Method — The Physical Layout of Theater Deemed Obvious

The First Circuit Court of Appeals reversed. It expressed the issue before it as whether, given the idea of an open-air drive-in theater, inventive faculty was required to carry it out. The court found that Hollingshead's arrangement of parking stalls, with their tiered layout and individual wheel riser ramps was obvious and lacking invention. In the words of the court:

This arcuate arrangement of parking stalls in a lot is obviously only an adaptation to automobiles of the conventional arrangement of seats in a theater employed since ancient times to enable patrons to see the performance while looking comfortably ahead in normal sitting position without twisting the body or turning the head.

* * *

[T]here is nothing inventive in adapting the old arcuate arrangement of seats in a theater to automobiles in a parking lot as the means to achieve horizontal pointing.[90]

Even with particular reference to Hollingshead's wheel riser ramps, the court refused to find any inventive contribution:

Making every allowance for viewing the patentee's contribution in the light of hindsight, it seems to us that grading the ground upon which an

---

[89] U.S. Patent No. 1,909,537, "Drive-In Theater," issued to Hollingshead August 6, 1932, claim 1.

[90] 174 F.2d at 552.

automobile is to be placed for the purpose of giving it the tilt desired would be the first expedient to occur to anyone who put his mind on the problem.[91]

In short, the First Circuit Court of Appeals was not impressed with the physical layout and arrangement of Hollingshead's drive-in movie parking lot. It found the layout to be an obvious expedient — an obvious adaptation of the theater as it had existed in the times of the ancient Greeks. That, however, is not why the *Drive-in Theatres* case is important. It is important because of how the court placed into perspective the District Judge's finding that Hollingshead's invention was "ingeneous . . . practical . . . a completely new idea."

The court disposed of Hollingshead's ingenuity by casting his true contribution — conception of the drive-in movie — into the realm of nonstatutory subject matter. Having disposed of his idea of a drive-in theater, his solution to the practical problem how to see the screen through the tiny windshield of a 1939 Packard, the court considered obvious:

> Furthermore, the art did not have to wait for years for Hollingshead to discover the means for making drive-in theaters practically useful. So far as the record discloses Hollingshead was the first to hit upon the idea of a drive-in theater and consequently he was the first to appreciate the problems inherent in designing a practical one. And it did not take him long to solve those problems. Thus while the art may have had to wait for years for the idea or conception of a drive-in theater, the finding of invention in devising the means for carrying out the idea cannot be buttressed by a finding that Hollingshead was the first to supply the solution for a long-standing problem.[92]

When read in the context of the district court's finding, the appellate court must have found the drive-in theater was unique and original. However, within the constraints of the patent law as it perceived them, the appellate court could not place patentable significance on the business method inherent in Hollingshead's creation. Of the district court's findings, the appellate court stated,

> Viewing the opinion, findings and conclusions of the court below as a whole it seems evident that that court was so impressed with the novelty, usefulness, and commercial success of out-door drive-in theaters that it failed to focus its attention sharply upon the means employed to make such theaters practically useful.[93]

[C] Understanding *Loew's Drive-in v. Park-in Theatres*

Was the Appellate Court correct in its reversal of the District Court's conclusion? To be sure, the Court was constrained by weighty precedent. Citing

---

[91] 174 F.2d at 553.

[92] *Id.* at 553.

[93] *Id.* at 553.

the U.S. Supreme Court decisions in *Rubber-Tip Pencil Co. v. Howard*[94] and *Miller v. Eagle Manufacturing*,[95] the court had to distinguish between abstract idea and its practical implementation:

> An idea of itself is not patentable, but a new device by which it may be made practically useful is.[96]

> * * *

> It is not the result, effect, or purpose to be accomplished which constitutes invention, or entitles a party to a patent, but the mechanical means or instrumentalities by which the object sought is to be attained.[97]

> * * *

> Thus a system for the transaction of business, such, for example, as the cafeteria system for transacting the restaurant business, or similarly the open-air drive-in system for conducting the motion picture theater business, however novel, useful, or commercially successful is not patentable apart from the means for making the system practically useful, or carrying it out.[98]

There was certainly nothing wrong in the court's logic, provided one can accept the court's conclusion that the arrangement of stalls and risers was an obvious expedient. Could Hollingshead have claimed his invention differently, so that the ingenious aspects seen by the district court would strengthen his claim? The district court had been impressed with this "completely new idea of giving entertainment to people who could not ordinarily go to the theater." What if Hollingshead had claimed the invention as a method? That would have forced the issue, for it would have been far more difficult to declare the method obvious, given that the prior art was an indoor theater. Of course, that would also have forced the court to grapple with whether a method of providing entertainment was statutory subject matter.

§ 6.19 Converting Bank Checks into Promissory Notes — *In re Wiechers*

Wiechers applied November 4, 1960, for a patent on a method of converting a bank check into a promissory note with installment payments. The patent examiner rejected Wiechers's application and Wiechers appealed to the Board of Patent Appeals, where he failed to win a reversal. Wiechers then appealed to the Court of Customs and Patent Appeals. A five judge panel heard the case,

---

[94] 87 U.S. 498 (1874).

[95] 151 U.S. 186, 14 S. Ct. 310, 38 L. Ed. 121 (1894).

[96] Citing Rubber-Tip Pencil Co., 87 U.S. 498 (1874).

[97] Citing Miller, 151 U.S. at 201.

[98] 174 F.2d at 552.

resulting in a 3-2 split affirming the Patent Office's denial to grant the patent. Judge Rich wrote the opinion for the majority. Judge Rich would, some 33 years later, author the opinion in *State Street Bank v. Signature Financial* that dispelled the myth that business methods were not patentable.

Wiechers's basic idea was to provide banking customers with a new type of account that had both check writing and consumer loan features. The bank provided each participating customer with standard checks, which could be used in the standard way to draw upon funds in the customer's checking account. If the customer wished to borrow money on a promissory note, instead of drawing funds from his or her checking account, the customer simply affixed the stamp to the face of the check. When the stamped check was presented to the bank for payment, the bank withdrew funds from the customer's line of credit and would begin charging periodic installment payments against the customer's checking account.

[A] Claims at Issue in *In re Wiechers*

A claim to any invention is like a three-dimensional object being captured on two-dimensional paper. There are often different vantage points that reveal the invention in slightly different ways. In this case, Wiechers chose to describe his invention both as a method of producing a promissory note and as a method of converting a check into a promissory note. Claims 9 and 10 are exemplary.

> 9. A method of producing and treating a promissory note, comprising the steps of recording a maximum credit limit between a bank and a customer for an account of said customer, preparing a stamp with a personalized notation thereon identifying said customer in the records of said bank and recording said notation on said records, writing a bank check in favor of a payee, affixing said stamp to said check and passing said check to said payee, presenting said check to said bank for payment and making said payment, and making a marking in said account in the amount of said check for debiting said amount to said customer on a periodic installment payment basis until said amount is paid by said customer.
>
> 10. A method of converting a bank check into a promissory note, comprising the steps of making a deposit of money and writing a checking account between a bank and a bank customer to have the latter obtain bank credit and utilize a bank check for acquiring the credit upon writing the check, preparing means bearing identification indicia such that said means can be applied to said bank check and distributing said means to said customer, recording said indicia in the records of said bank under the name of said customer, writing a bank check in favor of a payee, affixing said identification indicia to said check at the selection of said customer when the latter desires to make a promissory note out of said check, passing said check to a payee for payment from said bank, and writing an entry in said account for debiting said account with periodic installment payments if said means is affixed to said check and for debiting said account with one full payment if no means is affixed to said check.

[B] Judge Rich Deems Check Certifying Scheme Is Obvious

Against Wiechers's claims, the patent examiner had cited the prior art U.S. Patent No. 1,112,654 to Pfleiderer. That patent described a scheme for certifying checks whereby the account holder would attaching a stamp to the check. The bank issued stamps bearing the customer's account number in dollar denominations corresponding to the customer's deposit. The presence of the stamps on the check indicated to the payee that there was money on deposit to cover the amount of the check.

Wiechers had argued that the invention allowed a customer to write a conventional bank check and have it honored as either a conventional check or as a promissory note. This dual feature, he argued, made the invention patentably different from the prior art. Judge Rich dismantled Wiechers's argument, first noting that, by Wiechers own admission, the checks were conventional. He then observed that the customer really has two accounts with the bank, a regular checking account and a loan or installment credit account. No doubt millions of Americans at that time had both a checking account and an installment credit account, and all of them used conventional checks. Thus, the only difference was the indicia used in Wiechers's system to tell the bank to which account the instrument is intended to apply.

Judge Rich saw little patentable significance in Wiechers's adhesive stamp indicia. The physical device itself, an adhesive stamp, was found in the prior art Pfleiderer reference. As to the function of this device, Judge Rich characterized it as merely providing a way for the customer to tell the bank for which of its two accounts the instrument was intended. As to that function, Judge Rich stated, "it is apparent that the bank must know — as it always has to know when a customer has a plurality of accounts — which account is being drawn against."[99]

Wiechers argued that this was more than just having two accounts because the stamp allowed the check to be honored not as a check but as a promissory note. That changed the check into something new and different. Judge Rich rejected this argument:

> [A]ppellant refers to "honoring" a check with an indication affixed to it "as a promissory note." The terminology here seems confused. Such a check is not *honored* by the bank as a promissory note. It is honored *as a check* in that the money is given to the payee, just as Pfleiderer's check is honored. It is thereafter treated *by the bank* as though it were a note drawn by the customer in favor of the bank and it is the customer who has to "honor" it as such.[100]

From the above, note how Judge Rich carved out the steps that occur after the check is honored, namely that the bank treats the check as a note which the customer must honor. He then characterizes these later steps as merely a matter of contractual relations between the parties:

---

[99] *In re* Wiechers, 347 F.2d 608, 610 (C.C.P.A. 1965).
[100] *Id.*

This is merely a matter of the contractual relations between the bank and its customer with whom it has agreed to advance money when, as, and if called for by checks drawn on the loan account. The checks remain checks.[101]

Wiechers's claims did make reference to the periodic installment payment. That, Wiechers argued, made his invention different from the prior art. Rich rejected the patentable significance of this installment payment feature as well, stating:

> [We consider] . . . that claimed feature to be immaterial to patentability presumably for the same reason as the examiner, though he did not state his reason. Repayment of loans in installments, however, has become one of the commonest business practices in the United States long prior to appellant's 1960 filing date. We seem to recall that an excess of installment credit was considered to be one of the primary causes of the Great Depression which commenced in 1929.[102]

On this final note, Judge Rich proclaimed the invention was obvious and the patent was denied.

The ruling of the Court was by no means unanimous. Rather, the vote was merely three to two against the patent applicant. Judge Smith wrote the dissenting opinion in which he criticized the majority for analyzing the various aspects of the claim in piecemeal fashion. He saw the majority as remiss in failing to consider the claimed subject matter as a whole. Interestingly, Judge Smith notes that the Court was in agreement that this claim was to a business — but that the statutory subject matter question was not before them.

> Initially, it should be made clear that what Wiechers is claiming here is, beyond question, *a method of doing business.* I agree with the majority that the issue of patentable subject matter under 35 U.S.C. § 101 is not before us.[103]

Judge Smith also agreed with the majority that Wiechers's checks and checking accounts were conventional. He further agreed that the individual differences between Wiechers's claimed method and the Pfleiderer certified check were obvious. He disagreed, however, with how the majority dissected Wiechers's claim instead of considering the claimed subject matter as a whole.

> I am in complete agreement with the majority that the individual differences between appellant's claimed method and the certified check patented by Pfleiderer are all fairly conventional steps and, when considered individually and out of context, may be said to be ''obvious.'' But it is not obviousness of the *differences* which section 103 contemplates, but obviousness of the

---

[101] *Id.*
[102] *Id.*
[103] *Id.* at 611.

claimed subject matter *as a whole.* And the appellant's claimed subject matter as a whole is a unique method of doing business which involves short-circuiting that marvel of modern-day society, the installment charge account, and its ubiquitous companion, the credit card.

* * *

But when, as in the instant case, the art is devoid of the barest suggestion of the very *heart* of a claimed idea, I feel the step from what the art *does* show to the holding of *obviousness* is over-long, and one not countenanced by section 103.[104]

[C] Understanding *In re Wiechers*

The dissent found Wiechers's scheme ingenious; the majority found it obvious. That makes *In re Wiechers* a good test case for study. At the heart of this difference in views between majority and minority is the business arrangement between bank and customer. It is the method by which bank and customer agreed to act towards each other. Judge Rich characterized the arrangement as contractual:

> This is merely a matter of the contractual relations between the bank and its customer with whom it has agreed to advance money . . . The checks remain checks.[105]

Judge Smith, writing for the dissent, characterized this as a patentably ingenious scheme.

> Where, in the sum total of the prior art, including the conventional banking and business practices mentioned by the majority, is there even the slightest suggestion of such an ingenious scheme?[106]

Comparing these different viewpoints, note that Judge Rich and the majority were focusing on the check, the physical device used in practicing the business method. Judge Smith was focusing on the business method itself. Which viewpoint is correct? To answer this, it may be helpful to look back to the claims Wiechers was making.

Wiechers's claims were drafted as method claims; however, the focus was on the paper instrument, the check. Claim 9 read "a method of producing and treating a promissory note . . ." Claim 10 read "a method of converting a bank check into a promissory note . . ." Wiechers chose to characterize his invention as a method of producing a piece of paper having certain properties (agreed to behind the scenes between the bank and its customer). In the preamble of his

---

[104] *Id.* at 610-611.
[105] *Id.* at 610.
[106] *Id.* at 612.

claims, Wiechers did not directly assert that his invention was a new financial scheme.

Perhaps Wiechers's patent attorney chose to focus on the paper instrument because of the belief, widely held at that time, that business methods were not patentable subject matter. However, the focus on the paper instrument did have the unfortunate side effect of trivializing what Wiechers had conceived. It allowed Judge Rich to spend the bulk of his analysis discussing the check and how the stamp was simply a message to tell the bank when to use the loan account. Quite telling was Judge Rich's summation: "The checks remain checks."

Judge Rich was not willing to give patentable weight to the underlying business arrangement that dictated how the parties actually behaved when the stamp was used. Under that arrangement the bank would debit the loan account for the amount on the instrument and would begin extracting installment payments from the checking account. Judge Rich wrote of this arrangement that it was merely part of the contractual relations between bank and customer:

> The customer must repay the advances. We see nothing unobvious, or novel, in this. We agree with the board that this is merely a matter of "private arrangement between the bank and the customer" and not an unobvious difference from the reference.[107]

Thus, given the way Wiechers characterized his invention, the majority was justified in ruling the invention was obvious in view of the prior art. Yet this is a close case. Wiechers's claims also contain recitation of many steps of the underlying business method. That justified the minority in finding the invention was an inventive new business scheme. Perhaps Wiechers could have drafted stronger business method claims — focusing more on the contractual relationship between the bank and its customer. However, Wiechers can hardly be criticized for having failed to do so. In 1960 when his application was filed, the Patent Office and most of the patent bar considered it axiomatic that business methods were not statutory subject matter. The Patent Office Manual of Patent Examining Procedure (MPEP) at § 706.03 for a long time contained evidence of this:

> 706.03(a) Though seemingly within the category of process or method, a method of doing business can be rejected as not being within the statutory classes. See *Hotel Security Checking Co. v. Lorraine Co.*, 160 F. 467 (2nd Cir. 1980) and *In re Wait*, 24 U.S.P.Q. (BNA) 88, 22 C.C.P.A. 822 (1934).[108]

The Patent Office has since removed this language from its Manual.

---

[107] *Id.*

[108] U.S. Patent Office, *Manual of Patent Examining Procedure*, § 706.03(a) (August 1993).

§ 6.20 Codes for Processing Transactions at the Grocery Checkout Register — *In re Howard*

In 1959 Howard applied for a patent broadly titled, "Method and Apparatus for Handling Material." The patent application described a method of pricing merchandise in retail grocery stores in which the merchant put coded labels on grocery items designating to which type or category the item belonged. Thus, milk might be assigned to code 4454, bread to code 4667, and so forth. At the checkout register, the checkout clerk enters the code of each item through a keyboard attached to a centralized "converter." The converter has all of the item codes stored in memory along with an associated price for each item. The converter is linked to the cash register so that the price of each purchased item would be fed to the cash register and ultimately printed on the sales slip.

The Howard application included both method claims and apparatus claims. The patent examiner allowed the apparatus claims, but rejected the method claims as being drawn to nonstatutory subject matter. The examiner found that the method claims "do not relate to any art but are merely directed to business techniques."[109] The Board of Appeals affirmed and Howard appealed to the Court of Customs and Patent Appeals.

[A] Claims at Issue in *In re Howard*

Method claim 2 is illustrative of the subject matter the examiner and the Board of Appeals found to be a nonstatutory business method.

> A method of handling a large plurality of materials of varying identities comprising the steps of printing labels having visible coded indicia thereon, applying said labels to separate items at points of origin thereof with the same indicia upon each of the identical items only, applying said coded indicia to a memory system at a collection and distribution point of coded items, also applying local price information on said items to said memory system in correspondence with said coded indicia thereon, registering the coded indicia of each item distributed and comparing same with the indicia on said memory system to obtain the corresponding prices thereof, and printing the coded indicia and corresponding price of each item distributed at the point of distribution as a sales slip for items distributed.[110]

[B] Avoiding the Business Method Issue — The Board of Appeals Rules that Electronic Lookup of Product Codes Is Obvious

The Board of Appeals had rejected Howard's method claims on two grounds: first, that the claimed method was old and well-known; second, that the claims were drawn to a nonstatutory method of doing business. The Court of

---

[109] *In re* Howard, 394 F.2d 869, 870 (C.C.P.A. 1968).
[110] *Id.*

Customs and Patent Appeals addressed only the first issue, finding the claimed method unpatentable for lack of novelty. Having affirmed the Patent Office on this ground, the Court found it unnecessary to consider whether a method of business is statutory subject matter:

> Our affirmance of this ground of the rejection makes it unnecessary to consider the issue of whether a method of doing business is inherently unpatentable.[111]

The Court based its ruling, that the claims lacked novelty, entirely on judicial notice. The Court took judicial notice of practices dating back to the corner grocery store:

> It is a matter of common practice of wide notoriety, well within the ambit of judicial cognizance, for retail outlets to list by code or otherwise various items stocked for sale, together with the price assigned to each item, so as to enable the clerk or sales person to ascertain the charge to the customer. This practice relates back to the proverbial country merchant who has all but passed from the scene, but has his present-day counterpart in this context in the supermarket cashier who has a price list of advertised "specials" taped to his register.[112]

Howard's counsel had argued that the claimed method was different in that it used electrical means to look up the price of the item. The code stored in memory was compared with the code entered by the cashier through the keyboard allowing the converter to look up the price. Unfortunately, this argument carried little force because they conceded in their brief that "The claimed comparison may be done electrically or in any other way." The Court seized upon this admission and gave the electrical comparison no patentable significance.

Sitting on the panel for this decision was Senior District Judge Kirkpatrick from the Eastern District of Pennsylvania. As a trial judge, Judge Kirkpatrick was well acquainted with how facts are proved. It concerned him that the entire decision of the majority was based on judicial notice. He wrote:

> I do not see how, without any evidence, we can use the doctrine of judicial notice to find that a system exists which anticipates that of the application or is nearly like it as to make the application an obvious variation . . . Without some concrete evidence (of which there is none) I do not think that it is possible to find that the system of this application is old or that it is obvious under Section 103.[113]

---

[111] *Id.* at 872.
[112] *Id.* at 870.
[113] *Id.* at 872.

Judge Kirkpatrick did not disagree with the result, however. He belived the claims should have been rejected as being directed to an unpatentable business method.

> I would place the affirmance of the board's decision upon the ground that the application discloses merely a method of doing business and is therefore for an unpatentable invention.[114]

[C] Understanding *In re Howard*

Why was there no prior art in evidence upon which the invention could be rejected without resorting to judicial notice? That may have something to do with what resources the examiner had available — remember, the application was filed in 1959. While the Patent Office had, and still has, an excellent technical library, that technical library has science and technology as its focus. Finding a business method concept in a science and technology library, particularly without today's electronic searching tools, is next to impossible. Perhaps the examiner was simply unable to identify a good source for prior art.

This lack of prior art forced the Court to confront the business method issue, although the majority chose to duck it. Reading between the lines, the majority did not want to allow these method claims, but it did not want to endorse the business method argument either. So it used judicial notice to find the invention was obvious.

Reading this decision decades later, it is surprising to see a court use a list of prices taped to the cash register as prior art to reject claims to an electronic system — unless the claims were so very broad that they indeed read upon the taped list. Here the patent applicant was his own worst enemy. By arguing that the electrical comparison (price check lookup) "may be done electrically or in any other way," the applicant, in effect, conceded that his claims were so broad that they did cover the checkout clerk consulting a list to make the product code-price comparison.

§ 6.21 Performing Traffic Studies on Telephone Lines — *In re Waldbaum*

A discussion of what is statutory subject matter often leads to a more fundamental question, what is art. The U.S. Constitution states at Article I, Section 8, Clause 8 that Congress shall have the power "to promote the progress of science and useful arts." The constitutional term "useful arts" has been analyzed many times by the courts. Sometimes the term has been equated with "technological arts," a term that has seemed to help courts with the patentability of computer software, but one that is not as helpful in grappling with the patentabliltiy of business methods.

---

[114] *Id.*

The issue arose in *In re Waldbaum*[115] over a disputed claim to a computer method of analyzing data to determine how many number 1's a given data word contained. The method was useful in performing traffic studies on telephone lines. The Patent Office had rejected the Waldbaum invention as nonstatutory; the Board of Patent Appeals had affirmed. Waldbaum appealed to the Court of Customs and Patent Appeals.

[A] The Claims at Issue in *In re Waldbaum*

Claim 1 illustrates the scope of the invention Waldbaum sought to patent.

1. A method for controlling the operation of a data processor to determine the number of 1's in a data word; said data processor including a memory for storing data and instruction words at respective addresses; means for normally controlling the sequential execution of successively addressed instruction words; a plurality of registers; means for storing memory data words in said registers; means for performing logical operations on data words in said registers; and means responsive to the execution of a predetermined instruction word for examining the data word contained in a predetermined first one of said registers, changing the rightmost 1 in said first register to a 0 if said register contains at least one 1, controlling a transfer to the instruction word at a specified address if said first register contains all 0's, and storing in a predetermined second one of said registers the address of the following instruction word if said transfer is made; comprising the steps of:

> (1) controlling said storing means to store a memory data word whose number of 1's must be counted in said first register.
> (2) controlling the data processor to execute a series of identical ones of said predetermined instruction word, and
> (3) comparing the address of the first of the instruction words in said series with the content of said second register when a transfer is made during the execution of one of the instruction words in said series to derive the number of 1's in said data word.

[B] Mathematical Algorithm Deemed Statutory, Only to Be Reversed by Supreme Court

The Board had rejected Waldbaum's claims under 35 U.S.C. § 101 as a mental act of programming a mathematical solution to a mathematical algorithm. The Board reasoned that the manipulations required by Walbaum's method were already built into the prior art computer, thus the Waldbaum invention was not a new use of an old apparatus, but the same use implicit in the old apparatus. The Board considered the only difference in the Waldbaum method to be the

---

[115] 457 F.2d 997 (C.C.P.A. 1972).

interpretation in the mind accorded to the numerical data represented by the computer.

The Court of Customs and Patent Appeals reversed. It held that the patent statute did not place any special retraints on inventions characterized as a new use of an old apparatus. The Court interpreted the statutory subject matter requirements of 35 U.S.C. § 101 as encompassing the "technological arts," which included computers:

> With regard to the "mental steps" rejection, whether appellant's process is a "statutory" invention depends on whether it is within the "technological arts." The phrase "technological arts," as we have used it, is synonymous with the phrase "useful arts" as it appears in Article I, Section 8 of the Constitution . . . It is clear that appellant's process which is useful in the internal operation of computer systems, is within the "useful arts." . . . Appellant's process is therefore a statutory process within the meaning of 35 U.S.C. § 101.[116]

The Court's reversal did not stand, however. When the Court reversed the Board as noted above, it did so, in part, based on its reasoning in *In re Benson*.[117] However, the Court's decision in that case was reversed by the U.S. Supreme Court in *Gottschalk v. Benson*.[118] Following the Supreme Court decision, the Patent Office moved to vacate the Court's decision in *In re Waldbaum*, and the Court agreed. In view of *Benson* it held that the claims were not limited to any practical application, other than in connection with a computer, and therefore a patent on the claims would, in effect, be a patent on the algorithm itself.

Waldbaum argued that he was not claiming a computer program, but a "machine process."[119] Waldbaum sought to distinguish *Benson* in that Benson's claims had recited mathematical manipulations, whereas Waldbaum's claims recited machine steps. The Court rejected this argument. It noted that Benson's claims had recited a "re-entrant shift register" but the Supreme Court ignored those machine limitations in concluding that the Benson claim would wholly preempt the mathematical algorithm.[120]

[C] Understanding *In re Waldbaum*

In re Waldbaum contains part of the historic record of the struggle between the U.S. Patent Office and the Court of Customs and Patent Appeals over what should be done about patentability of computer software. *Benson* was unquestionably a setback for software patent advocates, a setback that lasted more than a

---

[116] *Id.* at 1003.
[117] 441 F.2d 682 (C.C.P.A. 1971).
[118] 409 U.S. 63 (1972).
[119] *In re* Waldbaum, 559 F.2d 611, 616 (C.C.P.A. 1977).
[120] *Id.*

decade. Ultimately, however, the Patent Office softened its position and permitted software patents to pass through the statutory subject matter gauntlet.

The business method patent raises one of the same issues present in *Waldbaum*, namely what is a useful art. The Waldbaum court, in its original decision, equated useful arts with technological arts. Judge Rich, who claims credit for having equated these two terms, had this to say about the issue:

> [T]he court, in applying the test for determining whether the claimed process is "statutory" under 35 U.S.C. § 101, which test is stated to be whether the process is within the "technological arts," says:
>
>> The phrase "technological arts," as we have used it, is synonymous with the phrase "useful arts" as it appears in Article I, Section 8 of the Constitution.
>
> As the originator of that "test" in *In re Musgrave*,[121] I hereby express my agreement with the above-quoted statement. The phrase "useful arts" which was written into the Consitution conjures up images of the Franklin stove, horse collars, and buggy whips. The term "technological arts" was selected in *Musgrave* as probably having a connotation in these times roughly equivalent to that which "useful arts" had in the eighteenth century. No new legal concept was intended . . . Now we have come full circle in pointing out that the intention all along has been to convey the same idea and to occupy whatever ground the Constitution permits with respect to the categories of patentable subject matter named in section 101.[122]

Are business methods within the "technological arts"? Certainly, business methods that sport the technological garb of the latest computer or internet fashion may be characterized as within the technological arts. However, to conclude that business method patents require some "technological" ingredient may be wrong. It all depends on how broadly the term "technological" is defined. The term technology comes from the Greek word "techne," meaning the skill one needs to practice an art, where "art" means made by the hand, that is, artificial. Technology thus spans a wide continuum of the artificial creations of humankind. It is not limited to what would today be considered high technology, or cutting edge technology, but rather includes even the mundane note card and postage stamp.

§ 6.22   Distribution of Recorded Audio — *In re Fox*

In 1967 Fox applied for a patent on a method of distributing recorded audio tapes. His application stated that the method was generally useful for educational purposes. Under Fox's scheme, a lecture would be recorded and dubbed onto several master tapes. The master tapes were then shipped to libraries where

---

[121] 57 CCPA 1352, 1367, 431 F.2d 882, 893, 167 USPQ 280, 289 (1970).
[122] 559 F.2d at 1003-1004.

students could make copies of the master tapes and take the copies home for study. When finished, the students returned the tapes to the library where the tapes were reused by students wishing to record and study different program material.

The Patent Examiner rejected Fox's claims on two grounds. The Examiner found that the subject matter was nonstatutory under 35 U.S.C. § 101. The Examiner also found that the claimed subject matter was obvious, without citing any prior art in support. Fox appealed and the Board of Appeals affirmed the Examiner's decision on both grounds. Fox then took his case to the Court of Customs and Patent Appeals. A five judge panel heard the case. Judge Rich wrote the opinion of the court. The court affirmed.

[A] The Claims at Issue in *In re Fox*

Fox's application had one independent claim and four dependent claims. The independent claim read as follows:

> A method of transmitting audio information from an origin location to a first remote location and then, in multiple copies of such audio information, to a plurality of further remote locations comprising the steps of:
>
>> (a) Recording at an origin location a plurality of programs of individual master tapes, one for each program, and packaging said individual master tapes in individual master tape cartridges;
>>
>> (b) Transmitting a plurality of said recorded programs in said master tape cartridges to a first remote location;
>>
>> (c) Selectively recording at said first remote location onto at least one tape in a slave cartridge one said recorded program from a selected one of said master tape cartridges at a recording speed greater than the audio playback speed of said recorded program;
>>
>> (d) Transporting said recorded slave tape cartridge from said first remote location to a further remote location;
>>
>> (e) Reproducing the audio information on said transported slave tape cartridge at said further remote location;
>>
>> (f) Transporting said slave tape cartridge back from said further remote location to said first remote location; and
>>
>> (g) Repeating steps (c) through (f) herein repeatedly with different programs.

[B] Judge Rich Finds Recorded Audio Distribution Scheme Is Obvious

Like the decision in *In re Howard*, discussed in § 6.20, the Examiner had found no prior art upon which to base the obviousness determination. Like his predecessor in *In re Howard*, Judge Rich relied upon judicial notice to affirm the Examiner's finding of obviousness. He did so, expressly declining to rule on the statutory subject matter issue.

Judge Rich was aided in his decision by the admission by Fox in his application that the recording devices used were of conventional design, and that this conventional hardware enabled wide dissemination of audio information "in an obvious manner." Judge Rich selected this quote from the Fox application to include in his opinion:

> The recording device is of conventional design and the structural details thereof are not necessary for an understanding of the method of the present invention. For purposes of this description it suffices to note that the recording device which is employed is capable of accommodating at least one, and preferably three, so-called slave tapes in cartridges for recording an audio program of a selected one of the master tapes and of achieving this recording at a fraction of the 30 minute play-back time of the program. This, in an obvious manner, enables wide dissemination of the audio information of the master tapes.[123]

Fox's counsel spent most of its energy arguing the statutory subject matter rejection. On the obviousness issue, counsel had simply argued that the applicant "is unaware of any system where individual master tapes are packaged in individual master tape cartridges." In essence, counsel argued that the distribution of individual *master* tapes was different from the prior art.

Judge Rich rejected this argument, stating, "We find no magic in the term 'master tapes.' They are simply tape recordings."[124] Thus, Judge Rich viewed the Fox invention as no different from the conventional tape lending library, where the library duplicates the tapes for a number of simultaneous borrowers.

[C] Understanding *In re Fox*

Fox lost round one by conceding in the patent application that, "This, *in an obvious manner*, enables wide dissemination of the audio information of the master tapes."[125] Since the recording equipment was also admittedly conventional, Fox's concession that wide dissemination was obvious left little upon which to hang his patentability argument. If there truly was a nonobvious distinction between Fox's master tapes and conventional tapes, he was not able to articulate it.

Judge Rich noted that Fox devoted a major portion of his brief to the "method of doing business" question. We know from his later decision in *State Street Bank v. Signature Financial Corp.*[126] that Judge Rich was not predisposed to favor the method of doing business rejection. However, this was certainly not the case for Judge Rich to make his point.

---

[123] *In re* Fox, 471 F.2d 1405 (C.C.P.A. 1973).

[124] *Id.* at 1406.

[125] *Id.* at 1406.

[126] State Street Bank v. Signature Financial Corp., 149 F.3d 1368 (Fed. Cir. 1998).

§ 6.23 Assigning Budget Categories to Bank Reports — *In re Johnston*

Anyone who has used a computer program to organize personal financial accounts — to create a budget or to generate records for income tax purposes — has certainly set up different categories to which expenditures may be assigned. Some categories, such as home mortgage interest or medical expenses may have income tax implications; other categories, such as utilities, groceries, and entertainment may have other budgetary implications.

In March 1967, Thomas R. Johnston filed a patent application seeking the right to exclude others from making automatic data processing equipment that would empower banks to create reports for its customers much like the reports today's personal financial software programs can do. Johnston's idea was to have the customer handwrite or machine print category codes on checks before tendering them to the payee.

When the bank received the check for payment, it recorded the account code in magnetic ink or special optical font on the face along with the amount of the check. Banks were already in the practice of recording check amounts on the face of each check, so that they could be read by automated processing equipment and input into their computer. Johnston's improvement allowed category information to be input into the computer at the same time. Thereafter, using this added information the bank could print reports for the customer, showing exactly how much was spent in each of the customer's assigned categories.

Little did Johnston know, in 1967 when he filed his application, that his invention would become the subject of a nine year tug-of-war stretching all the way to the U.S. Supreme Court. The final outcome? Johnston's claims were rejected.

[A] Claims at Issue in *In re Johnston*

Johnston's claims were to an apparatus, an automatic data processing equipment employing a digital computer. Claim 20 is exemplary. Be forewarned, this is a lengthy claim. Paragraph formatting has been added to aid readability.

> 20. A record-keeping machine system for financial accounts, said system comprising a data processor including a memory for storing combinatorially coded signal records;
>
> > a processor for combining and comparing the coded signals of said records;
> >
> > > input and output devices;
> > >
> > > and a control system for supervising and directing said other parts of said data processor to perform a certain sequence of operations;
> > >
> > > said memory including a storage file of a plurality of machine-readable records formed of coded combinatorial signals, a master

section of said file including a plurality of predeterminedly sequenced sub-files of said records, each said sub-file including a different signal group record for a certain one of a plurality of account identifications, a plurality of predeterminedly sequenced different signal group records for different record-keeping category codes individually associated with each of said account identification records, a plurality of different group records each for an amount total associated with a different one of said category records, and a plurality of totalizing operation records individually associated with certain pluralities of said category records and sequenced with the associated category records, a transaction section of said file including different pluralities of sub-files of said records, each of said transaction sub-files including one of said account identification signal group records, one of said associated category signal group records, and a separate transaction amount signal group record associated with each of said category records;

said control system including means for directing the processing of said files in accordance with the predetermined sequences of said master sub-files and of said category records of each sub-file, said processing directing means including means for locating in said master file section, for each of said transaction sub-files, that one of said category records corresponding to the transaction account identification record and the transaction category record and for augmenting the associated master amount total record by the amount records of the corresponding transaction sub-file; and

means for producing an output record for each of said master sub-files;

said output record producing means including means for identifying each of said master sub-file category records and for producing an output listing of the sub-file category codes and of the associated category amount total from said records, means for directing the accumulation of the amount totals of successive ones of said category records, and means for identifying each of said totalizing records and for producing an output record of said accumulation of totals of the preceding category records.

[B] The Board of Appeals Rules that Assigning Budget Categories Falls Within the Nonstatutory Liberal Arts

The Patent Examiner had rejected Johnston's claims under 35 U.S.C. § 112 and 35 U.S.C. § 102. However, the Board of Appeals did not consider the underlying basis of these rejections to be valid. Instead, the Board entered several new grounds of rejection, two of which were based on 35 U.S.C. § 112, but for different reasons than the Examiner had articulated. A third ground of rejection was under 35 U.S.C. § 103. A fourth ground was under 35 U.S.C. § 101.

The Board first found the claims failed to particularly point out and distinctly claim the invention, as required by the second paragraph of 35 U.S.C. § 112. It was the Board's view that Johnston's invention was not an automatic data processing apparatus — as stated in the preamble of Johnston's claims — but rather a broad system of keeping financial records. The Board saw Johnston's invention as "in effect, the relationship of a bank and its customers, not any particular configuration of business machinery."[127] Thus the claims were improperly drafted, in the Board's view, because they recited a business machine apparatus, instead of Johnston's true invention.

The Board also found the claims were indefinite, under the second paragraph of 35 U.S.C. § 112. The Board reasoned that portions of Johnston's apparatus were defined as means for identifying an account or type of transaction, and that the "identifying" function was performed in the mind of the user and not by the apparatus. In essence, the Board considered the record stored in computer memory to have no concrete "meaning" until the human mind gave it one.

The Board then applied the following precedent from *In re Foster*[128] to hold that the claims were indefinite because they encompassed a human being as part of the claimed apparatus:

> Judge Baldwin in the second *Prater* opinion made it clear that means-plus-function language such as that used in the present apparatus claim does not encompass the human being as the "means" or part thereof.[129]

The Board further rejected the claims as being obvious under 35 U.S.C. § 103. It relied on the prior art Dirks patent[130] that disclosed a networked data processing system for cost accounting, featuring different account numbers in which transactions were kept and updated.

Finally, the Board rejected the claims as being nonstatutory subject matter under 35 U.S.C. § 101. It stated that patentable subject matter can constitutionally flow only from the "technological arts." This invention, the Board ruled, was not within the technological arts, but rather within the "liberal arts, such as social or political sciences, humanities, music and art." Thus, it was not patentable subject matter.

[C] The CCPA Reverses the Board — Judge Rich Troubled by Decision

The Court of Customs and Patent Appeals reversed the Board. A three-to-two majority of the Court of Customs and Patent Appeals agreed that Johnston's claims should be allowed. In reaching this conclusion, the majority rejected the Board's finding that Johnston was claiming the "relationship of a bank and its customer." The majority stated quite simply, "Upon reading the appealed

---

[127] *In re* Johnston, 502 F.2d 765, 768 (C.C.P.A. 1974).
[128] 438 F.2d 1011 (C.C.P.A. 1971).
[129] 502 F.2d at 769.
[130] U.S. Patent No. 3,343,133, issued September 19, 1967.

apparatus claims we cannot perceive how any such 'relationship' is being claimed."[131]

The majority also rejected the Board's other § 112 finding that the human thought process was being claimed:

> Appellant's specification makes it quite clear that the claimed *apparatus* automatically performs the identifying operations referred to by the board; those operations are neither actually performed by a human being, nor can we imagine how they could be realistically performed by a human being.[132]

The court likewise did not accept the Board's finding that the invention was nonstatutory for being within the liberal arts, as opposed to the technological arts. The Constitution does not use the phrase "technological arts." Rather it uses the broader phrase "useful arts." In a prior decision, the Court of Customs and Patent Appeals had found useful arts and technological arts were synonymous.[133] Nevertheless, the court refused to place this computer apparatus outside the technological arts.

> With regard to the rejection under § 101, we cannot agree with the board that the apparatus of the appealed claims is not within the "technological arts." Record-keeping *machine* systems are clearly within the "technological arts." . . . Furthermore, we are not aware of, nor can we locate, any dictionary which would define a *machine* system as within the purview of the "liberal arts." [in the accompanying footnote] . . . We likewise do not agree with the board that banking is a "social science."[134]

The majority also disposed of the Board's § 103 rejection, finding that the claimed "category codes" were different from the "account codes" described in the Dirks prior art patent.

Judges Markey and Rich took exception with the majority view. Judge Markey believed that the claims should have been rejected on obviousness grounds (35 U.S.C. § 103); Judge Rich believed the claims should have been rejected on statutory subject matter grounds (35 U.S.C. § 101). Judge Markey's point was that the majority was treating the obviousness issue using the novelty test. He wrote:

> The majority opinion finds differences between the claimed subject matter and the prior bank data processing system. Thereupon, the majority opinion appears to treat the obviousness rejection below as though it were based on 35 U.S.C. § 102. There is no doubt, as the board recognized, that appellant's programmed computer differs from the conventional programmed computers previously employed in banks. But the question under 35 U.S.C. § 103 is

---

[131] *In re* Johnston, 502 F.2d at 770.

[132] *Id.*

[133] *In re* Waldbaum, 457 F.2d 997 (C.C.P.A. 1972).

[134] *In re* Johnston, 502 F.2d at 771.

whether appellant's programmed computers as claimed, i.e., his "record-keeping machine system," would have been obvious to those skilled in the art applicable to such record-keeping machine systems. In my view one skilled in the art, presented with the conventional machine system, would have found it obvious, without knowledge of appellant's disclosure, to have modified the system as set forth in the appealed claims.[135]

Judge Rich struggled with this case more than the other judges. He had authored the appellate decision that the U.S. Supreme Court had reversed in *Gottschalk v. Benson*.[136] This, perhaps, caused him to dig deeper into the *Benson* holding than the other judges. It is not easy being reversed by the Supreme Court, and Judge Rich had to accept the Supreme Court's view, even if he did not agree with it. Thus he felt the Court was compelled to apply the "thrust" of the *Benson* case to the claims of this case. *Benson*, in Judge Rich's view, had held that computer programs were not patentable. Inasmuch as the heart of Johnston's invention was a computer program, Rich concluded that the Court could not sustain Johnston's claims. He wrote:

> As the author of the opinion of this court in *Benson*, which was wholly reversed, I have not been persuaded by anything the Supreme Court said that we made a "wrong" decision and I therefore do not agree with the Supreme Court's decision. But that is entirely beside the point. Under our judicial system, it is the duty of a judge of a lower court to try to follow in spirit decisions of the Supreme Court — that is to say, their "thrust." I do not deem it to be my province as a judge to assume an advocate's role and argue the rightness or wrongness of what the Court has decided or to participate in what I regard as the inconsistent decision here, supported by a bare majority which tries in vain, and only briefly, to distinguish *Benson* by discussing form rather than substance and various irrelevancies like pre-*Benson* decisions of this court, the banking business, social science, and the liberal arts.[137]

Judge Rich then went on to make a plea to the Supreme Court to clarify *Benson*:

> I deem it to be the Supreme Court's prerogative to set the limits on *Benson*, which was broadly based. I hope it will do so. As John W. Davis, erstwhile outstanding Solicitor General, once said, "The first requirement of any judicial opinion is utter clarity."[138]

Unfortunately, insofar as this case was concerned, Judge Rich's plea went unanswered.

---

[135] 502 F.2d at 772.

[136] Gottschalk v. Benson, 409 U.S. 63 (1972).

[137] *In re* Johnston, 502 F.2d at 774.

[138] *Id.* at 774, citing Harbaugh, *Lawyer's Lawyer, The Life of John W. Davis*, Oxford University Press (New York: 1973), 108.

[D] The Supreme Court Has the Last Say in *Dann v. Johnston*—Patent Declared Invalid as Obvious

Johnston's victory in the appellate court did not last. The Commissioner of the Patents petitioned the Supreme Court for *certiorari* and the Supreme Court agreed to hear the case. Seven Justices took part in the decision and ruled unanimously that the Johnston invention would have been obvious to one of skill in this art. The Court reversed the appellate court solely on 35 U.S.C. § 103 grounds, without ruling on the 35 U.S.C. § 101 issue.

The Court agreed with Judge Markey's dissenting opinion. Justice Marshall, writing for the unanimous Court, stated:

> However, it must be remembered that the "obviousness" test of § 103 is not one which turns on whether an invention is equivalent to some element in the prior art but rather whether the differences between the prior art and the subject matter in question "is a difference sufficient to render the claimed subject matter unobvious to one skilled in the applicable art . . ." (quoting from Judge Markey's dissenting opinion).[139]

The applicability (or lack thereof) of *Benson* and the § 101 issue had been extensively briefed, both by the parties and by various *amici*. In a single sentence, the Court stated that it saw no need to address that question:

> Petitioner and respondent, as well as various amici, have presented lengthy arguments addressed to the question of the general patentability of computer programs. *Cf. Gottschalk v. Benson.* We find no need to treat that question in this case, however, because we conclude that in any event respondent's system is unpatentable on grounds of obviousness.[140]

The Supreme Court's resolution of Johnston's nine year struggle must have frustrated Judge Rich. He no doubt wished for some clarification of how far the Supreme Court was willing to extend *Benson*. The Court refused to say, only noting that the *Benson* decision was limited to its facts:

> In order to hold respondent's invention to be patentable, the CCPA also found it necessary to distinguish this Court's decision in *Gottschalk v. Benson*, . . . handed down some 13 months subsequent to the Board's ruling in the instant case . . . As we observed, "[t]he claims were not limited to any particular art or technology, to any particular apparatus or machinery, or to any particular end use." Our limited holding . . . was that respondent's method was not a patentable "process" as that term is defined in 35 USC § 100(b).[141]

---

[139] Dann v. Johnston, 425 U.S. 219, 228 (1976).

[140] *Id.* at 220.

[141] *Id.* at 224.

[E] Understanding *In re Johnston*

How could such a simple idea as recording category information on the face of a check, so that it could be input into a computer, attract so many differences of opinion? The views expressed by the various boards and courts who considered the issue covered a surprisingly wide range:

> Chapter 7: The invention encompasses human beings as part of an apparatus;

> Chapter 8: The invention lies within the domain of the liberal arts;

> Chapter 9: The invention is a method of conducting banking business;

> Chapter 10: The invention is a patentably inventive machine;

> Chapter 11: The invention is a nonstatutory computer program;

> Chapter 12: The difference between the invention and existing computerized banking and bookkeeping systems would have been obvious to one of ordinary skill in the art.

Clearly the Supreme Court had the last say, and it chose not to make this case into another landmark *Benson* decision. No doubt, much of the reason this case turned into a nine year tug-of-war was because of *Benson*. *Benson* had changed the course the Court of Customs and Patent Appeals had been following with regard to computer-related inventions. Everyone, litigants, amici, the Patent Board of Appeals, the Patent Commissioner, the Court of Customs and Patent Appeals each had their own views on what *Benson* truly stood for. It would take another nine years after *Benson* before the Supreme Court would finally give us the answer.[142]

§ 6.24 Assigning Priorities in Data Processing Systems — *In re Chatfield*

In 1972 Intel announced its 8008 microprocessor, its first commercial 8-bit microprocessor; Hewlett-Packard announced its HP-35 calculator, the first calculator to perform trigonometric functions to rival the slide rule; Nolan Bushnell introduced Pong, the early video game that launched the Atari video game company; and Glen F. Chatfield of Duquesne Systems, Inc. of Pittsburgh, Pennsylvania applied for a patent on an improved way of scheduling program operation in a multiprocessor computer environment.

When Chatfield made his invention, there were several known ways to allow several programs to share a piece of computer hardware. One was to have the Central Processing Unit (CPU) assign priorities to the programs, so that one program would always have priority over another. Another was to assign priorities to the programs themselves. The problem was that these approaches

---

[142] *See* Diamond v. Diehr, 450 U.S. 175 (1981).

lacked flexibility. Once a program began to run, it was not possible to alter the priority scheme.

Chatfield's solution dynamically evaluated and reassigned program priorities as the programs executed. He accomplished this by having the running program periodically halt, during which time usage data was analyzed to determine if a shift in priorities was warranted. In his patent, he characterized the invention as a new method of operating a computer system. The Examiner cited no prior art against Chatfield's claims. The Examiner rejected Chatfield's application as being drawn to nonstatutory subject matter. The Board of Appeals affirmed the Examiner's ruling and Chatfield appealed to the Court of Customs and Patent Appeals.

[A] The Claims at Issue in *In re Chatfield*

Chatfield's application had 14 claims. Claim 1 is exemplary:

1. A method of operating a computing system upon more than one processing program concurrently for improving total resource utilization, said computing system comprising at least one central processing unit, having a logic and main memory function and an interrupt capability, and a plurality of peripheral resources capable of functioning in parallel with the central processing unit, comprising steps for:

   (1) accumulating system utilization data for at least one processing program for at least one resource, said system utilization data comprising resource activity and/or resource degradation data;

   (2)(a) at spaced intervals interrupting the processing programs and analyzing the system utilization of at least one processing program;

   (2)(b) based on this analysis regulating resource access by assigning an individual resource access priority and/or preventing resource access altogether in an unlike manner to at least two resources for at least one processing program to increase thruput;

   (3) resuming the operation of the computing systems on the processing programs; and,

   (4) continually repeating steps (1) to (3).

[B] Chief Judge Markey Finds Data Processing Method Statutory, Judge Rich Dissents — Still Troubled by His Reversal in *Benson*?

The Court of Customs and Patent Appeals ruled to allow Chatfield's claims by a three-to-two majority. Judge Markey wrote the opinion for the majority. The principal issue that caused the three-to-two split was whether a computer program could be the subject of a patent. That issue arose because of the Supreme Court's 1972 decision in *Gottschalk v. Benson*. Judge Markey and the majority believed that Chatfield's claims were proper statutory subject matter because they were method claims within the literal terms of the patent statute and because they did not fall within the judicially created categories determined to be nonstatutory.

Although this case had nothing to do with business methods, per se, Judge Markey included business methods within the subject matter that had been judicially carved out as nonstatutory. Judge Markey wrote:

> Some inventions, however meritorious, do not constitute patentable subject matter, e.g., printed matter, *In re Miller*; methods of doing business, *In re Wait*; purely mental steps, *In re Prater*, *In re Musgrave*; naturally occurring phenomena or laws of nature, *O'Reilly v. Morse*; a mathematical formula and the algorithm therefor, *Gottschalk v. Benson*.[143]

In the above quote, Judge Markey cited the 1934 *In re Wait* decision for the proposition that business methods were not proper statutory subject matter. That case, which is discussed in § 6.16, had involved a system for transmitting stock offers and acceptance prices via electrical means. The patent claims were rejected for lack of novelty, not because they recited business method subject matter.

The minority, represented by Judges Rich and Lane, believed that Chatfield's claims could not be sustained because they were computer programs. Judge Rich wrote the minority position. Judge Rich took the position that the Supreme Court in *Benson* had placed a lid over all computer software subject matter, prohibiting it from being patented. The majority read *Benson* more restrictively to apply only to the specific facts presented in *Benson*. Judge Rich clearly was not happy about *Benson*. He had written the lower court opinion in *In re Benson*[144] which the Supreme Court had reversed. Perhaps wishing to goad the Supreme Court into clarifying its intentions in *Benson*, Judge Rich took the hard line view that until properly explained by the Supreme Court, or overridden by Congress, no computer software of any kind could be patented. Judge Rich wrote:

> My colleagues of the majority take a narrower view of *Benson* and arrive at an opposite result in these two new cases, as another majority did in *Johnston*. This, to me, signals an urgent need to settle the question of patent protection for software by higher authority than this court so that the Patent and Trademark Office, the Federal judiciary as a whole, and the data-processing industry (hardware and software both) may know what the law on software patentability is. It is a socioeconomic issue with an impact of considerable magnitude, particularly on the practical operation of the Patent and Trademark Office. It is obviously not going to be finally decided in this court. Two Commissioners (*Gottschalk* and *Dann*) have obtained review of our decisions in the Supreme Court and in both instances (*Benson* and *Johnston*) obtained reversals; but here we are in two more cases, three-to-two, reversing findings by the Office that program inventions are non-

---

[143] *In re* Chatfield, 545 F.2d 152, 157 (C.C.P.A. 1976).
[144] 441 F.2d 682 (1971).

statutory under § 101. It seems to me like taking the problem of school segregation to court on a case-by-case basis, one school at a time.[145]

[C] Understanding *In re Chatfield*

In re Chatfield stands as an historic mile marker along the all but forgotten road that first lead to software patentability. The decision has very little to say about the patentability of business methods, and what it does say is wrong. Judge Markey was making the point that there is some subject matter that is off limits to the patent system. In that, Judge Markey is correct. The Supreme Court has articulated that abstract ideas, laws of nature and natural phenomena lie outside the domain of the patent law. However, when the cases are carefully studied, one finds business methods have not been judicially ruled off limits to patent protection.

Admittedly, it is possible to express a business method concept so broadly that it amounts to nothing but an abstract idea. As the second circuit court of appeals stated back in 1908, ''advice is not patentable.''[146] That statement seems to hold true, even today. Business advice, medical advice, golf swing advice, all of these seem to lie in the domain of the abstract idea. They lie there until the advice is taken up and put into practice. Once that occurs, the abstract idea becomes a useful, concrete, and tangible result, which does constitute patentable subject matter.

§ 6.25   Controlling Oil Refinery Plants — *In re Deutsch*

On November 2, 1970, Murray L. Deutsch filed a patent application entitled, ''Multi-Unit Optimization.'' The application, which he assigned to Mobil Oil Corporation, described a computerized system and method for controlling multiple plants — for example, oil refineries at different geographic locations. His method involved a technique to optimize the performance of a system of plants, while at the same time optimizing the performance of each plant and each processing unit within each plant.

Control system engineers have long used closed loop control to optimize the performance of a system. Closed loop control uses a process called ''feedback'' to regulate what goes into the system, based on what is coming out. Using closed loop control, engineers can optimize the system and cause it to perform in a controlled and desired way. Deutsch devised a system employing several closed loops, nested inside each other like so many Russian matryoshka dolls. At the outer, system-level loop he envisioned a system of plants; within each plant he envisioned an inner, plant-level loop comprising a system of individual processing units. Each individual processing unit itself comprised an innermost unit-level control loop.

---

[145] *In re* Chatfield, 545 F.2d at 161.
[146] Hotel Security Checking Co. v. Lorraine Co., 160 F. 467, 469 (2nd Cir. 1908).

Deutsch's method continuously gathered material and energy cost data from the individual processing units and used that data in the inner-most loop, to optimize how each individual processing unit performed. Periodically the system would also gather cost data for the plant-level loop. Deutsch's method used this plant-level data to modify how the control systems for individual processing units were set. In this way, Deutsch optimized the performance of each individual plant. Plant-level data was collected less frequently than unit-level data. At a further less frequent interval, the system would gather cost data for the outer-most, system-level loop. The method used this data to modify how the plants were set up to function, thereby optimizing each plant in accordance with the needs of the system as a whole.

The Patent Examiner rejected Deutsch's claims as being drawn to nonstatutory subject matter under 35 U.S.C. § 101. The Board of Appeals affirmed the Examiner's decision and Mobil Oil Company appealed to the Court of Customs and Patent Appeals.

[A] The Claims at Issue in *In re Deutsch*

The claims on appeal were as follows:

1. The method of operating a system of multi-plants which produce finished products from material derived from a plurality of sources at fluctuating costs for delivery to markets of variable prices, each of said plants having a different, unique, cost function in producing the products, said method comprising operating computing apparatus to automatically perform the steps of:

    (a) establishing optimized control points for process units in a first plant,

    (b) substantially continuously feeding material and energy cost data, product price data, and variable process data to the optimized controls for said process units to maintain them at operating conditions dependent upon said data,

    (c) periodically modifying cost and process data fed to the control for said first plant to set operating points for said process units in said first plant dependent upon optimum operation of said first plant, and

    (d) periodically modifying cost and price data fed to the control for said system to set an operating point for said plant dependent upon optimum operation of said system.

    6. In the control of a system of multiunit plants which produce finished products from materials derived from a plurality of sources of supply wherein costs and market prices are subject to time variations and process variables are controllable, each of said plants operating at a different cost function, the method which comprises operating computing apparatus to automatically perform the steps of:

(a) closing an optimizing control loop on a process unit in a first plant,

(b) substantially continuously feeding material input functions, product output functions, and variable process functions to said control loop for maintaining said process unit at an optimum operating condition dependent upon said functions,

(c) periodically closing a control loop on said first plant to modify input and output functions fed to said control loop for said first plant periodically to establish an optimum plant operating point to set an operating point for said unit dependent upon optimum operation of said first plant, and

(d) periodically, but at less frequent intervals, closing an optimizing control loop on said system to modify cost and price functions fed to the control loop for said system to establish an optimum operating point and to reset operating points for said first plant for said process unit dependent upon optimum operation of said system.

8. In the control of a processing system including a plurality of plants which cooperate to produce finished products from materials derived from a plurality of sources of supply wherein costs and market prices are subject to time variations and process variables are controllable, each of said plants operating at a different cost function, the method which comprises:

supplying from at least one of said plants to another of said plants an intermediate product,

generating input signals indicating the input and operating parameters of each of said plants,

applying said input signals to plant-computer-controllers, one plant-computer-controller for each of said plants,

operating said plant-computer-controllers in response to said input signals to automatically produce an operating point signal for each of said plants,

operating each of said plants in response to the operating point signal from the respective plant-computer-controller,

operating a system computer-controller in response to inputs representing the cost and price of materials and finished products to automatically produce system control signals which optimize operation of said system, and

periodically applying said system control signals to said plant-computer-controllers to modify said operating point signals to produce optimized operation of said system as a whole.

[B] Explaining Supreme Court's Ruling in *Benson*, Chief Judge Markey Finds Refinery Control Method Patentable

Judge Markey wrote the opinion for the Court. They reversed. Because the Board had based its decision on its interpretation of *Gottschalk v. Benson*,[147] Judge Markey devoted a significant part of the court's opinion to that decision.

The Board had applied *Benson* based on the Court of Customs and Patent Appeals' prior decision in *In re Christensen*.[148] In that case, the court identified the point of novelty, determined that it was a mathematical equation and found that the claim would wholly preempt use of that equation. Judge Markey drew upon the court's prior decision in *In re Chatfield*[149] to explain that it was improper to zero in on the point of novelty and assess whether the novel element alone is statutory subject matter. The correct approach, Markey said, is to consider the claim as a whole. Of the *Chatfield* decision, he wrote:

> In *In re Chatfield* . . . the majority found the "fundamental rationale" of *Benson* to be that a method encompassing all practical use of a mathematical formula and the involved algorithm constituted nonstatutory subject matter. Chatfield's method, however, manipulated the machines of a computing machine system for improved efficiency in executing multiple processing programs. Any algorithm used in his method was incidental thereto.[150]

Judge Markey followed with the observation that Deutsch's invention was even more removed from *Benson* than *Chatfield*'s. Although Deutsch's method used certain "optimization techniques" that could be characterized as "mathematical," the method as a whole was directed to controlling the timing and sequence of how the nested control loops functioned to control an industrial process. Judge Markey wrote:

> Each of Deutsch's claimed methods, considered as an industrial process, is even further removed from that of *Benson* than was that of *Chatfield*. Each of Deutsch's methods is a method of operating a system of manufacturing plants. Though Deutsch's specification teaches that the claimed methods may be carried out by the use of known optimization techniques and apparatus, the claimed system control is exercised externally to a specific optimization or computing technique . . . the claimed invention lies in the timing and sequencing of control application, not in the control means ("optimization technique") itself.[151]

---

[147] 409 U.S. 63 (1972).

[148] 478 F.2d 1392 (C.C.P.A. 1973).

[149] *In re* Chatfield, 545 F.2d 152 (C.C.P.A. 1976).

[150] *In re* Deutsch, 553 F.2d 689, 692 (C.C.P.A. 1977).

[151] *Id.*

Deutsch's claims did not wholly preempt a mathematical algorithm, hence, the Court concluded it was improper to apply *Benson* and reject the claims under 35 U.S.C. § 101.

[C] Understanding *In re Deutsch*

In re Deutsch is interesting from a historical standpoint, for it represents one step in the appellate court's coming to grips with *Gottschalk v. Benson*. From a business method standpoint, the case is perhaps more interesting because it involves the collection and use of financial information (cost data). Courts of the 1920s, 1930s, 1940s, and even of the 1950s and 1960s would probably have addressed this application as a method of doing business. That is, of course, because the Patent Office during those earlier eras frequently rejected applications involving money as being nonstatutory business methods.

The *Benson* decision may have momentarily distracted the Patent Office from its self-proclaimed duty to reject business method patents. *Benson* had given the Patent Office a new basis for rejecting "all things computer." With such a seemingly powerful weapon in its quiver, why sling such a puny dart as the business method rejection. Rarely had that dart hit the mark anyway.

Benson remained at the forefront until 1981, when the Supreme Court decided *Diamond v. Diehr*.[152] Until that time, the business method rejection did not see a lot of action in the courts.

§ 6.26   Optimizing Sales — *In re Maucorps*

Maucorps filed an application for patent on December 27, 1974, entitled "Computing System for Optimizing Sales Organizations and Activities." Maucorps's concept was a computer model for a sales organization. The model determined the optimum number of times a sales representative should make a sales call on each customer over a period of time. The model also determined the optimum number of sales representatives a company should have and how they should be organized. Maucorps expressed his computer model as a fairly complex mathematical equation.

Maucorps had developed his model based on a South African sales organization. In South Africa, Maucorps determined, there should be no more than four sales representatives reporting to a single manager. It was commonly accepted that seven subordinates per manager was the proper number. Maucorps reasoned that the lower representative to manager ratio in South Africa was due to the greater mobility of representatives in South Africa.

Whether Maucorps was right or wrong about this, it set him thinking that there must be a way to determine the optimum number of subordinates based on objective criteria. Maucorps studied various model sales organizations and concluded that the control factor was the cost of maintaining a sales

---

[152] 450 U.S. 175 (1981).

representative. Maucorps called this control factor "gamma" (γ), and developed equations for different types of sales units based on this factor. Ultimately, Maucorps programmed a computer in Fortran IV to calculate these equations.

The Patent Office rejected Maucorps's invention, which had been expressed in apparatus form as a "computing system." The Office said the claims were drawn to nonstatutory subject matter. Maucorps appealed to the Board of Appeals, which sustained the Patent Examiner's ruling. Maucorps then appealed to the Circuit Court.

[A] The Claims at Issue in *In re Maucorps*

If ever a claim, on its face, begged to be considered as a mathematical algorithm it is this one. Maucorps's claim, reproduced below, includes a complex mathematical formula embedded in its very text.

A computing system for processing data to determine an optimum "coding," defined as the number of regular visits over a predetermined period of time, Pd, by a business representative to a client, to be selected for such client, comprising:

(a) means for calculating for each different value of x representing the coding of clients, a value for y representing the sales arising over said predetermined period of time from the representative's activity when x = 3 in accordance with the relation given by [a lengthy mathematical equation, omitted here]
in which

G_1 = M(0.5054 + 0.1930 log x)

K = 0.0581 M

D = Client's total demand over said period

M = "memory factor," i.e. retention after 24 hours, the values of x and the corresponding values of y defining "saturation curve" of sales;

(b) means for calculating, for the value of x = 1 representing the coding of clients, a value for y representing the sales arising over said predetermined period of time from the representative's activity in accordance with the relation by $y = D(1 - G_1)$;

(c) means for calculating for the value of x = 2 representing the coding of clients, a value for y representing the sales arising over said predetermined period of time from the representative's activity in accordance with the relation given by $y = D[(1 - G_1)^2 + K]$;

(d) means for calculating for each different value of x a value for y where [another lengthy mathematical equation, omitted here]
in which

C = direct cost over said period of a representative RE

a = additional visits/systematic visits factor

β = percentage of gross profits on sales

T = visiting times (days)/periods of time/representative

N = number of visits/day/representative

γ = "control factor," i.e. the cost C' over said period of controlling a representative divided by the direct cost C of that representative,

the values of x and the corresponding values of y defining a "minimum sales line" which intersects the saturation curve at a point defined as the "critical point", and

(e) means responsive to the output of said calculating means for selecting a value of x as a client's coding for which the value of y corresponding to the sales over said period arising from the representative's activity provides a "representative point" P having as coordinates x and y, such that said point P is above the minimum sales line and the saturation curve and as close to the latter as possible, each of said calculating means including electric circuits constituted so that when said electric circuits are in an activated state, values of y are automatically calculated upon receiving the necessary input data regarding the above-defined variables, and said value selecting means likewise including electric circuits constituted so that, when said selecting means is in an activating state, a value of x will be automatically selected upon said means receiving the necessary minimum sales line and saturation curve data.

[B] The Sales Optimizing Method Deemed a Nonstatutory Mathematical Algorithm

Maucorps had taken the position that the claimed subject matter was a statutory machine. Indeed the specification included a full hardware description of a digital computer, including RAM for storing the data for each calculation and ROM for storing the compiled machine language generated from Maucorps's Fortran program. Notwithstanding the computer hardware underpinnings, the Patent Examiner had found Maucorps's claims to be wholly preempting a mathematical algorithm as decried in *Benson*.

Judge Markey wrote the opinion of the Court. He agreed with the Examiner, and with the Board that had sustained the Examiner. Maucorps should not be permitted to achieve by indirection (apparatus claims) what he could not achieve directly (method claims). Maucorps's counsel had admitted in oral argument that the claims, if drafted as method claims, would not have represented statutory subject matter.

The problem with Maucorps's claims, as Judge Markey saw it, was they merely comprised "a solution for a set of equations wherein sets of numbers are computed from other sets of numbers."[153] Markey concluded that these claims wholly preempted the mathematical algorithm:

---

[153] *In re* Maucorps, 609 F.2d 481, 486 (C.C.P.A. 1979).

> [A]ppellant's claimed invention as a whole comprises each and every means for carrying out a solution technique for a set of equations wherein one number is computed from a set of numbers. Thus, appellant's claims wholly preempt the recited algorithms.[154]

This rendered the claims nonstatuory under the rule espoused by the Supreme Court in *Benson.*

[C] Understanding *In re Maucorps*

It is not difficult to see how Maucorps's claims fell into the "mathematical algorithm" trap of *Benson.* The case is of historical importance because it resolved whether *Benson* was limited to method claims alone. The Circuit Court ruled that *Benson* was not so limited.

Maucorps went to great lengths to characterize the invention as an apparatus, with the hope of avoiding the 35 U.S.C. § 101 nonstatutory subject matter rejection. Maucorps even went to the extent of describing in the specification that the computer program was embodied in read only memory (ROM). Perhaps this was an attempt to bolster the argument that the claimed computer system was a special purpose machine and not a general purpose computer performing a mathematical algorithm.

What was lacking in Maucorps's claim was an external tie to the real world. Had Maucorps recited a mechanism for effecting a physical state change, either before computation or after, then the nonstatutory subject matter rejection might have been avoided. As later court decisions would explain, the distinction is basically this: using one set of numbers to compute another set of numbers, without something more, is nonstatutory. The mere number-to-number translation lies within the domain of the abstract idea.

However, when the numbers correspond to measured physical objects or activities, the process can become statutory. Measurements of physical objects or activities are made and those measurements are transformed outside of the computer into data. Then when the first set of numbers is transformed into the second set of numbers, the second set of numbers comprise intangible representations of the physical objects or activities and the process is statutory subject matter.

§ 6.27 Cash Management Accounts — *Paine, Webber, Jackson & Curtis, Inc. v. Merrill Lynch, Pierce, Fenner & Smith, Inc.*

Thomas E. Musmanno applied July 29, 1980, for a patent on a new type of cash management account. He assigned his application, entitled, "Securities Brokerage — Cash Management System" to Merrill Lynch, Pierce, Fenner & Smith, Inc. The patent issued at U.S. Patent No. 4,346,442 on August 24, 1982. It covered a data processing system to administer a cash management account

---

[154] *Id.*

having three elements: a margin brokerage account, a money market account, and a credit card account. The integrated accounts worked as follows.

The subscriber could purchase stocks from the margin brokerage account, borrowing against the line of credit secured by the equity in that account. Proceeds from the sale of stock and stock dividends were automatically swept into the associated money market account, where the subscriber would earn interest on the money. The system automatically updated the credit limit of the credit card account, based on the cash value of the margin brokerage and money market accounts.

When the subscriber used the credit card, funds were extracted to cover the payment, first from the credit card account until its credit limit was exceeded, then from the money market account, and finally from the lendable equity in the brokerage account. The system provided for efficient use of funds because the subscriber did not incur the cost of a margin loan until all free credit cash balances and funds invested in the money market fund were fully utilized.

Paine, Webber, Jackson & Curtis, Inc. brought a declaratory judgment action against Merrill Lynch, seeking to have the patent declared invalid, unenforceable, and not infringed by its cash management system. Merrill Lynch countered by suing Paine, Webber for infringement. It also attempted to join Dean Witter Reynolds, Inc. as a third-party defendant, alleging that Dean Witter was also infringing the Musmanno patent.

Paine, Webber attacked the Musmanno patent on several grounds. First it contended that the patent was invalid under 35 U.S.C. § 101 because it was, in their view, "nothing more than the combination of familiar business systems, that is, a margin brokerage account, one or more money market funds, and a checking/charge account, which have been connected together so that financial information can be exchanged among them."[155] It argued that this was nothing more than a business method and system and that it should be struck down under § 101. Paine, Webber filed a motion for summary judgment on the § 101 issue. Dean Witter filed a motion urging that it had been improperly joined in this action.

[A] Claims at Issue in *Paine, Webber v. Merrill Lynch, Pierce*

1. In combination in a system for processing and supervising a plurality of composite subscriber accounts each comprising a margin brokerage account, a charge card and checks administered by a first institution, and participation in at least one short term investment, administered by a second institution, said system including brokerage account data file means for storing current information characterizing each subscriber margin brokerage account of the second institution, manual entry means for entering short term investment orders in the second institution, data receiving and verifying means for receiving and verifying charge card and check transactions from said first

---

[155] Paine, Webber, Jackson & Curtis, Inc. v. Merrill Lynch, Pierce, Fenner & Smith, Inc., 564 F. Supp. 1358, 1365 (D. Del. 1983).

institution and short term investment orders from said manual entry means, means responsive to said brokerage account data file means and said data receiving and verifying means for generating an updated credit limit for each account, short term investment updating means responsive to said brokerage account data file means and said data receiving and verifying means for selectively generating short term investment transactions as required to generate and invest proceeds for subscribers' accounts, wherein said system includes plural such short term investments, said system further comprising means responsive to said short term updating means for allocating said short term investment transactions among said plural short term investments, communicating means to communicate said updated credit limit for each account to said first institution.

* * *

3. A combination as in claim 1 or 2 where said updated credit limit generating means comprises means for accumulating the amount of charge card usage and checks for each subscriber, means responsive to said brokerage account data file means for generating a subscriber updated credit limit measured by the difference between the limiting residual subscriber brokerage account securities loan value augmented by the value of the subscriber's short term investment, decremented by the value of the subscriber's aggregate expenditures and funds required for brokerage account purposes, means for reporting said updated credit limit to said brokerage account data file means.

[B] Merrill Lynch Cash Management Account Deemed Patentable Subject Matter

Chief Judge Latchum of the Delaware District Court denied Paine, Webber's summary judgment motion. In doing so, he ruled that the subject matter of the Musmanno claims was statutory under 35 U.S.C. § 101. Paine, Webber had cited a string of cases that it suggested stood for the well accepted proposition that business methods are not statutory. The list of cases comprised:

- *Loew's Drive-in Theatres, Inc. v. Park-in Theatres, Inc.*[156]

- *In re Patton*[157]

- *Hotel Security Checking Co. v. Lorraine Co.*[158]

- *Berardini v. Tocci*[159]

---

[156] 174 F.2d 547 (1st Cir. 1949).
[157] 127 F.2d 324 (C.C.P.A. 1942).
[158] 160 F. 467 (2d Cir. 1908).
[159] 190 F. 329 (S.D.N.Y. 1911), *aff'd*, 200 F. 1021 (2d Cir. 1912).

• *United States Credit System Co. v. American Credit Indemnity Co.*[160]

These cases are discussed in preceding sections of this chapter. However, as noted in those sections, none of the above cases actually held that business methods are not statutory subject matter. In every case, the language to that effect was *dicta*.

Paine, Webber had further attacked the claims because the "means-plus-function" portions of the claims were supported in the specification, not by structure, but by method steps. This, they argued, meant that the claimed "system" could be performed by hand with the aid of paper, pencil and telephone. Paine, Webber urged that this was further proof that the invention was no more than a business method, having nothing to do with machinery, technology, process, manufacture, or composition of matter.

Merrill Lynch countered this argument, noting that 35 U.S.C. § 112, paragraph 6 does not place any special requirements on means-plus-function elements in order to support an apparatus claim. That statutory section provides:

> An element in a claim for a combination may be expressed as a means or step for performing a specified function without the recital of structure, material, or acts in support thereof, and such claim shall be construed to cover the corresponding structure, material, or acts described in the specification and equivalents thereof.

Judge Latchum said it was not necessary to resolve whether the invention was an apparatus or a process because labels are not determinative in a § 101 analysis. He cited *In re Maucorps*[161] and *In re Gelnovatch*[162] in support of this proposition. He then turned to the § 101 analysis. Judge Latchum broke his § 101 analysis into two sections, the first one assessing whether a mathematical algorithm was recited, as it might run afoul of the *Benson* prohibition against preempting mathematical algorithms; and the second one assessing whether this was a nonstatutory business method.

Judge Latchum devoted much time discussing the mathematical algorithm issue. After a thorough analysis, he concluded that the *Benson* Court had used the term "algorithm" in the specific sense of a procedure for solving a given type of mathematical problem. He noted that computer programs also perform "algorithms," however these are not necessarily solving mathematical problems, but rather defining a series of steps to be performed by the computer. Judge Latchum concluded that it was only the former type of "algorithm," the math-solving one, that evoked *Benson* issues.

The claimed invention here was a system for attending to the details of a brokerage system. Thus, Judge Latchum concluded that prohibition against

---

[160] 53 F. 818 (S.D.N.Y.), *aff'd*, 59 F. 139 (2d Cir. 1893).
[161] 609 F.2d 481 (C.C.P.A. 1979).
[162] 595 F.2d 32 (C.C.P.A. 1979).

preempting mathematical algorithms articulated by *Benson* was not applicable here. He wrote:

> [T]he Supreme Court in *Benson* used the term "algorithm" in a specific sense, "a procedure for solving a given type of mathematical problem." . . . Using this definition, this Court has carefully examined the claims in this case and is unable to find any direct or indirect recitation of a procedure for solving a mathematical problem. Rather, the patent allegedly claims a methodology to effectuate a highly efficient business system and does not restate a mathematical formula as defined by *Benson*. Nor are any of the recited steps in the claims mere procedure for solving mathematical problems. Accordingly, the claims do not recite or preempt an algorithm.[163]

To develop its "method of doing business" argument, Paine, Webber urged the Court to look at the "product" of the Musmanno patent. The "product" was the service that Merrill Lynch and others practicing the claimed invention provided to their customers. Cash management services did not belong, Paine, Webber argued, among the technological or useful arts.

However, in *In re Toma*,[164] the Court of Customs and Patent Appeals had already rejected the product-focused approach, as Judge Latchum was quick to point out. In *Toma*, the invention was a digital computer for translating Russian into English. Although languages were not within the technological arts, computers were. Thus the CCPA held that a computer-implemented language translation system was statutory subject matter, not withstanding that the end product was of value in the liberal arts.

Judge Latchum applied this reasoning to the Musmanno invention. He wrote of the Musmanno ('442) patent:

> The subject matter of the '442 patent claims are similar to the claims of the patents in *Toma*, *Phillips*, and *Johnston*. The product of the claims of the '442 patent effectuates a highly useful business method and would be unpatentable if done by hand. The CCPA, however, has made clear that if no *Benson* algorithm exists, the product of a computer program is irrelevant, and the focus of analysis should be on the operation of the program on the computer.[165]

On that note, the Court found the Musmanno claims to be statutory subject matter. It also found that Dean Witter had been improperly joined in the action and dismissed it from the suit.

---

[163] Paine, Webber, 564 F. Supp. at 1368.

[164] 575 F.2d 872 (C.C.P.A. 1978).

[165] Paine, Webber, 564 F. Supp. at 1369.

[C] Understanding *Paine, Webber v. Merrill Lynch, Pierce*

Judge Latchum's analysis today makes his conclusion look easy. Yet, the *Musmanno* patent was widely discussed among patent attorneys and many held the belief that the *Musmanno* subject matter was nonstatutory. That is, to a large part because it was in those days widely believed that there was, indeed, a fundamental prohibition against obtaining business method patents.

Illustrative of this is Paine, Webber's string citation in support of its representation that "courts do not hesitate to invalidate patents on the grounds that they merely describe business systems."[166] The cases Paine, Webber cited (listed at the beginning of the section above) did not hold patents invalid on account of being business systems. Most of them held the respective patents invalid for obviousness, a rejection having nothing to do with the statutory subject matter issue.

Was Paine, Webber's counsel simply remiss in citing these cases for the wrong proposition? Under the circumstances prevailing at the time, no. Indeed, the Patent Office cited several of the same cases in support of its *Manual of Patent Examining Procedure* § 706.03(a) admonishment against business method patents:

> 706.03(a) Though seemingly within the category of process or method, a method of doing business can be rejected as not being within the statutory classes. See *Hotel Security Checking Co. v. Lorraine Co.,* 160 F. 467 (2nd Cir. 1980) and *In re Wait,* 24 U.S.P.Q. (BNA) 88, 22 C.C.P.A. 822 (1934).[167]

The situation was simply this. The "rule" against business method patents had arisen out of *dicta*. It spread from case to case — always as *dicta* — until everyone ended up believing that business methods were not statutory. Paine, Webber thus fell into the trap of believing it was on solid footing — making the business method argument — when in fact it was standing in quicksand.

§ 6.28 Accounting Method — *Ex parte Murray*

Joseph C. Murray applied August 17, 1979, for a patent on an accounting method he had devised. On November 17, 1980, he filed a second application, a continuation-in-part of his first application, and allowed his first application to go abandoned. The Examiner rejected Murray's claims as being drawn to nonstatutory subject matter. Murray appealed to the Board of Appeals. The Board affirmed the Examiner's rejection. Murray did not appeal further, leaving his case one of the very few that had actually been rejected as a nonstatutory business method.

---

[166] *Id.* at 1365.
[167] U.S. Patent Office, *Manual of Patent Examining Procedure*, § 706.03(a) (August 1993).

[A] Claims at Issue in *Ex parte Murray*

Exemplary of Murray's claim was claim 12, which is reproduced below.

12. An accounting method utilizing a financial institution's documents, each having an account number and other information preprinted thereon, and a designated area for a user entry, comprising:

a. entering into said designated area one of a plurality of pre-selected identifying indicia to identify the nature and the purpose of the expenditure made with a particular document,

b. converting said user's pre-selected identifying expenditures into a periodic expense analysis statement, further including steps of:

c. sorting said expenditures in accordance with said user's account number and identifying indicia entered by said user in said designated area on said documents, and thereafter,

d. correlating, tabulating, and storing said sorted identified expenditures in sequence and operational function as predesignated by said financial institution for verification of the validity of said expenditures,

e. said periodic expense analysis statement having a plurality of vertically spaced columns, and wherein each of said columns is subdivided in accordance with the number of said identifying indicias,

f. entering all of one of said sorted expenditure indentifying indicias in one of said subdivisions and thereafter entering all of another of said sorted indicias in respective other subdivisions of said columns, until all of the indicias are entered,

g. debiting into adjacent vertically spaced columns the amount and entering other information related to said expenditure identified by said indicias, and positioning said debited amount and other information entries in alignment with said identifying indicias,

h. sub-totalling each of said amounts of said expenditures for like identifying indicia and entering said subtotals on said statement adjacent to said sorted indicia entered in said subdivided column,

i. totaling said subtotals and said totals on said statement and entering below said subtotals,

j. printing and issuing said expense analysis statement to said user.

[B] Accounting Method Deemed Unpatentable Method of Doing Business

The Board of Appeals began its analysis with an acknowledgment that a process is statutory under 35 U.S.C. § 101 unless it falls within a judicially determined category of nonstatutory subject matter. The Supreme Court had, for example, identified three broad categories that were not statutory subject matter: abstract ideas, laws of nature, and natural phenomena. However, here the Board stood on the premise that business methods were a judicially determined nonstatutory subject matter class. Quoting its predecessor review court, which in turn had relied on *In re Wait*,[168] the Board wrote:

> At issue here is whether the claimed accounting method defines such a judicially determined exception. As stated by our predecessor review court in *In re Chatfield*, "Some inventions, however meritorious, do not constitute patentable subject matter, e.g. . . . methods of doing business, *In re Wait*."[169]

Unfortunately for Murray, the Board misconstrued these cases. None of these cases had actually held a business method to be nonstatutory. *In re Wait* turned on lack of novelty; in *In re Chatfield* the applicant was awarded a patent. The statement quoted in *Chatfield*, in turn attributed to *Wait* had been *dicta*. Nevertheless, the Board proceeded to announce this invention was a method of doing business:

> We are convinced that the claimed accounting method, requiring no more than the entering, sorting, debiting and totaling of expenditures as necessary preliminary steps to issuing an expense analysis statement, is, on its very face, a vivid example of the type of "method of doing business" contemplated by our review court as outside the protection of the patent statutes. Accordingly, we will affirm the examiner's rejection of the claims as drawn to nonstatutory subject matter.[170]

At Murray's insistence, the board addressed the *Paine, Webber v. Merrill Lynch*[171] case. That case had held a similar accounting-related subject matter was statutory. The Board, however, distinguished Murray's claims from Musmanno's claims in *Merrill Lynch*. The Musmanno claims were directed to a "system," whereas the present Murray claims were directed to a "method." The Board thus said, "Whereas an apparatus or system capable of performing a business function may comprise patentable subject matter, a method of doing business generated by the apparatus or system is not."[172] In support of that statement, the Board cited *In re Johnston*.[173] Again, the Board misapplied its precedent. The CCPA in *In re Johnston* had found the invention to be statutory. Any language in that case was therefore *dicta*.

The Board did not stop with the business method rejection. It further rejected Murray's claims as being nonstatutory for preempting a mathematical algorithm, citing *Gottschalk v. Benson*.[174] The fact that Murray's claims included

---

[168] 73 F.2d 982 (C.C.P.A. 1934).

[169] *Ex parte* Murray, 9 U.S.P.Q.2d (BNA) 1819, 1820 (Bd. Pat. App. 1988).

[170] *Id.* at 1820.

[171] *Paine, Webber*, 564 F. Supp. 1358 (D. Del. 1983).

[172] Murray, 9 U.S.P.Q. at 1821.

[173] 502 F.2d 765 (C.C.P.A. 1974).

[174] 409 U.S. 63 (1972).

steps for "subtotaling" and "totaling," drew this subject matter into the nonstatutory domain. The remaining steps of the claim the Board construed as mere "preliminary data gathering steps" or "insignificant post solution activity."

[C] Understanding *Ex parte Murray*

Had Murray appealed the Board's decision to the circuit court, it seems unlikely that the court would have affirmed the statutory subject matter rejections. Aside from its reliance on *dicta* from the prior cases, the Board seemed to place too much emphasis on claim form over substance. It had distinguished Murray's method claims from Musmanno's apparatus claims, accepting that the latter were statutory whereas the former were not. However, in *In re Maucorps*[175] Judge Markey had admonished that "Labels are not determinative in § 101 inquires."

The Board's *Benson* analysis also leaves something to be desired. In assessing whether a claimed invention is statutory, it is improper to focus on one claimed mathematical step and ignore the remainder of the claim. Again, Judge Markey in *In re Deutsch*[176] explained that although mathematical steps may be recited, it is the claim as a whole that must be considered in making the statutory subject matter determination. If the mathematical algorithm was "subtotaling and totaling" as the Board had found, was Murray wholly preempting this mathematical calculation by his claim? The Board never truly addressed that question.

Had Murray appealed to the Circuit Court, would his claims have been granted? That seems doubtful, as well. The steps Murray recites in his claim appear to read upon conventional bookkeeping techniques. The Circuit Court could thus have ruled the claims unpatentable on obviousness grounds under 35 U.S.C. § 103. That, of course, is speculation, but the approach is not without precedent. The U.S. Supreme Court took a similar approach in *Dann v. Johnston*.[177]

§ 6.29 Medical Diagnostics — *In re Grams*

Ralph A. Grams and Dennis C. Lezotte filed an ambitious patent application on June 27, 1984, claiming a diagnostic method. Originally conceived as a medical diagnostic system, Grams and Lezotte stated in their application that the method could be used to diagnose any complex system, be it "electrical, mechanical, chemical, biological, or combinations thereof."[178] Their method was highly mathematical. The method involved first obtaining clinical data and

---

[175] 609 F.2d 481, 485 (C.C.P.A. 1979).
[176] 553 F.2d 689, 692 (C.C.P.A. 1977).
[177] 425 U.S. 219 (1976).
[178] *In re* Grams, 888 F.2d 835, 836 (Fed. Cir. 1989).

then analyzing that data to ascertain the existence and identity of an abnormality and possible causes thereof.

The Patent Examiner rejected their application under 35 U.S.C. § 101 as being either a mathematical algorithm or a business method, both of which the Examiner considered to be nonstatutory subject matter. The Board of Appeals affirmed the Examiner's ruling and Grams and Lezotte appealed to the Federal Circuit Court of Appeals.

[A]   The Claims at Issue in *In re Grams*

Grams and Lezotte's application contained 15 claims on appeal. Claim 1, presented below, is exemplary. Formatting has been added for readability. The bracket letters were added by the court to allow easier reference to particular steps of the method.

> 1. A method of diagnosing an abnormal condition in an individual, the individual being characterized by a plurality of correlated parameters of a set of such parameters that is representative of the individual's condition, the parameters comprising data resulting from a plurality of clinical laboratory tests which measure the levels of chemical and biological constituents of the individual [sic] and each parameter having a reference range of values, the method comprising
>
>> [a] performing said plurality of clinical laboratory tests on the individual to measure the values of the set of parameters;
>>
>> [b] producing from the set of measured parameter values and the reference ranges of values a first quantity representative of the condition of the individual;
>>
>> [c] comparing the first quantity to a first predetermined value to determine whether the individual's condition is abnormal;
>>
>> [d] upon determining from said comparing that the individual's condition is abnormal, successively testing a plurality of different combinations of the constituents of the individual by eliminating parameters from the set to form subsets corresponding to said combinations, producing for each subset a second quantity, and comparing said second quantity with a second predetermined value to detect a non-significant deviation from a normal condition; and
>>
>> [e] identifying as a result of said testing a complementary subset of parameters corresponding to a combination of constituents responsible for the abnormal condition, said complementary subset comprising the parameters eliminated from the set so as to produce a subset having said non-significant deviation from a normal condition.

[B]   Broadly Claimed Diagnostic Method Deemed a Nonstatutory Mathematical Algorithm

Judge Archer wrote the opinion of the court, which agreed that the claims did not recite statutory subject matter. Although the Board had rejected the claims as both mathematical algorithm and method of doing business, the court based its ruling only on mathematical algorithm grounds. Thus it expressly declined to rule on whether the claims were nonstatutory as a method of doing business.

Judge Archer began his analysis with a discussion of claim 1. He noted that step [a] requires the performance of clinical laboratory tests on an individual to obtain data for the parameters. The patent gave the example that the parameters could be sodium content. The remaining steps [b] to [e], Judge Archer noted, analyze the data gathered in step [a]. He found that these later steps were:

[I]n essence a mathematical algorithm, in that they represent "[a] procedure for solving a given type of mathematical problem." *Gottschalk v. Benson*[179]

Judge Archer discussed the mathematical algorithm in light of not only *Gottschalk v. Benson*, but also the later Supreme Court decisions on statutory subject matter, namely, *Parker v. Flook*,[180] *Diamond v. Chakrabarty*,[181] and *Diamond v. Diehr*.[182] He concluded that the decision in *Benson*, which was not overturned by the later decisions, required a finding that Grams and Lezotte's claims were wholly preempting a mathematical algorithm. As such, they represented nonstatutory subject matter.

In discussing *Flook* and its impact upon *Benson*, Judge Archer made this perfunctory statement about the patentability of business methods (*emphasis added*):

[E]ven though the application of an algorithm to data is a "process" in the literal sense, it is not one that is contemplated by section 101, i.e., it is "nonstatutory subject matter." Thus, mathematical algorithms join the list of non-patentable subject matter not within the scope of section 101, including *methods of doing business*, naturally occurring phenomenon, and laws of nature. *In re Sarkar. . . In re Chatfield*.[183]

The *In re Chatfield* case, cited by Judge Archer in support of his generalization about business methods, had held that a computer program to allow multiple computer processes to share hardware resources was statutory subject matter. The Chatfield invention was expressed as a method of operating a computer. In the *Chatfield* case, Judge Markey made a passing remark where he included business methods within the subject matter that had been judicially carved out as nonstatutory. Judge Markey cited *In re Wait*[184] for the proposition that business methods were not statutory. *In re Wait*, however, had not held this.

---

[179] *Grams*, 888 F.2d at 837.
[180] 437 U.S. 584 (1978).
[181] 447 U.S. 303 (1980).
[182] 450 U.S. 175 (1981).
[183] *Grams*, 888 F.2d at 837.
[184] 73 F.2d 982 (C.C.P.A. 1934).

In re Wait involved a system for transmitting stock offers and acceptance prices via electrical means, which the court rejected for lack of novelty. Thus both Judges Archer and Markey were citing *dicta* from earlier judges opinions.

Ultimately, Judge Archer rested his opinion on the mathematical algorithm rejection. Thus, the court made no ruling in this case that the claims were nonstatutory as business methods.

[C] Understanding *In re Grams*

The basic structure of Grams and Lezotte's claim was this:

> [a] perform tests, measure parameters;

> [b] compute a first number using the parameters;

> [c] compare the first number with a reference number;

> [d] perform more tests, measure different subsets of parameters and compute second numbers, compare second numbers to second reference numbers;

> [e] identify one of the subsets as corresponding to the abnormal condition.

The logic of steps [a] through [c] was to determine if an abnormal condition existed. If so, then steps [d] and [e] identified which parameters were responsible.

The algorithm expressed by their claim employed the basic scientific method: formulate a hypothesis, test the hypothesis through observation, refine hypothesis, test again. Grams and Lezotte had simply associated numbers with their observations so that their hypothesis could be tested using statistics. It was as if their claim had read:

> A method of testing for an abnormal condition comprising:

> applying the scientific method of forming a hypothesis, testing said hypothesis through measurement and refining said hypothesis based on said testing;

> whereby said testing step is performed by calculating and comparing numbers based on said measurement.

Seen in this light, the Grams and Lezotte claim had scope comparable to the far reaching claim of Samuel Morse, in which he claimed all use of electromagnetism for communication:

> I do not propose to limit myself to the specific machinery, or parts of machinery, described in the foregoing specifications and claims; the essence of my invention being, the use of the motive power of the electric or galvanic current, which I call electro-magnetism, however developed, for making or printing intelligible characters, letters or signs, at any distances, being a new

application of that power of which I claim to be the first inventor or discoverer.[185]

The *Grams* court, like the Supreme Court had in *O'Reilly v. Morse*, recognized that the Grams and Lezotte claims were overreaching. As demonstrated above, the claims effectively covered the scientific method where numeric values were used in testing the hypotheses. Although the Court might have rejected the claims as being obvious, it chose to base its decision on *Benson*. *Benson* was convenient because it had used a preemption analysis — Benson's claims preempted all use of the mathematical formula — that seemed to apply to Grams and Lezotte. Grams and Lezotte were preempting all use of numerical representation in utilizing the scientific method by their claims.

Grams and Lezotte had argued that their claims should not be rejected simply because a mathematical algorithm had been recited. Their proposition had some basis. In *Diamond v. Diehr*,[186] the Supreme Court had upheld a claim that included a mathematical algorithm among its recited elements. However, there was a difference between reciting a mathematical algorithm and entirely preempting a mathematical algorithm. Grams and Lezotte were far closer to the latter than was Diehr.

Grams and Lezotte's ultimate downfall was that their nonalgorithmic step, the data gathering in step [a], was expressed in very general terms with little support in the specification. Indeed, in an attempt to make this patent as broad as possible, the specification placed little specific emphasis on data gathering — to leave wider room for argument that the claims could cover any complex system, not just the human body. Judge Archer pointed this out in his opinion:

> The sole physical process step in Grams' claim 1 is step [a], i.e., performing clinical tests on individuals to obtain data. The specification does not bulge with disclosure on those tests . . .
> The specification also states that "[t]he invention is applicable to any complex system, whether it be electrical, mechanical, chemical or biological, or combinations thereof." From the specification and the claim, it is clear to us that applicants are, in essence, claiming the mathematical algorithm, which they cannot do under *Gottschalk v. Benson*.[187]

In re Grams contains a lesson for patent applicants — make the specification count. If the claims encompass a broadly applicable algorithm, be sure the data gathering inputs and the data using outputs to the algorithm are fully described in the specification. If the algorithm may be used in multiple different fields, explicitly show how the data would be gathered, and how the output would be used, in each different field of use. Doing so will make it more difficult to invoke the *In re Grams* analysis. Of course, it also helps to recognize when a

---

[185] O'Reilly v. Morse, 56 U.S. 62, 112 (1854).
[186] 451 U.S. 175 (1981).
[187] *In re* Grams, 888 F.2d at 840.

claim is encroaching upon an abstract idea, expressed as a mathematical algorithm or otherwise. Despite the apparent vacillation over computer programs, the Supreme Court has steadfastly held that abstract ideas are not patentable subject matter.

Some business method patents may involve mathematical algorithms. A method for picking stocks using mathematical calculations based on historical prices would be an example. Whether such patents are statutory or not may well depend on what data is gathered and how. Usually it is possible to draft a statutory claim that gives proper coverage of the invention, without wholly preempting the mathematical algorithm. Business methods involving purely abstract ideas are another matter.

§ 6.30   Competitive Bidding — *In re Schrader*

Rex Schrader and Eugene Klingaman filed an application on June 19, 1989, for a method of competitive bidding. They had observed that in a classic auction the total amount received by the auction house could depend on how the individual items were grouped for sale. For example, in an auction involving two or more contiguous tracts of land, tracts 1 and 2, the following bids might be received and recorded: *Bid 1* — $100,000 for tract 1 by bidder *A*; *Bid 2* — $200,000 for tract 2 by bidder *B*; and *Bid 3* — $250,000 for both tracts 1 and 2 by bidder *C*. The combination of bids that maximizes the revenue to the seller is bids *1* and *2*.

Schrader and Klingaman envisioned a system in which bidders could bid on both individual items and groups of items. All bids received were entered into a record and a "completion" was assembled, presumably by computer, identifying the bid combination that would yield the highest overall revenue. In an alternate form, the system could be used to minimize the overall price. The alternate form might be used, for example, in conducting competitive bids for defense contracts.

The Patent Office rejected Schrader and Klingaman's claims as being drawn to nonstatutory subject matter. The Board of Appeals affirmed, finding the claims were nonstatutory on the following three grounds:

> Chapter 13 — The claimed subject matter "falls within a judicially determined exception to a process set forth in § 101. The process involves only information exchange and data processing and does not involve a process of transforming or reducing an article to a different state or thing . . . ."[188]
>
> Chapter 14 — The claimed method "involves a mathematical algorithm or mathematical calculation steps, as the method includes a procedure for solving a given type of mathematical problem . . . ."[189]

---

[188] *In re* Schrader, 22 F.3d 290, 292 (Fed. Cir. 1994).
[189] *Id.*

Chapter 15 — "[T]he issues in the case relating to the § 101 rejection are analogous to the issues in *Ex parte Murray* . . ."[190]

[A] Claims at Issue in *In re Schrader*

Claim 1 was exemplary of Schrader and Klingaman's position. It is reproduced below with indentation added to improve readability.

1. A method of competitively bidding on a plurality of items comprising the steps of

identifying a plurality of related items in a record,

offering said plurality of items to a plurality of potential bidders,

receiving bids from said bidders for both individual ones of said items and a plurality of groups of said items, each of said groups including one or more of said items, said items and groups being any number of all of said individual ones and all of the possible combinations of said items,

entering said bids in said record,

indexing each of said bids to one of said individual ones or said groups of said items, and

assembling a completion of all said bids on said items and groups, said completion identifying a bid for all of said items at a prevailing total price, identifying in said record all of said bids corresponding to said prevailing total price.

[B] Placing Bids Lacks Physicality — Patent Denied as Nonstatutory

Judge Plager wrote the opinion of the court; Judge Newman wrote a dissenting opinion. The majority of the court agreed with the Board that Schrader and Klingaman's claims were nonstatutory. Judge Plager began his analysis with a focus on the mathematical algorithm. The mathematical algorithm had become the popular focus of many software-related inventions after the Supreme Court had ruled a certain mathematical algorithm was nonstatutory in *Gottschalk v. Benson*.[191]

Schrader and Klingaman had argued that the claim did not recite a mathematical algorithm, hence it was improper for the Board to analyze the claims as if a mathematical algorithm were involved. However, Judge Plager found that the claim term "assembling a completion" did involve a mathematical algorithm. He wrote:

This process, although expressed in general terms, is within or similar to a class of well known mathematical optimization procedures commonly applied to business problems called linear programming. Thus, a mathematical algorithm is implicit in the claim.[192]

---

[190] *In re* Schrader, 22 F.3d at 292.

[191] 409 U.S. 63 (1972).

[192] *In re* Schrader, 22 F.3d at 293.

Schrader and Klingaman next argued that even if a mathematical algorithm were present, their claims did not wholly preempt the algorithm because there was sufficient physical activity apart from the algorithm step (assembling a completion) to render the claims statutory. Their argument reflected the outcome of the court's decision two years before in *Arrhythmia Research Technology v. Corazonix Corp.*[193] In that case, the invention involved an apparatus and method for analyzing electrocardiograph signals. The claims recited a mathematical algorithm, but were held statutory because the input signals from the patient's heart were physically transformed into a value related to the patient's heart activity. The input signals and resultant output were not abstractions; rather they were related to physical conditions within the patient.

Regarding the Schrader and Klingaman claims, Judge Plager noted that there was nothing physical about bids, per se. He wrote:

> There is nothing physical about bids, per se. Thus the grouping or regrouping of bids cannot constitutie a physical change, effect, or result . . .
>
> The only physical effect or result which is *required* by the claim is the entering of bids in a "record," a step that can be accomplished simply by writing the bids on a piece of paper or chalkboard. For purposes of § 101, such activity is indistinguishable from the data gathering steps which we said in *In re Grams* . . . were insufficient to impart patentability to a claim involving the solving of a mathematical algorithm.[194]

Why the focus on physicality? Judge Plager explained that a process claim requires some kind of transformation or reduction of subject matter to be statutory under § 101. He reached this by reviewing the seminal Supreme Court decision in *Cochrane v. Deener* that articulated the rule that:

> A process is . . . an act, or a series of acts, performed upon the subject matter to be transformed and reduced to a different state or thing.[195]

That rule, Judge Plager found, was ultimately codified in the 1952 Patent Act. As Judge Plager explained, when Congress approved the addition of the term "process" to the categories of patentable subject matter in 1952, it incorporated the definition of "process" that had evolved in the courts.[196]

Thus, Judge Plager reached the conclusion in this case that Schrader and Klingaman's method (process) claims were not statutory because they did not involve any transformation of subject matter into a different state or thing.

The decision of the Schrader court was not unanimous, however. Judge Newman wrote a dissenting opinion in which she respectfully disagreed with the approach taken by Plager and the majority. Judge Newman believed the subject

---

[193] 958 F.2d 1053 (Fed. Cir. 1992).
[194] *In re* Schrader, 22 F.3d at 294.
[195] 94 U.S. 780, 787-788 (1877).
[196] *In re* Schrader, 22 F.3d at 295.

matter of Schrader and Klingaman's claims was statutory; but she believed the case should be remanded to determine whether the claims would pass muster under §§ 102 and 103. Judge Newman considered Schrader and Klingaman's claims as proper method claims involving more than mental steps, theories, and plans. Although the claims recited computational steps, she did not consider the claims to encompass an abstract mathematical algorithm. She reached this conclusion, noting:

> The Supreme Court has defined mathematical algorithm as a "procedure" or "formulation" for solving a particular mathematical problem. *Gottschalk v. Benson*, 409 U.S. 63 (1972). The only mathematical problem in Schrader's invention is identifying that combination of bids which yields the highest return, and he does not claim any particular procedure or formula for solving that problem. Neither the trivial procedure of adding up the returns on all permissible combinations of bids and selecting the combination with the highest return, nor some more elegant mathematical manipulation, is claimed. One must distinguish the answer to be found from the method of finding that answer. The latter might be a mathematical algorithm; the former is not.[197]

Judge Newman urged the court to separate the technology of the claimed invention from the nature of the subject matter to which the technology is applied:

> As stated in *In re Musgrave*, 431 F.2d 882, 893 . . . (C.C.P.A. 1970), a statutory "process" is limited only in that it must be technologically useful. A process does not become nonstautory because of the nature of the subject matter to which it is applied, or the nature of the product produced. *In re Toma*, 575 F.2d 872, 877-78 . . . (C.C.P.A. 1978).[198]

The majority had it wrong, according to Judge Newman. The majority had imposed a value judgment that data representing bid prices did not deserve the same patentability treatment as data representing electrocardiogram signals. That, Judge Newman maintained, was impermissibly tinkering with the scope of 35 U.S.C. § 101.

> The majority now imposes fresh uncertainty on the sorts of inventions that will meet the majority's requirements. All mathematical algorithms transform data, and thus serve as a process to convert initial conditions or inputs into solutions or outputs, through transformation of information. Data representing bid prices for parcels of land do not differ, in section 101 substance, from data representing electrocardiogram signals (*Arrhythmia*) or parameters in a process for curing rubber (*Diehr*). All of these processes are employed in technologically useful arts.

---

[197] 22 F.3d at n.1.
[198] 22 F.3d at 297.

By enlarging section 101 beyond its statutory scope, the majority implies that it is more desirable, from the viewpoint of social policy, to withhold the patent incentive from innovative activity such as that here illustrated. Such policy decisions have been made for a variety of reasons, as exemplified in the denial of patents on inventions relating to nuclear weapons, perpetual motion machines, and, until 1977, gambling machines.[199]

Judge Newman also took the opportunity to address the business method rejection made by the Board. She urged that the business method doctrine had no place in the determination of statutory subject matter under 35 U.S.C. § 101. Instead, Judge Newman favored using the novelty, obviousness, and formal disclosure and claiming requirements of the patent law (e.g., 35 U.S.C. §§ 102, 103, and 112) to assess whether a business invention deserves a patent.

The Board also relied on the "method of doing business" ground for finding Schrader's subject matter non-statutory under section 101. In so doing the Board remarked that the "method of doing business" is a "fuzzy" concept, observed the inconclusiveness of precedent, and sought guidance from this court. Indeed it is fuzzy; and since it is also an unwarranted encumbrance to the definition of statutory subject matter in section 101, my guidance is that it be discarded as error-prone, redundant, and obsolete. It merits retirement from the glossary of section 101.

* * *

I discern no purpose in perpetuating a poorly defined, redundant, and unnecessary "business methods" exception, indeed enlarging (and enhancing the fuzziness of) that exception by applying it in this case. All of the "doing business" cases could have been decided using the clearer concepts of Title 35. Patentability does not turn on whether the claimed method does "business" instead of something else, but on whether the method, viewed as a whole, meets the requirements of patentability as set forth in Sections 102, 103 and 112 of the Patent Act.[200]

[C] Understanding *In re Schrader*

The *Schrader* decision rests squarely at the dividing line between the algorithm-centric approach adopted by Federal Circuit's predecessor court (*Freeman-Walter-Abele*) and the claim-as-a-whole approach adopted by the Federal Circuit in *Arrythmia*, and later by the Court, *en banc*, in *Alappat*. *Schrader* was the last decision of the Federal Circuit Court of Appeals to fully embrace the *Freeman-Walter-Abele* test developed by its predecessor court (Court of Customs and Patent Appeals). The tension between the two approaches

---

[199] 22 F.3d at 297, and n. 2.
[200] *Id.* at 298.

may be seen by contrasting Judge Plager's majority opinion with Judge Newman's dissenting opinion.

By adopting the *Freeman-Walter-Abele* test, the majority focused on whether the Schrader invention was a mathematical algorithm. Thus the majority avoided the business method issue altogether. In the eyes of the majority, the Schrader invention lacked physicality. According to the majority, there was no physical transformation and that supported the Board's contention that the Schrader invention was just an abstract idea. This lack of physicality can be a troublesome point. District Judge Saris found similar shortcomings in the patent at issue in *State Street Bank*.

Judge Newman appears to be less concerned with physicality, for that issue receives scant, if any, treatment in her dissenting opinion. If physical transformation is required, as Judge Plager has stated, where is the physical transformation in Schrader's method claim? Judge Plager answers that question, finding the "bid entering" step to be the only one requiring a physical effect. Schrader's claim contained the following steps, in which the "bid entering" step has been emphasized:

Chapter 16 — identifying . . . related items in a record,

Chapter 17 — offering . . . items to a plurality of potential bidders,

Chapter 18 — receiving bids . . .

Chapter 19 — *entering said bids in said record,*

Chapter 20 — indexing each of said bids . . .

Chapter 21 — assembling a completion . . .

It could be argued that the indexing and assembling steps also involve similar physical effect. Indexing would presumably involve generating and entering additional data to relate each bid with the offered items. Assembling would similarly involve generating a report from the entered bids and index data, thus representing a further physical transformation of the bids stored in computer memory.

One might argue that moving around bids stored in memory is not the kind of physical transformation the patent law requires. That does not appear to be Judge Plager's point, however. In his criticism of the claim's lack of physical transformation, Judge Plager noted that the "bid entering" step could be carried out with pencil and paper.

> The only physical effect or result which is *required* by the claim is the entering of bids in a "record," a step that can be accomplished simply by writing the bids on a piece of paper.[201]

---

[201] *Id.* at 294.

Although use of pencil and paper are trivial in today's world, there is undeniably a physical transformation that takes place when pencil lead is adhered to paper surface. Thus, Judge Plager appears to be saying that some physical transformations are simply too commonplace or trivial to lend patentable significance. Indeed, he states this in his opinion:

> For purposes of § 101, such activity [writing bids on a piece of paper] is indistinguishable from the data gathering steps which we said in *In re Grams* . . . were insufficient to impart patentability to a claim involving the solving of a mathematical algorithm.[202]

As the later *State Street Bank* decision demonstrates, there is considerable merit in Judge Newman's dissent. Although there are several differences between *Schrader* and *State Street Bank*, her view appears to be winning favor in the Court.

---

[202] *Id.* at 294.

CHAPTER 7

E-COMMERCE TECHNOLOGY

§ 7.01   E-Commerce Technology Building Blocks

When sitting at one's computer and interacting with a popular Web site, it is easy to think of the Internet as a giant resource of information, software, and services. The Internet is actually a collection of thousands of individual computer networks. Chances are that the computer you will use tonight to check closing stock prices, or to order an on-line movie, is connected to an Internet service provider that is part of a regional computer network. The regional network communicates, in turn, with a larger network, which perhaps includes a high speed telecommunications pipeline or backbone that stretches from one city to another. Some of the nodes on this larger network may include satellite telecommunications links that allow it to communicate with similarly configured networks around the world.

What makes this global communication possible? There are several important building blocks. While the underlying technology is fairly complex, the concepts are surprisingly simple. The following sections will explain some of the basic building blocks used to construct many Internet-related inventions. Knowing about these basic building blocks will help when drafting claims to Internet and e-commerce inventions.

[A]   TCP/IP Protocol

The first building block is a common communication protocol. A protocol is simply a set of conventions governing how data is treated. It governs how data is formatted, how it is packaged for transmission, and how data travels over the network, how information is sent and received. Currently the TCP/IP protocol governs much of the information flow.

The TCP/IP protocol is actually two different protocols: the TCP protocol which governs how information is broken up into packets and the IP protocol which governs how those packets are routed to their proper destination over the Internet. TCP stands for Transmission Control Protocol. According to this protocol, information to be sent over the Internet is first chopped up into standard sized chunks, as you might chop up a carrot into slices when making a salad. Each chunk is called a packet. The computer doing the chopping attaches an error detection code, called a checksum, to each packet that may be used to test whether the information within the packet was garbled during transmission.

The other part of the TCP/IP protocol, the IP protocol, stands for Internet Protocol. According to the Internet Protocol, each packet created according to the TCP protocol is placed into an electronic IP "envelope," complete with electronic address label so that it may be shipped to its proper destination. Devices on the Internet, called routers, examine the address information on these electronic envelopes and route each packet in the direction of its proper destination. Because the Internet forms complex inter-linked connections of many smaller networks, there are often many paths that a packet can take from source to destination. Each router makes a split-second decision on which way to route each incoming packet, based on traffic congestion at that particular time.

This means that the individual packets of a given message may not necessarily all arrive in the same sequence in which they were sent. A first packet may be held up in a traffic jam in Tulsa, while the remaining packets may dash ahead by routing through Duluth.

As the packets arrive at their destination, they are reassembled in the proper sequence by using the IP address label information. Then, following the TCP protocol, the checksum for each packet is examined to validate that the packet was received intact. If the checksum calculated when the packet was sent does not match the checksum when the packet was received, the receiver requests the sender to retransmit the packet.

While the TCP protocol is used for most text and graphic image delivery, there are other protocols for special purposes. Discussed below in connection with streaming video and audio, the UDP protocol (User Datagram Protocol) serves the packet formatting function where retransmission of packets would be counterproductive. When sending a video clip, for example, it is better to simply skip a garbled packet, rather than delay the video clip until the garbled packet may be resent.

To speak TCP/IP protocol, a computer needs the appropriate software. Because of the Internet's popularity, most computers have the necessary software built in or bundled with the computer operating system. The software component needed to speak TCP/IP is sometimes referred to as a *socket* or a TCP/IP *stack*. Computers that are physically connected to a network, such as to a local area network or LAN, further require the appropriate hardware driver and software to enable the computer to communicate with the network card that provides the physical connection to the network. Hardware drivers are usually shipped with the network card. The network card may be an Ethernet card, for example. The TCP/IP stack or socket must be written to work with the specific hardware driver used by the network card.

The TCP/IP protocol has two important properties. First, it allows packets of information to carry their own routing information, so that intelligent devices, such as routers, can create electronic pathways on the fly. This property qualifies the Internet as a "packet switched" network, in which the switches (e.g., routers) are able to make split-second decisions about how to route information on a packet-by-packet basis. Contrast this with the conventional telephone system that has thus far predominated voice communication. The telephone system (call POTS, for "plain old telephone system") is a "circuit switched" network. The switches that determine the path a call will take are switched into a particular routing configuration at the outset of the call. This connection then persists for the duration of the call. Even if the call consists of packets of information, those packets have no way to change what route they will take. The circuit switched network has configured the route and the packets of information have no choice but to go along for the ride.

For persons using a standard POTS telephone line to connect to the Internet via a modem, the distinction between circuit switched networks and packet switched networks may seem confusing. You may ask, if my modem is using a regular phone line (circuit switched network connection), how is it able to communicate with the packet switched Internet? The answer lies with your local Internet service provider and with another set of protocols. Internet service providers that offer dial-up access over regular phone lines via modem have specially configured equipment to allow dial-in computers to become part of the Internet (only while connected to the service provider's equipment, however). The two most popular protocols for this purpose are SLIP, which stands for Serial Line Internet Protocol, and PPP, which stands for Point-to-Point Protocol. PPP is newer and supports the retransmission of packets if they are garbled during transmission.

When a dial-up connection is made between a computer and a local service provider using conventional POTS phone lines, the connection is made by a circuit switched connection between the respective modems of the computer and the service provider. This circuit switched connection remains active for the duration of the communication, just as any regular telephone circuit must remain active for the duration of a call. However, once the SLIP or PPP connections have been established, the information flowing over this connection will conform with Internet standards and may be routed as all other packets are routed over the Internet. Thus, while the packets of information originating from a dial-up connection are initially constrained to a single POTS phone line connection (to the service provider), once the packets reach the service provider they are placed on the Internet where they are free to travel by whatever route their address labels and local routers may dictate.

[B] Routers, Bridges, Hubs, and Gateways

Routers provide the all important switching function of sending information to the proper destination by the most efficient route. Routers are nodes on the network, typically specialized hardware devices, that read the address labels of each information packet and determine from those labels where to send each packet next. The typical router has two or more physical input/output ports that can both receive and send data. Within the router is a lookup table, called a routing table, that contains routing instructions that specify where incoming packets should be sent. The routing table specifies to which output port each incoming packet is sent, based on the packet's IP address. If the destination address specifies a computer on the local network to which the router is connected, the router will send the packet directly to the destination computer. If the destination address specifies a computer that is not on the local network, the router will send the packet to another router that is closer to the final destination.

Routers can have different degrees of intelligence. Some employ static routing tables that always route packets the same way, regardless of traffic conditions. More sophisticated routers have dynamic routing tables that can adjust routing paths based on traffic conditions at a particular time. If traffic is slow via one route, the router automatically selects a different route.

Routers perform the all important role of traffic cop, directing packets to their proper destination. However, you may encounter other devices with similar, but more specialized roles, called hubs, bridges, and gateways.

Hubs link groups of computers together so that they may communicate with one another. In a typical office network, a hub is what each individual computer is connected to. If you could follow the network cable coming out of the back of your office computer, trace it through the telephone-like wall jack behind the walls and along the ceiling or floor to the wiring closet or computer room, chances are the other end of that cable plugs into its very own port within a hub. The computer in the neighboring office connects to the hub in the same way. Thus when you want to send information from one computer to another, you do so through the mutual connection, the hub. If the local area network includes a file server, as many do, the file server would also be connected to the hub.

Bridges do for individual networks what hubs do for individual computers. Bridges link local area networks together. If your company or firm has more than one building or location, with separate local area networks at each building or location, these local area networks may be interconnected through a bridge. Typically a bridge will connect two networks of like protocol. Thus, if two local area networks both utilize TCP/IP protocol, they would be connected to each other through a bridge. If, however, the two networks do not employ like protocol, then some form of data translation may be required before the networks can be interconnected.

When such data translation is required, engineers call the bridging device a *gateway*. Gateways are often encountered when connecting between different hardware platforms, such as between personal computers and mid-sized business computers or between personal computers and mainframes. For example, a systems engineer might employ a gateway to connect a personal computer network to an AS/400 mid-sized computer network. Gateways are also encountered when connecting between different software platforms. For example, a systems engineer might use a gateway to connect a legacy Microsoft Mail (MS Mail) system with a more modern Internet e-mail system.

[C] The Domain Name System

According to the IP protocol, information packets are delivered to the proper destination based on the address attached to each packet. Addresses are specified numerically. By convention each numeric IP address is represented as four decimal numbers (each ranging from zero to 255) separated by periods or dots. Thus a numeric IP address might look like this:

165.55.128.60

Because it is difficult for people to remember such lengthy numeric addresses, the Internet uses a domain name system that associates an alphabetical name with each numeric address. Instead of numbers, words are used. Thus the corresponding alphabetical name might look like this:

res.ipsa.loquitur.com

Domain names are hierarchically configured. Reading the above name from right to left, the right-most word (.com) is the root and the preceding words are hierarchically related like parts of a plum tree. Using this analogy, the next-left word (.loquitur) would be the trunk; the next word (.ipsa) would be a branch; the left-most word (res) might be a plum.

How are the address words converted back into numerical IP addresses? Special computers called domain name servers translate the alphabetical domain names into numeric addresses. Each domain name server contains a lookup table that can translate at least some of the alphabetical names into numeric addresses. As you might expect, there are so many domain names in use today that no one domain name server can be expected to know them all. Fortunately, domain name servers are taught to ask when they don't know the answer. If a domain name server cannot translate an alphanumeric URL address into a numeric IP address, it asks another domain name server for assistance.

To illustrate, assume your computer is connected to a local area network (or to an Internet service provider that is part of a regional network). When you ask to communicate with the Internet site at the address, <res.ipsa.loquitur.com>, the domain name server of your local area network tries first to translate the address. If that address is part of your local area network (or part of the regional network known to your service provider) chances are good that your local area network's domain name server has an entry in its lookup table that will translate <res.ipsa.loquitur.com> into its numeric address, 165.55.128.60.

If your local domain name server does not know how to translate the alphabetical name, it will ask the domain name server associated with the root (.com) for assistance. Assuming the requested domain name has been properly registered, the root domain name server will know the addresses of both a primary and a secondary domain name server. These will in turn know the address of <res.ipsa.loquitur.com>. Thus, after checking with the root server, your local domain name server will contact either the primary or secondary domain name server to learn the address of the requested <res.ipsa.loquitur.com> site.

[D] Client-server Architecture

The TCP/IP protocols define how communication takes place over the Internet; these protocols define the low level information packaging. At a higher level, the domain name system defines an information delivery infrastructure; it defines how addresses are interpreted and how a network device looks up an address it doesn't know. At a still higher level is the client-server relationship. The client-server relationship is an architectural model that defines what roles two or more network devices will play when they communicate with one another.

The relationship revolves around service. A network device is called a client or a server, depending on the role it plays: clients utilize services; servers provide services. The range of services is limited only by human imagination. On the Internet today, many servers provide informational services, in the form of Web pages. Other servers provide transactional services, ranging from catalog sales to auctions. Still other servers provide access to databases. The U.S. Patent and Trademark Office, for example, maintains a server which clients may access to explore the issued patent literature on file in Patent Office databases. Still other servers provide entertainment services, including video and audio on demand.

Most servers on the Internet today are social introverts. They quietly sit on the Internet doing nothing unless engaged by a client to provide services. Thus a server consumes little or no communications resources unless a client asks that server for something that it can provide. Clients are another matter. Clients are chatty extroverts that initiate Internet conversation by asking servers to engage in dialogue with them. When the Internet slows down every afternoon, most likely it is the clients who are to blame — clients initiate, servers respond. Thus the more information clients request, the more the Internet pipelines fill up. Information technology engineers use the term ''pull technology'' when referring to this traditional use of the client-server relationship. Clients ''pull'' services from the servers.

That is not to say that the client-server relationship is limited to pull technology; it is not. Clever information technology engineers have devised schemes to allow servers to ''broadcast'' information to multiple clients at once, much like your local television or radio station broadcasts the news to multiple homes at once. They call this ''push technology.''

[E] Push Technology

With conventional pull technology, the user must take an affirmative action, such as clicking on a button or hyperlink, each time he or she wants a quantity of information to be shipped from the server. Thus, with pull technology, if the user wants to read the news each morning, the user would need to click the ''news'' button every morning. Push technology does away with the need for repeated clicking. With push technology, once the news channel is activated on the user's browser, the server will periodically ship up-to-date information without the user's further request.

Currently there is no standard by which push technology is implemented. Microsoft's web browser implements it one way; Netscape's web browser implements it a different way; third party client software applications, such as PointCast, Marimba, and BackWeb implement it in still different ways. Typical of most push technology is the concept of subscription. The user wishing to receive information by push technology accesses a server hosting a push site and sends the server a request to subscribe. The server then provides the user's browser or client application with the requested information. In addition, the server and client maintain a relationship whereby the server can notify the client if there is new content to be downloaded. In some forms of push technology, the server merely notifies that new information is available, the client must explicitly ask that the information be sent. In another form, the server will automatically ship the new information to the client without being asked.

Push technology defines an ongoing relationship between client and server. Periodically the client and server communicate automatically so that the latest information subscribed to may be pushed to the user's computer. While presumably either client or server could mediate this periodic exchange, typically the client performs this function. The client (or web browser) periodically contacts the server (automatically, without user involvement) to check for and download any pertinent updated information that it finds.

Because push technology allows information to be delivered without continued user interaction, it may be used to stockpile information on the user's computer for later viewing. Stockpiling of information on the computer's local hard disk can vastly improve the perceived speed at which information is delivered because the local hard disk operates at a much higher throughput than the typical Internet connection. For example, push technology could be used to download an entire Web site, complete with the full content of all hyperlinked pages. Once all information is pushed to the local hard disk, the user may view each page with perceived instant access.

[F] Cookies

The Internet is a cacophony of intermittent client-server relationships. Users log onto Web sites, obtain bits and pieces of information and then browse elsewhere, with no guarantee they will ever revisit the same site twice. Thus, it is not practical for each Internet server to save the state of every dialogue it has ever had with every client application that has accessed it. Most server sites are stateless. They provide information when requested and then promptly forget that the exchange ever occurred. Of course, some Internet transactions require a way to hold an ongoing dialogue between client and server. Cookies represent one of the most popular ways to accomplish this.

Cookies are pieces of data that are placed on the hard drive of a client computer, usually by a web server. The data stored within these cookies saves the particular state of a transaction, so that the transaction may be recalled and continued at a later date or time. When you visit a Web site to conduct an e-commerce transaction — buying a book through an Internet-based bookseller, for example — the browser software on your computer examines the URL address of the bookseller's server and then checks its cookie file to see if it has stored a cookie associated with that URL address. If so, the browser sends the cookie information to the bookseller's server, thereby "refreshing" the server's memory regarding the state of the previous transaction between your client application (browser) and the server. Because the server typically forgets everything it knows about you immediately after processing your request, the previous transaction "recalled" through the stored cookie may represent a selection you had made only seconds before. The cookie would store, for example, the identity of items you have just placed in your e-commerce shopping cart. If the browser finds no cookie associated with the URL address being accessed, the server may place a cookie inside the cookie file for use next time.

Cookies represent an important component of client-server interaction. They allow these inherently stateless communication mechanisms to save a dialogue context so that more complex interaction is possible.

[G] Streaming Audio and Video

When the Internet evolved from a text only medium to a text plus graphics medium, most modem connections were quite slow. Graphic images required far more data than text; thus graphic images took much longer to download through the slow modem connections. At first this was a significant problem. Web pages with rich graphic content took an unacceptably long time to download. Engineers attacked this problem using compression. Special graphic data formats were devised to compress graphic image files by selectively discarding data that was not critical to image quality. Today, the most popular compressed graphics formats are GIF and JPEG. Web browsers can read these file formats without special software.

Today, data compression techniques and faster Internet connections have largely eliminated the long wait that was once experienced when accessing static graphic content such as images, photographs, corporate logos, artwork, and computer icons. However, consumers always want more. With static graphic content problems mastered, consumers now want audio and video delivered to their Internet web browsers and, once again, they don't want to wait. Consumers want to enjoy an audio or video clip on a computer, the computer must be capable of performing the clip in real time, that is, at the same rate as the clip was originally recorded. Unfortunately, unless a very fast Internet connection is available, it may not be possible to deliver the audio and video data in real time.

Engineers have attacked this problem with buffers. They devised special client applications that would store incoming audio or video data in a special storage location or buffer within the computer's local memory. Once stockpiled, the data within the buffer would be sent to the media player while additional data would continue to flow into the buffer from the Internet. By sizing the buffer appropriately for the speed of the Internet connection, engineers have achieved smooth playback.

While buffers could provide the mechanism for real time playback, engineers quickly discovered that the error correcting retransmission feature of the TCP protocol was actually counterproductive. If an audio or video data packet was garbled during transmission, the TCP protocol required it to be resent. If enough packets were garbled, due to a poor telephone connection for example, the buffer might empty and the player would need to interrupt playback while the garbled packets could be resent. The solution to this problem was to use a different protocol, one that did not enforce error correcting retransmission. The protocol selected was the User Datagram Protocol or UDP. The UDP protocol works seamlessly with the IP protocol, simply replacing TCP while continuing to use IP for packet delivery.

§ 7.02   Secure Electronic Commerce

For electronic commerce to exist at all there must be trust. The consumer wishing to purchase a book over the Internet must trust the e-commerce channel enough to submit her credit card number, and must trust the merchant to fill her order. The merchant, receiving the order, must trust the consumer is who she says she is and that there are funds behind the proffered credit card number. Trust is not blind. Internet system developers have already constructed fairly complex systems to secure our trust and the valuable information we send over the Internet. A knowledge of these security systems is essential if we are to fully exploit electronic commerce patent protection.

There are perhaps four basic security issues underlying electronic commerce. They are:

1. Confidentiality — ensuring others cannot read your information, or steal it;
2. Integrity — preventing others from creating, deleting, or corrupting information you rely upon;
3. Regulation of Use — authorizing use of your information and preventing malicious damage or theft;
4. Non-repudiation of Contract — preventing participants from denying they took part in an electronic contract transaction. (Yes, that's my name, but I didn't sign this check.)

How have Internet system developers addressed these issues? In a word: cryptography. Developers have devised a panoply of encryption techniques that make the Internet more confidential, more secure, and more binding.

[A]   Cryptography Basics

It is fairly straightforward to understand how cryptography may be used to impose confidentiality. Every child who has ever played with a secret decoder ring understands how one letter can be exchanged for another, rendering a written message unreadable to anyone who does not have the ring needed to reverse the process. In cryptographic terms, the original message is called the *plaintext* message; the encrypted version is called the *ciphertext* message. The cryptographic algorithm used to perform encryption and decryption is called the *cipher*.

The childhood cipher is not terribly secure. One could begin to guess at the letter substitution list and eventually reverse engineer the entire decoder ring. Fortunately for the youthful cryptographer, most childhood messages, furtively passed in study hall or English class, are likely to be deemed unimportant by all but the most relentless young code breaker. Thus the decoder ring provides reasonably effective privacy for classroom communication. In short, the trial and error needed to break the code is not worth the effort.

Computers, however, can devour mindless trial and error tasks. Give the decoded message to a reasonably fast desktop computer and the decoder ring's secrets can be revealed in an instant. Assuming the decoder ring has been used for simple one-to-one letter substitution, a simple decryption algorithm, with prior knowledge of spelled words, seeks out all of the one letter words (a, I) and constructs hypotheses on which letters represent possible substitutes for those. The algorithm then moves on to two letter words (an, or, my) and constructs further hypotheses. Possibly some of these hypotheses will rule out some of the hypotheses previously made. If so, the incorrect hypotheses are discarded. By the time the algorithm has moved on to three letter words (the, cat, dog, Bob) and four letter words (Dave, coat, bill, boat) many of the ambiguities will have been resolved.

Clearly any code breaker with even rudimentary programming skill can break letter-for-letter substitution — decoder ring — encryption. One reason the decoder ring is so easy to decrypt is that encrypted words have the same number of letters as their unencrypted counterpart. The white space between words gives a valuable clue about the number of letters each word contains. We can make a marked improvement in the decoder ring's encryption scheme by treating white spaces as any other characters. Making that simple improvement greatly increases the decryption difficulty. To illustrate, the following sentence uses a decoder ring encryption technique to encrypt the beginning of a popular nursery rhyme:

bowrmfoamomjyeejqmjobc

mary had a little lamb

Hint: the encryption algorithm is based on the QWERTY keyboard layout.

The childhood decoder ring uses what cryptographers would call a *substitution cipher*. A simple substitution cipher, such as that employed by the decoder ring, replaces a character of the plaintext with one character of the ciphertext. A more sophisticated variation, called *homophonic substitution*, replaces a single character of plaintext with one or more characters of ciphertext. Thus, the letter 'a' could be encoded as 847523. A further improvement, called *polyalphabetic substitution*, involves using different sets of substitution ciphers, where the set chosen for use changes throughout the plaintext message. The set chosen could change with the position of each character of the plaintext, for example. This would be like using a different decoder ring for each letter of the message to be encoded. "Mary had a little lamb" would require 22 decoder rings if you count the spaces between words as letters. Clearly, long messages would require lots of decoder rings (cryptograhers would call these *keys*). Usually a polyalphabetic substitution cipher will use a fixed number of keys and will then recycle the keys after all have been used. If there were 26 one-letter keys, then every twenty-sixth letter would be encrypted with the same key. The cipher would be said to have a *period* of 26. The longer the period, the harder the cipher is to break.

A popular variation, called *polygram substitution*, encrypts blocks of characters as groups. Thus "AD2T" might become "M39Y," "AD3T" might become "9Y76" and so forth. The British used a polygram substitution cipher, called the Playfair cipher, during World War I.[1] Another popular variation, called a *running-key cipher*, uses one text to encrypt another. One could use the text of Macbeth to encode the text of this book. For example, each letter of the following quote from Macbeth could tell which decoder ring to use on each letter of this paragraph.

> Round about the cauldron go; In the poison'd entrails throw. Toad, that under cold stone Days and nights has thirty-one Swelter'd venom sleeping got, Boil thou first i' the charmed pot.

As you are probably beginning to see, the number of possible ciphering schemes is limited only by human ingenuity. Each of the above substitution ciphers involves replacing letters of plaintext with other letters. Another approach, which is particularly popular with Hollywood filmmakers, involves leaving the letters of the message intact, but shuffling their order around so that the message is obscured. Cryptographers call this a *transposition cipher*. A simple example involves arranging the plaintext message in a fixed width column written horizontally and then encrypting by writing out the ciphertext vertically.

Plaintext: E-COMMERCE PATENTS ENHANCE A PATENT PORTFOLIO

```
E—C O M M E R C E P A
T E N T S E N H A N C E
A P A T E N T P O R T F
O L I O
```

Ciphertext: ETAO-EPLCNAIOTTOMSEMENENTRHPCAOENRPCTAEF

Transposition ciphers produce good visual effects favored by movie producers. The next time you are watching a movie involving encrypted messages, see if you can spot a transposition cipher as the animated letters displayed on the decryption computer scurry about to exchange places in a columnar grid.

The preceding encryption techniques are mere child's play in comparison with the complex infrastructures computer scientists have developed to support electronic commerce. Several sections later in this chapter will describe some of the more popular encryption infrastructures. However, before moving on to these topics, we should explore how Internet systems engineers address the basic security issues — confidentiality, integrity, regulation of use, and non-repudiation of contract.

---

[1] D. Kahn, *The Codebreakers: The Story of Secret Writing* (New York: Macmillan, 1967).

[B] Using Cryptography to Address Security Issues

Encryption techniques play an important role in addressing each of the general security concerns described in the preceding section, namely, confidentiality, integrity, regulation of use and non-repudiation of contract. While it is fairly easy to see how encryption techniques apply to some of these concerns, it may not be readily apparent for others. The next subsections treat each of these security issues in greater detail.

[1] Ensuring Confidentiality Through Encryption

Using encryption techniques to ensure confidentiality is straightforward. Confidential information is encrypted at its source and thereafter stored and disseminated only in its encrypted form. Assuming the encryption cipher is not broken, the confidential information remains secure. The encryption technique changes the information itself, so that even if it is intercepted it cannot be read or used.

Clearly, if a message is to be passed in encrypted form, both message originator and intended message recipient need to know how to apply the encryption-decryption algorithm. Encryption algorithms typically perform an encryption transformation upon the plaintext message. Often the transformation is based on an independent data value known as the *encryption key*. Decrypting the message involves a reverse transformation, using an often related data value known as the *decryption key*. How to make both parties aware of these keys, without broadcasting them to eavesdroppers of the world, poses an interesting dilemma.

One logical, yet less than ideal, solution is to provide originator and recipient with the secret keys ahead of time. *Party A* wishing to send *Party B* an encrypted file, could pick up the telephone ahead of time and let *Party B* know, ''The secret password is FRED.'' That works reasonably well, so long as *A* and *B* can be sure no one is listening in on their telephone call. What if *A* and *B* meet through a public Internet Web site. How can *A* publish the key to *B* over this public Web site for *B* to use, without making it available to everybody? There are numerous solutions. An important one worth knowing about is the *public key cryptosystem*. A brief overview of one particular public key cryptosystem, the Diffie-Hellman cryptosystem, is described in the following subsection. The public key cryptosystem is but one of several cryptosystems in use today. Subsection [E] below describes more of them.

[a] Public Key Cryptosystems

Public key cryptosystems have been around since 1976, when the concept was first introduced by Whitfield Diffie and Martin Hellman of Stanford

University.[2] Instead of using a single encryption-decryption key as prior systems had done, Diffie and Hellman proposed a system that used a pair of related keys — one key for encryption and the other key for decryption. The encryption key, called the *private key*, is known only to the encrypting party and remains secretly guarded. The decryption key, called the *public key* is made freely available to the public. The private and public keys work in concert. Both are mathematically linked to the same underlying encryption scheme. Yet, the mathematical link is chosen so that even given knowledge of the public key, it remains infeasible to determine the private key. The original Diffie-Hellman algorithm used the mathematics of discrete logarithms as the link between the public and private keys. Discrete logarithms are very difficult to calculate as compared with the reverse process of calculating exponents. The original algorithm worked like this.

Albert and Betty agree on two large integers, n and g, where g is less than n but greater than 1. These two integers comprise the public key. They do not have to be secret. Albert then chooses a large random integer x and computes the following intermediate value:

$$X = g^x \bmod n$$

In case your math is a bit rusty, the *mod n* term in the above expression simply says to use base n arithmetic (we use base 10 — ten digit — arithmetic at the grocery store, computers use base 2 — two digit — arithmetic, any base will do for this calculation).

Betty, using the public key information computes a similar intermediate value:

$$Y = g^y \bmod n$$

Next an important exchange of information takes place. Albert sends X to Betty and Betty sends Y to Albert. Note that Albert keeps x, and Betty keeps y secret during the exchange.

Both parties, using the information just exchanged, compute a common key. Albert computes the key, using the information received from Betty along with the value x secretly retained, as follows:

$$k = Y^x \bmod n$$

Betty computes the same key, using the information received from A along with the value y secretly retained, as follows:

$$k' = X^y \bmod n$$

[2] W. Diffie and M. Hellman, "New Directions in Cryptography," 1976 IEEE Transactions on Information Theory, vol. IT-22, no. 6, pp. 644-645.

It takes a bit of mathematical reasoning to figure this out, but Albert's newly calculated key k, and Betty's newly calculated key k' are the same number. It turns out that both k and k' are equal to g^{xy} mod n. Thus Albert and Betty have, in effect, agreed upon a value to serve as an encryption-decryption key, without revealing it to any third party. Albert and Betty each computed the key independent of the other.

The Diffie-Hellman algorithm works because it is extremely difficult to extract the x and y values hidden within the intermediate calculations $X = g^x\,mod\,n$ and $Y = g^y\,mod\,n$ that were publicly exchanged. To do so requires calculating discrete logarithms, a process that is very difficult and computationally expensive.

You do not need to fully understand the mathematics of the Diffie-Hellman algorithm to get the gist of how public key cryptosystems work. Albert and Betty needed to agree on a common key that could thereafter be used to control encryption and decryption. Instead of picking up the phone and agreeing that "The secret password is FRED," they spoke in mathematical tongues. They communicated certain information over an insecure channel (the public key values n and g and the intermediate calculations X and Y) and then each independently figured out the secret password by mathematically manipulating the information they publicly exchanged.

[2] Ensuring Data Integrity Through Encryption

It should be self-evident, but encrypting the information does not protect the integrity of the information. Encrypting a confidential file does not prevent someone from deleting that file or corrupting the information in it. One straightforward way to guard against attacks on data integrity is to use a lock. Most computer users are familiar with the process known as "logging on" whereby the user must enter ID and password before he or she is given access to computer files and resources. The data stored in the computer may not be encrypted. However, the user ID and password creates the functional equivalent of a locked box. Without the user ID and password "key" the box cannot be opened and the contents cannot be revealed, deleted, or tampered with.

While user IDs and passwords are by far the most popular form of lock used today, there are other alternatives. Hardware locks, called *dongles,* are sometimes used to prevent unauthorized access to computer resources. The dongle is a hardware-based security device that attaches either to an input/output port of a computer, or to the computer bus. Often the dongle is designed to attach to the serial or parallel port of the computer, or to the PCMCIA (PC Card) slot. The dongle contains a hardware key that uses codes and passwords embedded in the key to control access to software applications or data. Even though a user may otherwise have full access to the computer, a software application protected in this way will not run unless the dongle is attached to the computer. The dongle is effective because it is very difficult to reverse engineer the embedded codes or keys embodied in its hardware.

A somewhat less secure variation of the dongle is the software key. Functionally similar to the dongle, a software key can be implemented on a diskette or CD-ROM. Software applications protected in this way will not run unless the key diskette or CD-ROM is inserted in the appropriate disk drive of the computer.

[a] Checksums

What if it is not possible to lock the computer with a key? For example, what if the information needs to be disseminated across an insecure channel — such as over the public Internet. How can data integrity be ensured when information cannot be stored in a locked box? Computer engineers have devised several solutions. One is to attach a *checksum* to the information being protected. The checksum is a value that has been calculated based on the content of the information it protects. To verify the integrity of the information, the user simply repeats the calculation process that produced the checksum in the first place and compares the result with the checksum value previously attached to the information. If the values match, this is a good indication that the information has not been tampered with.

By way of simple illustration, imagine that each letter of the alphabet is assigned a unique number, a = 1, b = 2, c = 3, etc. A simple checksum could be calculated by adding the numeric values of all letters in a given word and appending that value to the end of the word.

$$BAT \ [23] - that \ is, \ 2 + 1 + 20$$
$$MAT \ [34] - that \ is, \ 13 + 1 + 20$$
$$etc.$$

If one or more of the letters is altered (either deliberately, or because of faulty transmission) the checksum alerts us to the condition.

Originally, computer engineers used checksums to ensure that data being sent over noisy communication lines was received intact. For that purpose it did not much matter that the checksum value was readily discernible as part of the data stream. In its classic form, the checksum was not much of a security mechanism. Anyone with the savvy to identify the checksum (usually tacked on to the end of the data stream) could readily change the checksum value to hide the fact that data in the stream had been tampered with.

The answer was to encrypt the checksum value. Thus, unless the tampering party knew the encryption cipher, he or she could not manufacture a new checksum to hide the fact that the data stream had been tampered with.

[b] Data Hiding

Another technique for ensuring data integrity is *data hiding*. The data hiding technique involves embedding identification data — such as a digital watermark — directly within the information being protected. Usually great care is taken to mask the embedded data's presence. Thereafter, if the information data is tampered with, chances are good that the identification data will have been tampered with as well, particularly if it was cleverly hidden in the first place.

The technique is functionally quite similar to an old spy trick used by James Bond in one of Ian Flemming's popular novels. Plucking a hair from his head, James adhered it with saliva across the gap between two closed French doors of his hotel room closet. Later that evening, upon returning from a black tie event full of intrigue and mystique, James checked his closet doors. Finding the hidden hair intact he knew no one had tampered with the contents of his closet. That is data hiding in a nutshell.

Data hiding techniques have become quite popular with purveyors of audio and video content. Digital watermarks embedded in the audio, video, or still image data files show, by their presence, that the data files are authentic. Imperfect copies of the audio, video or image data content lack the authenticating watermark and may be spotted instantly as a bootleg copy. Why is this so? An example will demonstrate.

Assume we are dealing with an enterprising college student, Brent, who has an award winning collection of Metallica CD's he wants to share with his girlfriend over the Internet. Both are enrolled at the same university. State-of-the-art Internet access is not a problem. Although most of Brent's collection is pre-Napster, the most recent Greatest Hits CD in Brent's Metallica collection has an embedded ''Ride the Lightning'' watermark as hidden data in one of the CD's title track header files. Unaware of the hidden data, Brent copies his entire collection to the Internet, using a popular mp3 compression algorithm. As a result of the mp3 compression process, the ''Ride the Lightning'' watermark does not get reproduced in Brent's Internet copies.

Several months later, an offshore CD pressing house of dubious scrupples downloads all of Brent's Metallica collection and re-records them on CD's for distribution to non-U.S. markets. Although the re-recorded CD's are based on somewhat lower quality mp3 data files, the sound quality is still excellent to most listener's ears. The jewel case artwork is meticulously copied. By all outward appearances, these bootleg copies present themselves as authentic — except for the digital watermark. That was lost when Brent made his original copy for his girlfriend.

Eventually bootleg copies of the re-recorded Metallica CD's begin showing up in discount bins in the United States. How does Metallica's legal counsel prove to U.S. Customs that the imported Metallica products are bootleg? You guessed it — the absence of the hidden digital watermark is clear and convincing proof that these CD's are contraband.

The preceding example perhaps unduly focuses on the content provider — in this case, Metallica and its authorized distributors. The data integrity issue actually impacts both content purveyor and content recipient. Clearly, a purchaser of the bootleg Metallica CD would have good reason to be disappointed on learning that his or her purchase did not have the full sound quality of the original

CD—even if most audio equipment could not convincingly reproduce the difference.

In 1998, Congress passed the Digital Millennium Copyright Act, which made it illegal to tamper with or remove certain "copyright management information," identifying the copyrighted work or its author. Data hiding techniques could be used to embed such information, hence removal or tampering with such information might run afoul of the Act. The Digital Millennium Copyright Act is discussed in more detail in § 7.02[H][4].

[C] Regulating Use Through Encryption

Digital information and software can move from manufacturing source to end user destination through different distribution channels. The purveyors of digital information select which channel to use, based on what the information is and how it will be used. Often a fundamental trade-off between time and space dictates which distribution channel will be used. Time and space tug in opposite directions. The larger the information file, the longer it will take to deliver. Thus if the information must be delivered quickly, make the file size small—a constraint that implies the delivered content will be less rich in detail. If richly detailed content is desired, find ways to placate the user while he or she waits for delivery—because the large file size implies a long delivery time.

Because of the time-space trade-off, digital information of rich content was initially delivered on CD-ROM or diskette. Indeed, most computer software and digitally recorded music is still distributed via this channel. However, as the speed of the Internet continues to improve, expect to see the purveyors of rich content—computer software, digitally recorded music, and movies—switch to Internet distribution channels to supply their customers. Distribution over the Internet eliminates the manufacturing, shipping, handling, and middleperson expenses in delivering the information product. It also supports a pay-per-use business mode. Thus, expect to see rich content such as computer software, music, and movies being supplied via the Internet on a per use, rental basis.

How the information content provider gets paid when a user accesses its content is one half of the regulation of use issue: access control. Content providers need a way to control access to their content so that they can charge for it. Both parties, content provider and content user, need assurance that the other party is who he claims to be. That is the other half of the regulation of use issue: authentication or assurance of identity.

[1] Access Control

Traditionally, the most common form of access control has been the password or personal identification number (PIN). Most computer users have logged onto at least one computer system that required entry of a user ID and password before the system would grant access to the computer resources. Banks control access to their automated teller machines (ATM) in a similar way. The bank card contains the user ID embedded in the magnetic strip on the back of the card. The user then enters his or her PIN through a numeric keypad to gain access to the ATM services.

To protect systems where the user ID or password must be sent over a public communication channel, such as over the Internet, security experts employ encryption. The user ID, password, and any other sensitive information (e.g., credit card number) are encrypted before being sent and are then decrypted at the receiving end. There are various ways to impose encryption security. In Internet web-based systems, one way is to employ a *Secure Sockets Layer (SSL)*. The Secure Sockets Layer is a communication protocol developed by Netscape Communications Corporation. You can identify a Web site protected by SSL because the uniform resource locator (URL) address begins with ''https://'' (note the 's' appended to the hypertext transport protocol http designation).

Another way to impose encryption security in web-based systems is to utilize *Secure HTTP (S-HTTP)*. Secure HTTP addresses the same basic issues as SSL. Developed by Enterprise Integration Technologies in response to a business consortium request, Secure HTTP differs from SSL in where the security mechanisms are deployed. SSL places the security mechanisms in a separate session protocol layer above the Internet TCP protocol layer. S-HTTP is a security extension to the existing HTTP protocol. Both provide similar functionality.

[2] Assurance of Identity

When one signs a personal check, his or her signature provides assurance to the bank that the information on the check may be relied upon as an expression of the maker's intent to spend his or her hard earned money. Digital signatures provide the same function of assuring identity where digital information is involved. Digital signatures come in different flavors. Most use a form of public key encryption.

Basically, the digital signature is a digital data item or code that accompanies the digitally encoded message. The signature is encrypted using a private key encryption process known only by the originator. The encrypted signature is then decrypted by the recipient, using a public key made available to the recipient by the originator. The private key and public key are designed to work hand-in-hand. In effect, the public key contains the knowledge needed to ''undo'' the encrypting effect imparted by the private key. The private and public keys are selected to provide mathematically astronomical odds against someone else's public key being able to ''undo'' the originator's private key.

After decrypting the originator's signature, the recipient compares it with the originator's advertised identity, which may be sent in plaintext along with the digitally encoded message. If the decrypted signature and originator's identity match, the recipient can be assured that the message is genuine.

There are several popular digital signature systems in use today. In addition, many states and national governments have enacted, or are considering enacting, legislation governing the legal effect of digital signatures. Digital signature technology is extremely important to electronic commerce. Given the importance of electronic commerce to the global economy, it is certainly only a matter of time before digital signatures become a *de facto* standard. Adding to their importance, digital signatures also represent an essential component for creating binding legal documents, as we shall see.

[D] Non-repudiation of Contract

So far, the focus has been on information. Encryption techniques ensure confidentiality of information, ensure integrity of information and help regulate use of information. There is another issue, however, which focuses not on the information, but on the *originator* of the information. The issue may be termed *non-repudiation of contract*. It involves locking in the originator's commitment to an electronic transaction, so that the recipient can rely on the commitment without fear that the originator will later deny having made it. "Yes, that's my name on the signature line of the contract, but I deny having signed it."

Digital signatures serve as the principal safeguard against repudiation, just as conventional, handwritten signatures do for pen and ink contracts. For a digital signature to be effective it must do more than the simple check value indicia did to ensure data integrity, discussed in subsection [2] above. The digital signature must be capable of convincing a third party that the signer's purported identity is genuine. Thus, the digital signature must not be easily faked by a forger. Fortunately, computer scientists have developed several robust digital signature techniques that are very difficult to fake.

[1] RSA Digital Signatures

Three scientists, Ron Rivest, Adi Shamir, and Len Adleman, at the Massachusetts Institute of Technology, developed a clever encryption algorithm which they described in a paper published in 1978.[3] The algorithm has been widely used to implement digital signatures. The algorithm relies on using large prime numbers to fabricate the public and private keys used to encrypt the signature. Prime numbers are those which can be divided only by themselves and one (1, 2, 3, 5, 7, 11, 13, . . . 611, 957 . . .).

The algorithm works like this. Choose any integer number to serve as a *public exponent*, call it e. Then select two large prime numbers at random, call them p and q. Any prime numbers can be chosen with one restriction. When one is subtracted from the chosen prime number ($p-1$, $q-1$) it must not have any divisors in common with the chosen public exponent e. In other words, if the selected prime number minus one is divisible by five, and the public exponent is also divisible by five, choose a new prime number (or choose a different public exponent).

---

[3] R.L. Rivest, A. Shamir, and L. Adleman, "A Method for Obtaining Digital Signatures and Public-Key Cryptosystems," 1978 Communications of the ACM, vol. 21, no. 2, pp. 120-126.

Next, take the two prime numbers and multiply them together. This generates a whopping big number called the *public modulus*, call it *n* = *pq*. Take the public modulus and the public exponent together and treat them as the public key. In true public key encryption fashion, the public key is advertised to the world. The public key is useless, however, without a matching private key.

To generate the private key, a *private exponent*, call it *d*, is selected to meet the following criterion. When you multiply the private exponent with the public exponent and subtract one from the product (*de* − *1*), the resulting number must be divisible by both the first randomly selected prime number minus one (*p* − *1*), and the second randomly selected prime number minus one (*q* − *1*). The values of the public modulus and the private exponent are then used as the private key. The private key is kept secret and is used for decrypting.

What actually happens to all these public and private exponents, modulus and key values gets a bit mathematical. As it turns out, the public and private exponents (*e* and *d*) exhibit the important property that *d* functions as the inverse of *e* when modular arithmetic is used. A message, call it *M*, may be encrypted and decrypted by capitalizing on the following mathematical relationship:

$$(M^e)^d \bmod n = M \bmod n$$

The encryption process for message M involves computing $M^e \bmod n$. Anyone who knows the public key (*n* and *e*) can perform this operation. The decryption process of message M′ involves computing $M'^d \bmod n$. Only the party with access to the private key (*e* and *d*) can perform this operation.

Using the RSA algorithm to construct a digital signature, the message originator encrypts the message to be sent using the private key. The encrypted version of the message serves as the digital signature. The digital signature is then attached to the unencrypted message and both are sent to the recipient. The recipient uses the originator's public key to decrypt the signature, thereby reversing the decryption process used to create the signature in the first place. At this point, the decrypted signature will reveal the originally encrypted message, which should match the non-encrypted version which was also sent. If the two messages do not match, there is forgery afoot.

The RSA algorithm is not impenetrable. The security of the RSA algorithm hinges on the fact that, while it is easy to select two large prime numbers, it is not so easy to factor their product. Not easy, but possible nevertheless. In 1994, scientists and students collaborated over the Internet to successfully crack the encryption code of a public challenge made by RSA originators, Rivest, Shamir, and Adleman 17 years earlier. The public challenge was encrypted using a modulus of 129 digits (429 bits). One hundred twenty-nine digits cast a long computational shadow in 1997 when the challenge was originally made, but it eventually receded to conquerable proportions as the speed of computers rose higher.

The RSA algorithm is far from being defeated, however. Fortunately, the factoring problem at the heart of RSA's success grows astronomically larger as the number of modulus digits increases. Ron Rivest, one of RSA's founders, has estimated that in the year 2000 it will cost $25,000 in computational power to crack a 425-bit modulus and $25 million to crack a 619-bit modulus.[4] This suggests that a 512-bit modulus may not be secure for too long into the 2000s, but that a 1024-bit modulus (the next logical step up) should be secure for quite some time.

[E] Popular Cryptosystems for Electronic Commerce

If encryption algorithms or ciphers are the bricks and mortar of a secure electronic commerce Web site, then cryptosystems are the architectural plans describing how those bricks and mortar should be laid out. This subsection will describe some of the more popular cryptosystems used for electronic commerce. One of these cryptosystems, the Diffie-Hellman public key cryptosystem, has already been discussed in subsection [B][1][a].

Cryptosystems fall into two basic categories, symmetric cryptosystems and public key cryptosystems. Symmetric cryptosystems represent the older of the two categories. In a symmetric cryptosystem both communicating parties hold the same key to the encryption algorithm. The key is kept secret from everybody else. Public key cryptosystems represent a more recent improvement over symmetric cryptosystems. In a public key cryptosystem, the key to the encryption algorithm is calculated by both communicating parties using information that is made available to everybody. The encrypting and decrypting parties typically do not use the same values to calculate the key. The encrypting party calculates the ultimate key using a private key known only to him or her. The decrypting party calculates the ultimate key using a public key available to everybody and other information sent over a potentially public channel by the encrypting party.

[1] Symmetric Cryptosystems

Symmetric cryptosystems have been around since the early 1970s. A distinguishing feature of symmetric systems is both parties have the same key. The key is used to encrypt and it is used to decrypt. Thus symmetric cryptosystems function pretty much the same as the metal key to your front door. If you have two identical keys cut and give one to your spouse, then either you or your spouse can lock and unlock the front door.

[a] The Data Encryption Standard [DES]

The first symmetric cryptosystem to be placed in wide commercial use was IBM's Data Encryption Standard, or DES. It was developed in response to a request from the U.S. Department of Commerce in 1974. The federal government adopted it as a U.S. federal standard in 1977. The banking community later adopted it in 1981 as their standard as well.

---

[4] W. Ford and M. Baum, *Secure Electronic Commerce* (Prentice Hall, Upper Saddle River, NJ, 1997) p. 110.

The DES cyrptosystem works on messages that have been subdivided into blocks of 64 bits. DES is therefore called a *block cipher*. The standard uses a 56-bit key. Internally, the DES system takes a block of the plaintext message and passes it through 16 successive rounds of encryption in which the 64-bit block is split into two 32-bit halves. One half is then fed through a complex mathematical function, turning it into digital mincemeat. This mincemeat is then recombined with the other half using a Boolean exclusive-OR operation. The complex mathematical meat grinder used on the first half contains eight non-linear, table-specified substitutions known as S-boxes. After each of the successive 16 rounds, the first and second halves of the block swap places so that each 32-bit half is forced several times through the meat grinder.

The final output of the 16th round through the meat grinder serves as the encryption key, which is then applied to the plaintext message to encrypt it. The decryption process takes the same form, except that the selected portions of the key used internally for the 16 rounds are used in reverse order. Because the DES S-boxes contain non-linear components, it is exceedingly difficult to guess what the encryption key is by looking at the encrypted message.

The DES cryptosystem is not impenetrable, however. Any block cipher can be attached through a brute force approach of trying every possible key value until one works. Breaking the DES code would therefore be like trying every possible combination of numbers on your bicycle lock, until you discover the one combination that works. Clearly, the larger number of digits in the key, the more difficult the DES cipher becomes to break. The power of current computers has rendered the original 56-bit DES encryption key somewhat insecure.

You can probably guess the proposed solution. Increase the size of the key. One proposed way to accomplish this, called *Triple-DES*, increases the effective size of the key by using three DES keys in succession. Key One is used to encrypt the message, Key Two is used to decrypt the encrypted message from Key One. Key Three is used to encrypt the result from Key Two. If desired, Keys One and Three can be the same. Some experts believe that use of three keys in this fashion is many orders of magnitude stronger than DES. The current drawback to using this more secure approach is the computational burden it places on the computer system doing the encryption and decryption.

[b] SKIPJACK and the Clipper Chip

In 1993 the U.S. government proposed a new encryption scheme, known as SKIPJACK, that uses an 80-bit key to perform a 64-bit block cipher. The scheme has a cryptographic strength several orders of magnitude stronger than DES. How SKIPJACK works precisely is classified. SKIPJACK has gained considerable notoriety, not because of its improved strength, but because of a backdoor that was built into the system to allow law enforcement officials to break the SKIPJACK code. The backdoor could only be unlocked by a hardware device called the *Clipper Chip*. Anyone possessing the Clipper Chip could break SKIPJACK's powerful code and thereby read the plaintext message.

The U.S. government had planned to require a Clipper Chip to be installed in every computer, modem, telephone, and television sold in the United States. That way, law enforcement officials could activate the Clipper Chip and decode encrypted messages being sent to or from those devices. Public outcry against the plan was considerable. Many expressed concern that this Clipper Chip was a significant step towards the feared Big Brother is Watching You world described in George Orwell's novel, *1984*.

[c] Rijndael

Faced with the realization that the venerable DES encryption standard would eventually be undermined by increasingly fast computers, the U.S. government held a contest to find a replacement encryption scheme. Administered by the National Institute of Standards and Technology (NIST), the government launched the effort on January 2, 1997, with a formal call for cipher algorithms. The effort was named the Advanced Encryption Standard (AES) Development Effort.

The call stipulated that the AES would specify an unclassified, publicly disclosed encryption algorithm(s), available royalty-free, worldwide. In addition, the algorithm(s) must implement symmetric key cryptography as a block cipher and (at a minimum) support block sizes of 128-bits and key sizes of 128-, 192-, and 256-bits.

On August 20, 1998, NIST announced a group of 15 AES candidate algorithms at the First AES Candidate Conference (AES1). These algorithms had been submitted by members of the cryptographic community from around the world. At that conference and in a simultaneously published Federal Register notice,[5] NIST solicited public comments on the candidates. A Second AES Candidate Conference (AES2) was held in March 1999 to discuss the results of the analysis conducted by the global cryptographic community on the candidate algorithms. The public comment period on the initial review of the algorithms closed on April 15, 1999. Using the analyses and comments received, NIST selected five algorithms from the 15.

The AES finalist candidate algorithms were MARS, RC6, Rijndael, Serpent, and Twofish. MARS was submitted by IBM; RC6 by RSA Laboratories; Rijndael by two Belgian computer scientists, Joan Daemen and Vincent Rijmen; Serpent by European computer scientists, Ross Anderson, Eli Biham, and Lars Knudsen; and Twofish by Counterpane Internet Security, Inc. These finalist algorithms received further analysis during a second, more in-depth review period prior to the selection of the final algorithm(s) for the AES FIPS.

Until May 15, 2000, NIST solicited public comments on the remaining algorithms. Comments and analysis were actively sought by NIST on any aspect of the candidate algorithms, including cryptanalysis, intellectual property,

---

[5] Federal Register: September 14, 1998 (Volume 63, Number 177), pp. 49091-49093.

crosscutting analyses of all of the AES finalists, overall recommendations, and implementation issues.

Near the end of Round 2, NIST sponsored the Third AES Candidate Conference (AES3) — an open, public forum for discussion of the analyses of the AES finalists. AES3 was held April 13-14, 2000, in New York, NY, USA. Submitters of the AES finalists were invited to attend and engage in discussions regarding comments on their algorithms. Following the close of the Round 2 public analysis period on May 15, 2000, NIST studied the available information and ultimately selected Rijndael as the winner.

Rijndael is pronounced approximately like "Reign Dahl," "Rain Doll," or "Rhine Dahl." Daemen and Rijmen coined the name, as they report on their Web site, "because we were fed up with people mutilating the pronunciation of the names 'Daemen' and 'Rijmen.'"[6] Cryptically, they state that there are two messages in the preceding answer.

Rijndael has at its core an operation called the *round transformation*. Unlike some other symmetric encryption algorithms, in which bits within intermediate states are simply transposed unchanged to another position, based on something called the *Feistel Structure*. The Rijndael round transformation concentrates considerably more on mixing things up at the bit level. The Rijndael algorithm performs bit-level manipulation in a series of three distinct transformations, called *layers*. These transformations comprise a linear mixing layer, a non-linear layer, and a bitwise exclusive-OR addition layer. Within each layer, all bits are treated in a similar way. Specifically, the Rijndael round transformation comprises a non-linear byte substitution operation, row and column shifting operations, and a round key addition operation. Addition is performed at the bit level via an exclusive OR operation.

Based on this underlying three level manipulation, multiple rounds are performed using *round key*s that are extracted from an initially selected *cipher key*. The cipher key is selected and expanded into a linear array of four-byte words and the round keys are selected from this expanded cipher key.

After multiple rounds have been iteratively performed, the resulting ciphertext is very difficult to attack. Rijndael's creators attribute this to the fact that the cipher behaves in an asymmetrical way, despite the large amount of symmetry present in the algorithm. This is obtained by the round constants that are different for each round. The fact that the ciper and its inverse use different components practically eliminates the possibility of generating weak keys — a known problem with DES encryption. The non-linearness of the key expansion practically eliminates the possibility of finding equivalent keys.

Whether Rijndael will prove to be as resistant to attack as its creators claim remains to be seen. However, for the present, Rijndael is the U.S. government standard for symmetric encryption, hence it will likely be implemented in a wide variety of consumer and commercial products. Under the terms of the AES

---

[6] <http://www.esat.kuleuven.ac.be/~rijmen/rijndael/>

competition, Rijndael has been placed in the public domain. Its creators have represented that to the best of their knowledge Rijndael is not covered by any patents.

[F] Popular Security Systems for Sending Messages

The cryptosystems described thus far have multiple uses. They can be used to secure credit card transactions, encrypt confidential business information, lock and unlock proprietary software applications delivered over the Web, and for many other uses. These more sophisticated e-commerce uses aside, still probably one of the most important uses of encryption technology is for messaging security. For many, e-mail has replaced other forms of electronic communication, such as fax, and voice mail. Instead of sending a colleague a faxed document, many prefer sending the document as an e-mail attachment. The attachment can be sent in its native format, to allow the recipient to edit the document, or it can be sent as a portable electronic document (such as an Adobe Acrobat document) that can be viewed and printed, but not edited.

When it comes to e-mail security, there are several popular ways of providing it. Perhaps the most popular are those that are compatible with the popular MIME protocol used for encoding e-mail attachments. The MIME protocol, or Multipurpose Internet Mail Extensions protocol, is actually a set of specifications for structuring a message into body parts of different content types. Supported content types include text, images, audio, and completely encapsulated messages. Although MIME is not the only e-mail attachment encoding protocol in use today, it is by far the most popular. Chances are good that if you regularly receive e-mail attachments from colleagues, they have been encoded using MIME.

There are two prevalent systems for adding security to MIME-encoded documents. They have similar security goals but approach the problem in somewhat different ways.

[1] MIME Object Security Services (MOSS)

The MIME Object Security Services, or MOSS, was developed by the Internet Engineering Task Force (IETF) and completed in 1995. The IETF is an unofficial Internet standards body. The IETF Web site describes itself as follows:

> The Internet Engineering Task Force is a loosely self-organized group of people who make technical and other contributions to the engineering and evolution of the Internet and its technologies. It is the principal body engaged in the development of new Internet standard specifications. Its mission includes:
>   • Identifying, and proposing solutions to, pressing operational and technical problems in the Internet;
>   • Specifying the development or usage of protocols and the near-term architecture to solve such technical problems for the Internet;

• Making recommendations to the Internet Engineering Steering Group (IESG) regarding the standardization of protocols and protocol usage in the Internet;

• Facilitating technology transfer from the Internet Research Task Force (IRTF) to the wider Internet community; and

• Providing a forum for the exchange of information within the Internet community between vendors, users, researchers, agency contractors and network managers. The IETF meeting is not a conference, although there are technical presentations. The IETF is not a traditional standards organization, although many specifications are produced that become standards. The IETF is made up of volunteers who meet three times a year to fulfill the IETF mission.[7]

Providing security for messages can potentially involve three different cases: digital signatures, encryption, and digital signatures plus encryption. The MOSS specification provides specific solutions for the first two. The MOSS specification breaks a MIME-encoded message into two body parts, the first part containing any MIME content the originator wishes to send (text, image, audio, structured type, etc.) and the second part containing a digital signature and any control information needed by the recipient to verify the signature. In both digital signature and encryption cases, the body part ends up being passed through an encryption algorithm. In the digital signature case, the encrypted body part serves as the signature. In the encryption case, the encrypted body part serves as the encrypted message.

Both cases present a communication problem because the Internet messaging environment uses a text-based transport system with special characters to designate carriage returns [CR] and line feeds [LF]. The system is designed to carry character-encoded messages, not binary data. After the body part has passed through a potentially random encryption process there is no telling what it will look like, and no telling what has happened to the carriage returns and line feeds. After encryption, there is no way to tell whether the message was originally encoded as ASCII characters (used by personal computers) or as EBCDIC characters (used by IBM mainframes). Likewise, there is no way to tell whether lines of text within the message were terminated by carriage return plus line feed [CR + LF] or simply carriage return [CR] alone — both standards are in use today.

To address this ambiguity, the MOSS specification dictates that the first body part, containing the message content, must be *canonicalized* prior to encryption. The canonicalization process involves converting the content, regardless of its original encoding form, into text-based strings using an agreed upon set of character encoding rules dictated by the MOSS specification. It is necessary to follow these rules, otherwise the encrypted message or digital signature, computed from the message, may not be recognized by the recipient.

---

[7] <http://www.ietf.org/tao.html#What_Is_IETF>

To effect a digital signature under the MOSS specification, the originator's private key encrypts the canonicalized body part containing the message content. The result is stored in the second body part containing the signature. The message is stored in the first body part, unencrypted.

To effect encryption, the MOSS specification provides a somewhat more complex procedure. After putting the message part into canonical form, a random data encryption key is generated on the message and used to encrypt the message, which is stored in the second body part. The same random key is then further encrypted using an RSA public key to generate an encrypted key which then placed in the first body part of the MIME message. If you think about it, the MOSS specification for encryption actually uses a symmetric cryptosystem to encrypt the message. The random key is a symmetric key. The random key is used to encrypt the message and that same random key is needed to decrypt the message. The MOSS specification uses a public key cryptosystem to securely pass the symmetric key. The recipient's public key encrypts the symmetric key so that it may be safely passed in the first body part of the multipart MIME message.

[2]   S/MIME

The S/MIME specification was developed by a private group lead by RSA Data Security, Inc. RSA Data Security, Inc. owns several encryption algorithm and communications security patents, including several invented by Ronald Rivest. Rivest is one of the inventors of the basic RSA encryption algorithm. The basic patent on that algorithm was issued to the Massachusetts Institute of Technology.[8] The S/MIME specification adapts for MIME-compliant security a set of *de facto* public key encryption standards, known as *Public-Key Cryptography Standards* (PKCS). Also developed by RSA Data Security, Inc., these standards include a specification (PKCS #7) that defines the data structures and procedures used for digitally signing and encrypting other data structures. S/MIME simply applies those standards to the MIME body part.

As with the MOSS specification, described above, the MIME body part to be signed must first be canonicalized, to ensure proper passage through the character-based e-mail transport system. The body part, in canonical form, is then signed using the originator's public key, and treated as a PKCS #7 data structure under the PKCS standards. The PKCS standards specify use of an international standardized data type, known as *Abstract Syntax Notation One* (ASN.1) instead of character encoding used by MOSS. This requirement is largely an implementation detail. The ASN.1 encoded content is then further processed using *base 64 encoding*, so that the message will be able to travel, uncorrupted, through the character-based e-mail transport system.

In addition to supporting digital signatures, the S/MIME specification can be used to encrypt the body part of a message, just as MOSS was able to do. Except for ASN.1 encoding, the basic procedure for encrypting content under

---

[8] U.S. Patent No. 4,405,829, Rivest et al., Cryptographic Communications System and Method.

S/MIME is similar to that used by MOSS. The message is encoded using symmetric key encryption (both sender and recipient use the same key) and the symmetric key is sent to the recipient using public key encryption.

S/MIME, in its basic form, has one shortcoming over MOSS. The recipient of a S/MIME signed message must have an S/MIME-compliant e-mail reader. Otherwise the recipient will not be able to read the message, even if it has not been encrypted. To address this shortcoming, the S/MIME developers employ a multi-part message having two body parts. One body part contains the signature (generated using S/MIME) and the other body part contains the original message, unchanged by any S/MIME processes. A recipient of this multi-part message would be able to read the original message even if the e-mail reader is not S/MIME compliant and is thus not able to verify the digital signature.

[3] Pretty Good Privacy (PGP)

Pretty Good Privacy, or PGP as it is more often referred to, spread throughout the world like a virus after its creator, Phil Zimmerman, gave a copy to a friend who released it. The story of PGP's release is more fascinating than fiction.

Zimmerman, a computer scientist in Boulder, Colorado, developed PGP using publicly available information about various encryption techniques. Like many users of public information infrastructures, Zimmerman was concerned with security. The Internet was particularly vulnerable, Zimmerman recognized, because packets of information sent over the Internet cannot be sealed shut. Anyone who intercepts an e-mail message en route can open the packets and read the message. Zimmerman developed PGP to encrypt the contents of e-mail messages, so that even if the packets were intercepted and opened, they could not be read. Zimmerman planned to offer PGP to the public as shareware.

Zimmerman was friends with one Kelly Goen who lived in the San Francisco Bay area. Goen was also interested in computer security, so Zimmerman gave Goen a copy of his PGP program, including the source code and full documentation. Had history taken a different turn of events, Zimmerman and Goen might have experimented with PGP a bit, sending encrypted e-mail messages between themselves, and then moved onto other interests. PGP, like so many other shareware programs, might have been forgotten.

However, history did not take that turn. Indeed, the Internet was already exerting such a strong gravitational pull on history's direction that PGP found itself in the mainstream of Internet commerce, and Zimmerman found himself the subject of a U.S. federal investigation.

The U.S. Congress, too, was interested in computer security, although Congress was not looking at the issue the way Zimmerman and Goen did. Congress was concerned that strong encryption techniques might allow criminals and unfriendly foreign governments to use the public communication infrastructure to conduct criminal and subversive activities under the very noses of the FBI, undetected. In the wake of the Gulf War, in 1991, Congress introduced a bill,

Senate Bill 266, that would prohibit U.S. citizens from owning strong encryption products, such as PGP. The proposed new law treated all but the weakest forms of encryption as if they were high tech munitions. Owning a shareware PGP encryption program was like owning a nuclear rocket launcher as far as the proposed new law was concerned.

The Internet free speech cognoscente reacted to Senate Bill 266 as if a formerly benign information virus had just mutated into a deadly strain. Word of Senate Bill 266 spread instantly. The cognoscente's message urged citizens to "stock up on cyrpto gear while you still can." Zimmerman's friend Goen wasted little time. Reportedly with Zimmerman's permission, Goen began uploading PGP, source code and all, onto as many public bulletin boards as he could find. Fearing that the FBI was hot on his trail, Goen reportedly drove all over the Bay area with laptop computer and modem with acoustic coupler, using public telephones to upload PGP onto the Internet. Goen feared that the government would try to trace his efforts, seize his computer, and then shut down all bulletin boards that he had reached. Hence he restricted his uploading efforts to furtive moments at public pay phones and then moved quickly on. If enough copies of PGP could be put on the Internet before the federal authorities caught him, Goen believed that PGP would become so infused within the worldwide Internet community that the federal authorities couldn't suppress it. That is exactly what happened.

It took two years, but in February 1993, U.S. Customs officials paid Zimmerman a visit. Even though he had not personally uploaded PGP to the Internet, they charged him with exporting munitions without a license. The Colorado computer scientist found himself the subject of a federal grand jury indictment. The grand jury investigation of Zimmerman lasted three years. Ultimately, the government decided not to pursue the charges.

With all the surrounding controversy, one is inclined to conclude that PGP must be very powerful indeed. Although it does provide a fairly high modicum of privacy (pretty good would be an understatement), Zimmerman based PGP on currently available public key encryption technology. PGP is therefore quite similar to MOSS or S/MIME.

PGP's approach utilizes both symmetric and public key encryption. First, PGP generates a secret key for the symmetric encryption algorithm by using a random number that it generates. The random number is based on information supplied by the user. It then encrypts the message text and attaches the secret key to the end of the message. Once this has been done, PGP utilizes the RSA public key encryption algorithm to encrypt the secret key. Because the secret key can only be decrypted by the holder of the intended private key, only the recipient can decrypt.

PGP, at least in its original implementation, differed from MOSS and S/MIME in how the public keys were distributed. MOSS and S/MIME use recognized public key infrastructures to distribute everyone's public keys. PGP based its key management system on looser, *ad hoc* relationships among the users of PGP. Instead of employing a single trusted entity to manage and distribute the public keys, PGP public keys could be distributed by any PGP user — defined as being in the PGP "web of trust." Any PGP user could certify any other PGP user's public key. Thus, if a PGP user that you trust (the *trusted introducer*) introduces you to a PGP-using friend whom she trusts, your PGP system will accept public keys from this newly introduced PGP site.

Since its original introduction, PGP has undergone change; not at the insistence of the federal government, but in response to negotiations with a private entity. As we know, Zimmerman designed PGP using existing public key encryption technology. Specifically, he used the famous RSA encryption algorithm — a famous *patented* encryption algorithm. Rivest, Shamir, and Adleman patented their RSA algorithm and assigned the patent to the Massachusetts Institute of Technology (MIT). The patent, U.S. Patent No. 4,405,829, entitled "Cryptographic Communications System and Method" was issued September 20, 1983, based on a patent application filed December 14, 1977.

Although the details of the transaction are not publicly available, PGP was revised in its version PGP 2.6 and is currently being distributed by MIT. Reading between the lines, PGP must have required a license under the RSA patent, and Zimmerman and MIT worked out a deal whereby MIT would become the official distribution site for PGP. In an open letter posted on the Internet, Zimmerman made the following comment in support of this new relationship with MIT:

> MIT did not steal PGP from me. This was a joint venture by MIT and myself, to solve PGP's legal problems. It took a lot of manuevering by me and my lawyers and by my friends at MIT and MIT's lawyers to pull this off. It worked. We should all be glad this came off the way it did. This is a major advance in our efforts to chip away at the formidable legal and political obstacles placed in front of PGP; we will continue to chip away at the remaining obstacles.[9]

[G] Systems for Distributing Encryption Keys

All of the previously described encryption systems have relied upon keys. Symmetric encryption systems employ a secret key, shared by message originator and recipient. Public key encryption systems employ a pair of keys, a private one for the message originator and a public one for the recipient. Given the explosive popularity of e-commerce, there are undoubtedly millions of encryption keys being passed back and forth over the Internet. Who manages all these keys? That is a very good question. As it turns out, there are several systems currently in place for managing and distributing keys used by most e-commerce cryptosystems.

Managing encryption keys is a bigger problem than you might at first think. Key management involves at least eight different tasks:

---

[9] <http://www.virtualschool.edu/mon/Crypto/ZimmermannPGP2.6Myths.html>

1. Generation — the key must be created before it can be distributed. Typical key generation algorithms use random numbers based on some seed information provided by the user.

2. Registration — the key, once generated, needs to be associated with the maker or intended user. For instance, the public key used to communicate securely with John Doe must be registered in John Doe's name, somewhere and in some way that is considered trustworthy and reliable.

3. Distribution — users of the key need to obtain access to it. There are many ways to distribute the key to its intended user. Preferably, key distribution employs a trusted distribution authority. Encryption users want assurance that the key they have been given is indeed the key for their intended message recipient, and not some impostor.

4. Backup and Recovery — keys can get lost or damaged. The key management system should have some way to keep backup copies of keys should the original keys become lost or damaged. The backup and recovery system, itself, may need to be secure, to prevent keys from being stolen from the backup reserves.

5. Escrow — at times third parties may need access to the encryption keys of another. Escrow relationships usually arise by business relationship or contract. An employer, for example, might require a key escrow of its employees' keys, to allow it to access secure information in the event the employee leaves. A contracting party might require a key escrow to allow it to access secure information if the other party does not live up to the terms of the contract. Under certain circumstances, law enforcement officials under court order may employ key escrow to allow them to monitor secure communications.

6. Replacement or Update — periodically, encryption keys need to be replaced or updated to maintain security. Many encryption systems rely on mathematical manipulation of very large numbers. These systems are secure because it is computationally difficult to reverse engineer the mathematics employed — difficult but not impossible. Give a fast computer enough time and eventually most encryption schemes can be broken through brute force trial and error. Thus, replacing encryption keys on a periodic basis makes sense. If it would take a Pentium V processor two years to crack the code of a given key by brute force, changing the key every six months renders the system impervious to this attack.

7. Revocation — permission to an encrypted system may need to be retracted for a variety of reasons. Thus the key management system needs a way to revoke rights to previously distributed keys.

8. Destruction — sometimes merely revoking a key is not enough. It may be necessary to eliminate all traces that the key ever existed. For example, assume that a *version A* of a software system is encrypted using a first set of keys. *Version A* contains confidential source code.

Later *version B* is produced and is encrypted using a new set of keys. *Version B* is based on *version A* and thus contains most of the same confidential source code. The software system producer may want to destroy all record of the first set of keys, to prevent a hacker from discovering the confidential source code contained in *version A*.

Many of the popular key management systems are tied to particular encryption systems or standards, such as the DES symmetric key standard (discussed in subsection 7.02[E][1][a]), or the RSA public key system (discussed in subsection 7.02[D][1]).

[1]   Distribution of Symmetric Keys

Symmetric key technology originated prior to public key technology. Banks began using symmetric keys in the 1980s to protect the transmission of financial data. Characteristic of symmetric key technology, both communicating parties use the same key to encrypt and decrypt. Thus one party wishing to communicate securely with another party must find a way to securely pass the symmetric key to the intended recipient. How can this be done over a public communication channel?

The banking industry grappled with this question and ultimately devised a standard, known as ANSI X9.17,[10] that set forth an intricate set of key-within-key rules to promote secure key distribution. The ANSI standard was devised before the invention of public key encryption, so the standards committee proposed using symmetric keys to secure distribution of other symmetric keys. The result was a set of nested rules for protecting the key that protects the key. The standard defines three layers of keys:

1. Session keys — these are used to protect the basic financial data to be communicated.
2. Key-encrypting keys — these are used to protect the session keys, allowing the session key to be communicated securely.
3. Master keys — these are used to protect the key-encrypting keys, allowing key-encrypting keys to be communicated securely.

In the simplest point-to-point communication between one bank and another, for example, a master key might be delivered by courier in a highly physically secure way. The master key would be locked in the respective safes of both banks and used only as needed when updating the key-encrypting keys. Using the master key, key-encrypting keys would be exchanged between banks, and these key-encrypting keys would then be used each time a new session key was required. A new session key might be generated for each communication

---

[10] American National Standards Institute, ANSI X9.17: American National Standard for Financial Institution Key Management (Wholesale), 1985.

session between the two banks. The session key would encrypt and decrypt all wire transfer banking transactions conducted during that session.

The problem with such simple point-to-point communication is that banks typically work in larger cooperative groups. If every bank had to store and maintain unique master keys, key-encryption keys, and session keys with every other bank, the keymaster's pockets would literally be bulging with thousands or millions of electronic keys. To address this geometric explosion of keys the ANSI X9.17 standard devised a *key center configuration* in which a central key center manages the keys for each bank. Each bank exchanges a master key with the key center, but not with each other. The key center then distributes session keys to individual banks as needed to protect communication sessions between banks.

Another system for distributing symmetric keys is *Kerberos*. Kerberos was designed at the Massachusetts Institute of Technology (MIT) as a means for authenticating communication in client-server environments. Kerberos uses DES symmetric cryptography. See subsection 7.02[E][1] above for more about DES symmetric cryptography. In addition to one or more host servers and one or more client workstations, Kerberos employs an *authentication server*. Every client workstation and every host server shares a symmetric key with this authentication server, whose function is to hand out data items called *tickets* containing the session key needed for the secure communication.

The Kerberos process begins when a client communicates with the authentication server asking for a ticket. The authentication server furnishes the client with a ticket containing a DES session key that will be used to secure communication between that client and the host server the client wishes to communicate with. The session key embedded in the ticket is specifically constructed to be interpretable only by the target server the client wishes to communicate with. The client then passes the newly received key to the target server. The server uses the session key embedded in the ticket to encrypt the information requested by the client. It sends the encrypted information to the client, which is able to decrypt the message using the same session key.

[2] Distribution of Public Keys

Public key encryption was introduced in 1976 by Whitfield Diffie and Martin Hellman of Stanford University.[11] It was a radical departure from symmetric key encryption thinking. Public key encryption offered a way to exchange encryption keys over a public communication channel without the need to initially exchange a common master key. This initial exchange had been the Achilles heel of symmetric key encryption systems — how to exchange a common master key without allowing eavesdroppers to discover it.

Public key encryption solved this problem by exploiting the power of mathematics. Certain reversible mathematical functions are easy to compute in

---

[11] W. Diffie and M. Hellman, *supra*, note 2.

one direction, but very difficult to compute in reverse. For example, exponential functions and logarithmic functions compute the inverse of each other. It is comparatively easy to solve the exponential function and far more difficult to solve its inverse counterpart, the logarithmic function. If you can't remember the difference between exponents and logarithms, here is a brief refresher.

The *exponential function* is how you raise a number to a given power, like two squared (two raised to the power of two) is four. The following equation is an exponential function:

$$y = b^x$$

If you know the values of b and x, you can easily calculate the exponential value for y (assuming you remember how to raise a number b to a power x, e.g., $10^2 = 100$). If, however, you know the values of y and b and would like to calculate the value of x, you would do so by using the following *logarithmic function*:

$$x = \log_b y$$

Thus, if b is 10 and y is 100, x would be calculated as $x = \log_{10} 100 = 2$.

If the above refresher hasn't helped, no need to worry. The key point is that exponential functions and logarithmic functions compute the inverse of each other. The knot tied by one function is untied by the other. In comparative terms, one function is easy to perform, the other is difficult.

Public key encryption solves the key distribution problem by using one mathematical function to tie the encryption key into a Gordian knot that is extremely difficult to untie, unless you have the precise inverse mathematical function designed to untie it. The inverse mathematical function designed to untie the knot is generated by the intended *recipient* based on information published by the sender. Thus members of the public who may be eavesdropping on the encrypted key exchange will have no knowledge of the precise inverse mathematical function selected by the intended recipient and are forced to use brute force, trial and error tactics that would need years to untie the Gordian knot.

Public key distribution has proven to be quite convenient and flexible. In fact, many symmetric key encryption systems now use public key encryption to distribute the symmetric keys. This eliminates the need for cumbersome physical delivery of master keys, while allowing otherwise secure symmetric key systems to remain in operation.

[3] Certificates

While public key encryption is convenient, distributing public keys requires care. Because public keys are, as the name implies, public, it is possible to fool someone into believing he is having a secure communication with *party A*, when in fact the communication is with an impostor, masquerading as *A*. The impostor substitutes his public key for the party he is masquerading as. If there is no way to detect that the impostor's key is not genuine, parties wishing to communicate securely with *party A* can be fooled into communicating with the impostor.

To guard against such a masquerade attack, public key distribution systems employ *certificates*. In technical terms a certificate is a data structure that is digitally signed by a party whom the users trust. When a public key is linked to the certificate, users can check the certificate's digital signature, verify that it is genuine, and then be more assured that the public key is genuine as well.

There are several public key certification systems in popular use today. Certification is not a new concept. Certain legal documents have undergone certification for years. For example, imagine that your client wishes to apply to register a trademark in Brazil. The Brazilian government is willing to grant your client's wish, but first your client must submit a signed affidavit, attesting to certain facts regarding use of the trademark. You draw up the affidavit, your client signs it and you are about ready to put it in the mail to Brazil when you realize there is a problem. How is the Brazilian government going to know your client's signature is genuine? Upon investigation you learn of a procedure that will allow the Brazilian government to verify the genuineness of the signature. The procedure works like this.

Your client executes the affidavit in the presence of a notary. The notary attaches his or her signature to the document as witness. Then the affidavit is sent to the Secretary of State, which checks the authenticity of the notary's signature based on notarial registration information the office has on file. The Secretary of State appends its official seal to the affidavit, in effect vouching for the authenticity of the notary's signature. The affidavit then proceeds to the Brazilian consulate. The Brazilian consulate, having previously established diplomatic relations with various Secretaries of State throughout the country, is able to verify the genuineness of the Secretary of State's seal. The consulate applies additional seals and colorful ribbons to the affidavit, adding its certification that the document is genuine. By now the affidavit is dressed up like a military general, all very official looking. When the affidavit is finally sent to the Brazilian government, the consulate's seal, and the procedure by which it was placed on the document, acts as certification that all other signatures and seals are genuine.

Digital certificates work the same way. The most widely used public key certificate format is defined in the ISO/IEC/ITU X.509 standard. The certificate format has evolved through several versions (version 1 in 1988, version 2 in 1993, and version 3 in 1996). Each version added support for new security features, but the basic certificate concept remains the same. Like the affidavit submitted to the Brazilian government in the preceding example, the X.509 certificate contains a collection of information that the recipient can use to verify that the certificate is genuine. This collection of information is placed in pre-assigned fields to allow a computer to process it:

- Version — indicating which version of the X.509 standard the certificate complies with.

- Serial number — uniquely identifying and distinguishing this from all others issued by the certification authority.

- Signature — the digital signature of the certification authority.

- Issuer — providing the name of the certification authority against which the signature is checked.

- Validity — denoting the range of dates during which this specific certificate is valid.

- Subject — providing the name of the holder of the private key for which the corresponding public key is being certified.

- Subject public-key information — providing the value of the public key that this certificate is authenticating.

- Issuer unique identifier — optional additional information used to uniquely identify the issuer in the event the Issuer field is also used by another certification authority.

- Subject unique identifier — optional additional information used to uniquely identify the subject in the event the Subject field is also used by another entity.

The X.509 standard does more that merely define the certificate. It also defines how to address the eight issues discussed in subsection [G][1], namely:

1. Generation
2. Registration
3. Distribution
4. Backup and Recovery
5. Escrow
6. Replacement or Update
7. Revocation
8. Destruction

There are several commercial products currently being offered to support distribution of public keys. One of the more popular ones is VeriSign, distributed by VeriSign, Inc.[12] Another commercial source for public key encryption technology is RSA Security, Inc.[13] You should also check the MIT Web site for public key encryption information as well. MIT distributes the PGP technology discussed in subsection 7.02[F][3].

[H] Legislation Supporting Secure Electronic Commerce

Legislators at all levels of international, national, and state government have jumped on the e-commerce bandwagon. As a result, there are several major bodies of statutory law dealing with electronic commerce, in particular dealing with issues of encryption. This is a rapidly changing area of the law. You may wish to periodically check the Internet for updates that could affect the

---

[12] VeriSign, Inc., may be reached through its Web site at <verisign.com>.

[13] RSA Security, Inc., may be reached through its Web site at <www.rsasecurity.com>.

e-commerce with which you are dealing. The following will present some of the highlights of the current statutory law and proposed legislation.

[1] United Nations Model Law on Electronic Commerce

The United Nations Commission of International Trade Law drafted its Model Law on Electronic Commerce to show how to remove some of the barriers that prevent the spread of electronic commerce today. The Model Law was adopted by that Commission December 16, 1996. The Model Law provides, for example, that a suitably formatted data message can satisfy the requirement of a writing in contracts, or that a properly constructed digital signature may serve as a binding substitute for the conventional handwritten signature. The Model Law is not binding, rather a guideline for legislators wishing to draft binding law to facilitate electronic commerce. The Model Law could become binding law if adopted by a sovereign state, or if adopted as controlling by parties to a private contract. Although the Model Law has not the force of law, its influential presence will undoubtedly ease the transition to a global electronic economy.

Currently several countries and one U.S. state have adopted legislation based on the Model Law on Electronic Commerce. These countries include Australia, Bermuda, Colombia, France, Hong Kong, Ireland, Republic of Korea, Singapore, Slovenia, the Philippines, and the States of Jersey (Crown Dependency of the United Kingdom of Great Britain and Northern Ireland). In the United States, legislation based on the Model Law has been adopted in the state of Illinois.

In addition, the Model Law has served as the basis for other uniform legislation, including Canada's Uniform Electronic Commerce Act of 1999, and the Uniform Electronic Transactions Act of 1999, adopted by National Conference of Commissioners on Uniform State Law. Appendix A provides a full text of the Model Law on Electronic Commerce.

[2] Uniform Commercial Code, Article 4A — Funds Transfers

The Uniform Commercial Code that forms the backbone of commercial law throughout the United States has specific provisions that deal with electronic funds transfers between a bank and its customers. The drafters' comments to Article 4A explain that the article deals primarily with wholesale wire transfers:

> Article 4A governs a specialized method of payment referred to in the Article as a funds transfer but also commonly referred to in the commercial community as a wholesale wire transfer. A funds transfer is made by means of one or more payment orders. The scope of Article 4A is determined by the definitions of ''payment order'' and ''funds transfer'' found in Section 4A-103 and Section 4A-104.[14]

---

[14] U.C.C. § 4A-102.

Section 4A-103 defines a "payment order" as

[A]n instruction of a sender to a receiving bank, transmitted orally, electronically, or in writing, to pay, or to cause another bank to pay, a fixed or determinable amount of money to a beneficiary if:

> (i) the instruction does not state a condition to payment to the beneficiary other than time of payment,

> (ii) the receiving bank is to be reimbursed by debiting an account of, or otherwise receiving payment from, the sender, and

> (iii) the instruction is transmitted by the sender directly to the receiving bank or to an agent, funds-transfer system, or communication system for transmittal to the receiving bank.

Section 4A-104 defines a "funds transfer" as

[T]he series of transactions, beginning with the originator's payment order, made for the purpose of making payment to the beneficiary of the order. The term includes any payment order issued by the originator's bank or an intermediary bank intended to carry out the originator's payment order. A funds transfer is completed by acceptance by the beneficiary's bank of a payment order for the benefit of the beneficiary of the originator's payment order.

The article contemplates that the parties may use a "security procedure" such as one of the encryption algorithms discussed earlier in this section, as may be agreed to between the parties:

"Security procedure" means a procedure established by agreement of a customer and a receiving bank for the purpose of (i) verifying that a payment order or communication amending or canceling a payment order is that of the customer, or (ii) detecting error in the transmission or the content of the payment order or communication. A security procedure may require the use of algorithms or other codes, identifying words or numbers, encryption, callback procedures, or similar security devices. Comparison of a signature on a payment order or communication with an authorized specimen signature of the customer is not by itself a security procedure.[15]

The article allocates considerable risk upon the bank's customer by making payment orders effective, whether or not authorized by the customer, provided reasonable security procedures were in place and followed:

If a bank and its customer have agreed that the authenticity of payment orders issued to the bank in the name of the customer as sender will be verified pursuant to a security procedure, a payment order received by the

---

[15] U.C.C. § 4A-201.

receiving bank is effective as the order of the customer, whether or not authorized, if (i) the security procedure is a commercially reasonable method of providing security against unauthorized payment orders, and (ii) the bank proves that it accepted the payment order in good faith and in compliance with the security procedure and any written agreement or instruction of the customer restricting acceptance of payment orders issued in the name of the customer. The bank is not required to follow an instruction that violates a written agreement with the customer or notice of which is not received at a time and in a manner affording the bank a reasonable opportunity to act on it before the payment order is accepted.[16]

[3] Electronic Funds Transfer Act and Regulation E

The U.S. federal statutes and regulations now modify the impact on consumers of what U.C.C. Article 4A provides at the state level. Specifically, the Electronic Funds Transfer Act,[17] and its associated Regulation E,[18] define how the risk of loss is allocated in funds transfers involving certain consumer transactions, significantly softening the blow to consumers who find they have been the victim of electronic fraud.

Congress identified a number of deficiencies in e-commerce law being applied at the state level. Thus, it enacted the Electronic Funds Transfer Act for these stated reasons:

(a) Rights and liabilities undefined

The Congress finds that the use of electronic systems to transfer funds provides the potential for substantial benefits to consumers. However, due to the unique characteristics of such systems, the application of existing consumer protection legislation is unclear, leaving the rights and liabilities of consumers, financial institutions, and intermediaries in electronic fund transfers undefined.

(b) Purposes

It is the purpose of this subchapter to provide a basic framework establishing the rights, liabilities, and responsibilities of participants in electronic fund transfer systems. The primary objective of this subchapter, however, is the provision of individual consumer rights.[19]

The Act limits the risk of loss in electronic consumer transactions involving an "access device," which the Regulations define as "a card, code, or other means of access to a consumer's account, or any combination thereof, that may

---

[16] U.C.C. § 4A-202b.
[17] 15 U.S.C. §§ 1693 *et seq.*
[18] 12 C.F.R. pt 205.
[19] 15 U.S.C. § 1693.

be used by the consumer to initiate electronic fund transfers.''[20] As provided in Regulation E, consumer liability is limited to $50, provided the consumer notifies the bank of the unauthorized transaction within two days after learning of the loss or theft of the access device:

Sec. 205.6 Liability of consumer for unauthorized transfers.

(a) Conditions for liability. A consumer may be held liable, within the limitations described in paragraph (b) of this section, for an unauthorized electronic fund transfer involving the consumer's account only if the financial institution has provided the disclosures required by Sec. 205.7(b)(1), (2), and (3). If the unauthorized transfer involved an access device, it must be an accepted access device and the financial institution must have provided a means to identify the consumer to whom it was issued.

(b) Limitations on amount of liability. A consumer's liability for an unauthorized electronic fund transfer or a series of related unauthorized transfers shall be determined as follows:

(1) Timely notice given. If the consumer notifies the financial institution within two business days after learning of the loss or theft of the access device, the consumer's liability shall not exceed the lesser of $50 or the amount of unauthorized transfers that occur before notice to the financial institution.

(2) Timely notice not given. If the consumer fails to notify the financial institution within two business days after learning of the loss or theft of the access device, the consumer's liability shall not exceed the lesser of $500 or the sum of:

(i) $50 or the amount of unauthorized transfers that occur within the two business days, whichever is less; and

(ii) The amount of unauthorized transfers that occur after the close of two business days and before notice to the institution, provided the institution establishes that these transfers would not have occurred had the consumer notified the institution within that two-day period.

(3) Periodic statement; timely notice not given. A consumer must report an unauthorized electronic fund transfer that appears on a periodic statement within 60 days of the financial institution's transmittal of the statement to avoid liability for subsequent transfers. If the consumer fails to do so, the consumer's liability shall not exceed the amount of the unautho-

---

[20] 12 C.F.R. § 205.2(a)(1).

rized transfers that occur after the close of the 60 days and before notice to the institution, and that the institution establishes would not have occurred had the consumer notified the institution within the 60-day period. When an access device is involved in the unauthorized transfer, the consumer may be liable for other amounts set forth in paragraphs (b)(1) or (b)(2) of this section, as applicable.

(4) Extension of time limits. If the consumer's delay in notifying the financial institution was due to extenuating circumstances, the institution shall extend the times specified above to a reasonable period.

(5) Notice to financial institution.

(i) Notice to a financial institution is given when a consumer takes steps reasonably necessary to provide the institution with the pertinent information, whether or not a particular employee or agent of the institution actually receives the information.

(ii) The consumer may notify the institution in person, by telephone, or in writing.

(iii) Written notice is considered given at the time the consumer mails the notice or delivers it for transmission to the institution by any other usual means. Notice may be considered constructively given when the institution becomes aware of circumstances leading to the reasonable belief that an unauthorized transfer to or from the consumer's account has been or may be made.

(6) Liability under state law or agreement. If state law or an agreement between the consumer and the financial institution imposes less liability than is provided by this section, the consumer's liability shall not exceed the amount imposed under the state law or agreement.[21]

Note that Regulation E applies not just to credit card purchases but to any transaction effected using an "access device," which includes a code used to effect an electronic funds transfer. Thus the Act and its associated Regulation may apply to a variety of different forms of "digital cash" to make purchases over the Internet.

---

[21] 12 C.F.R. § 205.6.

[4] Digital Millennium Copyright Act

On October 12, 1998, the U.S. Congress passed the Digital Millennium Copyright Act which addressed the data integrity issue among other issues. The Act made it a crime to remove or tamper with identifying information embedded in a copyrighted work — such as a digital watermark added by the copyright owner to ensure integrity of the work. The Act calls such embedded identifying information *copyright management information.* The Act imposes civil and criminal penalties for tampering with copyright management information. The relevant excerpt from the Act is presented below. The definition of copyright management information appears in §1202(c) of the Act.

Sec. 1202. Integrity of copyright management information

(a) FALSE COPYRIGHT MANAGEMENT INFORMATION — No person shall knowingly and with the intent to induce, enable, facilitate, or conceal infringement —

(1) provide copyright management information that is false, or

(2) distribute or import for distribution copyright management information that is false.

(b) REMOVAL OR ALTERATION OF COPYRIGHT MANAGE-MENT INFORMATION — No person shall, without the authority of the copyright owner or the law —

(1) intentionally remove or alter any copyright management information,

(2) distribute or import for distribution copyright management information knowing that the copyright management information has been removed or altered without authority of the copyright owner or the law, or

(3) distribute, import for distribution, or publicly perform works, copies of works, or phonorecords, knowing that copyright manage-ment information has been removed or altered without authority of the copyright owner or the law, knowing, or, with respect to civil remedies under section 1203, having reasonable grounds to know, that it will induce, enable, facilitate, or conceal an infringement of any right under this title.

(c) DEFINITION — As used in this section, the term "copyright management information" means any of the following information conveyed in connection with copies or phonorecords of a work or performances or displays of a work, including in digital form, except that such term does not include any personally identifying information about a user of a work or of a copy, phonorecord, performance, or display of a work:

(1) The title and other information identifying the work, including the information set forth on a notice of copyright.

(2) The name of, and other identifying information about, the author of a work.

(3) The name of, and other identifying information about, the copyright owner of the work, including the information set forth in a notice of copyright.

(4) With the exception of public performances of works by radio and television broadcast stations, the name of, and other identifying information about, a performer whose performance is fixed in a work other than an audiovisual work.

(5) With the exception of public performances of works by radio and television broadcast stations, in the case of an audiovisual work, the name of, and other identifying information about, a writer, performer, or director who is credited in the audiovisual work.

(6) Terms and conditions for use of the work.

(7) Identifying numbers or symbols referring to such information or links to such information.

(8) Such other information as the Register of Copyrights may prescribe by regulation, except that the Register of Copyrights may not require the provision of any information concerning the user of a copyrighted work.

(d) LAW ENFORCEMENT, INTELLIGENCE, AND OTHER GOVERNMENT ACTIVITIES — This section does not prohibit any lawfully authorized investigative, protective, information security, or intelligence activity of an officer, agent, or employee of the United States, a State, or a political subdivision of a State, or a person acting pursuant to a contract with the United States, a State, or a political subdivision of a State. For purposes of this subsection, the term ''information security'' means activities carried out in order to identify and address the vulnerabilities of a government computer, computer system, or computer network.

(e) LIMITATIONS ON LIABILITY —

(1) ANALOG TRANSMISSIONS — In the case of an analog transmission, a person who is making transmissions in its capacity as a broadcast station, or as a cable system, or someone who provides programming to such station or system, shall not be liable for a violation of subsection (b) if —

(A) avoiding the activity that constitutes such violation is not technically feasible or would create an undue financial hardship on such person; and

(B) such person did not intend, by engaging in such activity, to induce, enable, facilitate, or conceal infringement of a right under this title.

(2) DIGITAL TRANSMISSIONS —

(A) If a digital transmission standard for the placement of copyright management information for a category of works is set in a voluntary, consensus standard-setting process involving a representative cross-section of broadcast stations or cable systems and copyright owners of a category of works that are intended for public performance by such stations or systems, a person identified in paragraph (1) shall not be liable for a violation of subsection (b) with respect to the particular copyright management information addressed by such standard if —

(i) the placement of such information by someone other than such person is not in accordance with such standard; and

(ii) the activity that constitutes such violation is not intended to induce, enable, facilitate, or conceal infringement of a right under this title.

(B) Until a digital transmission standard has been set pursuant to subparagraph (A) with respect to the placement of copyright management information for a category of works, a person identified in paragraph (1) shall not be liable for a violation of subsection (b) with respect to such copyright management information, if the activity that constitutes such violation is not intended to induce, enable, facilitate, or conceal infringement of a right under this title, and if —

(i) the transmission of such information by such person would result in a perceptible visual or aural degradation of the digital signal; or

(ii) the transmission of such information by such person would conflict with —

(I) an applicable government regulation relating to transmission of information in a digital signal;

(II) an applicable industry-wide standard relating to the transmission of information in a digital signal that was adopted by a voluntary consensus standards body prior to the effective date of this chapter; or

(III) an applicable industry-wide standard relating to the transmission of information in a digital signal that was adopted in a voluntary, consensus standards-setting process open to participation by a representative cross-section of broadcast stations or cable systems and copyright owners of a category of works that are intended for public performance by such stations or systems.

(3) DEFINITIONS — As used in this subsection —

(A) the term "broadcast station" has the meaning given that term in section 3 of the Communications Act of 1934 (47 U.S.C. 153); and

(B) the term "cable system" has the meaning given that term in section 602 of the Communications Act of 1934 (47 U.S.C. 522).

[5] Digital Signature Legislation

The following legislation with digital signature provisions were introduced in the 106th Congress. Any differences between measures passed in the Senate and the House will be resolved through Conference Committee.

[a] H.R. 439, Paperwork Elimination Act of 1999

The Act provides for the acquisition and use of alternative information technologies that provide for electronic submission, maintenance, or disclosure of information as a substitute for paper and for the use and acceptance of electronic signatures. It is designed to minimize the burden of Federal paperwork demands on small business. The bill passed in the House by voice vote 413-0 on February 9, 1999, and was sent to the Senate where it was referred to the Committee on Governmental Affairs on February 22, 1999. According to the House Committee on Small Business press release on February 3, 1999, the Paperwork Elimination Act of 1999 goes further than the measure (Title XVII, GPEA) attached to the FY 1999 Omnibus Appropriations Bill by requiring the Federal Government to communicate electronically with those people who choose electronic submission, rather than relying on the longer term schedule provided by Title XVII.

[b] Millennium Digital Commerce Act

The Senate Commerce, Science and Transportation Committee held a hearing on May 27, 1999 on S.761, Millennium Digital Commerce Act. This proposed legislation builds on the GPEA and is designed to promote the use of electronic signatures in business transactions and contracts. It preempts State law on an interim basis until uniform State law on electronic signatures is in place.

[c] Electronic Signatures in Global and National Commerce Act

The Subcommittee on Telecommunications, Trade and Consumer Protection of the House Committee on Commerce held a hearing on June 9, 1999, on H.R. 1714, Electronic Signatures in Global and National Commerce Act. This proposed legislation provides for the acceptance of electronic signatures and records in interstate and international commerce and for the acceptance of electronic signatures and records by the securities industries.

[d] Digital Signature Act of 1999

H.R. 1572, Digital Signature Act of 1999, was introduced by Representative Bart Gordon on April 27, 1999, and was referred to the House Committee on Science. This proposed legislation calls for the Director, National Institute of Standards (NIST), in consultation with industry, to develop digital signature infrastructure guidelines and standards to enable Federal agencies to utilize digital signatures in a manner that is sufficiently secure to meet the needs of those agencies and the general public. A key objective is to ensure that digital signature technologies that agencies deploy are interoperable. The measure also calls for the establishment of a National Policy Panel for Digital Signatures composed of government, academic and industry technical and legal experts on the implementation of digital signature technologies.

[e] Internet Growth and Development Act of 1999

H.R. 1685, Internet Growth and Development Act of 1999, was introduced by Representative Rick Boucher on May 5, 1999, and was referred to the House Committee on Commerce and the House Committee on Judiciary. This measure provides for recognition of electronic signatures for the conduct of interstate and foreign commerce, restriction of transmission of unsolicited electronic mail advertisements, authorizes the Federal Trade Commission to prescribe rules to protect the privacy of users of commercial Internet Web sites and promotes the rapid deployment of broadband Internet services.

§ 7.03   XML — The *Lingua Franca* of the New Internet?

At the turn of the millennium, XML is poised to become the *lingua franca* of the Internet. Will it replace the more primitive, but widespread, HTML? Bill Gates thinks so. Speaking to a technical audience at the WinHEC[22] 2000 conference in Anaheim, California, on March 26, 2001, Microsoft founder Bill

---

[22] Windows Hardware Engineering Conference.

Gates telegraphed his vision of the computing future, in which XML will play a key role.

> I believe that we're entering into a new era, an era where a new approach to how applications are created is taking place, and a way of looking at the PC for a broader set of tasks. Every time we move from one era to the next, the kinds of things we did with the PC are expanding. The era that we're entering now is one where the key protocol is XML, allowing the richness of the server and the richness of the client to be complementary. Bill Gates

§ 7.04   What Is XML?

XML stands for Extensible Markup Language. A markup language is a system of commands that communicates how to treat selected elements of a text or data set. Usually the markup commands are written directly on or embedded in the text or data set being marked up. When your fourth grade teacher scrawled red ink all over your history paper about Abraham Lincoln, those red ink comments were a form of markup language, communicating how to write the history paper next time. XML is a particularly powerful markup language because it defines a language with which you can describe other languages. Imagine if your fourth grade teacher's red ink comments first taught you ancient Chinese, and then proceeded to correct your paper in the language of Genghis Khan.

[A]   XML's Predecessor HTML

When Tim Berners-Lee invented the World Wide Web, he employed a markup language, called Hypertext Markup Language (or HTML), to communicate which words within a page of text would function as hot buttons or hyperlinks to take the reader instantly to other pages of text.

Today, most are familiar with the concept behind HTML and how it controls the appearance of millions of Web pages on the Internet. HTML markup commands are embedded in a text document and these commands are interpreted by the web browser as instructions on how to display the associated text. For example, the following simple HTML commands (<h1> and </h1>) instruct the browser to display the enclosed text as a level 1 heading:

<h1>Frequently Asked Questions about the
Extensible Markup Language</h1>

Note that according to the HTML syntax, the text to be treated as a level 1 heading is bracketed by the HTML commands. Thus, reading from left to right, as the browser would, one first encounters the <h1> command. This tells the browser that, until told otherwise, everything that follows is to be displayed in the browser's "level 1" heading style. Thus the text, "Frequently Asked Questions about the Extensible Markup Language" will be displayed in a larger, more prominent font. How does the browser know when to turn the level 1 heading style back off? That command comes at the end of the heading, the </h1> command. There are many HTML markup commands in popular use today. More commands are added with each new release of web browsers.

When Tim Berners-Lee first began developing the Web in the late 1980s, his HTML command set had few of the sophisticated display features to which today's Web users have grown accustomed. It did not, for example, have the ability to define a Web page into independently scrollable frames. Currently one of the most challenging aspects of designing Web pages using HTML is that the language continues to grow. The leading browsers, Internet Explorer and Netscape, often interpret HTML commands in slightly different ways; and even within a single browser family there are many different generations or versions still in use. Internet Explorer 3.0 does not handle all of the HTML commands that Explorer 4.0 does, and so forth. Thus, the Web designer faces the challenge of developing user-friendly, engaging Web pages that will run, at least reasonably well, under a range of different browser technologies.

[B] What Makes XML Different?

HTML describes the layout and appearance of the Web page. HTML gives Web designers a way to place font changes, frames, title headings and hypertext links in a text file that is destined to become a Web page. XML does something more. It allows Web designers to place *structured data* in that text file. You can think of structured data as a data having associated labels or categories. Spreadsheets, address books, configuration parameters, financial transactions, and technical drawings comprise structured data.

Perhaps you have seen what appears to be a table of spreadsheet data in an HTML-generated Web page. By all outward appearance, the table *looks* like a spreadsheet, with data arranged in rows and columns. However, on deeper inspection you would find that the appearance is only skin deep. If you were to copy and paste that table into a spreadsheet program, you would be disappointed to find that none of the column and row structure was carried over during the cut and paste. Why is this? It is because HTML merely defines how the data *looks* on the Web screen. It defines nothing with regard to the underlying structure of the data, that is how the columns and rows actually relate to one another. See Figure 7-1.

In contrast, if you took an identically appearing table that was generated using XML and tried to copy and paste it into a spreadsheet, you would be much happier with the result. As shown in Figure 7-2 the column and row information is preserved. This is because XML allows the underlying structure of the data to be described, in this case the identity of the column and row headings of the table. Note that these column and row headings can be anything the table creator dreams up. XML does not itself define these specific headings; it gives the table creator the tools to define them.

Data Displayed using HTML:

	Monday	Tuesday	Wednesday
Shopping	yes	no	no
Laundry	no	yes	no
Dusting	no	no	maybe
Organizing	no	no	never

Cut and Paste into a Spreadsheet
(with unacceptable results)

Copied to Spreadsheed (structure is lost)

	Monday	Tuesday	Wednesday
Shopping	yesnononoyesnononomaybenononever		
Laundry			
Dusting			
Organizing			

FIGURE 7-1:   EXAMPLE OF TABLE USING HTML

Data Displayed using XML:

	Monday	Tuesday	Wednesday
Shopping	yes	no	no
Laundry	no	yes	no
Dusting	no	no	maybe
Organizing	no	no	never

Cut and Paste into a Spreadsheet
(with good results)

Copied to Spreadsheet (structure is retained)

	Monday	Tuesday	Wednesday
Shopping	yes	no	no
Laundry	no	yes	no
Dusting	no	no	maybe
Organizing	no	no	never

FIGURE 7-2:   EXAMPLE OF TABLE USING XML

§ 7.05   Why Use XML?

As the above example illustrates, XML gives the Web designer the freedom to pass structured data to Web page users. This seemingly small change has a rather profound impact on what can be built using Web technology. By being able to pass structured data from Web server to client, the system designer can shift the computational load from the server to the client applications. Because the server now has less to do, it can handle more clients without any apparent drop in speed.

How would this work? A stock trader's information system, illustrated in Figure 7-3, demonstrates the concept. The client, which might be a day trader wanting to obtain the latest stock quotes along with detailed historical information, asks for data by filling out a form in the browser. The server sends back the results, coded using XML. The client browser then optionally asks for a style sheet that describes how the data should be displayed. The style sheet could be served by the server that supplied the stock data, or it could be served by any other server. The style sheet is served and the client browser then applies it to the XML data, to format the data and display the results. In this case, the browser uses the data to generate and render graphs showing historical stock prices over the timeframe specified by the user.

Note that in the preceding example, the client browser did all the formatting work. The server merely sent the raw data in XML format and the client browser did the rest of the work. If the user wanted a graph of a different timeframe, the browser would be capable of generating it, without requesting assistance from the server.

Just as HTML Web designers have had to struggle with a population of different browsers of different version levels, so too are XML designers hampered. Currently, only higher level browsers support XML. For example, Internet Explorer 5.0 does, but older versions do not. Thus some system designers may choose not to use XML simply because their client base has not migrated to the higher level browsers.

§ 7.06   Alternatives to XML — Middleware Technologies

While Bill Gates's acknowledgment of the importance of XML undoubtedly carries considerable weight, there are other technologies that provide power similar to XML. Many of these fall within a category of software known as *middleware*. Middleware is the software glue that can bind two or more disparate software systems together. Middleware is often used, for example, to integrate otherwise disparate legacy mainframe applications and more modern Web-based applications.

When a large corporation such as General Motors wishes to make data available to its suppliers from a legacy COBOL mainframe database, chances are a middleware application will be used. Built long before the Web even existed, legacy applications do not know how to speak XML, or even HTML for that matter. To bridge between the past and the present, a middleware application is

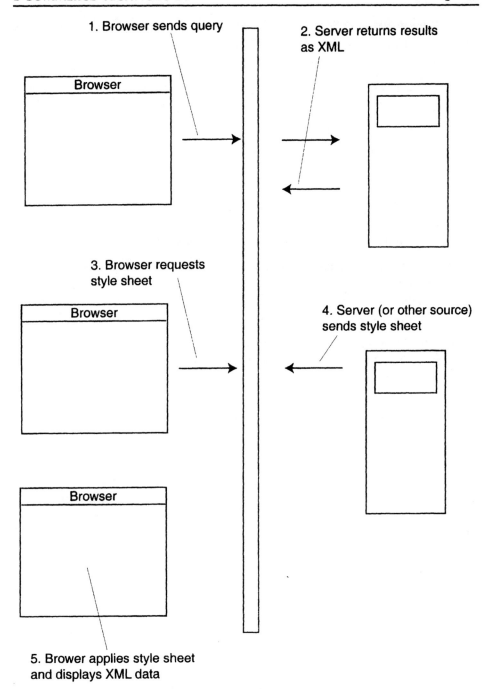

1. Browser sends query

2. Server returns results as XML

3. Browser requests style sheet

4. Server (or other source) sends style sheet

Browser

Browser

Browser

5. Brower applies style sheet and displays XML data

FIGURE 7-3:   STOCK TRADER'S INFORMATION SYSTEM

deployed that knows how to talk to both the mainframe application and modern Web browser applications. The middleware application knows how to ask for information in a language the mainframe understands and how to translate the mainframe applications answers into a language the Web browser understands.

There is a multitude of different middleware technologies available today, no doubt driven by the Internet's voracious appetite for data of all descriptions. Here are a few middleware contenders you may encounter:

- ODBC — Open Database Connectivity, an application program interface defined by Microsoft in which requests for data are processed on the server side, featuring multi-platform support.

- JDBC — Java Database Connectivity, an application program interface defined by Sun Microsystems in which requests for data are processed on the server side.

- CDML — Claris Dynamic Markup Language, a middleware variation on the XML theme in which the client asks for data using proprietary tags; the server retrieves data and converts output to standard HTML prior to sending.

- JSP — Java Server Pages, a middleware technology which uses XML-like tags and scriptlets written in the Java programming language to encapsulate the logic that generates the content for the page.

- ASP — Microsoft Active Server Pages, a server side scripting environment that allows a script, such as a Visual Basic script, to be embedded in the HTML code for processing by the active server page. The active server page processes the script and returns the result to the browser.

§ 7.07   Components of an XML System — Learning the Jargon of XML

XML is a language, and like any language it has its own grammar and syntax. For example, XML defines a syntax used to subdivide and hierarchically structuring a text document into its component parts. This book, for example, has a hierarchical structure of chapters, sections, subsections and paragraphs. XML could be used to describe these hierarchical layers.

Unlike HTML, XML tells nothing about how the data is to be displayed. Thus specifying the typeface and font size of the text document is beyond the reach of XML. In an XML implementation, how information is displayed is relegated other mechanisms, such as style sheets, which describe how the data should be formatted and displayed.

[A] Elements

XML defines something called the *element* as the fundamental building block. Elements are delimited by Start and End XML tags.

[B] XML Tags

XML Tags communicate to the browser information about the data element with which it is associated, as well as the meta-data about what the data is. The

XML tag typically includes a user-defined tag name, and optionally some additional attribute information.

[C] Tag Names

XML Tag Names can be any user-defined name. This is quite a departure from HTML, which allows only a limited set of predefined names (such as <h1> and </h1> illustrated in the example above).

[D] Attributes

Where XML tags need to specify initial conditions or initial values, the tag will contain attribute fields, typically an ordered pair that assigns a value to a name, using the syntax name = "value." For example, the following attribute description would specify an initial height value:

height = "140"

[E] XML Application

This term does not mean what you might suspect. In much of the XML literature, it refers to the application of XML technology to a particular domain, such as chemistry, accounting, inventory.

[F] XML Grammar (also XML Vocabulary)

The set of all possible documents that a given group of elements can define. For example, such grammar might encompass database-dependent grammars, database-independent grammars and structure and layout grammars.

[G] Document Type Definition [DTD]

This is an optional format description telling which elements are allowed in certain contexts. The document type definition (or DTD) may be embedded in the XML document itself, or stored in a separate file. When the XML parser within the browser reads an XML document, it checks to see if the DTD is complied with.

[H] Style Sheet

The style sheet is a document separate from the XML document that describes how the data should be formatted and displayed. Thus the style sheet performs the function relegated to the HTML tags in an HTML-based Web page. There are several different style sheet technologies available. These include:

- CSS — cascading style sheets
- XSL — extensible style language

- XSLT — a part of the extensible style language used to perform transformations on objects

- XSL-FO — a part of the extensible style language used for formatting objects

- JavaScript

- VBScript — Visual Basic Script

[I] CGI

CGI or Common Gateway Interface refers to a standard for interfacing external applications with information servers, such as HTTP or Web servers, to provide dynamic content.

CHAPTER 8

PURE BUSINESS MODEL PATENTS

§ 8.01   Pure Business Model Patents

With all the interest in the Internet, it is not surprising that many business model patents today have an e-commerce flavor. However, the patent laws do not limit business model patents to e-commerce technologies. Indeed, as many of the business model patents of the 1800s demonstrate, the U.S. Patent Office has long been granting patents on business methods that can be practiced with simple devices, such as printed business forms (*See* §§ 6.03, 6.09), coupon bond forms (*See* § 6.02), and restaurant menus (*See* § 6.08).

You might be tempted to cast out these early examples as irrelevant today. After all, these "low technology" business patents — now all more than 100 years old — were taken out in simpler times, when steamboats and buggy whips were the norm. Indeed, 100 years ago, mass production using interchangeable parts was still being perfected. Back then, a patent on a business method practice using simple paper forms was a plausible proposition, but certainly not today. You might understandably argue that in this world of broadband communication, wireless Internet, and commerce at the speed of light, there is no longer a place for simple, low-tech business innovations. You would be wrong.

A case in point is the kanban card. When Toyota Motor Company in the 1970s grabbed significant U.S. market share from the U.S. car makers, it did so using the kanban card. The kanban card, a simple card with production or shipping information written on it, drives Toyota's highly effective just-in-time manufacturing process that allowed it to rapidly become a major worldwide automotive manufacturer.

§ 8.02   Just-in-time Production

Today the term kanban has become essentially synonymous with just-in-time manufacturing, or at least strongly tied to it. It was not always that way. In the Japanese language, the word kanban means simply a sign, signboard, poster, or billboard. Originally the term probably had an advertising connotation. Were you to have wandered the streets of Kobe, or Osaka, or Tokyo in the 1950s no doubt you would have seen many kanban in storefront windows. Perhaps you might also have had the fortune to spot a kanban-musume working inside a small shop. Musume, means "daughter," and kanban-musume is a pretty daughter who attracts boys to her parents' shop.

If you were to visit a manufacturing plant in Japan today, you would most certainly see hundreds of kanban cards, encapsulated in plastic sleeves and alligator clipped to products being manufactured, to wire baskets holding component parts awaiting assembly, and to shipping pallets being loaded by forklift onto trucks. How these kanban cards proliferated first in Japan and then throughout the manufacturing world is an interesting story and provides a fine example of why the low-tech business model should never be cast aside as unworthy of patent protection.

To best understand the significance of the kanban card it will help to briefly trace the history of mass production using interchangeable parts. U.S. inventor

Eli Whitney, perhaps more famous for his invention of the cotton gin, developed the concept of mass production using interchangeable parts around the turn of the 19th Century.

[A] The History of Mass Production Using Interchangeable Parts

[1] Eli Whitney Invents Manufacturing with Interchangeable Parts

Whitney invented mass production using interchangeable parts after a ten year legal struggle to collect royalties on his wildly successful cotton gin from which he emerged penniless. Although Whitney's cotton gin had been a technological success and had changed the face of the American South, Whitney had great difficulty getting plantation owners to pay licensee fees for his invention. When forced to sue to collect royalties, he found the courts — all located in cotton growing states where the cotton gin was used — favored the "public good" and awarded Whitney paltry royalties.

Weary from his legal battles and desperate for money, Whitney, now age 39, took a gamble that his inventive prowess might save him. Wary of having further dealings with the likes of the plantation owners, Whitney bid on a federal government contract to produce ten thousand musket rifles at the very competitive price of $13.40 each. Whitney had no factory and no machinery either, but the federal government awarded him the contract, mostly on his reputation as a brilliant inventor. Whitney did not let them down.

Whitney's competitors manufactured rifles one at a time, relying on skilled craftsmen to work raw metal into rifles that, like sculpture, were each one of a kind. Parts from one rifle would not fit on another rifle, and no one expected them to. Whitney realized that skilled craftsmen would always be in short supply, and unwilling to work for him until he was well established. Therefore, Whitney focused his inventive genius on the problem and concluded that he needed to develop a manufacturing technique that would allow unskilled laborers to produce rifles of equal quality to those manufactured by the skilled metalworker. Whitney reasoned that if he could make batches of component parts that were all identical, then unskilled laborers could be trained to assemble the parts into finished rifles.

He first developed templates or patterns to be used as a guide in forming the shape of each part. No doubt his inspiration for this first step was the dressmaker's pattern by which cloth was cut prior to sewing a garment. To make cutting the metal shapes easier and more repeatable, Whitney invented the rotary saw by filing teeth into the circumference of a wheel. Some of the rough cut parts would need further refinement. To perform this task Whitney invented the milling machine. His basic design for that machine remains virtually unchanged to this day.

With his newly invented tools, Whitney then developed a manufacturing system whereby unskilled labor could fabricate interchangeable components and then assemble those components into finished rifles. Whitney ultimately fulfilled his federal government contract. It took him eight years to perfect his manufacturing techniques and deliver the ten thousand rifles. Thereafter he easily filled his next order for 15,000 rifles in two years.

[2] Henry Ford Develops the Assembly Line

Henry Ford refined and exploited mass production in the production of the Model T automobile, which first rolled off the assembly line in 1908. The Model T was Ford's 20th design. It exploited not only interchangeable parts, but interchangeable workers. Ford designed his manufacturing processes for the Model T with the division of labor in mind. In the past, even though interchangeable parts were used, manufacturing was done by skilled craftsmen who obtained the tools and materials they needed from the company tool room, repaired them if necessary, and then proceeded to build the entire automobile from the ground up. In contrast, Ford's Model T workers each had a single and repetitive job to do. As the partially assembled vehicle rolled down the assembly line, each worker simply added his or her assigned part to the work in progress. If tools needed repairing, someone else would do it.

Ford's innovative assembly line mass production technique changed the commercial landscape in countless ways. It made trucks and automobiles economically affordable for the average citizen. That, in turn, created tremendous demand for better roads. With better roads came freedom to travel any time, any place. Interstate commerce, once the exclusive domain of the railway companies, was now accessible to anyone with a car or truck.

There was a problem with Ford's assembly line mass production technique, however. The technique was good at repetitive production, but it was not very nimble. To work best, the assembly line needed an ample supply of parts at each assembly station. This meant having warehouses of component parts at the ready. Care was taken not to let the reserves of parts drop too low, for if any reserve got depleted the manufacturing line would stop. Because the assembly line technique relied on repetition, each work station was set up to assemble one specific part onto the vehicle. The entire assembly line was thus configured to turn out one model of vehicle only. Using Ford's assembly line approach it would have been very difficult to manufacture different types of vehicles from the same assembly line. That would have required stopping the line to change tooling and to swap component parts of one model for the other. Stopping the line meant stopping Ford's money making machine, which clearly he did not want to do.

Ford's answer to manufacturing multiple vehicle models would have been to set up multiple assembly lines, one for each model. There is nothing inherently wrong with that. However, as demand for vehicle models fluctuates, some lines may be producing more vehicles than required and others less. The realities of supply and demand go directly to the bottom line. When more vehicles are produced than required, the manufacturer must give dealers special incentives to take the vehicles off their hands. The cost of such incentives eats into the manufacturer's profits. When fewer vehicles are produced than required, the manufacturer loses profit opportunities.

[3] Taiichi Ohno Introduces the Kanban Card to Achieve Just-in-time Manufacturing

During the 1950s, the U.S. automobile manufacturing companies dominated the U.S. automobile market. In 1955, for example, Ford, General Motors, and Chrysler had 95 percent of the market. Six vehicle models accounted for 80 percent of all cars sold.[1] At that time, the Toyota Motor Company, under the direction of Taiichi Ohno, Toyota's chief production engineer, was experimenting with a new, just-in-time technique based on a simple kanban card. The U.S. automobile manufactures would later learn of this technique, after the oil crisis in the Fall of 1973 placed their gas guzzling products on the unwanted list, giving nimble Toyota the opportunity to plant a foothold on American soil.

The Toyota Motor Company was founded in 1937 out of a family run business of the Toyoda family. (The family name, Toyoda, means abundant rice field. The coined name, Toyota, although similar sounding, has no meaning.) The Toyota Motor Company struggled to survive its first 15 years. It nearly folded in 1949, when sagging sales forced it to lay off much of its workforce. In the spring of 1950, Eiji Toyoda, nephew of founder Kiichiro Toyoda, made a three month pilgrimage to Ford's Rouge plant in Detroit to study Ford's assembly line manufacturing methods. At that time the Ford Rouge plant was turning out over twice the number of vehicles in a single day than Toyota had produced in its past 13 years.

What Eiji Toyoda saw in Detroit was a well-oiled manufacturing machine running at full tilt. Eiji's uncle, Kiichiro had pronounced that Toyota would need to catch up with America in three years. Otherwise, he predicted, the Japanese automobile industry would not survive. Standing in the Ford Rouge plant, Eiji Toyoda knew that it was going to take longer than that. He knew it would be economically infeasible to copy the American assembly line system to the last detail. Americans turned out few models in high volume. Japan needed many models in smaller volume. Certain improvements would have to be made.

The improvements came from the Toyota's chief production engineer, Taiichi Ohno. Interestingly, Ohno's inspiration came not from America's assembly lines, but from America's grocery stores. At that time in Japan, most people commuted by train and then walked home from the train station, or rode bicycles. Shopping for groceries was an everyday affair. Few owned cars to carry home a large trunk full of groceries. Most people would simply stop in one or two small markets to pick up a few items needed for the evening meal on the way home. The American supermarket, with aisles and aisles of produce of every description, begot wonderment in most Japanese people, Ohno concluded.

---

[1] Womack, James P., et al., *The Machine That Changed the World, the Story of Lean Production* (HarperCollins Publ., 1991).

In America's supermarkets Ohno saw a just-in-time process at work. Fresh produce would spoil if left out on display for too long. Therefore, by monitoring what was sold at the checkout line, grocers knew precisely when to restock their shelves and what quantities to order for tomorrow. This just-in-time system worked well. It allowed grocers to offer numerous products, including competing brands, in a very efficient, shopper-friendly way. The tiny grocery stores he shopped at in Japan could not afford to carry this wide variety of merchandise.

In a stroke of inspiration Ohno recognized that the just-in-time concept of American supermarkets might just be the way for Toyota Motor Company to compete. Toyota, like other Japanese auto makers, needed to offer many different vehicle models, in order to build a sufficient customer base to stay in business. The American auto companies did not have this problem because their market was considerably larger and they already had a substantial market share of customers that were satisfied with the two or three models being offered. If multiple models could be produced just-in-time, like grocers stocked the shelves in American supermarkets, Toyota might be able to stay alive and grow, Ohno reasoned.

While Ohno's logic appears simple in hindsight, cabbages and cars are very different products. It was by no means readily apparent at first how Ohno was going to establish a just-in-time flow within an automobile manufacturing plant. Indeed, at first, many thought Ohno was crazy. Stocking parts bins, just-in-time, like grocer's shelves seemed exceedingly risky. If any bin goes empty, even for a moment, the plant shuts down and valuable momentum is lost.

Ohno solved the problem with kanban cards. If an assembly line was going to operate without warehouses full of parts to keep the parts bins full, the flow of production would need to be very carefully monitored. Computers are good at many things, but Ohno wanted something more simple, more fundamental. He wanted a system workers in a plant could understand and operate for themselves. When he thought about American grocery stores, he saw information about sales being captured at the checkout line and then immediately used to send a signal to the stocker to replace, on the shelves, what had just been sold. In the stockroom behind, he envisioned a similar information gathering and signaling operation to alert the buyer when it was time to place another order for lettuce.

To implement this concept in the automobile plant, Ohno envisioned using cards bearing only the key information: what, when, and how many. As he thought about the problem more, he was able to categorize the key information as (1) pickup information, (2) transfer information, and (3) production information. He devised simple cards to bear this information and had the workers in the machine shop attach the cards to the parts being manufactured and to the pallets containing the parts in transit. By 1953, he had this kanban system working in the machine shop. It allowed workers to manufacture and deliver parts just-in-time to the upstream manufacturing processes.

Just as in any large company, it took considerable persuading to change the culture to accept kanban, but Ohno was ultimately successful. By 1962, the entire Toyota Motor Company was operating just-in-time using kanban cards to pass information throughout the Toyota nerve system. Changing the culture of the company to accept the just-in-time way of thinking had been difficult, but Ohno had a powerful ally. Ohno reports in his book that Kiichiro Toyoda, Toyota's founder, once told Ejii Toyoda, Toyota's president, "that in a comprehensive industry such as automobile manufacturing, the best way to work would be to have all the parts for assembly at the side of the line just in time for their use."[2] Thus Ohno already had management on his side.

[B] Principles of Kanban

Simply making kanban cards and attaching them to everything will not, by itself, achieve just-in-time production. As with any endeavor, it is possible to botch the job if the system is not understood or is not being carefully followed. Just-in-time offers the significant advantage of eliminating time-consuming, resource-wasting warehousing operations. However, it also offers the significant risk that if the kanban system is not well designed, or if the kanban cards are not obeyed, supplies may run dry and the manufacturing line will stop.

Ohno reduced the basic kanban system to these five rules, which, if followed, would ensure a smooth flowing just-in-time system:

1. The later process picks up the number of items indicated by the kanban at the earlier process;
2. The earlier process produces items in the quantity and sequence indicated by the kanban;
3. No items are made or transported without a kanban;
4. Always attach a kanban to the goods; and
5. Defective products are not sent on to the subsequent process. (The result is 100% defect-free goods.)
6. Reducing the number of kanban increases their sensitivity.[3]

These rules contain some hidden features we should discuss here. Note that kanban carries messages between earlier and later processes, and that the kanban travels with the goods. Ohno recognized that the goods being manufactured were inherently going to move from workstation to workstation. By attaching the kanban cards to the goods, the cards move from workstation to workstation, essentially for free. Thus the information channel between earlier and later processes is essentially free.

To understand how these rules might work, assume that you work in a litigation department that generates highly confidential documents. To ensure the documents are not read by unauthorized eyes, you have designed the documents to self-destruct when exposed to light or x-rays. Specifically, the documents are

---

[2] Ohno, Tiichi, *Toyota Production System — Beyond Large-Scale Production* (Productivity Press, 1988), p. 75.

[3] *Id.* at 30.

recorded on sensitive photographic plates that become unreadable a few microseconds after being exposed to light.

From time to time you may need to ship documents abroad. To avoid self-destruction, special x-ray tight postage-prepaid containers are required. The containers are very expensive and are frequently updated to match ever changing postage fees of the international shipping companies. Your office manager, always looking to cut cost, would like to avoid having a stockroom full of these expensive containers, particularly because they are used sporadically and regularly go out of date when the postage fees change. It takes three days to order new containers from the supplier, but your customers or clients cannot tolerate a three day delivery delay. Thus to eliminate the storeroom, you need a just-in-time system.

Imagine that the documents being shipped are generated by an earlier process, say by a graphics technician. The graphics technician gets his or her order from the litigation department that is investigating the merits of a case. To ensure that the special shipping container is ready when needed, a worker in the litigation department sends the supply ordering clerk a kanban card indicating that an x-ray proof shipping container will be needed. The card specifies specifically what documents are to be placed in that shipping container. The ordering clerk obtains the shipping container from the supplier and attaches the card received from the litigation department. In this way, that specific shipping container is earmarked for a specific set of documents.

Meanwhile, the graphics technician produces the documents according to the litigation department's instructions. The instructions are provided on another kanban card. The card specifies exactly what is to be produced and to whom the documents will be shipped once produced. The kanban specifies that shipment shall be made using a special x-ray proof container. When the graphics technician finishes producing the documents he or she attaches the kanban from the litigation department and forwards the finished goods to the shipping clerk in the mail room.

The shipping clerk receives the documents and notes from the kanban that a special x-ray proof shipping container is required. Because the ordering clerk has already obtained the needed container and attached to the container is a kanban specifying what is to be placed in it, the shipping clerk simply matches up the container kanban with the documents kanban and sends the package on its way, using the address indicated on the documents kanban.

Ohno's second rule states that the earlier process must produce not only the quantity specified by the kanban, but must also follow the sequence specified. What does this mean, exactly? In our hypothetical litigation department, assume that shipping containers come in different sizes and bear different amounts of prepaid postage, say, small ($12.50), medium ($14.50), and large ($18.00). Assume that the litigation department generates the following document sequence: Monday — one small package; Tuesday — one small package; Wednesday — one large package.

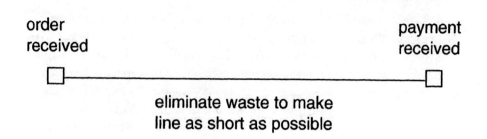

order
received

payment
received

eliminate waste to make
line as short as possible

FIGURE 8-1:   OHNO'S STRAIGHT LINE PHILOSOPHY

The litigation department would thus end up sending kanban cards to the ordering clerk in the following sequence: small, small, large. The ordering clerk would obtain the containers following the same sequence, small, small, large. Thus, when the shipping clerk receives the document set generated Monday, a small shipping container will be waiting. When the clerk later receives the set generated Wednesday, a large shipping container will be waiting.

Note how the kanban cards function to provide both transportation information (the shipping clerk will not ship without a kanban to specify the address) and production information (the graphics technician will not produce photographic plates without instruction from the litigation department). The system prevents overproduction of photographic plates, and also prevents over-stocking of shipping containers. Following Ohno's fifth rule, the graphics department will not transfer a defective photographic plate to the shipping department, and the ordering clerk will not send a defective shipping container to the mailroom. This ensures 100% defect-free shipments.

Ohno's sixth rule, reduce the number of kanban, tells us how to minimize waste. In our example, if the ordering clerk gets ten shipping containers ahead of the game when the project is suddenly cancelled, those ten containers are likely to become waste. If the ordering clerk were merely one container ahead when the project halted, there would be only one container wasted. Reducing the number of kanban by carefully engineering the process flow does certainly reduce potential waste. However, it also reveals problems with the system in a fairly profound way: the production line stops. Ohno treats this as an advantage and not a shortcoming. His philosophy is that all waste should be identified and eliminated. Why mask waste by providing built-in slack in a stockpiling warehouse? Figure 8-1 sums up Ohno's philosophy nicely.

How does one eliminate waste? The key is to first identify it. That takes a little ingenuity. Waste is everywhere, but we become so habituated to it that it is often overlooked. Shuffling papers within your in-basket, without actually working on anything represents waste in the office environment of which most of us are guilty. Storage of intermediate products in a warehouse, or transporting goods to a place other than their final destination are also examples of waste. That's right, the stack of magazines that have been sitting in the corner of your office for the last year, or the boxes of files cluttering the hall next to your partner's secretary's desk represent waste. A more efficient process would have

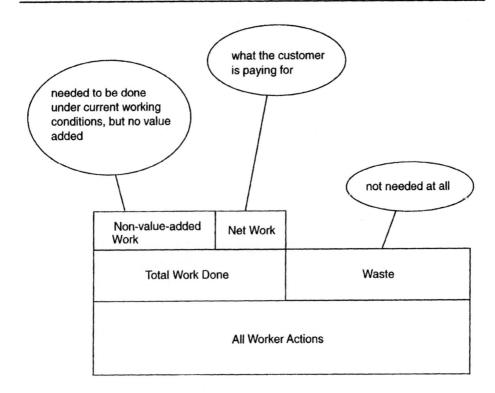

FIGURE 8-2: OHNO'S WASTE DIAGRAM

whisked those magazines and boxes away to more useful storage locations, or to their final destinations.

As illustrated in Figure 8-2, waste represents part of the total effort expended by every worker, but it is not the whole picture. Some effort expended may be necessary under current working conditions, even though it adds no value to the finished product. Perhaps those boxes sitting next to your partner's secretary's desk, which you thought were merely in intermediate storage, are actually undergoing a slow but steady auditing process. That's right, that section of the floor next to the secretary's desk is actually a horizontal staging area for document processing. Didn't you know that?

Ohno would call such activity non-value-added work. Working on the floor takes extra effort, and causes coworkers to work around the clutter, but it may be the only way to accomplish the required auditing task under current working conditions. A redesigned workspace might allow for the boxes to be located more conveniently elsewhere. Better yet, a redesigned process might audit the files before they are placed in boxes, thus eliminating the need to even route them to the secretary for review.

[C] Patenting Just-in-time Business Models

Toyota exploited just-in-time manufacturing to become one of the world's largest automobile manufacturers. In many respects, it was the simple kanban card that allowed Toyota to gain a foothold in American markets during the 1970s oil crisis. Low-tech as the kanban card was, the innovative manufacturing processes and business methods it unlocked were of immeasurable value to Toyota.

So did Toyota patent the kanban? Initially, no. Ohno introduced the supermarket system in the Toyota machine shop as early as 1953. Thus, the most fundamental aspects of the kanban system probably entered the public domain without patent. However, eventually Toyota's patent attorneys did recognize the patentable value in the kanban system and began filing patents to protect those aspects that had not fallen into the public domain. As with many Japanese companies, Toyota first applied for patents in Japan, and then sought protection elsewhere in the world, including in the United States, under the Paris Convention.

The Paris Convention allows an application filed in the patent office of a member country (such as Japan) to serve as the basis for prosecution of a patent in the patent office of another member country (such as the United States). If filed within one year of the first filed application, the later application is given a priority filing date of the first filed application.

Toyota's first U.S. patent to reference the kanban system was filed in the U.S. Patent Office on August 19, 1991. The U.S. application was based on a Japanese parent application, filed in Japan on August 23, 1990. Entitled "Production Schedule Making Method," the patent describes an improvement to the basic kanban technique, most applicable to scheduling delivery of component parts for manufacturing, where the truck delivery schedule is taken into account. The abstract and first claim of this patent are reproduced below.

Abstract:

A method for making a production schedule of an instant process which produces a plurality of types of products and supplies the products to a plurality of following processes by trucks. A truck delivery schedule including a number of truck deliveries and times is taken into account when the production schedule of the instant process is made. On the production schedule, a production order of the products is determined, and a stocking schedule also is made on the production order schedule.

Claims:

1. A method for making a production schedule for an instant process that produces a plurality of types of products and supplies the products to a plurality of second processes by trucks, the method comprising the steps of:
   recording a second process schedule which includes numbers of products needed by the second process with respect to respective product types and respective days;
   recording a truck delivery schedule from the instant process to the second process and information included in a card called a *KANBAN* which is carried by each truck between the instant process and the second processes and which includes information about delivery types;

determining a product shipment schedule of the instant process based on the second process schedule, the truck delivery schedule, and the *KANBAN* information;

recording a working condition and a production condition of the instant process;

determining a production schedule of the instant process based on the product shipment schedule, the working condition of the instant process, and the production condition of the instant process; and

determining a production order of the products to be produced at the instant process based on the production schedule of the instant process.

Just to be clear that this is indeed a low-tech patent, the term "computer" appears only once in the specification and not in the claims. The use of the term appears to be gratuitous, in that the computer is never substantively used:

A KANBAN is here defined as an instruction card which is carried between the instant process and the following processes by trucks and which includes an instruction or information about production and delivery of products. This entered information is stored in a memory of a computer.[4]

There have been a number of kanban patents issued in the United States since the above Toyota patent was taken out. Table 8-1 below shows the state of the issued U.S. patent kanban art. As of February 2001 there were 22 issued patents that made explicit reference to kanban.

TABLE 8-1
Kanban Patents Issued in United States Since 1994

Count	U.S. Patent No.	Title of Patent
1	6,005,571	Graphical user interface for managing security in a database system
2	5,993,041	Production controller for facility group work start
3	RE36,360	Method for determining flexible demand in a manufacturing process
4	5,963,919	Inventory management strategy evaluation system and method
5	5,937,197	Updating of electronic performance support systems by remote parties
6	5,930,763	Method of and system for order amount calculation
7	5,918,054	Distributed electronic performance support systems
8	5,778,386	Global view storage management system for semiconductor manufacturing plants

---

[4] U.S. Patent No. 5,278,750, issued January 11, 1994, to Kaneko, "Production Schedule Making Method, col. 4, ln. 30."

TABLE 8-1. CONTINUED

Count	U.S. Patent No.	Title of Patent
9	5,751,580	Fuzzy logic method and system for adjustment of priority rating of work in process in a production line
10	5,655,086	Configurable electronic performance support system for total quality management processes
11	5,612,886	Method and system for dynamic dispatching in semiconductor manufacturing plants
12	5,544,348	Simulation method and apparatus
13	5,528,489	Method of and apparatus for scheduling parts deliverer running and method of managing parts deliverers
14	5,442,561	Production management system and its application method
15	5,440,480	Method for determining flexible demand in a manufacturing process
16	5,285,392	Parallel manufacturing system
17	5,278,750	Production schedule making method
18	5,233,533	Scheduling method and apparatus
19	5,204,821	Parts supply instruction apparatus
20	5,193,065	System for requisitioning and distributing material in a manufacturing environment
21	4,759,123	Method and apparatus for producing electronic devices which each include a memory
22	4,676,090	Press machine with an adjustable shut height

§ 8.03   Model for Reducing Financial Risk

The credit card industry, quite understandably, views each of us as a financial risk. That is not to say that each of us is a bad financial risk, but that there is some degree of risk associated with each credit card that a company like Visa or MasterCard issues. To control that risk, Visa International developed a business model with a fairly simple purpose: to receive credit card transaction data, score each transaction based on a model, and identify those transactions that scored below a predetermined threshold indicative of poor financial risk.

Presumably what made Visa's business model different was its use of a model reflecting "transaction patterns" to assess risk. The patent specification describes the model as being based on a neural network built upon patterns observed in a known data set with known results (i.e., historical scoreable transactions, the associated account, and the known risk level). The abstract and two representative claims from Visa's patent on this business model are presented below. Claim 1 is a method claim covering the method for predicting financial risk. Claim 26 is a system claim covering certain components of the business model.

[A] Financial Risk Prediction Systems and Methods Therefore — U.S. Patent No. 6,119,103, Basch, et al., Issued: September 12, 2000

[1] Abstract

A computer-implemented method for predicting financial risk, which includes receiving first transaction data pertaining to transactions performed on a first financial account. The first financial account represents a financial account issued to a given account holder by a first account issuer. The method further includes receiving second transaction data pertaining to transaction performed on a second financial account different from the first financial account. The second financial account represents a financial account issued to the given account holder by a second account issuer different from the first account issuer. There is further included scoring the first transaction data and the second transaction data based on a preexisting model to form a score for the account holder. Additionally, there is included transmitting, if the score is below a predefined financial risk threshold, the score to one of the first account issuer and the second account issuer.

[2] Representative Claims

1. A computer-implemented method for predicting financial risk, the computer-implemented method comprising:

receiving transaction data pertaining to a plurality of transactions for a credit account, said transaction data including at least one of a transaction type and a transaction amount for each transaction of said plurality of transactions;

scoring said transaction data, wherein scoring said transaction data includes scoring a transaction pattern ascertained from said transaction data based on a preexisting model to form a score for said credit account, the transaction pattern being indicative of a pattern associated with the plurality of transactions for the credit account, said transaction pattern being arranged to include events that impact the financial risk;

determining when said score is below a predefined financial risk; and

transmitting said score to an account issuer of said credit account when it is determined that said score is below the predefined financial risk.

2. A financial risk prediction system, the financial risk prediction system comprising:

a receiving mechanism, the receiving mechanism being arranged to receive scoreable transaction data from a first data source, the scoreable transaction data being associated with a first credit account issued to a given account holder by an account issuer;

an authenticator, the authenticator being in communication with the receiving mechanism, the authenticator being arranged to authenticate scoreable transaction data received by the receiving mechanism, wherein the scoreable transaction data includes transaction data from the first credit account and transaction data from a second credit account issued to the given account holder;

a scoring mechanism, the scoring mechanism being arranged to generate a score associated with the given account holder, the score being arranged to indicate a financial risk level of the given account holder, wherein the scoring mechanism is arranged to generate the score using the scoreable transaction data in conjunction with a predictive model;

an evaluator, the evaluator being arranged to compare the score against a predefined financial risk threshold; and

a transmitter, the transmitter being arranged to transmit the score to the account issuer.

§ 8.04   Model for Improving Janitorial Services

It is hard to believe that such basic functions as housekeeping would be the subject of a business model patent, but that is, in fact, the case. Virginia Commonwealth University obtained such a patent on November 7, 2000. Entitled, "Method of Managing Contract Housekeeping Services," the patent recites the following steps:

- store selected housekeeping service-specific tasks as variables;

- define a set of performance metrics;

- allow the person grading performance to select a subset of the variables and input an actual performance score for each of the subset;

- the system then outputs a grade, which is then used to make decisions about management of the staff performing the housekeeping services.

In essence, the invention is a grading system in which the housekeeper's duties are broken down into component tasks represented by the system as "variables." It is easy to imagine the inspiration behind the University's janitorial improvement model. As the following excerpt suggests, managing custodial services for an entire university is a multifaceted exercise:

> A comprehensive list of different housekeeping service tasks is created to include all of the different types of tasks that may be performed. The list would include all the different tasks necessary to maintain an entire campus or series of different buildings. Possibly, the tasks would cover all the necessary service tasks for just a single building or portion thereof. For instance, a business office can be expected to have different housekeeping requirements than academic classroom buildings.[5]

The abstract and a representative claim from this patent are presented below.

---

[5] U.S. Patent No. 6,155,943, issued November 7, 2000, to Minder, "Method of managing contract housekeeping services."

[A] Method of Managing Contract Housekeeping Services — U.S. Patent No. 6,144,943, Minder, Issued: November 7, 2000

[1] Abstract

The invention provides a method and apparatus including a computer system for managing contract housekeeping services so as to improve the quality and value of the housekeeping services received. The method invention comprises a series of actions in order to generate a grade representative of the quality of housekeeping services. This grade is then used to make at least one decision regarding the management of housekeeping services. The management having an impact upon the physical appearance and maintenance of a given facility.

[2] Representative Claim

1. A method of using a computer system to manage contract housekeeping services comprising:

storing in said computer system a plurality of variables describing specific housekeeping service tasks;

defining a plurality of possible performance criteria scores probative of the plurality of variables;

receiving in said computer system user input operative to select a subset of said plurality of variables;

receiving in said computer system actual performance criteria scores for the selected subset of variables;

using said computer system to process the actual performance criteria scores and to output a grade representative of the quality of housekeeping services;

using the grade in making at least one decision regarding management of housekeeping services, the management having an impact upon the physical appearance and maintenance of the given facility.

§ 8.05 Service Business Management System

Janitorial services are not the only thing being automated. At least one enterprising tradesperson has determined that the building trades can also be managed more efficiently through the use of computer or microprocessor-based business management systems. U.S. Patent No. 6,216,108, which issued to Mark R. LeVander on April 10, 2001, illustrates the level at which the U.S. Patent Office has been awarding patents on business systems.

The patent describes a computerized system that determines the minimum job price, and generates a contract proposal. The minimum job price is based on such things as minimum labor rate and job parameter data, including a factored-in return on investment equity in the service business. The abstract and several representative claims are reproduced below.

413

[A]   Service Business Management System — U.S. Patent No.
      6,216,108, LeVander, Issued: April 10, 2001

[1]   Abstract

The invention provides a microprocessor, an input device for entering job
costing data and job parameter information, an output device such as a printer
to generate contract proposals and/or management reports. A display device
is preferably also provided to display the minimum labor rate calculated by a
program executed on the microprocessor from the job costing data. Memory
is preferably also provided to store at least the minimum labor rate for use in
generating management reports.

[2]   Representative Claims

1. A service business management system comprising:
a microprocessor;
an input device for entering job costing data, job parameter information, and a proposal price to said microprocessor;
a program executing on said microprocessor which calculates a
minimum labor rate from the job costing data, and calculates a minimum job
price from the minimum labor rate and the job parameter information;
a display device for displaying the minimum job price;
an output device for generating a contract proposal having a proposal
price entered after review of the displayed minimum job price; and
wherein the minimum labor rate calculation includes as factors, the
equity investment in the business (EI) and a desired rate of return on the
equity investment (ROE).
2. The system of claim 1 wherein the minimum labor rate calculation
also includes the parameters:
average direct labor wages (AW), annual direct labor hours (DLH),
workmen's compensation insurance rate (WCIR), and estimated overhead
expenses (OE).
3. The system of claim 2 wherein the minimum labor rate (MLR) is
calculated according to the formula: [lengthy mathematical equation, omitted
here]
4. The system of claim 1 in which the job parameter information
comprises material costs and hours of labor.
5. The system of claim 1 in which the job parameter information
comprises measurements of the job site.
6. The system of claim 5 in which the minimum job price is calculated
from the job site measurements and an industry standard cost per unit
measurement.
7. The system of claim 5 in which the minimum job price is calculated
from the job site measurements and industry standard materials usage per unit
measurement.
8. The system of claim 5 in which the minimum job price is calculated
from the job site measurements and industry standard labor time per unit
measurement.

§ 8.06   Microsoft Marketing to Influential Rumormongers

When you are a company with the resources of Microsoft, there is practically no limit on how technology may be brought to bear on business problems. A case in point is a system developed by Microsoft for inferring who out there are the "influential rumormongers" using data and resources on the Internet. Why Microsoft wants such information and how it intends to use that information is suggested by the following passage from its U.S. Patent No. 6,151,585:

> The present invention concerns methods and apparatus for analyzing resource usage data, such as Internet web site usage data for example, to infer users who communicate with, and influence, other users. Marketing information may then be targeted to such influential users with the hope that they will pass on such information to, and influence, other users.

The system described in the patent is actually fairly sophisticated. It extracts information about usage patterns from "cookies" which many web systems store on users' computers. The system extracts data from these cookies and constructs a *directed graph* representation of each user, which can then be analyzed to infer which users are most likely to influence the buying decisions of others. Based on information stored within your computer, Microsoft may have ascertained that you are an influential rumormonger.

[A]   Details of the Microsoft Information Gathering System

The following excerpt from the patent describe some of the details of the Microsoft system. The subheadings correspond to those of the patent.

[1]   Directed Graph Representation of Information Dissemination on the Internet

> The dissemination of information may be depicted as a directed graph . . . .
> The vertices may represent users. Users may be individual people, an individual computer, a company, a local area network, a group, etc. In the context of the Internet, the vertices may correspond to Internet users who have visited an Internet Web site. The numbers assigned to the edges of the graph correspond to a probability that the user corresponding to the originating vertex will communicate with, and influence, the user corresponding to the destination vertex.
>   [U]ser A communicates with, and influences, user B 90 percent of the time and user C 5 percent of the time. User D always communicates with, and influences, users A and B, and communicates with, and influences, user C 95 percent of the time. User E communicates with, and influences, user F 50 percent of the time. User A takes an action, apparently influenced by no one (i.e., a "ghost user") 5 percent of the time. Similarly, user C takes an action, apparently influenced by no one (i.e., a "ghost user") 10 percent of the time. Such "communications" may be direct (e.g., in person, via mail,

e-mail, telephone, video phone, etc.) or indirect (e.g., via hearsay conversation, television, print media, Internet posting, etc.)

Note that users may influence each other. For example, referring to users A and C . . . user A communicates with and influences user C 5 percent of the time as stated above. However, as shown in the dashed line, user C may also communicate with and influence user A 3 percent of the time. Note also that the probabilities associated with vertices entering or exiting any node need not add to 1.0.

Once the values of the vertices (i.e., users) and the values of the edges (i.e., probability that one user influences another) of the directed graph are known, "influential rumormongers" can be determined from the directed graph using known greedy, graph covering, algorithms for example. Other public or proprietary algorithms may also be used to determine "influential rumormongers" from directed graphs. The challenge thus becomes to generate a directed graph with appropriate values for the vertices and edges.

[2] Determining Values of a Directed Graph; Determining Vertices

In the context of the Internet, users can be uniquely recognized, though not necessarily identified, by so-called "cookies". More specifically, when a user "visits" an Internet site, a cookie is generated by the resource server at the Internet site, provided to the user's computer, and stored at the user's computer. Users may be identified if they voluntarily provide identifying information (e.g., name, address, Internet address, e-mail address, etc.). Alternative means for recognizing or identifying users may be used.

[3] Determining and/or Inferring Edge Values

There are a number of ways to gather data for determining the values of edges in the directed graph. The most direct way may be to ask users, by means of web-based questionnaires for example, (a) who they communicate with, (b) who they influence, (c) who communicates with them, and/or (d) who influences them. Unfortunately, this way of data gathering relies on users to volunteer information. Another way to gather data for determining the values of edges in the directed graph may be to monitor "chat rooms" so that it can be determined who communicates with whom. "Newsgroups" may be monitored to identify those who start "threads" (i.e., a sequence of related communications). E-mail forwarding information can be used to determine who forwarded a communication to whom. Unfortunately, it is difficult to gather this data. Moreover, some users may have an expectation of privacy in their chat room, newsgroup, and e-mail communications, even if the content of their communications is not monitored. Such expectations of privacy may lead to legal or voluntarily placed limitations on such data gathering. Furthermore, although this data may indicate whom communicates with whom, it does not indicate user influence.

Finally, a "referred from" embedded link may be used to inform an Internet site where a visiting user "came from", i.e., the preceding Internet site that the user visited.

Thus, gathering "explicit" information to identify "influential rumor-mongers" is difficult. The present invention uses resource usage log data to infer data of a directed graph from which "influential rumormongers" may be determined.

The fact that a user requests a resource after another already requested that same resource does not necessarily mean that they were influenced by the first requester. However, such influence may be inferred. For example, once an Internet user has visited an Internet Web site, they may tell others about it. This may take place directly, via an in-person conversation, mail, telephone, video call, e-mail, for example, or indirectly, via a hearsay conversation, television, the print media, or a posting on an Internet Web site.

[4] The Use of E-Mail as a Means to Disseminate Information

As discussed above, the use of e-mail over the Internet has dramatically increased over the past five years and is expected to increase. Moreover, most modern e-mail programs permit other information in electronic form (such as a word processing document, or a JPEG image for example) to be "attached" to the e-mail. Hence, a wealth of information may be disseminated quickly, and relatively inexpensively, via e-mail.

[5] Greedy (or Covering) Algorithm for Determining "Influential Rumormongers" from Directed Graph

[I]n the directed graph the vertices may represent users and the edges may represent a probability that a user corresponding to a vertex at the destination of the edge will request a resource within a "memory length" time period of a request for that resource by a user corresponding to the vertex at the origin of the edge. Influence may be inferred from these probabilities. Hence, "influential rumormongers" can be derived from such a graph with known edge values using a greedy algorithm to "cover" the graph. Although such algorithms are known to those skilled in the art, their operation in this context is briefly described for the reader's convenience.

Initially, a first user (also referred to as a "rumor starter") that maximizes the total number of people that will request a resource (or "know the rumor") within a time period (e.g., an "observation length" time period) is determined. Next, a second user (or "second starter") is determined that maximizes the number of users, that have not yet requested the resource, that will request the resource (or "know the rumor") within the observation length time period. This process continues for a predetermined number of rumor starters. The observation length time period may be set to be equal to the "memory length" time period, although it may be less than or greater than the "memory length" time period.

[B] Representative Claims

1. A computer-implemented method for ascertaining an influential rumormonger from amongst a plurality of users through a directed graph from resource usage log data including records having (i) user information,

(ii) resource identification information, and (iii) resource request time information, the method comprising steps of:

a) inferring a corresponding measure of influence between each of a plurality of pairs of said users from the usage log data based on (i) the user information, (ii) the resource identification information, (iii) the resource request time information, and (iv) a memory length parameter so as to define a plurality of influence measures, wherein, for each of said pairs of users, one of the users in said each pair appears to exhibit influence, as reflected in said corresponding influence measure, over action taken by another one of the users in said one pair;

b) generating the directed graph having, for said each one of the pairs of users, first and second vertices corresponding to the one and the other one, respectively, of the users and having an edge connecting the first and second vertices and associated with the corresponding influence measure; and

c) determining the influential rumormonger from amongst the users and the influence measures provided in the directed graph.

2. A device for inferring an influential rumormonger from amongst a plurality of users from resource usage log data including records having (i) user information, (ii) resource identification information, and (iii) resource request time information, the device comprising:

a) a graph edge determination facility for determining a corresponding influence measure between each of a plurality of pairs of said users based on the user information, the resource identification information, the resource request time information, and a memory length parameter so as to define a plurality of influence measures, wherein, for each of said pairs of users, one of the users in said each pair appears to exhibit influence, as reflected in said corresponding influence measure, over action taken by another one of the users in said one pair; and

b) an influential rumormonger determination facility for determining the influential rumormonger based on the corresponding influence measures between all the pairs of said users, the users, and an observation length parameter.

§ 8.07   Business Model for Strategic Management of Lawsuits

Lest you might think for a moment that the legal profession is oblivious to business model patents, think again. Here is a patent by Heckman, Rickerson, Kauffman, and Zaremski that describes a client-server system to automate litigation strategy decisions. Basically, the system comprises three main entities, a few databases, and some program code with which the system is glued together. The main entities include a service provider and two classes of subscribers, client subscribers and law firm subscribers.

The service provider maintains a resource containing historical legal decision information. Presumably Lexis/Nexis or Westlaw would fulfill this requirement. The service provider also maintains an evidence database, into which law firms and clients can store their case-specific information. A "baseline" database contains templates of different types of legal procedures, presumably in the form of checklists and formbook-like documents. The baseline database associates a legal cost with each procedure. This cost data can be used to estimate the cost of a given legal proceeding, allowing the law firm and client to make a cost-benefit analysis. The service provider also gives the users access to other third party databases, if needed.

Client and law firm subscribers can log-in to the service provider, to input information, conduct preliminary forecasts of tasks, and, based on case-specific data, compute projected costs and modify tasks based on actual costs. Based on case-specific data, the service provider can also supply users with automatic reminders and suggested courses of action when certain predetermined conditions occur (e.g., discovery cutoff dates, deposition dates, hearing dates, etc.)

The abstract and a representative claim are presented below.

[A] Legal Strategic Analysis Planning and Evaluation Control System and Method — U.S. Patent No. 5,875,431, Heckman, et al., Issued: February 23, 1999

[1] Abstract

This invention is directed to a strategic planning control system, and more particularly to a computer-based, closed-loop legal strategic planning system and method having iterative convergence to an optimal strategy and dynamic tracking of current prevailing legal climates. The system of this invention includes a computer-generated legal strategy for streamlining the legal process by converting it from a traditional task-oriented system to a process-oriented system. By so doing, predetermined objectives and tasks are defined according to a disciplined time schedule, cost targets are defined, and deliverables agreed upon prior to beginning the legal process. A key aspect of the system and method of this invention is a series of computer programs which provide a strategic planning template outlining the objectives and tasks, and their associated timing. The template is case category and case type specific and presents the "best practices" strategic process from which to launch a legal action. The "best practices" are taken from previously concluded well managed cases having a similar case category and case type as the instant case, and which have been identified as paradigms. Three closed-loop control systems are integrated into the system and method of this invention for dynamically monitoring and measuring legal cost reporting and billing, for dynamically monitoring and measuring attainment of objectives and milestone tasks, and for dynamically measuring and controlling the deliverables derived from the timely completion of the legal objectives and tasks. These control systems have special features for maximizing the likelihood of a desired legal outcome, increasing legal productivity, minimizing the cost to achieve that outcome.

[2] Representative Claim

1. An outcome-oriented, closed-loop legal strategic planning and control system comprising in operative combination:

419

a) a service provider having a first computer system, said first computer system having a central processing system, a memory storage means, and a telecommunications means, said first computer system further comprising:

i) a historical legal case database, said historical legal database containing attribute data on completed paradigm legal cases,

ii) an evidentiary resources database containing therein information to be used in the preparation and prosecution of a legal action,

iii) a baseline template database having therein baseline templates for the execution of a legal process, said baseline templates further comprising objectives, tasks, and activities predetermined to provide the maximum likelihood for a desired legal outcome, and costs and schedules associated with said objectives, tasks and activities, and

iv) a service pass-through means to permit access to third party databases, said third party databases containing information supplemental to information contained in said historical case database and said evidentiary resources database;

b) at least one client service subscriber, said service subscriber having at least one second computer system, said second computer system having a central processing system, a memory storage means, a display means, and at least one of a telecommunications means, optical disk drive, magnetic disk drive, and flash memory;

c) at least one law firm subscriber, said law firm subscriber having at least one third computer system, said third computer system having a central processing system, a memory storage means, a display means, and at least one of a telecommunication means, optical disk drive, magnetic disk drive, and flash memory;

d) a computer readable program code system accessible by said first computer and said second computer, said computer code further comprising:

i) a first computer readable program code subsystem for configuring said second and said third computer systems to perform in any operational order at least one of accepting the keystroke entry of a user identification and password by the user of said computer readable program code system, accepting by user keystroke entry the selection of a task to be performed by said second or third computer system including at lest one of the activation of a second or third computer readable program code subsystem, interfacing to telecommunications for downloading, uploading and for access to external computer readable database and computer applications, and for providing automatic reminders or suggestions to the user upon the occurrence of a set of predetermined conditions,

ii) a second computer readable program code subsystem for configuring said second and said third computer systems for at least one of entry of a first set of case-specific data and a second set of law firm-specific data, for determining and for providing a preliminary forecast of tasks and activities based upon entry of said case specific data, for providing projected costs for said tasks and activities, for permitting modification of said tasks, activities, and costs, for permitting entry of actual costs and completion dates of said tasks and activities, for permitting downloading and uploading of said actual costs and completion dates to said third computer system, for permitting transfer of updated baseline templates from said baseline template database of said first computer system to said memory storage means of said second computer system, and for transferring completed case information to said historical database of said first computer system,

iii) a third computer readable program subsystem for configuring said second and said third computer systems to permit access to at least one of said evidentiary resources database of said first computer system, and to said third party database.

§ 8.08 Buying and Selling Capacity in the Semiconductor Manufacturing Market

As a patent holder IBM commands worldwide respect. When it is awarded a patent that purports to change the way the semiconductor manufacturing market buys and sells capacity, the world takes notice. On May 29, 2001, the U.S. Patent Office awarded IBM U.S. Patent No. 6,240,400, to Chou, et al. The patent describes what is essentially an electronic marketplace in which players can buy and sell capacity, while retaining their anonymity, if desired. This excerpt from that patent describes IBM's motivation for developing the system and some of its advantages:

> Semiconductor manufacturing is a very capital-intensive industry. The cost of building a wafer fabrication (fab) facility is usually more than $1 billion. Therefore, semiconductor manufacturers want to increase the utilization of their existing facilities. However, fluctuations in demands for chips may result in a mismatch between manufacturing capacity and demand, thereby disadvantageously resulting in excess manufacturing capacity in some companies and a shortage in others. Currently, under such conditions, companies may delay order deliveries or incur very large losses due to idle capacity. These capacity shortages at the fabrication facility can affect manufacturers that use chips as raw materials in their products. Delayed deliveries greatly increase requirements for safety stocks, and can also result in considerable lost sales for a company.
>
>   * * *
>
> We have now discovered a novel method and system which can enable semiconductor manufacturers and downstream product manufacturers and users of chips to trade semiconductor manufacturing capacity, thereby creating a situation that can result in benefits for all parties. Some advantages of such a method/system include:
> * an efficient utilization of semiconductor manufacturing capacity;
> * a preservation of anonymity of the players in this market, and hence a preservation of competitive information a company may not want to divulge;
> * a creation of a common marketplace where all the players can trade semiconductor manufacturing capacity simultaneously, thereby obviating one-to-one negotiations and their attendant high search costs;
> * a flexibility to change capacity commitments in line with customer demands from time to time.

[A] **Method and System for Accommodating Electronic Commerce in the Semiconductor Manufacturing Industry — U.S. Patent No. 6,240,400, Chou, et al., Issued: May 29, 2001**

[1] Abstract

A method for accommodating electronic commerce in a semiconductor manufacturing capacity market. The method comprises the steps of identifying a plurality of players in the semiconductor manufacturing capacity market, each of which players can solicit capacity in semiconductor manufacturing capacity market; providing a neutral third-party, the neutral third party and the plurality of players configured in a hub arrangement for communicating with each of the plurality of players in semiconductor manufacturing capacity trades; and realizing an open market conditionality between each of the plurality of players and the neutral third party so that the semiconductor manufacturing capacity supplied by one or more of the players can be bought and sold among the players; and, the neutral third party can preserve anonymity of each of the plurality of players soliciting semiconductor manufacturing capacity.

[2] Representative Claims

1. A computer implemented method for accommodating electronic commerce in a semiconductor manufacturing capacity market, the method comprising:

1) identifying a plurality of players in the semiconductor manufacturing capacity market, each of which players can solicit capacity in semiconductor manufacturing capacity market,

2) providing a neutral third-party, the neutral third-party and the plurality of players configured in a hub arrangement for communicating with each of the plurality of players in semiconductor manufacturing capacity trades; and

3) realizing an open market conditionality between each of the plurality of players and the neutral third party so that:

i) the semiconductor manufacturing capacity supplied by one or more of the players can be bought and sold among the players; and

ii) each of the players may individually select at any time during said market conditionality, one of the following regardless of selection of the other players: identifying themselves to the other players and preserving their anonymity among each of the plurality of players soliciting semiconductor manufacturing capacity.

2. A method according to claim 1, wherein said plurality of players is selected from a group consisting of semiconductor manufacturers, semiconductor design companies, semiconductor manufacturing capacity bulk buyers and sellers, bulk buyers and sellers of semiconductors, and manufacturers of products that incorporate semiconductor chips.

3. A method according to claim 1, wherein a solicit comprises ascertaining a market price of the semiconductor manufacturing capacity.

4. A method according to claim 1, wherein a solicit comprises specifying the availability of the semiconductor manufacturing capacity.

5. A method according to claim 1, wherein said semiconductor manufacturing capacity comprises specifying attributes of said semiconductor manufacturing capacity including a type of semiconductor, a size of wafer, a number of wafer starts, a set of processes required for manufacturing these wafers, and a time period during which these processes need to be used.

6. A method according to claim 1, wherein preserving anonymity by the neutral third party comprises the neutral third party not revealing a player's identity to anyone.

7. A method according to claim 4, wherein specifying available semiconductor manufacturing capacity comprises connotating a price at which any of said plurality of players is willing to buy.

8. A method according to claim 4, wherein specifying available semiconductor manufacturing capacity comprises connotating a price at which any of said plurality of players is willing to sell.

9. A system for accommodating electronic commerce in a semiconductor manufacturing capacity market, the system comprising:

1) means for identifying a plurality of players in the semiconductor manufacturing market, each of which players can solicit capacity in the semiconductor manufacturing market;

2) means for providing a neutral third-party, the neutral third-party and the plurality of players configured in a hub arrangement for communicating with each of the plurality of players in semiconductor manufacturing capacity trades; and

3) means for effecting an open market conditionality between each of the plurality of players and the neutral third party so that:

i. the semiconductor manufacturing capacity supplied by one or more of the players can be bought and sold among the players; and,

ii. each of the players may individually select at any time during said open market conditionality one of the following regardless of selections of the other players: identifying themselves to the other players and preserving anonymity of each of the plurality of players soliciting semiconductor manufacturing capacity.

§ 8.09 Music Popularity Rating

To lighten things up from the last example, the business model presented here is a system for rating music. Top 40 radio stations have been doing that for years. The difference here is the system rates music by having a selected group of listeners rate a collection of music "hooks," which are essentially the toe-tapping riffs, catchy melodies, and clever refrains that make popular songs popular. Through field surveys, listeners are asked to rate their favorite hooks. The ratings are then tabulated and used to identify which songs are likely to be the most popular. Armed with this information, radio stations can adjust their program content to increase their listener base.

On June 15, 1999, the Patent Office awarded U.S. Patent No. 5,913,204 to Thomas L. Kelly for an invention entitled, "Method and Apparatus for

Surveying Music Listener Opinion About Songs.'' The abstract and several representative claims from that patent are reproduced below. Note, in particular, some of the dependent claims which appear to add conventional details, such as making a follow-up telephone call to ensure the listener has completed the survey.

While rating of songs based on rating of hooks may be the primary difference between this system and conventional listener preference surveys, the patent defines the term ''hook'' to include even the entire song, as this excerpt from the patent demonstrates:

> The music medium contains a predetermined number of representative portions of songs or song hooks which are to be reviewed by the select listeners. Each song hook is identified by code, for example, by introducing each song hook by number only. The length of each song hook varies depending on the type of music being surveyed. For example, contemporary, country, rock and pop songs generally require a hook of about 7 to 10 seconds to allow the listener enough time to recognize the song. Other types of music such as classical require a hook of about 60 seconds to allow the listener enough time to recognize the song. Further, if brand new or unfamiliar music is being tested, the hook may comprise the entirety of the song.

[A] Method and Apparatus for Surveying Music Listener Opinion About Songs — U.S. Patent No. 5,913,204, Kelly, Issued: June 15, 1999

[1] Abstract

A method and apparatus for surveying and reporting listener opinion of a list of songs such as the songs comprising a radio station music library. A group of music listeners is selected from whom individual listener opinions are recorded. A home music preference test kit is fielded to the select listeners for use in each listener's residence. Data from each music preference test kit is then collected, compiled and tabulated to reflect a mean or average listener opinion for each song in the list. The home music preference test kit includes instructions for completing the test and returning the survey form, a music medium having a preselected number of song hooks representing each song in the library of the radio station. The kit also includes a survey form for recording the listener's opinion of each song based on the song hooks and an honorarium for agreeing to properly complete a survey form.

[2] Representative Claims

1. A method of surveying and reporting listener opinion of a list of songs from the music library of a radio station including the steps of:

a) selecting a group of music listeners from which individual listener opinions are to be recorded;

b) fielding from a survey taker a home music preference test kit including a music audio medium containing a number of song hooks to the select listeners at each listener's residence;

each listener listening to a list of music or music song hooks,

entering his preferences as to the song hooks on a test sheet without the presence of a survey interviewer;

returning the completed test sheets to the survey taker,

c) collecting and compiling data from each music preference test kit test sheet answered by each listener;

d) tabulating the test sheet data in a computer to reflect listener opinion for each song in the list,

reporting the tabulated data to the radio station, and modifying the music programming of the radio station based on the results of the survey.

2. The method of claim 1 wherein the step of selecting a group comprises defining a geographic market from which the music listeners are selected.

3. The method of claim 2 wherein the step of selecting a group comprises contacting and interviewing randomly-selected individuals within the geographic market to determine their compliance with specific listener criteria.

4. The method of claim 3 wherein the step of contacting randomly-selected individuals comprises random telephone dialing in the geographic market using random digit dialing utilizing known telephone prefixes or exchanges.

5. The method of claim 3 wherein the step of contacting randomly-selected individuals comprises publicly advertising in the geographic market.

6. The method of claim 3 wherein the step of interviewing randomly-selected individuals comprises asking individuals a series of questions to determine if the individual meets the specific listener criteria.

7. The method of claim 6 including the step of asking perceptual questions to gather general impressions about issues that may impact a radio station's marketing strategy.

8. The method of claim 3 wherein the step of selecting a group further includes recruiting the individuals complying with the specific listener criteria.

9. The method of claim 8 including the step of recruiting individuals from a demographic age group, a household income range, or a particular racial or ethnic background.

10. The method of claim 8 wherein the step of recruiting individuals comprises inviting the individual to participate in a home music survey.

11. The method of claim 10 including the step of asking the individuals control questions to verify data received from the individual.

12. The method of claim 2 wherein the step of selecting a group comprises telephone contacting individuals on a list sample who meet the specific listener criteria.

13. The method of claim 1 wherein the step of fielding a home music preference test kit comprises delivering the test kit to the select listeners, said test kit including instructions for completing the test survey, a music medium containing a number of song hooks, a survey form, and an honorarium.

14. The method of claim 13 including the step of providing a control device to verify the data received from the select individual.

15. The method of claim 14 including the step of including verification checkpoints in the survey.

16. The method of claim 13 including the step of placing a follow-up telephone call to the select listener to confirm arrival of the home music preference test kit at the select listener's residence.

17. The method of claim 16 including the step of instructing the select listener that the completed test should be returned within a predetermined short duration.

18. The method of claim 17 including the step placing a second follow-up telephone call to the select listener to confirm that the home music preference test was completed and returned.

§ 8.10   Mediating Purchase Transactions Over a Network

On March 2, 1999, the U.S. Patent Office awarded to Microsoft U.S. Patent No. 5,878,141 for a computerized purchasing system. The system has a straightforward purpose: to mediate purchase transactions by keeping track of which merchants accept which payment methods and by further keeping track of which personal payment methods each purchaser uses.

The system uses two databases, (a) a purchaser database that stores a list of purchasers and their associated personal payment methods, and (b) a merchant database that stores a list of merchants and the accepted payment methods of each. The system processor receives purchase requests, identifies merchant and purchaser and then computes the "intersection" between the merchant's accepted payment methods and the purchaser's available payment methods. In essence, the "intersection" matches the credit card the purchaser holds to the credit card the merchant will accept.

The motivation behind this Microsoft invention is provided, in part, by background description contained in the patent. The following excerpt is illustrative:

> In a transaction for the purchase of goods and/or services, the purchaser typically has the ability to pay for the items using any one of many different payment methods. For instance, consider the familiar situation where a purchaser in a department store wishes to buy an article of clothing. The purchaser can pay for the clothing article with cash or by check. Alternatively, the purchaser might wish to use a credit card or a debit card. Indeed, it is not uncommon for the purchaser to carry many payment options in the form of cash, a checkbook, a bank debit card, as well as many different kinds of credit cards, including cards issued by the merchants themselves (e.g., a Sears.RTM. charge card or a Nordstroms.RTM. charge card), bank issued credit cards (e.g., a SeaFirst Bank VISA.RTM. credit card or a Bank of America MasterCard.RTM. credit card), an organization-related credit card (e.g., United Airlines Mileage Plus First Card.TM. or an IEEE credit card), and association credit cards (e.g., Discover.RTM. and American Express.RTM.).

The department store, on the other hand, might only accept a few of these forms of payment, such as cash, local checks, its own charge card, and American Express.RTM., while not accepting other forms of payment. The department store often posts these accepted forms of payment at the point-of-purchase counter.

During the purchase transaction of the clothing article, the purchaser mentally takes note of the forms of payment accepted by the department store. The purchaser then tenders payment using a suitable payment method. If the purchaser chooses to pay with a personal check, the sales clerk performs an authentication process. The clerk only accepts the check if it is local, if the clerk recognizes the person writing the check, or if the person presents another piece of identification (e.g., a credit card or driver's license) to verify the authenticity of that person who is offering the check.

In the event the purchaser tenders a credit card to pay for the clothing article, the sales clerk performs a check to verify that the purchaser has sufficient funds in the credit card account and has not exceeded the spending limit imposed by the issuing institution. This is typically done by passing the purchaser's credit card through a magnetic-stripe card reader, such as a Verifone.RTM. system, that is located at the point-of-purchase counter to electronically read the purchaser's account information contained in the magnetic stripe on the credit card. The purchaser's account information is validated on-line with the card issuer with respect to the purchaser's account balance and spending limit. Assuming that the verification process returns a normal status, the sales clerk accepts the tendered credit card and consummates the purchase.

The complexity of a purchase transaction increases when moved from the point-of-purchase context, where the purchaser and merchant are face-to-face, to a remote purchase context, where the merchant and purchaser are separated from one another. For example, consider another familiar transaction where a purchaser wants to buy a new lamp from a mail order catalog. The purchaser places an order for the lamp over the telephone or through the mail. The purchaser might use a credit card, enclose a check, or simply wait to be billed at the end of the month. The merchant takes an assumed risk that the ordering consumer is legitimate and that payment will be forthcoming, and based upon that assumption, ships the new lamp to the purchaser.

In these transactions, the merchant accepts a fairly high risk of not being paid (compared to other types of sales transactions) because the purchaser does not present a credit card or sign a credit card receipt. The purchaser can deny that the transaction ever occurred, leaving the merchant with the burden of proving that a transaction took place. To meet this burden, the merchant typically tries to show that the purchaser signed for receipt of the product.

The patent issued with 55 claims. The following provides the abstract and a representative sample of those 55 claims.

[A] Computerized Purchasing System and Method for Mediating Purchase Transactions Over an Interactive Network — U.S. Patent No. 5,878,141, Daly, et al., Issued: March 2, 1999

[1] Abstract

A computerized, electronic purchase mediating system includes a purchaser database having a list of purchasers and a merchant database having a list of merchants. The purchaser database stores information about each purchaser including a set of personal payment methods that the purchaser could use to purchase goods and/or services. Similarly, the merchant database stores information about each merchant including a set of accepted payment methods that the merchant would accept for sale of the goods and/or services. The purchase system also includes a processor coupled to the purchaser and merchant databases. The processor receives a purchase request and accesses the merchant database according to a merchant identified in the purchase request to retrieve the set of accepted payment methods which corresponds to that merchant. The processor also accesses the purchaser database to retrieve the set of personal payment methods which corresponds to the identified purchaser. The processor then computes an intersection of these two sets to derive a common set of any available payment method that is both accepted by the merchant and can be used by the purchaser for purchase of the goods and/or services. The purchaser is presented with the purchase amount and the common set of available payment methods to choose a most preferred form of payment. Upon selection, the processor consummates the sale and signs a digital signature with the purchaser's permission via password verification to ensure for the merchant that a completed transaction has occurred.

[2] Representative Claims

1. An electronic purchase mediating system comprising: a purchaser database having a list of purchasers, the purchaser database also storing a set of many personal payment methods for corresponding ones of the purchasers whereby an individual purchaser could use any one of the personal payment methods in that purchaser's corresponding set to purchase goods and/or services;

a merchant database with a list of merchants, the merchant database also storing a set of many accepted payment methods for corresponding ones of the merchants whereby an individual merchant is willing to accept any one of the accepted payment methods in that merchant's corresponding set for sale of the goods and/or services;

a processor coupled to the purchaser and merchant databases, the processor also being coupled to receive a purchase request for goods and/or services, the purchase request identifying a merchant and a purchaser;

the processor accessing the merchant database according to the merchant identified in the purchase request to retrieve the set of many accepted payment methods which corresponds to that merchant, the processor also accessing the purchaser database according to the purchaser identified in the purchase request to retrieve a set of many personal payment methods which corresponds to that purchaser; and the processor computing an intersection of these two sets to derive a common set of any available payment method that is both accepted by the merchant and can be used by the purchaser for purchase of the goods and/or services.

2. An electronic purchase mediating system as recited in claim 1 wherein:

the purchase request further includes a purchase amount; the purchaser database includes personal purchase allowances for associated purchasers, each personal purchase allowance being imposed by the purchaser to prevent an expenditure in excess of the personal purchase allowance; and

the processor evaluates whether the purchase amount contained in the purchase request exceeds a personal purchase allowance associated with the identified purchaser.

3. An electronic purchase mediating system as recited in claim 1 wherein:

the purchase request further includes a purchase amount; the personal payment methods of the purchasers have associated spending limits, each spending limit being imposed and maintained by an institution that administers the payment method to prevent an expenditure in excess of the spending limit; and

the processor communicates with the institution to evaluate whether the purchase amount contained in the purchase request exceeds a spending limit of any available payment method in the common set.

4. An electronic purchase mediating system as recited in claim 1 wherein:

the purchase request further includes a purchase amount; the purchaser database includes account balances for corresponding ones of the personal payment methods for related purchasers; and

the processor evaluates whether the purchase amount contained in the purchase request exceeds an account balance of any available payment method in the common set.

5. An electronic purchase mediating system as recited in claim 1 wherein:

the purchaser database includes unique signing keys for creating digital signatures for corresponding ones of the purchasers; and

the processor creates a digital signature on behalf of the identified purchaser to authorize the purchase of the goods and/or services.

§ 8.11   Administration of Life Insurance Business

Occasionally the Patent Office will issue a patent on a system that is designed to mirror, or comply with, a state or federal law or regulatory requirement. As discussed in Chapter 2, the patent-in-suit in the *State Street Bank v. Signature Financial* litigation was criticized by State Street Bank as having been designed merely to take advantage of certain provisions of the U.S. tax laws.

On September 8, 1998, roughly two months after the Federal Circuit decision in *State Street Bank*, the U.S. Patent Office issued U.S. Patent No.

5,806,042 on a system to manage a Bank Owned Life Insurance (BOLI) plan that was configured to ensure compliance with regulatory guidelines regarding the maximum premiums that could be charged under law. The system would purportedly balance the financial needs of the bank with the complex legal regulatory guidelines.

The abstract and a representative claim from the patent are presented below.

[A] System for Designing and Implementing Bank Owned Life Insurance (BOLI) with a Reinsurance Option — U.S. Patent No. 5,806,042, Kelly, et al., Issued: September 8, 1998

[1] Abstract

The present invention involves a computer software and hardware system which smoothly integrates the following functions into an integrated computer-based system for designing and administering a BOLI plan for national banks under current federal and state guidelines and financial market constraints. The system includes determining the highest BOLI premium permitted under OCC Banking Circular 9651, determining insurable interest requirements by accessing a database with the appropriate state's insurable interest guidelines, generating performance estimates for the BOLI plan and allocating premium amount by business unit and employee. The system also ensures that the BOLI plan is in compliance with the regulatory requirements for the business unit. In addition, the system reinsures the BOLI plan through a captive insurance company of the financial organization, obtaining policy values for the captive insurance company. Other aspects of the system include verifying, reconciling, consolidating and reporting policy values for the financial organization, and performing administrative procedures for the BOLI plan of the financial organization.

[2] Representative Claim

1. A system for designing and administering a bank owned life insurance plan for a financial organization, comprising:

a. means for determining the highest bank owned life insurance premium permitted under Office of the Comptroller of the Currency banking guidelines

b. means for determining insurable interest requirements by accessing a database with the appropriate state's insurable interest guidelines;

c. means for generating performance estimates for said bank owned life insurance plan, wherein said performance estimates are used to ensure compliance with risk-based capital requirements;

d. means for allocating premium amount by business unit and employee, and for calculating regulatory requirements for said business unit and employee, and for ensuring that said bank owned life insurance plan is in compliance with the regulatory requirements for said business unit;

e. means for calculating maximum risks and premium values that can be reinsured through a captive insurance company of said financial organization;

f. means for said captive insurance company to obtain policy values;

g. means for verifying, reconciling, consolidating and reporting policy values for said financial organization; and

h. means for performing administrative procedures for said bank owned life insurance plan of said financial organization.

§ 8.12 Determining Insurance Rates Based on Geographic Location

The previous example was for a business method for managing a life insurance business. This next example is also from the insurance industry. Specifically, the example comes from U.S. Patent No. 6,186,793 to Brubaker, which issued February 13, 2001. The patent describes a methodology for determining insurance rates for different geographic locations. Essentially, the method involves selecting a number of geographic points for which insurance rate values may be calculated, based on selected and weighted data. Then rates for the intermediate locations are calculated by interpolation. All rates are then used to create a three-dimensional surface or map reflecting the expected cost to cover future insurable events.

The abstract and several sample claims from the patent are presented below.

[A] Process to Convert Cost and Location of a Number of Actual Contingent Events Within a Region into a Three Dimensional Surface Over a Map that Provides for Every Location Within the Region Its Own Estimate of Expected Cost for Future Contingent Events — U.S. Patent No. 6,186,793, Brubaker, Issued: February 13, 2001

[1] Abstract

A method of establishing the insurance rate at a desired location comprising the steps of selecting a plurality of predetermined points, calculating the insurance rate at each predetermined point, and interpolating among the values of insurance rates at predetermined points adjacent to the desired location to calculate the insurance rate at the desired location.

[2] Representative Claims

1. A method that utilizes and transforms historic insurance data such as losses, policyholder location, coverage provided, and other factors affecting likelihood of loss into a three dimensional surface over a map that provides a uniquely estimated rate value for every location desired, comprising the steps of:

selecting a plurality of predetermined points such that every location for which rate values are desired is located at or among the predetermined points;

determining a rate value for each predetermined point by using data uniquely selected and weighted from other locations;

interpolating among the rate values at the predetermined points to determine the rate values at desired locations which are located among the predetermined points; and

specifying the rate values calculated at the predetermined points and, by formula, the rate values at desired locations among the predetermined points to create a three dimensional surface over a map that provides a uniquely estimated rate value for every location desired.

2. The method of claim 1, wherein the plurality of predetermined points are uniformly spaced from each other.

3. The method of claim 1, wherein the step of calculating the rate values at each predetermined point comprises the step of identifying relevant information affecting the likehood of loss.

4. The method of claim 3, wherein the step of calculating the rate values at each predetermined point further comprises the step of assigning relative weights to the historic insurance data.

5. The method of claim 4, further comprising the step of adjusting the rate values at each predetermined point to reflect actuarial credibility.

§ 8.13   The Never Ending Subscription

One of the many patents generated by the Walker Digital think tank was this one for an open-ended subscription service. Note, the representative claims include means-plus-function claims, method claims, article of manufacture claims, and system claims.

[A] Method and Apparatus for Providing Open-ended Subscriptions to Commodity Items Normally Available Only Through Term-based Subscriptions — U.S. Patent No. 6,014,641, Loeb, et al., Issued: January 11, 2000

[1] Abstract

A system for providing an open-ended subscription to commodity items normally available on a term basis includes a central agent that serves as the front-end for commodity suppliers. This central agent maintains databases containing information associated with a group of commodity items and their sales. Using these databases, the central agent produces subscription records to provide open-ended subscription services to its customers, while support-ing the term-based subscriptions of the commodity suppliers.

[2] Representative Claims

1. A system for managing subscriptions to commodity items normally available through renewable term-based subscriptions, comprising:

means for receiving customer orders for open-ended subscriptions to said commodity items, said open-ended subscriptions expiring only upon request of respective customers;

means for storing said received customer orders;

means for receiving from suppliers subscription information for said commodity items;

means for storing said subscription information;

means for generating supplier orders for renewable term subscriptions to said commodity items based upon said stored customer orders and said stored subscription information; and

means for transmitting said supplier orders to respective suppliers of said commodity items.

8. A method of managing subscriptions to commodity items normally available through renewable term-based subscriptions, comprising the steps of:

receiving customer orders for open-ended subscriptions to said commodity items, said open-ended subscriptions expiring only upon request of respective customers;

storing said received customer orders;

receiving from suppliers subscription information for said commodity items;

storing said subscription information;

generating supplier orders for renewable term subscriptions to said commodity items based upon said stored customer orders and said stored subscription information; and

transmitting said supplier orders to respective suppliers of said commodity items.

15. An article of manufacture for causing a computer to manage subscriptions to commodity items normally available through renewable term-based subscriptions, comprising:

means for causing a computer to receive customer orders for open-ended subscriptions to said commodity items, said open-ended subscriptions expiring only upon request of respective customers;

means for causing a computer to store said received customer orders;

means for causing a computer to receive from suppliers subscription information for said commodity items;

means for causing said computer to store said subscription information;

means for causing a computer to generate supplier orders for renewable term subscriptions to said commodity items based upon said stored customer orders and said stored subscription information; and

means for causing a computer to transmit said supplier orders to respective suppliers of said commodity.

28. A system for ordering magazines through open-ended magazine subscriptions, comprising:

a first database storing subscription information relating to a plurality of renewable term subscriptions for magazines from respective publishers;

a second database storing order information relating to customer orders for selected ones of said magazines on an open-ended basis;

a third database storing publisher information relating to each of said plurality of publishers; and

means for selecting information from said first, second, and third databases to create subscription orders for initiating renewable term magazine subscriptions from respective publishers to respective customers.

§ 8.14   Separating Product Pricing from Product Distribution

Buy the chosen product through our central controller and get a lower price. That is how this Walker Digital invention might be advertised. According to the patent disclosure, the invention enables a manufacturer to legally and effectively control the customer price of goods and services, while preserving the profitability of the retailer. The system is said to enable a manufacturer to establish a "private" price between himself and the customer. According to the patent disclosure, this manufacturer-controlled price can be set legally and without upsetting either the normal pricing structure or profit margin of the retailer. Because the profitability of and hence good working relationship is maintained with the retailer, the retailer remains highly motivated to sell the manufacturer's products. The patent specification asserts that the invention creates a new retail paradigm:

> By effectively separating out the pricing of goods from the distribution, sale, and customer support for those goods, the present invention presents a new retail paradigm favorable to the manufacturer while motivating and using the best skills of the retailer. Further, the present invention provides such benefits and advantages using technologies available to all participants in the retail process.

The patent issued with 132 claims. The abstract and a representative claim are provided below.

[A] Systems and Methods Wherein a Buyer Purchases a Product at a First Price and Acquires the Product from a Merchant that Offers the Product for Sale at a Second Price — U.S. Patent No. 6,249,772, Walker, et al., Issued: June 19, 2001

[1] Abstract

Systems and methods are provided wherein a buyer purchases a product at a first price and acquires the product from a merchant that offers the product for sale at a second price, the second price being different from the first price. Transaction information associated with the buyer and the merchant is received. Information that allows the buyer to acquire the product from the merchant in exchange for providing payment of an amount based on the first price, such as by providing payment to a central controller, is transmitted. According to one embodiment, the central controller provides payment of an amount based on the second price to the merchant.

[2] Representative Claim

1. A computer-implemented method, comprising:
receiving transaction information, the transaction information being associated with (i) a buyer who has arranged to purchase a product from a central controller at a first price established between the buyer and the central

controller and (ii) a merchant that offers the product for sale at a second price, different from the first price, without offering the product for sale to buyers at the first price; and

transmitting information that allows the buyer to acquire the product from the merchant in exchange for the buyer providing payment to the central controller of an amount based on the first price,

wherein the central controller provides payment of an amount based on the second price to the merchant.

§ 8.15   Upgrading Your Airline Seat

Imagine logging onto an Internet site prior to departure and making an offer to purchase an upgrade to first class for an amount of your choosing. In effect, you would be bidding against other customers for the premium seat. If airlines did adopt such a policy, it could represent a substantial source of revenue. According to the specification of this Walker Digital patent, there are practical reasons why this is not already being done today:

> Airlines recognize that there is a large source of incremental revenue that may be obtained from existing passengers willing to purchase upgraded tickets for available premium-class seats at a favorable price. There is currently no effective way, however, for an airline to receive an offer from a customer for an upgraded seat or other premium service at a particular price set by the customer, below the airline's published fare. Thus, due primarily to operational concerns, airlines are not maximizing their at-the-gate revenue opportunities for highly valued perishable services, such as available premium seats. Specifically, the airlines want to keep the gate area for boarding flights as simple and fluid as possible. Thus, airlines are unwilling to place complex or judgment-based systems at the gate, which can delay flights, frustrate passengers and increase anxiety of operating personnel who are already under significant time pressure to get the flight boarded and pushed back from the gate.
>
> In addition, there is currently no effective way for the airline to be confident that if the airline accepts the customer's offer, the customer will book the upgraded ticket without using the information to ascertain the airline's underlying level of price flexibility, which, if known to an airline's competitors or customers, could dramatically impact the airline's overall revenue structure. Thus, airlines typically provide upgraded seats to preferred customers, such as frequent flyers, for free or in exchange for upgrade coupons purchased by such customers in advance, or allow available premium-class seats to fly empty.

The patent purports to provide a solution. The abstract and a representative claim are presented below. The patent issued with 84 claims.

[A] **Automated Service Upgrade Offer Acceptance System — U.S. Patent No. 6,112,185, Walker, et al., Issued: August 29, 2000**

[1] **Abstract**

An automated service upgrade offer acceptance system is provided for receiving (i) reservations for a selected category of assigned services, such as seating, and (ii) offers for upgraded services, such as an upgrade of an initially selected category of seating to a preferred seating category, from confirmed customers. The automated service upgrade acceptance system permits customers to submit offers for a number of upgraded services or upgrade offer items, including offers for an upgrade of an initial category of seating to a different seating category, including premium seats within a given category of seating, as well as other premium services, such as priority for special meals or drinks, priority for receipt of luggage upon deplaning, and discount companion tickets. The automated service upgrade acceptance system allows a customer to place a binding offer for an upgraded offer item, should the item become available. Offers for upgraded services can be accepted by the automated service upgrade acceptance system at any time during a seller-defined offer acceptance period, from the time of making a reservation, up to a predefined expiration period. The automated service upgrade acceptance system processes the received offers for upgraded services, at one or more intervals until an offer acceptance period expires, to determine whether to accept or reject each offer for an upgraded service and thereafter notify the customers of any revised seating assignments. The received offers for each different upgrade offer item are preferably processed in a predefined sequence, such that offers for the highest categories of seating are processed first. The automated service upgrade acceptance system can enhance the value of offers in accordance with seller-defined criteria for preferred customers, such as frequent flyers, or in accordance with a promotional offer.

[2] **Representative Claim**

1. A method for processing reservations for a selected category of assigned service, said method comprising the steps of:

receiving a reservation from a customer for said selected category of service at a specified price from among a plurality of service categories;

receiving an offer from said customer for a change of said selected category of service to a preferred category of service at a price defined by said customer;

evaluating said offer based on predefined offer acceptance criteria; and

indicating said offer as acceptable if said offer meets said predefined offer acceptance criteria.

CHAPTER 9

PRIOR ART

§ 9.01 The Importance of Prior Art

Prior art is important to all patents. It is axiomatic that no patent can cover the prior art, no matter how broadly drafted its claims are. Thus, you often look to the prior art when construing a patent claim to determine what the patent does not cover. This is certainly true for business model patents.

During the 1970s and 1980s many attorneys believed software could not be patented, and counseled their clients accordingly. Few software applications were filed. Thus, in the 1990s, when more and more software patent applications were being filed, the Patent Office had scarcely any prior art software patents to which it could refer. Without an adequate collection of prior art, the Patent Office granted quite a number of software patents that software developers consider to be invalid. History seems to have repeated itself, for the same criticism is currently being lodged against business method and business model patents.

Today the Patent Office is making great strides in improving its prior art collection. Much of this collection is the result of applications for business method and business model patents, and information disclosed to the Patent Office by patent applicants. It is in your interest to help the Patent Office by submitting pertinent prior art of which you are aware. Aside from the fact that 37 C.F.R. § 1.56 requires such disclosure, by submitting prior art you help strengthen the patent system. It also strengthens a client's patent position, for a patent is presumed to be valid over all art considered by the Patent Office.

Finding the most pertinent prior art is another matter. Often a client will be aware of pertinent prior art, such as prior products, journal articles, textbooks, and the like. The remainder of this chapter concentrates on different ways of searching for prior art.

§ 9.02 Patent Classification System

Traditionally patent examiners have searched for prior art by hand. This entails combing through the vast collection of U.S. patents, foreign patents, and literature that is kept at the Patent Office. Copies of United States patents are kept in rooms with floor-to-ceiling compartment shelves, called *shoes*. There are millions of United States patents in these shoes. To a lesser extent, foreign patents and literature are also placed in these shoes. Today patent examiners also have computerized text searching tools to help search for prior art. The paper copy shoes are still maintained and occasionally used.

Difficult as the hand searching task may be, it is not nearly as difficult as it would be without the Patent Classification System. All United States patents are assigned to one or more classes and subclasses, according to the invention claimed. Using this classification system the examiner can usually find pertinent prior art, by identifying the relevant classes and subclasses, and then by quickly leafing through each patent in the pertinent subclasses by hand.

Anyone can search for prior art in the same fashion. All that is required is an understanding of the Patent Classification System and a road map of the Patent Office—in order to find where the relevant classes and subclasses are physically located in the Patent Office. The following sections will give you an understanding of the Patent Classification System. For a road map of the Patent Office, you will have to wait until you get to the Patent Office and then ask. Patent Office employees are constantly moving things around to make room for the ever-increasing patent collection.

§ 9.03   Patent Classification Procedures

When the Patent Office was established in 1790 there was little need for a patent classification system. Under the 1790 Patent Act only 57 patents were granted, hence it was feasible to search the entire body of issued patents by hand. Later, under the 1793 Patent Act, there was still little need for a classification system, for under that act Congress did away with patent examination altogether. Questions of novelty, utility, and scope of the patent grant were left to the courts. During that era, patents were not printed and were only available to the public in manuscript form in the files of the State Department.[1]

The need for a patent classification system arose in 1836, when Congress reinstituted an examination system and established the Patent Office to administer the system. Since its inception, the Patent Office has tried various classification systems, but finding one that works has proved quite difficult. One early system (1837) was a patent list divided into 21 classes.

Class 21, entitled ''Miscellaneous'' was the problem. It had no discernible order or basis and bore no relationship to the other classes. Every new technology (and many patents do involve new technology) ended up being classified under Miscellaneous. Later, in 1868, the Patent Office dropped the Miscellaneous category and adopted a class-subclass system. Problematically, the classes and subclasses were arranged alphabetically, so they still bore no relationship to one another.

Eventually the alphabetical arrangement was replaced by an outline arrangement, which shows relationships among classes and subclasses by hierarchical position. The current classification system uses such an arrangement. However, even the outline arrangement has problems with inconsistency and ambiguity arising from differences in subjective judgment. Deciding where a new subclass should be added in the positional hierarchy, the classifier must use subjective judgment. Naturally, different classifiers see classification relationships differently and therefore inconsistencies and ambiguities creep in.

To demonstrate, note the ambiguity in Subclasses 2, 11, and 28 in the classification of Artificial Fuels, established in 1920. A Subclass Apparatus appears in three different places:

1   ARTIFICIAL FUEL
2   Apparatus

---

[1] U.S. Dep't of Commerce, *Development & Use of Patent Classification Sys.*, Libr. of Cong. Catalog No. 65-62235, at viii (1966).

10 Briquetting
11 Apparatus
27 Peat
28 Apparatus

Today, there is a *Manual of Classification* that dictates how patents are assigned to a class and subclass. The Patent Office provides classification information in both paper copy and Internet Web-based content. Many business-related inventions are classified in Class 705. For the most up-to-date information, consult the Patent Office Web site at <www.uspto.gov>. The *Manual of Classification* is structured according to formal principles of the United States Patent Classification System. These principles govern where a patent should be classified and when a new class or subclass should be created.

[A] Principles of Patent Classification System

The present Patent Classification System is built on 24 principles or rules. Knowing these rules will help you decipher the *Manual of Classification*. Here are the rules.

[1] Utility as a Basis of Classification

Patents are classified principally in terms of utility. Focus is upon the function, effect, or product of a process or apparatus. Specifically, it is the direct, proximate, or necessary function, effect, or product, and not some remote or accidental use or effect. By grouping together patents on the basis of function, effect, or product, the classification system tends to collect together similar processes or means that achieve similar results by the application of similar natural laws.[2]

This means that when searching for prior art, think about how the invention functions and search subclasses that concentrate on those functions.

[2] Proximate Function as a Basis of Classification

Function is used as the basis of classification where a single causative characteristic can be identified and which requires essentially a single unitary act. Take a drilling machine, for example. It is designed to perform a single act, namely, to drill. Simple, single function inventions like this are classified according to the function they perform.[3]

Thus, if searching for prior art to an invention that has a well-defined, single causative characteristic, such as a business-related invention involving postage metering, look to find a subclass on point, for example, Class 705, Subclass 60, "Postage Metering System."

---

[2] *Id.* at 3.
[3] *Id.*

[3]   Proximate Effect or Product as a Basis of Classification

The effect or product is used as the basis of classification where a single causative characteristic cannot be identified. In a complex system there may be many separate and diverse operations taking place. The result is not the utility of any one of the operations, but rather the contribution of each. For example, a shoemaking machine performs many diverse operations; it cuts, glues, and stitches to make the resulting shoe. Complex inventions like this are classified according to the product or effect produced.[4]

Often business systems involve many diverse operations. Thus you can expect many business-related patents to be classified according to the effect produced, for example, Class 705, Subclass 8, "Allocating Resources or Scheduling for an Administrative Function."

[4]   Structure as a Basis of Classification

The structural features (such as configuration or physical make-up) of the invention may be used when the subject matter is so simple that it has no clear functional characteristics. Tubular stock material is an example. Tubular stock can be used to make fence posts, plumbing pipes, or motorcycle engine cylinders. It has no single clear functional characteristic upon which it may be subdivided. Thus its configuration or physical make-up is the only characteristic that can be used to differentiate it, that is, round, hollow, made of steel, and so forth.[5]

Business technology has probably not yet evolved to the point where it may be considered simple. Thus this rule presently has little applicability. In the future, if object-oriented agent-based e-commerce components, for example, achieve a generic business reusability, it may very well be that agents will someday be treated like tubular stock material. This rule may someday be useful.

[5]   Basis of Classification Applicable to Chemical Compounds and Mixtures or Compositions

Chemical compounds are classified on the basis of chemical constitution regardless of utility. Mixtures or compositions are classified on the basis of disclosed utility for the particular material.[6]

Obviously, this rule has primarily to do with chemistry. It is included here for completeness.

[6]   Analysis as a Prerequisite to System Development

The current classification system was created (and continues to evolve) to accommodate actual inventions, not hypothetical ones. This means that the

---

[4] *Id.*

[5] *Id.* at 4.

[6] *Id.*

claimed disclosure of a patent to be classified is first analyzed and then classes and subclasses are created to match, if they do not already exist. It is a pragmatic approach. Patent Office classifiers do not attempt to create theoretical classes and then try to fill them.[7]

[7] Patents Grouped by Claimed Disclosure

All patents are classified according to the subject matter defined by the claims. This is an important rule, and this may not be what you would expect. Because the primary purpose of the classification system is to help patent examiners find anticipatory subject matter, you might think that patents should be classified according to the subject matter disclosed in the specification. Such an approach does not work well, for it produces multiple classifications for identical subject matter.[8]

Using the disclosure to classify (as opposed to the claim), circuit boards, for example, might be classified under televisions, under garage door openers, under computers, and under room thermostats, simply because in each case the patentee happened to disclose that a conventional circuit board is used. This would not be a useful classification result.

By focusing instead on the claimed subject matter, the classification system places emphasis on what the inventor believes to be new. Under the patent law, 35 U.S.C. § 112 requires the applicant to fully teach how to make and use the claimed invention, thus the patent specification is usually most detailed surrounding the claimed subject matter. Classifying inventions based on claimed subject matter thus collects the most meaningful specification disclosure under one classification. As an important side benefit, the classification system has made a powerful tool in infringement investigations.

[8] Patents Diagnosed by Most Comprehensive Claim

Many times a patent will contain several claims, each of which covers the disclosed subject matter to different degrees. Under the current patent classification system, a patent is classified according to the claim that encompasses the largest amount of the disclosed subject matter. This is not the broadest claim—from a claim drafting standpoint—but rather, this is the most comprehensive claim, that is, the claim that includes the most disclosed elements. Simply put, a patent is assigned to the class that best fits the claim with the most disclosed elements.[9]

To illustrate how classification based on disclosed claim elements works, consider the following example. Assume an e-commerce patent discloses a computer program having a graphical user interface, a joystick pointing device

---

[7] *Id.* at 4.
[8] *Id.*
[9] *Id.*

interface, a quicksort module, and a relational database, and presents the following two claims:

1. A computer-implemented business system comprising,
    a quicksort module; and
    at least one relational database coupled to said quicksort module.
10. A computer-implemented business system comprising,
    a relational database having a joystick pointing device interface;
    a graphical user interface coupled to said relational database; and
    a quicksort module coupled to said relational database.

Claim 10 would probably be chosen for classification of this example patent because claim 10 encompasses more of the claimed subject matter. Thus the original patent copy is placed with other patents containing elements comparable to claim 10. If the quicksort module happens to contain unusual features, or if the patent gives a particularly helpful disclosure of the quicksort module, the classifier may place a cross-reference copy of the patent in the class where other quicksort modules are classified.

[9] Exception to Claimed Disclosure Principle for Assigning Patents to Specific Class

Classification is not an exact science. There are always exceptions. There are four exceptions to the above rule that claimed disclosure is the basis of classification. These exceptions are:

1. When the claimed disclosure is directed to an old or "exhausted" combination, a claimed subcombinational feature, rather than the claimed combination, may be used as the basis for classification.[10] (For example, a very large group of patents for an automobile all disclose a body (A), a motor (B), a transmission (C), wheels (D), and the most comprehensive claim in each patent is ABCD; the differences among most of the patents resides in variations of the individual elements A, B, C, and D. The classifier may place the original patent in a subclass covering subcombination ABC, if that is the salient combination);

2. When the claimed disclosure is directed to an article defined by the material from which it is made, the composition of materials may be used. (For example, a claim reciting "A razor comprising composition r, s, and t" would not be classified under "razors" but rather under the composition r, s, and t);

3. When the claimed disclosure is directed to a process of using a composition, the composition may be used. (For example, a nominal mechanical process, such as refrigeration or coating, may be classified

---

[10] *Id.* at 4-5.

according to the composition of refrigerant or coating materials used); and

4. When the patent *claims* the subject matter of a subcombination subclass and *discloses* subject matter of a combination subclass, but when the disclosed combination subclass is classified subordinate to (indented under) the claimed subcombination subclass, the subordinate subclass is used. This exception takes some explaining. To illustrate, assume the following hypothetical arrangement of subclasses:

  . . Data structure

  . . . with sorting means

  . . . with parametcrizcd tcmplate means

In the above example, if the claim simply recites "data structure" but the disclosure includes the "sorting means" then the patent may be classified under the "with sorting means" subclass, because the "with sorting means" subclass already exists, indented under the subclass which would otherwise control.

Note that when conducting infringement investigations, consider subordinate subclasses of pertinent subclasses to avoid missing patents that have been classified according to this exception.

[10] Exception to Claimed Disclosure Principle for Patent Assignment Between Subcombination Subclass and Indented Combination Subclass

When a patent subclass has indented under it a combination subclass that includes as a subcombination thereof the subject matter of the parent subclass, a patent disclosing the subject matter of the combination subclass, but claiming only the subject matter of the subcombination subclass, is assigned to the indented combination subclass.[11]

The foregoing is essentially how the Patent Office explains the rule. Here is the translation. Assume the following classification exists:

  . . Subclass ABC

  . . . Combination ABCD

Even though the claim recites ABC, if the patent discloses ABCD, then the patent is classified under ABCD.

[11] Exhaustive Division—Miscellaneous Subclass

Despite the bad experience with Miscellaneous Class 21 in the 1830s, this rule provides that classes may have a miscellaneous subclass.[12] This is done to insure that the subdivisions or subclasses within every class are exhaustive. The

---

[11] *Id.*

[12] See § 9.03.

miscellaneous subclass is provided to ensure that the class may receive any future invention that falls within the scope of the class.

The provision for a miscellaneous subclass does not mean that the problems experienced in the 1830s have been solved. By its very nature, a miscellaneous subclass is easy to fill up haphazardly because assigning a patent to a miscellaneous subclass requires less work than creating a new subclass. The work savings is only temporary, for eventually the miscellaneous subclass becomes bloated and must be reclassified.

This happened in the computer and software arts during the 1980s. The miscellaneous Subclasses 200 and 900 of Class 364 grew at such an exponential rate with the explosion of the computer industry, that searching these classes was a daunting task. In November 1991, both subclasses were abolished and new Class 395 was created.[13] Then later, the 700 Series was created to replace Class 395. Currently business method patents are classified within the 700 Series, specifically in Class 705. This suggests that the Patent Office currently considers business-related inventions as part of the computer art. That could, of course, change if the business-related art grows sufficiently beyond the computer arts boundary.

[12] Exhaustive Nature of Coordinate Subclasses: Combinations to Precede Subcombinations

Coordinate subclasses are each exhaustive of the classification characteristic for which the subclass title and definition provides. Thus, in coordinate relationships, combinations including a detail should always precede subcombinations to the detail, per se.[14]

Coordinate subclasses refer to subclasses on the same hierarchical outline level. This requirement means that subclasses on the same level must be mutually exclusive.

[13] Indentation of Subclasses

As shown in the example reproduced below, the class schedule is arranged with certain subclasses appropriately indented. Hierarchy is further denoted by dots or periods preceding the subclass names.[15] The indenting format takes some getting used to, because the subclass numbers are not necessarily in numerical order. Example (from Class 705):

70 . . Home banking
71 . . Including key management
72 . . Verifying PIN

---

[13] Classification Order No. 1377 (U.S. Dep't of Commerce, 1991).

[14] U.S. Dep't of Commerce, *Development & Use of Patent Classification Sys.*, Libr. of Cong. Catalog No. 65-62235, at 5 (1966).

[15] *Id.* at 5-6.

73 . . Terminal detail (e.g., initializing)
74 . . Anonymous user system
75 . . Transaction verification
76 . . Electronic credential
77 . . Including remote charge determination or related payment system
78 . . . Including third party
79 . . . Including a payment switch or gateway
80 . Electronic negotiation
1 AUTOMATED ELECTRICAL FINANCIAL OR BUSINESS PRAC-
TICE OR MANAGEMENT ARRANGEMENT
2 . Health care management (e.g., record management, ICDA billing)
3 . . Patient record management
4 . Insurance (e.g., computer implemented system or method for writing
insurance policy, processing insurance claim, etc.)
5 . Reservation, check-in, or booking display for reserved space

[14] Diverse Modes of Combining Similar Parts

The classification system recognizes and provides for diverse modes of combining the same or similar parts or steps to obtain functionally (and possibly structurally) unrelated combinations.[16]

This rule essentially mirrors the rule that elements can be combined in different, nonobvious ways to result in different patentable combinations. The classification system should be expected to distinguish between these different inventions.

[15] Relative Position of Subclasses

The indented structure used by the current classification system follows four simple rules.

1. Characters deemed more important for purposes of search generally appear first.
2. Subclasses based upon effect or special use usually precede those based upon function or general use.
3. Subclasses that are directed to variants of a concept are usually indented under the subclass directed to such concept, or otherwise preceding the same.
4. Subclasses directed to combinations of the basic subject matter of the class, with means having function or utility unnecessary to the basic subject matter usually precede subclasses devoted to the basic subject matter.[17]

---

[16] *Id.* at 6.
[17] *Id.*

[B] Each Class and Subclass Must Be Defined

Each class and subclass is defined in a detailed statement setting forth the metes and bounds of the area of subject matter that each covers.[18] It is often helpful to read these descriptions, which can be found on the Patent Office Web site.

[1] Tentative Definition

A tentative or preliminary definition of a class is written shortly after the class is created. It is understood that the tentative definition may be modified later, if necessary.[19]

[C] Explanatory Notes for Class or Subclass Definition

In many instances, explanatory notes relating to excluded subject matter, the explanation of some term or expression used in the definition, and clarifying statements may be appended to the definition.[20] These can be found by consulting the Patent Office Web site.

[1] Search Notes for Class or Subclass Definition

To supplement or take the place of cross-referencing, search notes may be provided to indicate other classes or subclasses that are directed to analogous or related subject matter.[21] The *Index to Classification* is a useful guide in locating analogous or related subject matter. The *Index to Classification* is published in soft cover book form and also as part of the Patent Office Web site.[22]

[D] Cross-referencing

Nearly every patent discloses subject matter that is classifiable in a different class or subclass than that which provides for the subject matter of the controlling claim. Sometimes the additional subject matter is separately claimed in some claim other than the controlling claim. Sometimes the additional subject matter is disclosed, but not claimed.[23] In either case, the patent may be assigned to one or more cross-reference classes or subclasses, as discussed below.

---

[18] *Id.*

[19] *Id.* at 6.

[20] *Id.* at 7.

[21] *Id.*

[22] <www.uspto.gov>. See also § 9.03.

[23] *Supra*, note 14 at 7.

[1] Cross-referencing Claimed Disclosure

A patent may have several claims that would be assigned to different subclasses, if found separately in different patents, either in the same class or in different classes. In this instance, the classifier is obligated to cross-reference the patent in those other subclasses, unless search notes are provided to lead a searcher to the other subclasses.[24] This means you should consider the search notes, for they could lead to art you may not otherwise uncover.

[2] Cross-referencing Unclaimed Disclosure

A patent may have disclosure of technology that is not claimed, and yet that patent may be cross-referenced in other subclasses (within the discretion of the classifier). Generally, the classifier will cross-reference a patent if the disclosure is novel (in the classifier's best judgment) and if the disclosure is of sufficient detail and clarity to be useful as a reference. The classifier will probably not make a cross-reference when an appropriate search note is appended to the subclass eligible to receive the cross-reference, instructing a searcher that the subclass containing the original copy of the patent must be searched.[25]

[E] Superiority Among Classes

When a patent has only one claimed disclosure, classification is straightforward. However, there are times when a patent includes disclosures to diverse inventions. In every case, a patent is assigned, as the "original" copy, to one and only one subclass. It may, of course, be cross-referenced elsewhere. Deciding which class and subclass to make the "original" patent assignment can be difficult. Here are the rules that the Patent Office classifier applies (in the order listed) to select the single disclosure that will control assignment.

1. The most comprehensive claimed disclosure governs.
2. Order of superiority of statutory subject matter categories is as follows:
    (a) process of using a product;
    (b) product of manufacture;
    (c) process of making a product;
    (d) apparatus;
    (e) material used.
3. When, and only when, the above two rules fail to solve the question of the controlling class, the relative superiority of types of subject matter is used, based on the following list:
    (a) subject matter relating to maintenance or preservation of life is superior to subject matter itemized in (b)-(d) below.

---

[24] *Id.*
[25] *Id.*

(b) chemical subject matter is superior to electrical or mechanical subject matter.

(c) electrical subject matter is superior to mechanical subject matter.

(d) dynamic subject matter (that is, relating to moving things or combination of relatively movable parts) is superior to static subject matter (that is, stationary things or of parts nonmovably related).[26]

[F] Superiority Within a Class

When different subclasses of the same class are involved, the patent is usually assigned to the one of several subclasses defined to receive the several claimed inventions that stands highest in the schedule of subclasses.[27]

§ 9.04 Classification of Business-related Patents

Most business-related patents are currently classified in U.S. Class 705. From a review of the major subclass headings alone, it is clear that this class covers an expansive waterfront. The following is an excerpt from the Class 705 hierarchy as of June 30, 2000. Only major headings and top level subclasses are included here.

[A] Class 705: Data Processing: Financial, Business Practice, Management, or Cost/Price Determination

50	BUSINESS PROCESSING USING CRYPTOGRAPHY
51	. Usage protection of distributed data files
60	. Postage metering system
63	. Utility metering system
64	. Secure transaction (e.g., EFT/POS)
80	. Electronic negotiation
1	AUTOMATED ELECTRICAL FINANCIAL OR BUSINESS PRACTICE OR MANAGEMENT ARRANGEMENT
2	. Health care management (e.g., record management, ICDA billing)
4	. Insurance (e.g., computer implemented system or method for writing insurance policy, processing insurance claim, etc.)
5	. Reservation, check-in, or booking display for reserved space
7	. Operations research
12	. Voting or election arrangement
13	. Transportation facility access (e.g., fare, toll, parking)
14	. Distribution or redemption of coupon, or incentive or promotion program
15	. Restaurant or bar
16	. Including point of sale terminal or electronic cash register

---

[26] *Id.* at 8.

[27] *Id.*

26 . Electronic shopping (e.g., remote ordering)
28 . Inventory management
30 . Accounting
35 . Finance (e.g., banking, investment or credit)
400 FOR COST/PRICE
401 . Postage meter system
412 . Utility usage
413 . Fluid
414 . Weight
417 . Distance (e.g., taximeter)
418 . Time (e.g., parking meter)
500 MISCELLANEOUS (e.g., BY GENERIC OR NONELECTRICAL COMPUTING)

What the above listing shows is that many of the patents currently classified in Class 705 are machines or electronic systems used in various well-known industries. There are, for example several subclasses devoted to postage meters. Postage meters would hardly be considered controversial subjects for patent. The Amazon.com one-click patent is classified in Subclass 26, which relates to electronic shopping. The Priceline.com name-your-own price patent is classified in Subclass 5, which relates simply to reservations. It is difficult to tell from the subclass titles that there may be controversial patents lurking therein.

The way to crack the Class 705 nut is to conduct assignee searches within the class, to determine where the patents of similarly situated companies are classified. To illustrate, say that you are searching for prior art relating to banking. While Subclass 35 relating to "finance" does encompass banking, such a search may be too limited. However, if you first run a search to identify all of the patents of, say Citibank, you may begin to see where patent examiners actually classify things.

[B] Samples of Some Citibank Patents and Classes to Which They Are Assigned

6,226,623 Global financial services integration system and process

Classes: 705/35; 709/217; 709/218; 709/230; 709/238

6,205,436 Trusted agents for open electronic commerce where the transfer of electronic merchandise or electronic money is provisional until the transaction is finalized

Classes: 705/65; 705/41; 705/64; 705/80

6,188,993 System and method for creating and managing a synthetic currency

Classes: 705/37; 705/35

6,175,921 Tamper-proof devices for unique identification

Classes: 713/173; 380/279; 380/283; 705/67; 713/156; 713/159; 713/168

6,154,527 Interactive voice response system

Classes: 379/88.18; 379/93.12

6,131,810 Integrated full service consumer banking system and system and method for opening an account

Classes: 235/379; 235/380; 705/41; 705/42; 902/5; 902/14; 902/22; 902/25

6,122,625 Apparatus and method for secure transacting

Classes: 705/65; 235/379; 235/380; 235/381; 235/382; 705/35; 705/39; 705/43; 705/44; 705/67; 705/68

6,112,190 Method and system for commercial credit analysis

Classes: 705/38; 705/35; 705/36; 706/47; 706/50; 706/60; 706/61

6,098,053 System and method for performing an electronic financial transaction

Classes: 705/44; 705/35

6,088,686 System and method to performing on-line credit reviews and approvals

Classes: 705/38; 235/375; 235/379; 235/383; 705/35; 705/39

* * *

5,930,764 Sales and marketing support system using a customer information database

Classes: 705/10; 705/35

5,920,848 Method and system for using intelligent agents for financial transactions, services, accounting, and advice

Classes: 705/42; 705/33

5,920,629 Electronic-monetary system

Classes: 705/69; 235/379; 705/41

In the Citibank example above, most of the classifications fall within Class 705. In addition to that class, Classes 235, 375, 379, 380, 706, 709, 713, and 902 are also represented. Those classes cover related subject matter, as follows:

235 Registers

379 Telephonic Communications

380 Cryptography

706 Data Processing, Artificial Intelligence

709 Electrical Computers, Multiple Computer or Process Coordinating

713 Electrical Computers, Support

902 Electronic Funds Transfer

[1] Patent Classes to Which the Above Patents Are Assigned

The above excerpt of Citibank's patent portfolio fall into a fairly diverse number of patent classes and subclasses. Here is the list:

Class 705 (Business), Subclass:

10 Market analysis, demand forecasting or surveying

33 Checkbook balancing, updating or printing arrangement

35 Finance (e.g., banking, investment, or credit)

36 Portfolio selection, planning or analysis

37 Trading, matching or bidding

38 Credit (risk) processing or loan processing (e.g. mortgage)

39 Credit (risk) processing or loan processing including funds transfer or credit transaction

41 Funds transfer having programming of portable memory device (e.g., IC card, "electronic purse')

42 Remote banking (e.g., home banking)

43 Remote banking including Automatic Teller Machine (ATM)

44 Funds transfer requiring authorization or authentication

64 Secure transaction (e.g. EFT/POS)

65 Secure transaction including intelligent token (e.g., electronic purse)

67 Secure transaction including authentication

68 Secure transaction including balancing account

69 Secure transaction including electronic cash detail (e.g., blinded, divisible or detecting double spending)

80 Electronic negotiation

Class 235 (Registers), Subclass:

375 Systems controlled by data bearing records

379 Banking systems

380 Credit or identification card systems

381 Credit or identification card systems with vending

382 Credit or identification card systems permitting access

383 Mechanized store

Class 379 (Telephonic Communications), Subclass:

88.18 Interacting voice message system

93.12 Sales, ordering or banking system

Class 380 (Cryptography), Subclass:

279 Key distribution center

283 User-to-user key distributed over data link (i.e., no center)

Class 706 (Artificial Intelligence), Subclass:

47 Knowledge representation using rule-based reasoning system

50 Knowledge representation having specific management of knowledge base

60 Expert system shell or tool

61 Knowledge acquisition by a knowledge processing system

Class 709 (Multiple Computer or Process Coordinating), Subclass:

217 Remote data accessing

218 Remote data accessing using interconnected networks

230 Computer-to-computer protocol implementing

238 Computer-to-computer data routing

Class 713 (Computer Support), Subclass:

156 Central trusted authority provides computer authentication by certificate

159 Central trusted authority provides computer authentication by certificate, including intelligent token

168 Particular communication authorization technique

173 Particular communication authorization technique having intelligent token pre-loaded with certificate

Class 902 (Electronic Funds Transfer), Subclass:

5 Providing security with means to verify identity of user

14 Terminal with cash dispenser

22 Register Transaction Terminal (e.g., point of sale terminal)

25 Specific identifier (e.g., bank card)

The above list is interesting because it represents a very good cross-section of technology involved in e-commerce. You might not suspect that the first listed class, "235 Registers," is actually the granddaddy of all the computer and software classes. That is because the computer and software arts trace back to the adding machine. That remarkable device of the 1800s finds itself classified in a class now titled, "Registers" Class 235.

It is interesting, however, that Class 235 began with a simpler purpose, to classify things with springs. The earliest patent in Class 235 was U.S. Patent No. 551 for a Carriage Spring issued to Morton on January 9, 1838. During that era, Class 235 also included such diverse items as a Molasses Faucet and a Cooking Stove. Presumably the common element in these devices was a spring that made it work. Someday, someone may make a similar observation about business method patents—that they were originally classified as part of the "computer" art because many of the "early" inventions contained computers that made the method work.

The "Registers" Class evolved from springs to cash registers over the next two decades. In 1839, the Patent Office issued U.S. Patent No. 1,325 for an Odometer. It was classified under "Registers." Later, adding machines would adopt the odometer-style readout for showing the sum of numbers being added. Shortly thereafter, in 1843, the Patent Office granted U.S. Patent No. 2,907 for a Slide Rule. The following year, it issued U.S. Patent No. 3,574 to J. Hatfield for a Machine for Calculating Interest, issued May 6, 1844. The following year it issued U.S. Patent No. 4,632 for a Calculator, to Vanderveer, on July 14, 1846 and U.S. Patent No. 4,902 for a Calculating Machine, to McLelman, on December 22, 1846. The software age was on its way.

The "Registers" Class 235 continued to grow over the next one hundred years. Today it contains patents on cash registers, voting machines, weapon system computers, pocket calculators, and even bar codes. The electronic digital computer was invented in 1945 (see main text § 1.03[A][3]). In 1951, when Bell Labs was issued its patent on an "Error-Detecting and Correcting System,"[28] the Patent Office classified this software invention under "Telegraphy" Class 178. It also cross-referenced the patent under "Registers" Class 235 and under "Coded Data Generation" Class 341.

Searching for prior art is more an art form than a science. The above example shows how widely distributed the art of a single business, such as a bank, can be. The following sections present a searching strategy that can be helpful in searching for prior art. The searching strategy is a starting point. Be sure to tap your client's knowledge of the art, as there may be many repositories of prior art in newspapers, journals, university theses, and ongoing business practices that simply will not show up in a structured search of the U.S. Patent Office prior art collection.

---

[28] U.S. Patent No. 2,552,629, issued May 15, 1951.

§ 9.05   Conducting a Prior Art Search—Defining the Scope of Search

Conducting a prior art search begins with defining the scope of the search. Just as the previous banking example showed, the field of search spanned many disciplines. Every prior art search will necessarily include certain bodies of knowledge and exclude others. Deciding what to include and what to exclude can be difficult because it ultimately depends on the level of skill in a particular art. What bodies of knowledge are persons of skill in the art being searched normally exposed to? What subjects are presented at conferences these persons attend? What books are on their bookshelves? These are some of the questions to ask in defining the scope of a prior art search. Of course, to answer these questions one must first grapple with the more fundamental question, who is one skilled in the art?

[A]   Identifying One Skilled in the Art

One of ordinary skill in the art is a hypothetical person. That may seem surprising, inasmuch as inventors whose names appear on patents are real people and not hypothetical at all. However, as real people, each is an individual, with individual interests, intellect, upbringing, education, and life experiences. No two inventors are alike, even in a well-defined field. What may be within the skill of one individual within that field may not be within the skill of another in that same field. Thus the person of ''ordinary skill'' is, of necessity, a legal fiction. It is an amalgam of the many persons who practice their craft in that field. If statistical analysis could be used to describe a legal fiction, the person of ordinary skill would probably fall under that famous bell-shaped curve. Excluded from that bell-shaped curve would fall, on the one extreme, a thousand monkeys randomly typing Shakespeare, and on the other extreme, the Einstein, Newton, Mozart geniuses to whom everything is obvious.

Regarding the 1,000 monkeys, one statistician has calculated that 1,000 monkeys seated at typewriters, for all practical purposes, will *never* type ''To be or not to be, that is the question.'' He calculates that even if one were to employ all monkeys in the known universe (17 billion galaxies, each containing 17 billion habitable planets, each planet with 17 billion monkeys) and set them typing away at a rate of one line per second for 17 billion years, the chances that all monkeys will fail is 99.999999999995%.[29] These numbers suggest that innovation is not a random act.

Thus, it is unlikely the prior art searcher will ever need to explore the random acts of monkeys. On the other hand, most patent attorneys have encountered the overly-erudite person to whom everything is obvious.

The U.S. Patent Office has promulgated its standard for defining who is one of ordinary skill in the art. Those standards, set forth in the *Manual of Patent*

---

[29] <http://www.nutters.org/monkeys.html>.

Examining Procedure at § 2141.03, are based on the case law. The *Manual* defines ordinary skill as follows.

[1] Manual of Patent Examining Procedure—Defines Ordinary Skill

The following is an excerpt of the pertinent section of the Manual of Patent Examining Procedure, describing how the Patent Office defines the level of ordinary skill.

2141.03 Level of Ordinary Skill in the Art

FACTORS TO CONSIDER IN DETERMINING LEVEL OF ORDINARY SKILL

Factors that may be considered in determining level of ordinary skill in the art include (1) the educational level of the inventor; (2) type of problems encountered in the art; (3) prior art solutions to those problems; (4) rapidity with which innovations are made; (5) sophistication of the technology; and (6) educational level of active workers in the field: *Environmental Designs, Ltd. v. Union Oil Co.*, 713 F.2d 693, 696, 218 U.S.P.Q. (BNA) 865, 868 (Fed. Cir. 1983), cert. denied, 464 U.S. 1043 (1984).

The "hypothetical 'person having ordinary skill in the art' to which the claimed subject matter pertains would, of necessity have the capability of understanding the scientific and engineering principles applicable to the pertinent art": *Ex parte Hiyamizu*, 10 U.S.P.Q.2d (BNA) 1393, 1394 (Bd. Pat. App. & Inter. 1988). (The Board disagreed with the examiner's definition of one of ordinary skill in the art (a doctorate level engineer or scientist working at least 40 hours per week in semiconductor research or development), finding that the hypothetical person is not definable by way of credentials, and that the evidence in the application did not support the conclusion that such a person would require a doctorate or equivalent knowledge in science or engineering.)

References which do not qualify as prior art because they post-date the claimed invention may be relied upon to show the level of ordinary skill in the art at or around the time the invention was made: *Ex parte Erlich*, 22 U.S.P.Q. (BNA) 1463 (Bd. Pat. App. & Inter. 1992).

SPECIFYING A PARTICULAR LEVEL OF SKILL IS NOT NECESSARY WHERE THE PRIOR ART ITSELF REFLECTS AN APPROPRIATE LEVEL

If the only facts of record pertaining to the level of skill in the art are found within the prior art of record, the court has held that an invention may be held to have been obvious without a specific finding of a particular level of skill where the prior art itself reflects an appropriate level: *Chore-Time Equipment, Inc. v. Cumberland Corp.*, 713 F.2d 774, 218 U.S.P.Q. (BNA) 673 (Fed. Cir. 1983).

ASCERTAINING LEVEL OF ORDINARY SKILL IS NECESSARY TO MAINTAIN OBJECTIVITY

The importance of resolving the level of ordinary skill in the art lies in the necessity of maintaining objectivity in the obviousness inquiry: *Ryko Mfg.*

Co. v. Nu-Star, Inc., 950 F.2d 714, 718, 21 U.S.P.Q.2d (BNA) 1053, 1057 (Fed. Cir. 1991). The examiner must ascertain what would have been obvious to one of ordinary skill in the art at the time the invention was made, and not to the inventor, a judge, a layman, those skilled in remote arts, or to geniuses in the art at hand: *Environmental Designs, Ltd. v. Union Oil Co.*, 713 F.2d 693, 218 U.S.P.Q. (BNA) 865 (Fed. Cir. 1983), cert. denied, 464 U.S. 1043 (1984).

[B] Planning the Field of Search

Whether one is conducting a prior art search as a precursor to filing a patent application, or as an assessment of validity of an issued patent, the steps are essentially the same. Begin by ascertaining the level of ordinary skill in the pertinent art and defining the field of search on that basis. Then devise a search methodology that allows you to systematically uncover and explore the art spanning your field of search. Finally, select the references that are potentially material to patentability, or that suggest leads for further exploration.

Getting started is often the most difficult part. An ill-conceived start may leave important prior art undiscovered, or may send one off to plow through material that ultimately proves barren. Ascertaining the level of ordinary skill and defining the field of search will often go hand-in-hand. As you investigate the level of ordinary skill, you will discover resources to add to your field of search, and vice versa. Of course, previous experience with the particular art is invaluable—knowing where to look separates the successful from the unsuccessful.

So how do you assess the level of skill and devise your field of search when you have no prior experience with the particular art? A good place to begin is to ask the experts. You may be fortunate to have access to experts in the field, such as the inventor for whom the patent application is being drafted, or the expert witness who will be testifying at trial. If so, ask these experts where they go to learn more about their field of expertise. Ask them to define who they think would be a person of ordinary skill in their art.

[1] Identifying Search Field Through Seminars and Symposia

If you do not have direct access to experts, you may nevertheless consult them through their writings. For example, say you are researching an invention that will monitor nightly news broadcasts around the world to predict which stocks will be in play the following day. The inventor believes that the markets react to news reports in subtle ways, based on whether the news report is basically positive or negative. The invention captures this information from broadcasts in all major languages around the world and then predicts which stocks to buy and sell.

One possible inventive aspect is how the system automatically extracts meaning from syntactic content of the news stories. Instead of merely performing keyword spotting, look for words that suggest the story will have a "positive" or "negative" impact. The invention determines the language being used and then looks at the sentence structure and grammar used in the stories to gain understanding of whether a positive or negative vibe is being expressed with regard to a particular subject.

In constructing a prior art search strategy for this invention, you could begin by simply doing a keyword search through the U.S. Patent Office Web site—searching terms such as "syntax," "syntactic," "news," and "stock market." However, the results are likely to be quite hit-or-miss. A more powerful approach would be to first identify various communication channels through which the experts in this field exchange views with one another. One such channel might be conferences and symposia, and the technical papers presented at these. Trade publications represent another excellent channel. Each of these channels can be identified by conducting Internet searches.

What you are looking for is not necessarily the inventive concept, per se. Rather, you are looking for leads. What terminology do the experts use to describe the techniques involved in the invention? When you do perform keyword searches through published patent databases, these can be quite helpful in searching the keywords the experts are most likely to use. Who has authored papers in this field, and what companies or universities sponsor their work? Armed with this information, you can conduct focused "inventor name" and "assignee" searches to discover if any key players have obtained patents in the field.

If you discover a particularly promising author, find out what else that author has written, and what other conferences or symposia the author has attended. If you discover a promising conference paper or article, check the bibliography to see if there are other even more promising articles. Also explore the co-authors. Sometimes co-authors work together regularly and may be involved in very similar work. You may find other pertinent papers written by one of the co-authors. Other times co-authors may have joined forces to publish this particular paper. They may come from more diverse backgrounds. Where there are co-authors, see if you can separate which authors are the more widely published from those who are relatively new to the field. The widely published authors may help you discover a broad spectrum of relevant articles, possibly with other co-authors.

[2] Using Ph.D. Theses to Expand the Field of Search

Interestingly, the newly published author may be an even more important find. If the newly published author is a recent Ph.D. graduate, there is a good chance that his or her Ph.D. thesis will be relevant. Many companies hire recent Ph.D. graduates to boost their research and development expertise in growing technologies. Not infrequently, a newly hired Ph.D. graduate's work for the company will build upon or extend his or her Ph.D. thesis. A good resource for locating Ph.D. theses is through the UMI Company (formerly University

Microfilm, Inc.).[30] The company maintains a Web site at <www.umi.com> through which searches for recently published Ph.D. theses can be identified and obtained.

[3] Identifying Appropriate U.S. Patent Classes and Subclasses

The patent classification search listed above presupposes that you know in which classes and subclasses to look. That can be a rather tall order. How do you know in which patent classes and subclasses to look? There are at least two techniques. Often you will want to use both.

One is to discover the proper classes and subclasses by first finding at least one relevant patent and then identifying where that patent is classified and, if necessary, also finding what field of search the examiner used when that patent was examined. The first page of each issued patent identifies both where it is classified, and what field of search the examiner used during examination.

Another technique is to study the taxonomy or classification principles of the *Patent Office Manual of Classification*. Currently many business inventions are classified with the software prior art. The software prior art classification system has grown quite complex. The Patent Office undertook a major reclassification campaign following the explosive growth in software patent filings in the mid-1990s. Hence, software patents are far better classified than they were in 1994. Subsection [a] below shows the state of the organization of the software patent classification system as of January 2001, and how it evolved from a simple subclass in the computer art, called "Miscellaneous."

[a] A Taxonomy of the U.S. Patent Office Software Prior Art Collection

Many e-commerce patents involve prior art that is classified with the software patent art collection. That collection exists in many different places. Most of the newly issuing software patents are classified in the 700 series of the U.S. classification system. However, if you consult the search notes to those classes, you will find there are many other places to look for software patents.

Figure 9-1 is a chart that gives an overview of the current taxonomy of the software patent prior art and how the collection evolved from 1989 to 2001. A lot changed between 1989 and 2001. In 1994 for example, Class 395 was the primary class where software patent art was classified. In 2001, the 700 Series, comprising eight separate data processing classes, all but replaced Class 395. The business-related inventions fall within the middle of the 700 Series (Class 705). The expansion of the software patent prior art collection from a single class to multiple classes over this seven year period substantiates the unprecedented growth of the computer arts over that period.

---

[30] UMI Company, 300 North Zeeb Road, P.O. Box 1346, Ann Arbor, MI 48106. Phone: 800-521-0600 or 313-761-4700; Fax: 313-761-9836; Web site: <http://www.umi.com>.

FIGURE 9-1: EVOLUTION OF SOFTWARE PRIOR ART (1989-2001)

CLASS	1989	1990	1991	1994	1995	1996	1997	1998	1999	Jan 2001 web site
340	Communications, electrical	same	same	same	same	same	same	same	same	same
364	Electrical computers and data processing systems	same	same	same	same	same	same	same	dropped from index	Data Processing Control Systems, Methods or Apparatus
365	Static information storage and retrieval	same	same	same	same	same	same	same	same	same
369	Dynamic information storage and retrieval	same	same	same	same	same	same	same	same	same
370	Multiplex communications	same	same	same	same	same	same	same	same	same
371	Error detection/correction and fault detection/recovery	same	same	same	same	same	same	same	same	dropped in favor of 714
375	Pulse or digital communications	same	same	same	same	same	same	same	same	same
379	Telephonic communications	same	same	same	same	same	same	same	same	same
380	Cryptography	same	same	same	same	same	same	same	same	same
382	Image Analysis	same	same	same	same	same	same	same	same	same
455	Telecommunications	same	same	same	same	same	same	same	same	same
395	not yet	Information processing system organization	same	same	same	same	same	same	same	
700	not yet	same	same	same	same	same	same	same	Data Processing: Generic Control systems or specific applications	same
701	not yet	same	same	same	same	Data Processing: vehicles, navigation and relative location		same	same	same
702	not yet	same	same	same	same	same	Data Processing: Measuring, calibrating or testing		same	same

(continued)

461

FIGURE 9-1 (continued)

	Category									
703	Data Processing: structural design, modeling, simulation and emulation	same	same	same	same	same	same	same	same	not yet
704	Data Processing: Speech signal processing, linguistics, language translation and audio compression/decompression	same	same	same	same	same	same	same	same	not yet
705	Data Processing: Financial, business practice, management, or cost/price determination	same	same	same	same	same	same	same	same	not yet
706	Data processing: artificial intelligence	same	same	same	same	same	same	same	same	not yet
707	Data Processing: database and file management, data structures, or document processing	same	same	same	same	same	same	same	same	not yet
708	Electrical Computers: arithmetic processing and calculating	same	same	same	same	same	same	same	same	not yet
709	Electrical Computers and digital processing systems: multiple computer or process coordinating	same	same	same	same	same	same	same	same	not yet
710	Electrical computers and digital data processing systems: input/output	same	same	same	same	same	same	same	same	not yet
711	Electrical computers and digital data processing systems: Memory	same	same	same	same	same	same	same	same	not yet

FIGURE 9-1 (continued)

712	not yet	same	same	same	same	same	same	same	Electrical computers and digital data processing systems: processing architectures and instruction processing
713	not yet	same	same	same	same	same	same	same	Electrical computers and digital data processing systems: Support
714	not yet	same	same	same	same	same	same	same	Error detection/correction and fault detection/recovery
716	not yet	same	same	same	same	same	same	same	Data Processing: design and analysis of circuit or semiconductor mask
717	not yet	same	same	same	same	same	same	same	Data Processing: Software development, installation and management
902	Electronic Funds Transfer	same	same	same	same	same	same	same	same

[C] Assessing Whether the Identified Subclass Is Pertinent

Before simply searching the subclasses identified through relevant patent or taxonomy analysis, it is wise to do a spot check. Select a few patents in the subclass and try to determine why it was classified there. The detailed rules by which patents are assigned to a particular class are discussed in § 9.03 above. Basically, however, patents are classified by the subject matter they *claim*, not by the subject matter they *disclose*.

Also, do a focused keyword search in the identified subclass to see if you get a reasonable number of hits, suggesting that the subclass may be relevant. Remember, although patents are classified by claimed subject matter, it is not necessarily the claimed subject matter for which we are searching. Often a patent specification will include descriptive content beyond what is claimed. This descriptive content may be quite relevant to your particular search.

[D] Identifying Additional Subclasses by Taxonomy Study

With Class 705 as the primary focal point for a business-related patent search, we can now examine the classification taxonomy to determine if there are other subclasses that should be searched. The Patent Office usually groups related subject matter together, often in indented, outline form. Thus, if Subclasses 242 and 256 are potentially relevant, the neighboring subclasses may also be potentially relevant.

In addition, it may also be helpful to look at the field of search used by the examiner in examining our identified patent. Although the examiner of the identified patent was searching for different claimed subject matter, the field of search could point us to additional places that may be worth searching.

Does that mean we should now jump in and start searching all of these subclasses? We could, but there is one other thing to check. In addition to maintaining class listings and class definitions, such as the definitions of Subclasses 242 and 256 reproduced above, the *Manual of U.S. Patent Classification* also includes valuable search notes. The *Manual* can be examined on-line at the U.S. Patent Office Web site.[31] Figure 9-2 contains an excerpt from the search notes for Class 705. The notes tell us something about how the class evolved, how it is currently organized, and whether there are any related classes worth taking a look at.

[E] Conducting the Search—Using Leads to Access the Issued Patent Collection

Once you have explored some of the seminars, symposia, and technical papers, you are ready to construct a patent search. The U.S. Patent Office Web

---

[31] <www.uspto.gov/web/offices/ac/ido/oeip/taf/moc/index.htm>.

FIGURE 9-2: CLASS 705 SEARCH NOTES—EXCERPT

Class 705

DATA PROCESSING: FINANCIAL, BUSINESS PRACTICE, MANAGEMENT, OR COST/PRICE DETERMINATION

Class Definition:

This is the generic class for apparatus and corresponding methods for performing data processing operations, in which there is a significant change in the data or for performing calculation operations wherein the apparatus or method is uniquely designed for or utilized in the practice, administration, or management of an enterprise, or in the processing of financial data.

This class also provides for apparatus and corresponding methods for performing data processing or calculating operations in which a charge for goods or services is determined

SCOPE OF THE CLASS

1. The arrangements in this class are generally used for problems relating to administration of an organization, commodities or financial transactions.

2. Mere designation of an arrangement as a 'business machine' or a document as a 'business form' or 'business chart' without any particular business function will not cause classification in this class or its subclasses.

3. For classification herein, there must be significant claim recitation of the data processing system or calculating computer and only nominal claim recitation of any external art environment. Significantly claimed apparatus external to this class, claimed in combination with apparatus under the class definition, which perform data processing or calculation operations are classified in the class appropriate to the external device unless specifically excluded therefrom.

4. Nominally claimed apparatus external to this class in combination with apparatus under the class definition is classified in this class unless provided for in the appropriate external class.

5. In view of the nature of the subject matter included herein, consideration of the classification schedule for the diverse art or environment is necessary for proper search.

REFERENCES TO OTHER CLASSES

SEE OR SEARCH CLASS:

177, Weighing Scales, 25.11 for a computerized scale.

186, Merchandising, various subclasses for customer service methods and apparatus in a variety of areas including banking, restaurant and stores.

235, Registers, various subclasses for basic machines and associated indicating mechanisms for ascertaining the number of movements of various devices and machines, plus machines made from these basic machines alone (e.g., cash registers, voting machines), and in combination with various perfecting features, such as printers and recording means. In addition, search Class 235 for various data bearing record controlled systems. Search subclasses 375-386 for a system having a detail of a record-sensing device in combination with a system utilized for banking, determining credit, maintaining an inventory, access control, vending, voting, time or operations analysis and having no more than a nominal recitation of a computer or data processing arrangement. Search subclasses 7+ for cash register; and subclass 61 for mechanically computing a cost/price ratio. Note that a nominally claimed record or card sensor is considered to be a peripheral of the data processing system.

283, Printed Matter, various subclasses for business forms and methods of using such forms.

307, Electrical Transmission or Interconnection Systems, various subclasses for generic residual electrical transmission or interconnection systems and miscellaneous circuits.

340, Communications: Electrical, various subclasses for residual electrical communication systems, 825.30 for communication details including authorization, vending, credit and access control; and see related classes elsewhere.

* * *

site at <www.uspto.gov/patft> provides an excellent search tool that may be accessed at no cost. One effective way to use this tool is through a series of separate searches, each designed to find the most pertinent prior art but coming from different vantage points. A typical series of searches might include, for example:

- assignee searches—using companies, universities and institutions identified in the literature search to be active in the field;

- inventor name searches—using authors identified in the literature search;

- keyword searches—using terminology learned from the published literature;

- patent classification searches—using keywords as above but limited to specific patent classes and subclasses.

- pertinent patent searches—taking most pertinent patents and exploring both the prior art cited in those patents and also the later patents making reference to those patents

§ 9.06   The Internet as Prior Art

When did the first Internet software patent appear on the scene? The answer depends on how one defines the term Internet. The Internet can be defined as a global, packet-switched computer network that supports a wide range of services from e-mail to World Wide Web to streaming video. As such, the Internet has roots in much earlier technology that forms the prior art of this now ubiquitous mode of communication.

To have an Internet, computers need to be able to talk to one another. For many, this still implies a modem. The first commercial modem, called the Dataphone, was developed around 1960. It worked much like the modems of today, converting digital computer data into analog signals that could be sent over conventional telephone lines. By 1964, computers were communicating with each other in commercial business systems, such as the SABRE airline reservation system developed by IBM for American Airlines. The utility of networked computers spread quickly. In 1970, Citizens and Southern National Bank of Valdosta, Georgia, installed the first automated teller machine.

These applications of networked computers are interesting, but they still fail to suggest that someday a giant network would span the globe and become the central nervous system of society. The first hint of such an Internet probably occurred in around 1970, when the Department of Defense established the first nodes of its ARPANET computer network. ARPANET is heralded by many as the origin of the Internet. The first ARPANET network linked computers at the University of California in Santa Barbara, at UCLA, at SRI International, and at the University of Utah. While the Internet had military origins, it was not long before commercial networks sprang up. Telnet, the first packet-switched network to convey commercial civilian traffic, began operation in 1975.

The development of packet-switched computer network technology continued throughout the 1970s and early 1980s. This technology took a giant leap towards the modern-day Internet in 1985, when the National Science foundation formed NSFNET. This newly formed network linked computer centers at Princeton University, Pittsburgh, University of California in San Diego, University of Illinois in Urbana-Champaign, and Cornell University. Other similar networks were soon formed at other computer centers. The U.S. government also reassigned sections of its ARPANET to the NSFNET, allowing these centers to contribute as well. By linking these important computer centers, the National Science foundation created an incubator that germinated the ideas of some of the best computer science minds of that day. The Internet, as we know it today, was soon to arrive.

The original NSFNET transferred data packets at 56 kilobits per second. That action limited the practical uses of the early Internet, but it did not limit innovation. The slow throughput forced computer scientists to develop techniques, such as data compression techniques, to squeeze as much functionality as they could from the limited bandwidth. Those techniques are still in use today. Fortunately, however, bandwidth was soon to increase. In 1987, ARPA awarded a contract to Merit Network, Inc., IBM, and MCI to develop network nodes around the country that used much faster communication lines. This became the computer network backbone that links most U.S.-based Internet sites today. The original backbone transferred data packets at 1.5 megabits per second. In 1992, the network was upgraded with faster T-3 lines to support traffic at 45 megabits per second.

The World Wide Web began in 1990, as the brainchild of Tim Berners-Lee, a research scientist at the CERN high-energy physics laboratory in Geneva, Switzerland. Berners-Lee developed the HyperText Markup Language, or HTML, that is currently used to define how Web pages display their content and how pages are connected through hypertext links. Berners-Lee later founded the W3 Consortium to coordinate the efforts of many to build what we now know as the World Wide Web.

If you conduct a patent search through the U.S. Patent Office patent database looking for the terms such as Internet, TCP/IP, hypertext, HTML, it is interesting what you will find. The word "internet" first appeared in the text of an issued U.S. patent in 1984. This was one year before the National Science Foundation formed the NSFNET. Who obtained this early patent? A company called Telebit Corp. The patent, U.S. Patent No. 4,438,511, was entitled "Packetized Ensemble Modem." The patent described a high speed modem. The patent used the term "internet" in the context of the internet protocols, stating, for example:

> There are simply no high speed digital data modems of the prior art which are compatible with packet switched networks (not to mention any that are compatible with multiple packet switching internet protocols) and which

operate in both synchronous and asynchronous modes concurrently through signal multiplexing.[32]

The term "TCP/IP" first appeared in an issued U.S. patent two years later in 1987. The patent was assigned to Vitalink Communications. That patent, U.S. Patent No. 4,706,081, was entitled, "Method And Apparatus For Bridging Local Area Networks."

Given that the World Wide Web was not founded until 1990, one would expect to find few patents exploiting hypertext prior to this date. For the most part, this is true. The term "hypertext" first appeared in an issued U.S. patent in 1989, in a patent assigned to Bell Communications Research. The patent, U.S. Patent No. 4,839,853, was entitled, "Computer Information Retrieval Using Latent Semantic Structure." The patent dealt primarily with an information searching technique using statistical analysis of semantic structures. The term "hypertext" was mentioned in passing as one of the ways computers were improving how information could be stored, organized, and accessed. In fact, the term "hypertext" appears in several issued U.S. patents that predate Berners-Lee's creation of the Hypertext Markup Language or HTML. That is not to say that these earlier patents describe the World Wide Web brainchild of Berners-Lee, but they do show that even the seminal work on important breakthroughs exist in a sea of related thinking.

The first patents to actually specify "HTML" as the implementation language for a hypertext system were issued several years later, in 1996. Three such patents issued that year. The first to issue was U.S. Patent No. 5,530,852, assigned to Sun Microsystems, with the rather lengthy title of, "Method For Extracting Profiles And Topics From A First File Written In A First Markup Language And Generating Files In Different Markup Languages Containing The Profiles And Topics For Use In Accessing Data Described By The Profiles And Topics." Also issued in 1996 were U.S. Patent No. 5,590,250, entitled, "Layout Of Node-Link Structures In Space With Negative Curvature" assigned to Xerox Corporation; and U.S. Patent No. 5,572,643, entitled, "Web Browser With Dynamic Display Of Information Objects During Linking," issued to David Judson, an individual.

Table 9-1, below, shows the explosive growth of the Internet, as reflected in the popularity of such terms as "internet," "TCP/IP," "hypertext," "HTML" and "World Wide Web" appearing in issued U.S. patents. As an interesting point of comparison, the table also includes the considerably more recent markup language XML, which some suggest will eventually replace HTML.

---

[32] U.S. Patent No. 4,438,511, "Packetized Ensemble Modem," issued March 20, 1984.

TABLE 9-1
Number of Patents (by year) Which Use Various Internet Terminology

	TCP/IP	Internet	HTML	hypertext	XML	World Wide Web
1980	0	0	0	0	0	0
1981	0	0	0	0	0	0
1982	0	0	0	0	0	0
1983	0	0	0	0	0	0
1984	0	1	0	0	0	0
1985	0	0	0	0	0	0
1986	0	0	0	0	0	0
1987	2	1	0	0	0	0
1988	0	0	0	0	0	0
1989	2	12	0	1	0	0
1990	2	10	0	2	0	0
1991	6	7	0	7	0	0
1992	11	11	0	11	0	0
1993	28	21	0	17	0	0
1994	68	59	0	21	0	0
1995	90	79	0	32	0	0
1996	215	209	3	42	0	5
1997	279	426	26	92	0	57
1998	708	1818	299	391	1	505
1999	1077	3577	748	725	9	1073

§ 9.07   The Early Tributaries of Electronic Commerce

Long before there was an Internet, there were private networks designed to support commerce. In the 1960s, in fact, companies began exchanging information electronically over what was termed the Electronic Data Interchange (EDI). In that era the digital computer was but a teenager. The computer landscape was dominated by mainframes of 1950s design and minicomputers of 1960s design. EDI provided a standardized way for exchanging electronic documents. Because digital computers were still quite expensive (the personal computer was still 20 years away) only large corporations took advantage of EDI. Today, however, EDI has matured into several industry standards, including the ANSI X.12 standard which is used primarily in North America, and the UN/EDIFACT standard which is used worldwide.

Roughly in parallel with the development of EDI, banks developed a system for electronic funds transfer or EFT. Today, the federal government has adopted EFT as the means to pay federal employees and make tax refunds. EFT was adopted as part of the Debt Collection Improvement Act of 1996.

A characteristic of both EDI and EFT systems is that both sending and receiving parties communicate with each other using a predefined protocol. The data sent between sending and receiving computers are formatted and packaged in a predefined way. That, of course, implies that both sending and receiving computers must be specially programmed to handle data in the same way. The logical next evolutionary step was be to remove the need for both systems to speak a common protocol. That was accomplished by providing a centralized clearinghouse system, or a middleware system, to which both sending and receiving computers could connect to exchange data. The middleware system performed any necessary data conversion, allowing the sending and receiving systems to exchange information even though they could not directly speak with one another using a common protocol. Exemplary of such a middleware system is the following patent issued to Shavit.

[A] Example of Electronic Marketplace Patent

If you are looking for prior art relating to a marketplace system that performs a middleware function, consider U.S. Patent No. 4,799,156 issued to Shavit, et al., entitled, "Interactive Market Management System." The patent was issued January 1989, based on an application filed October 1, 1986. The abstract of the Shavit patent is presented below. The Shavit patent is currently classified in U.S. Class 705/26. It is cross-referenced in Classes 705/28; 705/39; 705/40; 705/42; 705/44. These classes may prove fruitful in conducting prior art searches involving electronic marketplace technology.

[1] Abstract

A system for interactive on-line electronic communications and processing of business transactions between a plurality of different types of independent users including at least a plurality of sellers, and a plurality of buyers, as well as financial institutions, and freight service providers. Each user can communicate with the system from remote terminals adapted to access communication links and the system may include remote terminals adapted for storage of a remote data base. The system includes a data base which contains user information. The data base is accessed via a validation procedure to permit business transactions in an interactive on-line mode between users during interactive business transaction sessions wherein one party to the transaction is specifically selected by the other party. The system permits concurrent interactive business transaction sessions between different users.

CHAPTER 10

CLAIMING BUSINESS MODEL AND E-COMMERCE INVENTIONS

§ 10.01　Legal Requirements

Patent claims are the most focused upon part of the patent because the patent claims define the metes and bounds of the patent grant. In both patent prosecution in the Patent Office and patent litigation in the courts, the claims are of paramount importance. Therefore, as a patent practitioner, give a great deal of thought to the drafting of claims.

Patent claims were not always this important. One hundred years ago the metes and bounds of the patent grant were defined by the specification. In those days, the patent claim might simply state omnibus claim language such as: "I claim the invention as substantially shown and described." Do not try to get an omnibus claim like the above allowed today. Patent Examiners are trained to reject the omnibus claim as nonstatutory under 35 U.S.C. § 112.[1] The exception to this prohibition is design patents. Design patent claims are directed to the ornamental appearance of a design as shown in the patent drawings. Thus claims in design patents can and should state "The ornamental design, as shown and described."

[A]　Requirements of § 101

The statutory starting point for drafting any claim is 35 U.S.C § 101 that states what is patentable subject matter. Under § 101, there are four main categories of subject matter for which utility patents can be granted. They are:

1. process
2. machine (apparatus)
3. article of manufacture
4. composition of matter.[2]

In addition to these categories, the patent laws contain separate statutory provisions for plants[3] and ornamental designs.[4]

These are the only statutory classes of patentable subject matter. No matter how novel, inventive, or clever a person's creation or discovery may be, if it does not fall within one of the statutory classes, the creation or discovery is not patentable. Examples include scientific principles, laws of nature, naturally occurring articles, and printed matter.

[B]　Nonart Rejections

Patent Examiners are trained to watch for a number of claiming errors, for which they issue nonart rejections under 35 U.S.C. § 112. Knowing these nonart rejections in advance will help you draft claims to avoid them.

---

[1] U.S. Dep't of Commerce, Manual of Pat. Examining Proc. § 706.03(h) (Mar. 1994).

[2] 35 U.S.C. § 101.

[3] *Id.* at § 161-164.

[4] *Id.* at § 171-173.

The nonstatutory subject matter rejection is more fully discussed in Chapter 4. 35 U.S.C. § 101 treats everything as nonstatutory, except "process, machine, [article of] manufacture, or composition of matter."[5] From the examiner's perspective, it is often easier to identify a claim to nonstatutory subject matter by what it *is*, rather than by what it is *not*. Hence, the *Manual of Patent Examining Procedure* lists and briefly discusses the following three categories of nonstatutory subject matter:

1. Printed matter
2. Naturally occurring article
3. Scientific principle.[6]

At one time methods of doing business were included in the above list. However, in recognition of the principles outlined in the *State Street Bank* decision that category has been removed.

In addition, to these categories, the *Manual of Patent Examining Procedure* specifies that a claim should be rejected as nonstatutory if the claimed invention lacks utility.[7] This is a rejection under 35 U.S.C. § 101 because that section reads, "Whoever invents or discovers any new and *useful* process, machine, manufacture, or composition of matter . . . [emphasis added]." The lack of utility rejection may be given if the invention, as disclosed, is inoperative. In this instance, the specification would presumably also be rejected under 35 U.S.C. § 112. Perpetual motion machines, for example, are rejected as lacking utility.[8]

[C] Printed Matter Rejections

The printed matter rejection applies when the claim seeks to protect the mere arrangement of printed matter on a page.[9] Software Patent Examiners will sometimes use this rejection when the claim seeks to protect the arrangement of information displayed on a computer screen.

It is often possible to overcome the printed matter rejection by focusing on the apparatus or method responsible for generating the computer screen display alleged to be printed matter. Often you will find a combination of elements that is patentable subject matter. See *In re Miller* which held that, "The fact that printed matter *by itself* is not patentable subject matter, because nonstatutory, is no reason for ignoring it when the claim is directed to a combination."[10] In some instances the printed matter may be responsible for producing physical results, as

---

[5] *Id.* at § 101.

[6] *Supra*, note 1 at § 706.03(a).

[7] *Supra*, note 1 at Form Paragraph 7.04 and § 706.03(p).

[8] *Supra*, note 1 at § 706.03(p).

[9] *In re* Miller, 164 U.S.P.Q. (BNA) 46 (C.C.P.A. 1969); *In re* Jones, 153 U.S.P.Q. (BNA) 77 (C.C.P.A. 1967); *Ex parte* Gwinn, 112 U.S.P.Q. (BNA) 439 (Bd. App. 1955).

[10] *In re* Miller, 418 F.2d 1392, 1396 (C.C.P.A. 1969) (emphasis in original).

in an optical instrument for producing a diffraction grating or a semiconductor chip mask. If so, then the printed matter rejection can be overcome.[11]

[D] Naturally Occurring Article Rejections

The naturally occurring article rejection applies when the claim seeks to protect a thing occurring in nature, which is substantially unaltered and hence not a "manufacture." A shrimp with the head and digestive tract removed is an example.[12]

It would seem unlikely that a business model invention will ever be rejected as a naturally occurring article. The rejection is usually given when a naturally occurring article is treated in a particular way, for example, soaking an orange in a borax solution to inhibit mold.[13] It is therefore conceivable that a naturally occurring article may be treated using a software-controlled process. In such a case, if the claim is to the processed article, a naturally occurring article rejection may be given. To avoid such a rejection, focus the claim on the treatment process.

[E] Method of Doing Business Rejections

The method of doing business was long treated as nonstatutory subject matter.[14] That treatment changed abruptly with the Federal Circuit Court of Appeals Decision in *State Street Bank & Trust Co. v. Signature Financial Group*.[15] In that important decision, Judge Rich made it clear that the Federal Circuit does not embrace the so called business method exception.

> We take this opportunity to lay this ill-conceived exception to rest. Since its inception, the "business method" exception has merely represented the application of some general, but no longer applicable legal principle, perhaps arising out of the "requirement for invention" — which was eliminated by § 103. Since the 1952 Patent Act [which added § 103], business methods have been, and should have been, subject to the same legal requirements for patentability as applied to any other process or method.[16]

The method of doing business rejection is an interesting one. Prior to the *State Street Bank* decision, few applicants attempted to secure patent protection for business methods, per se. Instead, they sought protection for the software systems that made the business method practical. Thus the software patent served as the

---

[11] *See In re* Jones, 373 F.2d 1007 (C.C.P.A. 1967).

[12] *See* American Fruit Growers v. Brogdex, 283 U.S. 1 (1930).

[13] *Ex parte* Grayson, 51 U.S.P.Q. (BNA) 413 (Bd. App. 1941).

[14] *See* Hotel Sec. Checking Co. v. Lorraine Co., 160 F. 467 (2d Cir. 1908); *In re* Wait, 24 U.S.P.Q. (BNA) 88 (C.C.P.A. 1934).

[15] 33 F.3d 1526 (Fed. Cir. 1994).

[16] *Id.* at 1602.

vehicle for obtaining patent protection on what was otherwise believed to be nonstatutory business methods.

Perhaps one of the first to exploit this practice was Merrill Lynch. Indeed, until the *State Street Bank* decision, many considered Merrill Lynch's Musmanno patent to be the high water mark of this practice.

Merrill Lynch obtained U.S. Patent No. 4,346,442 (the Musmanno patent)[17] for its cash management account (CMA) system on August 24, 1982. The CMA system combines three financial services commonly offered by financial institutions and brokerage houses, including a brokerage security account. So far, that system does not sound like patentable subject matter. The Musmanno patent further includes certain data processing methodology and apparatus to effect the CMA system. Merrill Lynch sued Paine, Webber[18] for infringing the Musmanno patent, and the Delaware District Court held that the claims recite statutory subject matter and not an unpatentable method of doing business. The following is an example of a Musmanno claim:

1. In combination in a system for processing and supervising a plurality of composite subscriber accounts each comprising

a margin brokerage account, a charge card and checks administered by a first institution, and participation in at least one short term investment, administered by a second institution,

said system including brokerage account data file means for storing current information characterizing each subscriber margin brokerage account of the second institution, manual entry means for entering short term investment orders in the second institution,

data receiving and verifying means for receiving and verifying charge card and check transactions from said first institution and short term investment orders from said manual entry means,

means responsive to said brokerage account data file means and said data receiving and verifying means for generating an updated credit limit for each account,

short term investment updating means responsive to said brokerage account data file means and said data receiving and verifying means for selectively generating short term investment transactions as required to generate and invest proceeds for subscribers' accounts, wherein said system includes plural such short term investments,

said system further comprising means responsive to said short term updating means for allocating said short term investment transactions among said plural short term investments, communicating means to communicate said updated credit limit for each account to said first institution.[19]

---

[17] Musmanno, Securities Brokerage Cash Management Sys., U.S. Pat. No. 4,346,442 (Aug. 24, 1982), assigned to Merrill Lynch, Pierce, Fenner & Smith, Inc.

[18] Paine, Webber v. Merrill Lynch, 564 F. Supp. 1358 (D. Del. 1983).

[19] *Supra*, note 17, claim 1.

[F] Scientific Principle Rejections

The scientific principle rejection, the *Manual of Patent Examining Procedure* states, applies when a claim seeks to protect "[a] scientific principle, divorced from any tangible structure."[20] An example is Samuel Morse's claim 8 that covered:

> [T]he use of the motive power of the electric or galvanic current, which I call electro-magnetism, however developed, for making or printing intelligible characters, letters, or signs, at any distances . . .[21]

The Supreme Court rejected that claim as trying to preempt all use of the scientific principle of electromagnetism.[22]

[G] Omnibus and Single Means Claim Rejections

Nonstatutory subject matter is not the only nonart rejection the Patent Examiner is trained to watch for and reject. There is the omnibus claim — "a device substantially as shown and described" — which Examiners are trained to reject for failing to particularly point out and distinctly claim the invention as required in 35 U.S.C. § 112. The *Manual of Patent Examining Procedure* refers to the omnibus claim as a nonstatutory claim, although this terminology should not be confused with nonstatutory subject matter (35 U.S.C. § 101).[23]

Related to the omnibus claim is the single means claim, which Examiners are also trained to reject. The single means claim is one step removed from the omnibus claim. It recites the ultimate function of the invention in a single recitation. For example:

> A Fourier transform processor for generating Fourier transformed incremental output signals in response to incremental input signals, said Fourier transform processor comprising incremental means for incrementally generating the Fourier transformed incremental output signals in response to the incremental input signals.

This was the claim to which Gilbert P. Hyatt insisted he was entitled, when prosecuting his application and appeal per se. The Federal Circuit Court of Appeals disagreed.[24]

The single means claim is an attempt to claim function or result, without being limited to any particular structure. This is impermissible, not because functional language is used, but because the claim has only functional recitation and is therefore too broad. Courts seem to have little difficulty finding some

---

[20] U.S. Dep't of Commerce, Manual of Pat. Examining Proc. § 706.03(a) (Mar. 1994).

[21] O'Reilly v. Morse, 56 U.S. 62 (1853).

[22] *Id.* at 86.

[23] *Supra*, note 20 at § 706.03(h).

[24] *In re* Hyatt, 708 F.2d 712 (Fed. Cir. 1983).

rationale for invalidating claims which are too broad by claiming only the desired result. See, for example, *O'Reilly v. Morse*,[25] which is a classic case of claiming too broadly.

[H] Functional Language

Functional language, unsupported by structure, may get a claim rejected. This is not to say that functional language may never be used. 35 U.S.C. § 112 permits functional language to be used in a claim:

> An element in a claim for a combination may be expressed as a means or step for performing a specified function without the recital of structure, material, or acts in support thereof, and such claim shall be construed to cover the corresponding structure, material, or acts described in the specification and equivalents thereof.[26]

Thus, it is not functional language per se that is impermissible; it is using functional language to claim too broadly by claiming only the end result. Examiners are trained to watch for this occurrence. The *Manual of Patent Examining Procedure* instructs Examiners to reject "[a] claim which contains functional language not supported by recitation in the claim of sufficient structure to warrant the presence of the functional language in the claim."[27] The example given of a claim of this character is the claim found in *In re Fuller*.[28] The claim reads: "A woolen cloth having a tendency to wear rough rather than smooth."

Many patent attorneys use functional language in "whereby" clauses. There is nothing wrong with using a whereby clause, provided you understand what it can and cannot do. Patent Examiners are trained to treat the whereby clause as a non-distinguishing recitation. The *Manual of Patent Examining Procedure* informs examiners of the holding in *In re Mason*, which held that the functional "whereby" statement does not define any structure and accordingly does not serve to distinguish.[29]

In litigation some courts also give the whereby clause no weight, if the clause expresses only a necessary result of the previously described structure or method.[30] Conversely, a whereby clause can sometimes help make the claim more understandable. Hence there are times when there may be good reason to use the whereby clause. Properly used, the whereby clause should not contain the only recitation of a patentably distinguishing feature of the invention; rather, it should sum up the function, operation, or result that necessarily follows from the previously recited structure or method.

---

[25] 56 U.S. 62 (1853).

[26] 35 U.S.C. § 112.

[27] *Supra*, note 20 at § 706.03(c).

[28] 1929 C.D. 172, 388 O.G. 279 (1929).

[29] 114 U.S.P.Q. (BNA) 740 (C.C.P.A. 1957); *supra*, note 20 at § 706.03(c).

[30] *In re* Certain Personal Computers, 224 U.S.P.Q. (BNA) 270 (Ct. Int'l Trade 1984).

[I] Unduly Broad Claims

Whether a claim is unduly broad depends on the nature of the art and on what is disclosed in the specification. When the art produces predictable results, such as in much of the mechanical and electrical arts, broad claims may be properly supported by the disclosure of a single species.[31] However, when the art produces unpredictable results, such as in much of the chemical arts, broad claims may not be supported by the disclosure of a single species.[32] This is because in arts such as chemistry it is not obvious from the disclosure of one species that other species will work or what those other species might be.

Software inventions usually produce results that are predictable, hence, broad claims may properly be supported by disclosure of a single preferred embodiment. This should be taken with a grain of salt, however, for there are classes of problems to which software may be applied that defy predictability.

Predicting the weather is an example. A supercomputer software invention that predicts the onset of hurricanes 24 hours in advance may not similarly predict the onset of hurricanes one year in advance. Likewise, it may not predict drought. Hence a broad claim to weather prediction would not be supported by disclosure of a single 24 hour hurricane predictor. Why this is so is the subject of James Gleick's book, *Chaos Making A New Science*,[33] that explains that some phenomena, like weather, are extremely "sensitive to initial conditions." A colorful way to state this sensitivity is that the flitting of a butterfly's wings in China will radically alter wind conditions halfway around the world.

[J] Vague and Indefinite Claim Rejections

There are a number of other nonart rejections amounting to rejections of poor claim drafting style that Examiners look for. These include rejections that the claim is vague and indefinite,[34] that the claim is incomplete,[35] that the claim is too wordy or prolix,[36] that the claim is a mere aggregation of elements because there is no claimed cooperation between elements.[37]

Usually the allegedly vague or ambiguous claim can be corrected by choosing different terms to describe the invention. Sometimes an Examiner will assert that a claim is indefinite because the word "or" has been used. Use of the word "or" does not necessarily render a claim indefinite, although in some instances it can. If the word "or" separates two different elements, the *Manual of*

---

[31] *In re* Cook & Merigold, 169 U.S.P.Q. (BNA) 298 (C.C.P.A. 1971); *In re* Vickers, 61 U.S.P.Q. (BNA) 122 (C.C.P.A. 1944); *supra*, note 20 at § 706.03(z).

[32] *In re* Dreshfield, 1940 C.D. 351, 518 O.G. 255 (1940); *In re* Sol, 1938 C.D. 723, 497 O.G. 546 (1938); *supra*, note 20 at § 706.03(z).

[33] James Glick, *Chaos Making New Science* (1987).

[34] U.S. Dep't of Commerce, Manual of Pat. Examining Proc. § 706.03(d) (Mar. 1994).

[35] *Id.* at § 706.03(f).

[36] *Id.* at § 706.03(g).

[37] *Id.* at § 706.03(i).

Patent Examining Procedure instructs Examiners to reject the claim as being indefinite. "Alternative expressions such as 'brake or locking device' may make a claim indefinite if the limitation covers two different elements."[38] However, the *Manual* also explains that if the word "or" separates alternate expressions of the same element, the claim should not be rejected as indefinite. "If two equivalent parts are referred to such as 'rods or bars,' the alternative expression may be considered proper."[39]

Software inventions may involve the word "or" as an operator in a Boolean logic expression, such as: If A or B then do C. In this case, the Patent Examiner should recognize the word "or" as part of a single Boolean expression, that is, as alternate expressions of a single claim element. As such, the claim is quite precise and definite, and should not be rejected. Nevertheless, if an examiner cannot be persuaded to retract a rejection of the word "or" you may be able to get around the issue by restructuring the claim as follows: Do C if at least one of A and B is met.

[K] Incomplete Claim Rejections

When the Examiner rejects a claim as being incomplete, this usually means that the claim omits essential elements or steps, or that the claim omits the necessary structural cooperative relationship of elements or steps.[40] Examiners are taught that this rejection should focus on the elements or steps that are essential to novelty or operability and that greater latitude should be given to elements or steps that are not essential.[41]

Aggregation is a related rejection, based on lack of cooperation between the elements of a claim. The example given in the *Manual of Patent Examining Procedure* is: A washing machine associated with a dial telephone.[42] As can be seen here, the term "associated with" fails to distinguish a dial telephone coupled to remotely control a washing machine, from a dial telephone temporarily placed upon a washing machine.

[L] Duplicate Claims and Double Patenting

Examiners not only review individual claims for integrity, they also review the entire set of claims presented for integrity. If two or more claims are exact duplicates, or if they are so close in content that they cover the same thing, Examiners are trained to give a duplicate claim or double patenting rejection.[43] Duplicate claims can occur by simple clerical or drafting error, often the result of inattentively using a word processor copy and paste function. Examiners may

---

[38] *Id.* at § 706.03(d).
[39] *Id.*
[40] *Id.* at § 706.03(f).
[41] *Id.*
[42] *Id.* at § 706.03(i).
[43] *Id.* at § 706.03(k).

also find claims to be duplicates where the wording of the claims differ only in subject matter that is old in the art.[44]

37 C.F.R. § 1.75(b) provides that "[m]ore than one claim may be presented, provided they differ substantially from each other and are not unduly multiplied." Undue multiplicity is thus a nonart rejection that the Examiner can make when, in the Examiner's judgment, the number of claims is "unreasonable in view of the nature and scope of applicant's invention and the state of the art."[45]

§ 10.02 Claim Drafting Process

Draft the claims first. That is what many experienced patent attorneys advise. There is good reason for this. The claims define the metes and bounds of the invention. The claims put the invention into perspective with the prior art. The claims are the words that the Patent Examiner spends the most time reviewing, and the claims are the instrument construed by the court in litigation. The claims are important and should therefore be given primary attention. Starting with the claims helps you focus on the invention.

There are those who prefer to draft the specification first, and then draft the claims. They will tell you that writing the specification first helps them figure out what the invention is. There is nothing fundamentally wrong with this approach, but it can be considerably less efficient. Consider this analogy.

Before painting his famous mural on the ceiling of the Sistine Chapel, Michelangelo undoubtedly made sketches. Edgar Degas did this also, before he rendered his impressionistic pastels for which he is famous. Indeed, most artists make sketches and studies before undertaking a major work. (There are always exceptions, like artist Jackson Pollock, who exploited chaos and randomness in his swirling drip paintings. Pollock spread his canvas on the floor and spilled paint on it. He then selected and framed the good parts.)

Patent application drafting is an art form, more constrained than painting, but nevertheless an art form. It is quite natural to make sketches or studies of important inventive aspects before undertaking the patent application in earnest. Those who draft the specification first are actually making written sketches or studies in an effort to find the invention. Thus, whether you draft the claims first, as advocated here, or draft the specification first, treat early efforts as sketches or studies, and be prepared to throw them away, just as the artist makes a sketch, learns from it, and then discards it. That is why it is more efficient to draft the claims first. Claims comprise fewer words; it is easy to draft a claim, study it, and discard it in favor of a better claim. It is much harder to draft a complete specification and then discard it, after learning that you missed the invention. So many pages of effort go into drafting the specification that many are strongly tempted to keep the specification that has been drafted, even though it may not focus precisely on the claimed invention.

---

[44] *Id.*, citing *Ex parte* Whitelaw, 1915 C.D. 18, 219 O.G. 1237.
[45] *Supra*, note 34 at § 706.03(l).

There is no one proper way to express the metes and bounds of an invention. An invention is not one-dimensional. It is multidimensional — like sculpture. There is no one vantage point from which to look at an invention. There are many. Claims should be drafted to describe the invention from several different vantage points, thereby capturing a better image of the multidimensional nature of the invention.

§ 10.03   Finding the Invention

How do you go about finding the invention? Begin by assessing what are the commercially important features. What are the features that will motivate purchasers of a product to select the invention over the prior art? Having these features in mind while drafting will help keep the claim focused on what the inventor wants "to exclude others from making, using or selling . . . throughout the United States."[46]

Next, identify the essential claim elements that are required to make the invention. Often you can identify the essential claim elements by studying the component parts (or subassemblies of component parts) found in the inventor's preferred embodiment. Identify those parts that are essential to the operation of the invention. In this endeavor, software inventions and hardware inventions are no different. Each can be broken down into component parts or modules or into subassemblies or submodules. The objective is to identify the essential component parts. Sifting through all the component parts, the essential ones are often those that make the commercially important features possible, or those that a competitor must employ to compete.

Having identified the essential component parts, devise claim element abstractions to represent the essential component parts. These abstractions are the essential claim elements that are required to make the invention, and that will ultimately become the elements of the claim you are drafting.

An essential component part and an essential claim element seem quite similar. Is there a difference? The answer is yes. The component part identifies part of the inventor's implementation or physical embodiment of the invention. The claim element more broadly identifies a class of component parts capable of making the invention.

Object-oriented software developers use a similar notion to write software. They model a solution to the problem to be solved, first by identifying classes of objects they believe are needed to solve the problem, and second by selecting specific objects from these classes to build the actual software embodiment. The specific objects selected are called "instances" of the class and the act of selecting such objects is called "instantiation." Knowing this terminology and the analogy between claim drafting and object-oriented software development may help if you are working with an inventor who is familiar with object-

---

[46] 35 U.S.C. § 154.

oriented programming. If you are fortunate, you may find that the software invention was developed using object-oriented techniques, in which case, there may be good design documentation that will virtually track the claim drafting process, step-by-step.

Having identified the essential claim elements, or at least those which at this stage you believe are essential, the next step is to identify the relationships between these elements. Often one element will be functionally "connected" to one or more other elements. The connection may be a direct or indirect physical attachment connection, for example, *element A* may be "coupled to" *element B*. The connection may be an information flow connection, such as, *element A* may "supply data (numbers, values, and so forth) to" *element B*; or *element A* may "supply a physical signal" to *B*. There are no doubt many other types of connections.

No element exists without some relationship to at least one other element. In your analysis, if an element floats alone, unconnected by any relationship with the other elements, chances are the floating element is not essential and should be discarded. If discarding the floating element destroys your perception of the essence of the invention, then put the element back in, and work harder at a deeper understanding of how that element is related to the other elements. A deeper understanding of the elusive relationship will often reward you with new insights and a new vantage point from which to see the invention.

For a handy checklist of things to look for in finding the invention, see § 12.10[B].

[A] Drawing Claim Diagrams

In identifying the essential claim elements and the relationship between those elements, it is helpful to make diagrams. These diagrams are similar to the diagrams that Patent Examiners make from the claims in preparation for searching the prior art. System engineers use a diagram known as the entity relationship diagram or ERD, which can be employed to diagram claims. The diagramming technique is quite simple. Draw a circle (or any preferred shape) for each essential element. Label the inside of the circle with a broad or generic term that aptly describes that element. Next draw connecting lines between elements that are related in a way that is important to the invention. Place labels adjacent to or on the lines to describe the nature of the particular relationship. That is essentially it, although there is a little extra detail that you may want to add to the diagram, if desired.

When appropriate, add arrowheads to the relationship lines, to show the direction of data flow or to show a dominant-subservient relationship. A claim element modifier (such as an adjective or a participle phrase), not important enough to be represented as an element itself, can be shown as an unterminated, cat whisker line originating from the element it modifies. Figure 10-1 shows a sample claim diagram. The claim corresponding to this diagram is taken from

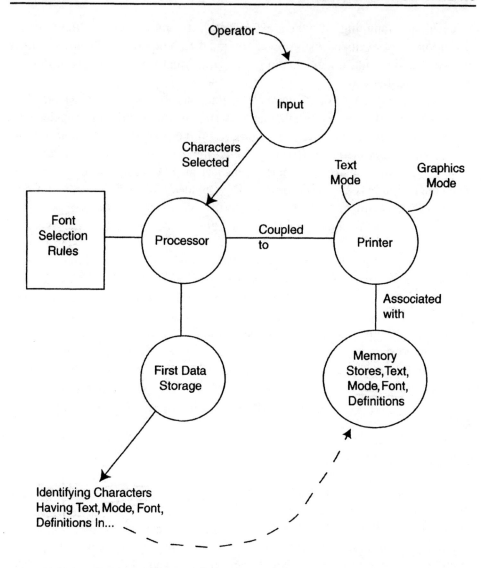

FIGURE 10-1:   A SAMPLE CLAIM DIAGRAM

claim 1 of U.S. Patent No. 5,233,685, originally assigned to WordPerfect Corporation.

The diagram does several things. First, it serves as a shorthand notation for your thoughts. It is quicker to draw a diagram than it is to draft prototype claim language. The diagram allows you to think about claim elements and relationships, without worrying about what words to use. Second, the diagram helps identify nonessential elements and limitations, and elements that are floating — unrelated to the rest of the claim. Finally, the diagram is easy to comprehend at a glance, hence you can set it aside and forget about it, and then later pick it up and be instantly reminded of your claim drafting thoughts.

[B] Working with Inventor's Drawings

Many times a claim diagram can be extracted from one of the inventor's drawings or from one of the patent drawings. This stands to reason — a drawing of a preferred embodiment of the invention would be expected to show all of the essential claim elements. However, drafting a claim to show every component of the preferred embodiment usually yields a narrow "picture claim." Thus, the inventor's drawing or the patent drawing should not be thoughtlessly used as a claim diagram.

Rather, starting with the inventor's drawing or the patent drawing, try to group together certain component parts that provide a common function. Draw a box around these component parts and label the box to denote the common function. Do this for the other component parts as well. For example, if the software embodiment has several data structures and processes pertaining to data acquisition, group these data structures and processes together under the heading "data acquisition system," for example. Similarly, if there are several data structures and processes for supplying input and output to the user, these may be grouped together under the heading "user interface."

In grouping together functionally related component parts, keep in mind that not all groupings necessarily represent essential claim elements. In the above examples, it may well be that the "data acquisition mechanism" and the "user interface," while required for a working product, are not essential elements of the invention.

[C] Working with Computer Source Code

With some e-commerce inventions, you may find the inventor has no drawings, only source code. When confronted with this situation, you may have to "reverse engineer" the source code to produce the claim diagram. Ideally, you should do this with the inventor's help, because deciphering someone else's source code is a *very* time-consuming task.

In reverse engineering the source code, do essentially the same thing you would do if supplied with drawings. Identify the data structures, processes, subroutines, procedures, and modules that have a common purpose or function and then draw an appropriately labeled box to represent these component parts. Often it can be helpful to start by identifying all component parts that make up the user interface (that through which the user supplies input to the program and through which the program replies to the user), and by identifying all component parts that make up the operating system interface (that through which the program attaches to and invokes services of the operating system, for example, disk storage, file management, keyboard, and mouse support). These two interfaces are frequently not essential elements of the invention, yet they are nearly always present in a working software product. By identifying and separating out these two interfaces, it is easier to examine the remaining elements for those that are essential to the invention.

§ 10.04   Testing Claims for Proper Scope

Having identified the essential elements, and having found the relationships between these elements, the next step is to test the model claim to see if it is of proper scope. Do this step by testing the claim to determine if there are any undue limitations, and also to make sure the model claim does not read on the prior art. At this stage, it is not necessary to reduce the model claim to writing. The claim diagram alone will suffice.

Begin by identifying which claim elements are themselves new. Also identify if any of the relationships between elements are new. Often, none of the elements alone will be new, but the combination of elements and their relationship is new. By identifying which elements and relationships are new, you begin to focus on the essence of the invention. If you find that you have included several elements that are not new, then possibly these old elements should be lumped together under a collective or generic name. Doing this helps avoid unnecessary limitations.

What if none of the elements themselves are new, and none of the relationships are new either? This means you have not yet found the invention; it is hiding. In identifying the essential elements at the outset, you may have chosen element names that are now hiding the invention. Try to determine which element or elements contain the new or inventive matter. Then break these elements into two or more component elements, and define the relationships between them, so that the invention is no longer hidden.

The model testing process takes an iterative hierarchical approach. Combine multiple related old elements into a more generic single element; break apart new and inventive matter into more specific, separate elements. In the end you have a model claim, in diagram form, which shows quite clearly what is new.

[A] Checking Claims Against Prior Art

It is usually a good idea to further test the model by attempting to read the model, as a whole, upon the known prior art. Take the known prior art, such as the inventor's own prior art product, or a competitor's prior art product, or the prior art patents developed during a patentability search, and see if each and every element of your model claim is found in a single prior art device or reference. If so, then an assumption made while preparing the model claim is incorrect. The elements identified as new are not new.

Do not despair. It is better to know that the claim is not of proper scope now, (before you file the application) while it can be corrected without a costly amendment, responding to a Patent Office rejection. Try to identify how the invention differs from the prior art and add at least one claim element or limitation that will highlight the difference.

If it is not apparent what element to add, try to determine what makes the invention better than the prior art. For example, the invention may operate more efficiently than the prior art; the invention may have features that the prior art does not have; the invention may be easier and less expensive to manufacture; or the invention may be easier to use. Once you have determined how the invention differs from or improves upon the prior art, devise a claim element or combination of elements that are necessary to effect this difference or improvement.

[B] Designing Around Claims

Claim drafting is an iterative art. Therefore, after adding the necessary claim elements to distinguish the invention from the prior art, go back and test the remaining elements to see if there are any that are unnecessary. One good way to test for unnecessary limitations is to pretend to be a competitor trying to design around the claim. Are there any elements that can be eliminated? If so, revise the claim model and eliminate those elements that a competitor can eliminate. Are there any elements that can be replaced with substantially nonequivalent elements? For these, try to devise different claim elements that are more generic and therefore cover both the originally contemplated element and the nonequivalent substitute. The objective is to make the claim devilishly difficult to design around.

[C] Considering Fall Back Positions

Once you have constructed a claim model that highlights what is new about the invention, distinguishes the invention from all prior art of which you are aware, and is devilishly difficult to design around, there is one more thing to do. Think about your fall back position. How will you revise the claim if it is rejected by the Patent Office over art of which you are not aware?

Naturally, until you know about the prior art, you cannot possibly predict with certainty how to revise the claim to avoid the art. However, since most claims are rejected in the first action, expect this one will be too. It is therefore wise to have some subject matter in the specification that can be later added to the independent claims.

[D] Drafting and Refining Claims

There are several ways of drafting actual claim language, once you have the claim model worked out. If you are fairly skilled using a word processor, you may want to draft the claim language directly on the screen using your word processor. This can be a fairly efficient way to work, even if you are not a fluent typist, because drafting the claim language involves a great deal more thinking than it does typing.

Decide first on the form of the claim, for example, apparatus, method, Jepson. This will dictate how the claim preamble should be worded. However, do not consider initially what to put in the claim preamble. It is usually easiest to build the claim preamble after the claim element language has been drafted. Using the claim model diagram as a blueprint, start by drafting language to describe each of the elements and the relationships between elements. Be sure to organize the elements so that elements supplying an antecedent basis for other elements appear first in the claim. To illustrate, see Figure 10-1 that shows a claim model diagram. A rough draft claim might be something like this:

A computer system for word-processing, comprising:

input responsive to operator-input characters;

a printer capable of printing in text mode and in graphics mode;

a memory associated with said printer for storing text mode font definitions;

a data store for identifying characters having text mode font definitions in said memory;

a processing coupled to said input and to said data store and to said printer for comparing operator-input characters with said characters in said data store and for causing said printer to print the operator-input characters according to a predetermined set of font selection rules.

Note in the above rough draft example, the preamble is nothing more than a place holder. The antecedent details have not been added in the preamble at this point. Similarly, the individual claim elements are also lacking details. Integration of the claim elements is not yet tight, and the claim does not yet distinctly represent the invention.

Work with the rough draft claim language, tightening the phrasing and supplying missing antecedent bases, until the claim reads, start to finish, without ambiguity. Having done this, the finished claim might be something like this:

1.   A computer system for performing word-processing operations, the computer system comprising:

(a)   input responsive to operator commands enabling an operator to specify any of a plurality of characters for inclusion in a document being created or edited;

(b)   printer capable of printing characters in either text or graphics mode;

(c)   memory associated with the printer for storing a text mode definition for each of a subset of the plurality of characters that may be specified by the operator using the input;

(d)   a first data store identifying each character having a text mode definition in the memory;

(e)   processor coupled to the input, the first data store and the printer programmed to compare each character selected by the operator with information stored in the first data store to determine whether, for each specified character, a text mode definition exits in the memory and,

(i) if said definition exists, sending data to the printer identifying the character; or

(ii) if said definition does not exist, taking alternative action comprising sending data identifying an alternative character to the printer for printing in graphics mode or sending data identifying said alternative character in a substitute font to the printer for printing in text mode.

[E] Narrow Claims

There is more to claim drafting strategy than drafting a single independent claim. A utility patent with only one claim qualifies under the old adage as "putting all your eggs in one basket." While having at least one broad claim is certainly desirable, it is far better to cover the spectrum with broad, intermediate, and narrow claims. This may entail having different sets of independent claims, each of different scope. It may also entail having claims of different types, for example, apparatus claims, method claims, Jepson claims, Markush claims, means plus function claims, and so forth. Nearly without exception, this will entail having dependent claims.

Why have intermediate and narrow claims, and specifically, why have dependent claims? Simply, these claims serve as insurance against broad claim invalidity. All claims issued by the Patent Office are presumed valid over all prior art considered by the Patent Examiner. However, when there is a great deal of money at stake, defendants in patent infringement litigation go to great lengths to find closer prior art than the Patent Examiner considered — to prove that the claims in litigation should never have been granted.

Because broad claims cover more intellectual territory, they are easier to invalidate. To invalidate any claim under 35 U.S.C. § 102, you must show that the claim reads on (covers, element by element) a single prior art device or reference. Broader claims read on more, and they are therefore more vulnerable to being invalidated. Therefore, while most patent prosecution attorneys strive for the broadest claim allowable, most patent litigation attorneys prefer the narrowest claim that covers the accused. Litigation attorneys prefer a narrow claim because a narrow claim is harder to invalidate and because a narrow claim, read on the accused, demonstrates how closely the accused has "copied" the invention.

[F] Dependent Claims

A good way to draft a claim of narrower scope is to draft a dependent claim. A dependent claim is one that incorporates by reference or "depends" from an independent claim. Object-oriented computer programmers will recognize the dependent claim as a child claim that "inherits" all of the elements and limitations of the parent claim — just as a rowboat inherits all properties of the class, watercraft. Incorporating all of the limitations of the independent claim parent, the dependent claim is, by definition, narrower than the independent claim parent. This narrowing inheritance relationship can cascade through several generations; hence one dependent claim can be based on another dependent claim, which in turn can be based on still another claim, perhaps an independent claim. The one dependent claim at the bottom of the cascade would be said to incorporate all of the elements and limitations of the independent base claim and any intervening claims.

There are typically two ways to draft dependent claims. You can amplify, by adding an additional limitation to an existing claim element of the parent claim; or you can augment, by adding a new claim element to the parent claim. The mechanics for each of these is quite simple. To amplify, use a ''wherein'' phrase:

> 2.   The invention of claim 1 wherein said balloon is blue.

To augment, use a ''further comprising'' phrase:

> 3.   The invention of claim 1 further comprising a second balloon.

It is also proper to cascade one dependent claim from another. Thus, you could add the following additional dependent claims, based on the ones presented above:

> 4.   The invention of claim 2 further comprising a second balloon.
>
> 5.   The invention of claim 3 wherein said second balloon is blue.

By the time you are done, you can have all permutations of one and two (blue and nonblue) balloon embodiments covered, as illustrated.

Dependent claims can get quite narrow quickly, when cascaded. This is usually not good practice. It is better practice to base most dependent claims directly upon an independent base claim. In that way, each dependent claim can amplify or augment a single feature or aspect of the invention, in a narrower sense, without bringing in unnecessary limitations.

§ 10.05   Software Claim-drafting Templates

Every invention is different. That is the claim drafter's starting point. More precisely, every patentable invention represents a novel, nonobvious improvement over what existed before it. Thus, rarely does study of a claim of a particular prior art patent shed much light on how to draft a claim for a new invention. Indeed, the patent claim for the new invention must represent something *different* than the prior art claim represents.

In this context, it seems clear that one should not draft a patent claim for a new software invention by starting with the claim of the closest prior art reference and then modifying it. Doing that is one sure way to increase the likelihood that your new claim will be rejected by the Patent Office under 35 U.S.C. § 103. That statute codifies that patents are not granted merely for making obvious modifications to the existing art. Yet, when dealing with unfamiliar territory, many attorneys feel the temptation to start with the claim of someone else (presumably someone more experienced) and then modify it. This will almost always lead to problems.

By analogy, it is far simpler to revise an existing contract than it is to draft a new contract from scratch. However, as contract lawyers know, always draft your own agreement; do not merely modify the other party's — that way, you remain in control of all the countless subtleties.

A patent claim is a contract with the government. It makes sense, then, to follow the contract lawyer's adage: always draft the contract yourself. Always draft the claim yourself. Do not modify the claim of someone else, who may indeed represent your client's present or future adversary.

Does this mean that you can learn nothing by studying the claims of others? Certainly not. Issued patents provide an excellent source of knowledge — not only technological knowledge, but claim-drafting knowledge as well. Simply understand that no single patent claim should be used as a starting point in drafting a new claim. Instead, use issued patent claims to discover claim-drafting techniques or solutions to claim-drafting problems with which you may be struggling. You will find that, like technological problems, the same claim-drafting problems recur over and over, each time in a different inventive context.

Recognizing that the same problems seem to crop up again and again, software developers now use "design patterns," a concept borrowed from architecture, to group together problems that share a common solution. To understand the power of design patterns, look around for the many different uses of the arch, or the tiled roof, in building architecture. Although the appearance may differ widely from use to use, the functions of these well-chosen patterns remain the same.

Claim-drafting patterns work the same way. Subsections § 10.05 [A]-[C] present several claim-drafting patterns in the form of claim-drafting templates. These templates show how to solve some recurring software patent claim-drafting problems. These come up frequently in e-commerce inventions and in business model inventions that employ computer or internet components.

[A] Claim-drafting Template for Real-time Processes

Speed and storage capacity are two scarce computer commodities. At the metaphysical level these commodities represent the very essence of the universe: time and space. It is not surprising, then, that software developers will spend substantial resources exploiting these commodities. Simply stated, in the computer industry faster is better, and you never have enough memory.

When it comes to speed, nearly every software developer must, from time to time, play beat-the-clock. However, some software developers do it for a living, earning them the moniker of real-time programmers.

Real-time programming is a term that defies definition. You will encounter some software inventors who use the term in a precise way to refer to processes that must fully execute, start to finish, between hardware interrupt cycles. You will encounter other software inventors who use the term in a more general way to refer to processes that occur interactively, as opposed to in batch. Patent

Examiners will often reject a claim in which novelty hinges on the term ''real-time,'' simply because the term ''real-time'' may have no precise meaning set forth in the claim or explained in the specification.

To include a ''real-time'' recitation in a claim, you may want to define a time frame, preferably one that ties to a physical condition within the system. You may then recite any real-time process as one that executes within that prescribed time frame. Think of a stopwatch whose hand makes one complete revolution in one second. If that one-second revolution defines the time frame for this system, then a real-time process might be one that starts and finishes before the hand makes one complete revolution. Although the stopwatch here is simply an example to make the point, try to find something akin to the stopwatch in every real-time process and use it to define the time frame.

The following claim template demonstrates how the real-time process may be recited. Both apparatus and method claim templates are provided. The phrases in brackets are optional or intended to show how you might incorporate the template into your existing claim. Remember, these templates contain concepts, not magic words. In using these concepts, choose words that best fit the particular invention at hand.

An apparatus claim might go something like this:

1.   An apparatus, comprising:

[Element A];
[Element B];
a timing element that defines a predetermined time interval [based on some physical condition or demonstrable state of the system being claimed or its environment];
a real-time element [processor, engine, etc.] that performs XYZ [the real-time process] fully within said predetermined time interval;
[Element C].

While a method claim might look like this:

2.   A method, comprising:

[Step A];
[Step B];
establishing a predetermined time interval [based on some physical condition or demonstrable state of the system being claimed or its environment];
performing XYZ [the real-time process] fully within said predetermined time interval;
[Step C].

As noted in these templates, you may wish to express the predetermined time interval in terms of some physical condition or demonstrable state of the system being claimed or its environment. The physical condition or state may be tied to one of the other elements or steps recited in the claim. If so, then that element or step should be recited first, to provide proper antecedent basis for the time interval recitation. Often, the physical condition or state may be tied to some entity in the environment in which the invention is intended to operate. In such case, you may want to recite that environmental entity.

For example, say the invention is a tool for making up-to-the-minute investment decisions. You might use the real-time template this way:

1.   A computer-implemented apparatus for making investment decisions, comprising:

an input processor coupled to a source of investment data;

a timing process associated with said input processor that generates data indicative of marketplace investment intervals;

a fuzzy logic processor communicating with said input processor for making buy/sell investment decisions, said fuzzy logic processor supplying a control parameter to said input processor, said control parameter in turn causing said input processor to selectively obtain investment data for use in making said investment decisions,

wherein said fuzzy logic processor updates said control parameter using a fuzzy XYZ relationship and supplies the updated control parameter to said input processor, all within a single one of said marketplace investment intervals.

[B]   Claim-drafting Template for Iterative Processes

Computers excel at repetition. Unlike its human creators, a computer makes little distinction between the command, "Do this once," and the command, "Do this a thousand times." The process by which computers will so readily loop through the same operation a thousand times is called *iteration*. Software developers express iteration in several ways. Here is one simple example:

```
Let X = 0
Repeat the following steps until X is 1000:
Begin
Y = m*X + b
X = X + 1
End
```

Actually, there are several different kinds of iteration that are worth recognizing. Each has its own purpose. You will encounter all types in practice. Nicholas Wirth, a world-famous computer scientist and educator, developed the computer language Pascal to teach his students basic programming concepts such as iteration. Although the Pascal language is no longer in widespread use, its syntax is easily understood, making it still one of the best languages for expressing basic computer software concepts.

The Pascal language defines at least three different kinds of iteration, each with its own unique twist. In the following discussion, keywords printed in boldface are terms defined by the Pascal language. Other computer languages may use different syntax.

The first iteration type uses the **while** statement. The form of this statement is:

> **while** condition **do**

To illustrate this form of iteration, consider this example.

> **while** daylight **do**
> **begin**
> read temperature sensor;
> look up correction factor in table;
> update processor input;
> **end**

The **while** statement causes this process to loop repeatedly or iterate until the condition (daylight) becomes false. What distinguishes this first iteration type is its ability to skip the process entirely if the condition is never true. If it is not daytime when this **while** procedure is invoked, none of the steps between **begin** and **end** ever get performed.

The second iteration type uses the **repeat** statement. The form of this statement is:

> **repeat**
> statement 1;
> statement 2;
>   . . .
> statement n
> **until** condition.

To illustrate this form of iteration, consider this example.

> **repeat**
> read temperature sensor;
> look up correction factor in table;
> update processor input;
> **until** nighttime.

Unlike the **while** statement, the **repeat** statement executes its included statements at least once, even if the **until** condition turns out not to be true. Thus, in the preceding example, the read, look up, and update steps are performed once (the first pass through) even if it is already nighttime.

The third kind of iteration uses the **for** statement. The form of this statement is:

> for control variable : = a **to** n **do**
> **begin**
> statement 1;

statement 2;

. . .

statement n
end

To illustrate this form of iteration, consider this example.

for n:= 1 to 1000 **do**
begin
read temperature sensor;
look up correction factor in table;
update processor input;
end

This program will execute for a prespecified number of iterations and the number of iterations does not depend on the effect of statements within the loop. In this example, the variable n serves as a counter that the program automatically increments each time the begin-end loop is performed.

You may encounter each of these iteration kinds in claiming an e-commerce or business model invention. The following claim templates demonstrate how each of these forms of iteration may be recited. Both apparatus and method claim templates are provided. The phrases in brackets are optional or intended to show how you might incorporate the template into your existing claim. Remember, these templates contain concepts, not magic words. In using these concepts, choose words that best fit the particular invention at hand.

[1]   The *While* Loop Template

A sample of the **while** loop template might look like this:

1.   A method, comprising:

[Step A];
[Step B];
while X is true [or false] iteratively performing:
[Step C],
[Step D], and
[Step E]
[Step F].

2.   An apparatus, comprising:

[Element A];
[Element B];
[iterator] element that iteratively performs CDE while X is true [or false];
[Element F]

To illustrate use of the template, consider the following claim for a method of accessing information over the Internet.

1. A method of accessing information from Internet Web pages having hypertext links to other Web pages for off-line viewing, comprising:

connecting a computer to a service provider;
running a browser on said computer;
prompting the user to terminate the running of said browser to end an information-accessing session;
while said browser is running, iteratively performing the following steps a-c:
a. accessing a Web page and storing the information content of said Web page in said computer;
b. identifying on said Web page at least one hypertext link to a different Web page;
c. accessing said different Web page and storing the information content of said different Web page in said computer;
disconnecting said computer from said service provider; and displaying said stored information content.

[2] The *Repeat-until* Loop Template

A sample of the **repeat-until** loop template might look like this:

1. A method, comprising:

[Step A];
[Step B];
[Step C];
[Step D]
iteratively repeating steps B, C, and D until X is true [or false]
[Step F].

2. An apparatus, comprising:

[Element A];
[Element B];
[iterator] element that performs CDE and iteratively repeats CDE while X is true [or false];
[Element F]

To illustrate use of this template, consider the following claim for a method of accessing information over the Internet.

1. A method of accessing information from Internet Web pages having hypertext links to other Web pages for off-line viewing, comprising:
a. connecting a computer to a service provider;
b. running a browser on said computer;

    c.   prompting the user to terminate the running of said browser to end an information-accessing session;

    d.   accessing a web page and storing the information content of said web page in said computer;

    e.   identifying on said web page at least one hypertext link to a different web page;

    f.   accessing said different web page and storing the information content of said different web page in said computer;

    g.   iteratively repeating above steps d-f until the running of said browser is terminated;

    h.   disconnecting said computer from said service provider, and

    i.   displaying said stored information content.

Comparing this **repeat-until** loop example with the preceding **while** loop example, you will note that steps d-f in the **repeat-until** example are required to be performed at least once. In contrast, the comparable steps a-c in the preceding **while** example may never be performed, if the browser is terminated before the iterative steps commence.

[3] The *For* Loop Template

The **for** loop can also be equivalently expressed using **if-then** terminology discussed in §§ 10.05[B][5]. You may find the **if-then** terminology more natural.

A sample of the **for** loop template might look like this:

1.   A method comprising:

> [Step A];
> [Step B];
> setting X [a counter] to an initial condition;
> for X less than [not equal to] Y [the terminating condition], performing the following:
> [Step C],
> [Step D],
> [Step E], and
> incrementing X;
> [Step F].

2.   An apparatus comprising:

> [Element A];
> [Element B];
> [iterator] element having a data structure for storing a counter initialized to a predetermined value, said [iterator] element being operable to iteratively increment said counter and to perform CDE for counter values less than [not equal to] Y [the terminating condition] [Element F].

To illustrate this template, consider the following example.

1.   A method of accessing information from Internet web pages having hypertext links to other web pages for off-line viewing, comprising:

>   connecting a computer to a service provider;
>   running a browser on said computer;
>   establishing a counter for storing values used in performing iteration;
>   setting said counter to an initial value and prompting the user to input a terminal value;
>   for counter values not equal to said terminal value, iteratively performing the following steps a-d:
>>   a.   accessing a web page and storing the information content of said web page in said computer;
>>   b.   identifying on said web page at least one hypertext link to a different web page;
>>   c.   accessing said different web page and storing the information content of said different web page in said computer;
>>   d.   incrementing said counter;
>   disconnecting said computer from said service provider; and
>   displaying said stored information content.

This **for** loop example is quite similar to the **while** loop example presented in § 10.05[B][1]. The **for** loop differs in that the initial and terminal conditions are more explicitly set forth. Also note that by specifying details about the terminal value, you can make the claim dictate whether the iterated steps must be performed at least once, or whether they may be skipped entirely. Do this by specifying the relationship of the terminal condition to the initial condition, as follows.

If the claim recites that the terminal condition is *greater than* the initial condition, then the iterative steps are performed at least once. In contrast, if the claim recites that the terminal condition is *greater than or equal to* the initial condition, then the iterative steps could be skipped (if the user selects a terminal condition equal to the initial condition).

[4]   The *If* Statement Iterator

Another useful variation on the iteration theme is the **if** statement. The keyword **if** in Pascal (and also in other computer languages) means much the same thing as in a standard English sentence. "If it is sunny, play golf, otherwise stay inside and work," would be coded in Pascal something like this:

>   **if** sunny
>   **then** golf
>   **else** work.

As in preceding sections, the keywords in boldface are defined terms in the Pascal language.

Important to note about the **if** statement is how the stated condition (in this case, "sunny") alters the flow of the program logic. The condition operates like a switch. If the condition is true, do one thing; if false, do another thing. Software developers sometimes refer to this feature of the **if** statement as performing "flow control" because the statement alters the logical flow of the program as railroad yard switches alter the flow of an approaching train.

When the switch is thrown defines an essential difference between the **if** statement and the other iteration forms discussed next. The **if** statement tests the condition **before** either succeeding option is performed. Thus it is possible that in a given situation, the **if** statement will do nothing at all. Consider this example:

if Saturday
then
begin
mow lawn;
go shopping;
wash car;
end
else
do nothing.

In this example nothing gets done, except on Saturday. After testing the condition, Saturday, program flow branches to the **else** do nothing statement for every day except Saturday. Although the **else** branch is shown here for clarity, most computer languages permit this "do nothing" statement to be omitted because "do nothing" is implied where no instruction is given.

[5] The *If-then* Loop Template

The **if-then** loop can also be equivalently expressed using **for** loop terminology discussed in § 10.05[B][3]. You may find the **if-then** terminology more natural. **If-then** statements can be used for many purposes besides iteration, including many examples of flow control, where program control follows different tracks depending on some switching condition. The template presented here demonstrates iteration.

1.   A method comprising:

[Step A];
[Step B];
setting X [a counter] to an initial condition;
if X is less than [not equal to] Y [the terminating condition], then performing the following:
[Step C],
[Step D],
[Step E], and
incrementing X;
[Step F].

2.   An apparatus comprising:

[Element A];
[Element B];
[iterator] element having a data structure for storing a counter initialized to a predetermined value, said [iterator] element being operable to iteratively increment said counter and to perform CDE if said counter value is less than [not equal to] Y [the terminating condition]
[Element F].

To illustrate this template, consider the following example.

1. A method of accessing information from Internet web pages having hypertext links to other web pages for off-line viewing, comprising:

connecting a computer to a service provider;
running a browser on said computer;
establishing a counter for storing values used in performing iteration;
setting said counter to an initial value and prompting the user to input a terminal value;
if the counter value is not equal to said terminal value then iteratively performing the following steps a-d:
    a.   accessing a web page and storing the information content of said web page in said computer;
    b.   identifying on said web page at least one hypertext link to a different web page;
    c.   accessing said different web page and storing the information content of said different web page in said computer;
    d.   incrementing said counter;
disconnecting said computer from said service provider; and
displaying said stored information content.

[C] Recursive Processes

Iteration, which involves repetition of a sequence, is not the only way to exploit the power of a computer. Indeed, software developers discovered long ago that some problems are very tedious to solve using iteration. Another approach, called *recursion*, exploits a different, divide-and-conquer strategy. Recursion involves subdividing a difficult problem into smaller, easier-to-solve problems that structurally resemble the big problem. If the smaller problems are still too difficult to solve, they may be subdivided into still smaller problems, ad infinitum, until the solution becomes easy.

Not all problems can be solved by recursion. Recursion requires that the problem meet these three criteria:

1. It must be possible to decompose the original problem into simpler instances of the same problem.

2. Once each of these simpler subproblems has been solved, it must be possible to combine these solutions to produce a solution to the original problem.

3. As the large problem is broken down into successively less complex ones, those subproblems must eventually become so simple that they can be solved without further subdivision.[47]

To better understand recursion, consider the task of looking up a word in the dictionary. Say the word is "sphinx." Clearly, using an iterative approach, starting at the beginning of the dictionary with the letter A and checking every entry *seriatim*, will be far too slow. A recursive approach is much faster.

Deciding to proceed recursively, you open the dictionary to the very middle and determine if "sphinx" is in the first half or the last half of the book. In this case, "sphinx" is in the last half, so you forgo further browsing of the first half and concentrate on the last half. Applying the very same divide-and-conquer technique, you subdivide the last half of the dictionary into halves, thereby opening the dictionary to the page beginning with the word "rumpy." From "rumpy," you determine that "sphinx" is in the latter half, so you forgo browsing the first half and proceed as before. Subdividing again takes you to the page beginning with "Themis." The word you seek, "sphinx," is now in the first half of this latest subdivision. You continue in this same fashion. After only a few more recursions you find your word "sphinx."

In Pascal-like pseudocode, a recursive process might look like the following.

```
procedure solve (problem);
begin
if problem is easy then
solve problem directly
else
begin
break problem into sub-problems (P_1, P_2 . . .)
solve(P_1); solve(P_2); . . .
reassemble solutions
end
end.
```

In this example, note that the steps "solve(P_1); solve(P_2); . . ." are themselves performed using the **procedure** "solve (problem)" defined by the pseudocode. It may seem strange that a procedure may call upon itself, but that is what occurs. The resulting code nests within itself like Russian matryoshka dolls,

---

[47] Eric S. Roberts, *Thinking Recursively* (John Wiley & Sons, 1986), p. 8.

sometimes many levels deep. The computer has no problem handling such requests because it has something called a control stack where it keeps a list of all unfinished tasks. Thus, when the innermost task is completed, the computer simply consults the control stack to get the numbers it needs to resume where it left off the next outermost problem.

[1] Recursion Template

A recursion process might look something like the template below:

1.   A method comprising:

[Step A];
[Step B];
defining a recursive procedure that includes steps a-d:
    a.   [Step C]
    b.   [Step D]
    c.   using said recursive procedure;
    d.   [Step E]
performing said recursive procedure.

2.   An apparatus comprising:

[Element A];
[Element B];
recursive processing element that performs ABC at least in part by recursively using said recursive processing element; and
[Element D].

Claiming a recursive algorithm can be tricky, particularly with the method claim, because you may encounter the antecedent basis difficulties. The following example shows how a recursive method claim may be constructed. As with the preceding examples, these templates are intended to show concepts, not magic words. In using these concepts, choose words that best fit the particular invention at hand.

1.   A method of generating a word list for a text containing a plurality of words, comprising:

storing said plurality of words as a binary tree having a root node and a plurality of offspring nodes that define left and right subtrees, said root node and said offspring nodes each being associated with one of said words;
defining a traverse procedure that includes the steps of:
    a.   visiting a node and printing its associated word;
    b.   recursively using said traverse procedure upon the left subtree of said visited node,
    c.   recursively using said traverse procedure upon the right subtree of said visited node;

applying said traverse procedure upon said root node such that said root node is the first visited node.

§ 10.06   Statutory Subject Matter Issues with Internet Patents

The statutory subject matter question is largely settled. Software patents can constitute statutory subject matter[48] as can business methods.[49] Nevertheless, some courts still have statutory subject matter concerns, particularly where the claims have broad-reaching implications. A case in point is *AT&T v. Excel Communications.*

In 1982, AT&T agreed to divest itself of local telephone companies as part of the consent decree in *United States v. AT&T.*[50] Since that time, local phone companies (local exchange carriers) and AT&T have operated independently. Under the "equal access" directive of the decree, callers must have the ability to presubscribe their telephones to a long-distance carrier (interexchange carrier) other than AT&T. Before the advent of "equal access," all interstate long-distance calls dialed on a 1+ basis were routed by the local exchange carrier to AT&T. A caller could reach another interexchange carrier only by dialing a seven-digit phone number, as well as a lengthy identification code, prior to dialing the called number. "Equal access" enabled callers to select a carrier other than AT&T to provide them long-distance phone service on a simple 1+ dialing basis. Thus any selected carrier could become the caller's "primary interexchange carrier" or PIC.

On May 6, 1992, AT&T inventors, Doherty, Lanzillotti, and Paulus, applied for a patent on an invention they termed, a "Call Message Recording For Telephone Systems." The system allowed AT&T to offer lower long-distance billing rates if both the calling party and the called party subscribed to AT&T as their primary interexchange carrier. The system worked by adding a PIC indicator to the standard exchange message. The PIC indicator identified which primary interexchange carrier the called party had selected. Using this PIC indicator information, and knowing the PIC of the calling party, AT&T could determine if both parties used AT&T as their primary long-distance carrier and could provide favorable billing treatment.

The patent issued July 26, 1994, with 41 method claims. Claim 1 is exemplary:

> 1. A method for use in a telecommunications system in which interexchange calls initiated by each subscriber are automatically routed over the facilities of a particular one of a plurality of interexchange carriers associated with that subscriber, said method comprising the steps of:
>   generating a message record for an interexchange call between an originating subscriber and a terminating subscriber, and

---

[48] *In re* Alappat, 33 F.3d 1526 (Fed. Cir. 1994).

[49] State Street Bank & Trust, Co. v. Signature Financial Group, 149 F.3d 1368 (Fed. Cir. 1998).

[50] 552 F. Supp. 131 (D.D.C. 1982).

including, in said message record, a primary interexchange carrier (PIC) indicator having a value which is a function of whether or not the interexchange carrier associated with said terminating subscriber is a predetermined one of said interexchange carriers.

Excel Communications is one of a number of resale carriers that contracts with facility owners to route their subscribers' calls through the facility-owners' switches and transmission lines. Excel began offering services with reduced rates if both parties used Excel as their primary interexchange carrier. AT&T sued Excel Communications for infringement in the U.S. District Court for Delaware. Excel moved for summary judgement on the grounds that the patent did not recite statutory subject matter.

Judge Robinson of the District Court of Delaware ruled in Excel's favor. She found that the claims implicitly recited a mathematical algorithm. Judge Robinson was mindful of the holding in *Diamond v. Diehr*,[51] that a claim drawn to statutory subject matter does not become nonstatutory simply because it uses a mathematical formula, computer program or digital computer. However, AT&T's claims, in Judge Robinson's view, lacked substance. Quoting from the *Diehr* decision, Judge Robinson wrote:

> Although the first element of these claims, generating a message record, is not "new" [fn omitted] but a standard practice in the industry, the court is directed to look at the claims as a whole to determine whether the process claimed "is performing a function which the patent laws were designed to protect (e.g., transforming or reducing an article to a different state or thing). . ."
>
> In this regard, the '184 patent can be described as claiming an invention whereby certain information that is already known within a telecommunications system (the PICs of the originating and terminating subscribers) is simply retrieved for an allegedly new use in billing. Characterized as such, the §101 inquiry employed, e.g., in *In re Grams*,[52] would dictate a finding of nonpatentability: Where "the only physical step [in the claim] involves merely gathering data for the algorithm," the subject matter is unpatentable.[53]

Judge Robinson thus granted Excel's motion for summary judgment, holding the patent invalid under 35 U.S.C. § 101.

AT&T appealed to the Federal Circuit Court of Appeals and that Court reversed. Judge Plager delivered the opinion. In it, he drew upon the Court's prior decisions in *Alappat*[54] and *State Street Bank*[55] which had both reached the conclusion that the presence of a mathematical algorithm or mathematical

[51] 450 U.S. 185 (1981).

[52] 888 F.2d 835, 839 (Fed. Cir. 1989).

[53] AT&T Corp. v. Excel Communications, Inc., 1998 WL 175878 (D. Del. 1998).

[54] *In re* Alappat, 33 F.3d 1526 (Fed. Cir. 1994).

[55] 149 F.3d 1368 (Fed. Cir. 1998).

formula in a claim did not *ipso facto* render the claim nonstatutory under § 101. However, both *Alappat* and *State Street Bank* involved apparatus claims. As such, the claims recited physical structure. However, the claims at issue in this case were method claims — no physical structure was directly recited. Thus, arguably, the claims did not require the transforming or reducing of an article to a different state or thing. Was Judge Robinson correct? Did the lack of a physical transformation render the claims nonstatutory?

Judge Plager answered this question, no. First, he made it clear that § 101 is to be applied the same, no matter whether apparatus or method claims are involved:

> In both *Alappat* and *State Street Bank*, the claim was for a machine that achieved certain results. In the case before us, because Excel does not own or operate the facilities over which its calls are placed, AT&T did not charge Excel with infringement of its apparatus claims, but limited its infringement charge to the specified method or process claims. Whether stated implicitly or explicitly, we consider the scope of § 101 to be the same regardless of the form — machine or process — in which a particular claim is drafted.[56]

As to the physical transformation, which District Judge Robinson had found lacking in AT&T's claims, Judge Plager clarified that physical transformation is not a requirement under 35 U.S.C. § 101. He reached this conclusion by carefully analyzing the words of the Supreme Court in *Diamond v. Diehr*:[57]

> Excel argues that method claims containing mathematical algorithms are patentable subject matter only if there is "physical transformation" or conversion of subject matter from one state into another. The physical transformation language appears in *Diehr* . . . and has been echoed by the court in Schrader[58] . . .
>
> The notion of "physical transformation" can be misunderstood. In the first place, it is not an invariable requirement, but merely one example of how a mathematical algorithm may bring about a useful application. As the Supreme Court itself noted, "when [a claimed invention] is performing a function which the patent laws were designed to protect (*e.g.*, transforming or reducing an article to a different state or thing), then the claim satisfies the requirements of § 101" . . . (emphasis added). The "e.g." signal denotes an example, not an exclusive requirement.[59]

If a physical transformation is but one indicia that statutory subject matter is present, are there any others? Judge Plager answered this question by reference to

---

[56] AT&T Corp. v. Excel Communications, Inc., 172 F.3d 1352, 1357 (Fed. Cir. 1999).

[57] 450 U.S. 185 (1981).

[58] *In re* Schrader, 22 F.3d 290, 294 (Fed. Cir. 1994).

[59] AT&T Corp., 172 F.3d at 1358-1359.

the Court's earlier decision in *Arrhythmia*[60] in which the presence of "transformed" data could be an indicia of statutory subject matter, if the data had a specific meaning that gave a useful, concrete and tangible result, as opposed to a mere mathematical abstraction.

> The finding [in *Arrhythmia*] that the claimed process "transformed" data from one "form" to another simply confirmed that Arrhythmia's method claims satisfied § 101 because the mathematical algorithm included within the process was applied to produce a number which had specific meaning — a useful, concrete, tangible result — not a mathematical abstraction.[61]

The result reached by the Court in this case left hanging the prior decisions in *In re Schrader*[62] and *In re Grams*.[63] In both of those cases, the court had relied upon the *Freeman-Walter-Abele* test, which involved the two-step process of first, analyzing the claims to determine if a mathematical algorithm was recited and, second, assessing whether the claims as a whole preempted all use of the mathematical algorithm. The *Freeman-Walter-Abele* test thus focused sharply on the mathematical algorithm and did not require a court to assess whether a useful, concrete, and tangible result ensued. Because the *Schrader* and *Grams* courts had applied the *Freeman-Walter-Abele* test and had not assessed whether a useful, concrete, and tangible result ensued, Judge Plager rejected the *Schrader* and *Grams* cases as "unhelpful."

It would seem that any computer program that "transforms" data, which arguably all programs do, would constitute statutory subject matter under Judge Plager's analysis. To prevent the jump to that conclusion, Judge Plager revisited the court's decision in *In re Warmerdam*,[64] a decision which he authored. In that case, the Court had rejected as nonstatutory Warmerdam's method claims for generating bubble hierarchies through the use of a particular mathematical procedure. The Court found that the claimed process did nothing more than manipulate basic mathematical constructs, and concluded that "taking several abstract ideas and manipulating them together adds nothing to the basic equation."[65]

The decision in Warmerdam is troubling because the Warmerdam "bubble" data structure was described as being useful for controlling the motion of objects and machines to avoid collision with other moving or fixed objects. That would seem to place the subject matter within the domain of the useful, concrete, and tangible result. However, as Judge Plager noted, the *Warmerdam* Court concluded on the facts that the claims did not encompass such useful subject

---

[60] Arrhythmia Research Technology, Inc. v. Corazonix Corp., 958 F.2d 1053 (Fed. Cir. 1992).
[61] AT&T Corp., 172 F.3d at 1359.
[62] 22 F.3d 290 (Fed. Cir. 1994).
[63] 888 F.2d 290 (Fed. Cir. 1994).
[64] 33 F.3d 1354 (Fed. Cir. 1994).
[65] *Id.* at 1360.

matter but rather the claims encompassed the manipulation of basic mathematical constructs. About *Warmerdam* Judge Plager wrote:

> The court found that the claimed processes did nothing more than manipulate basic mathematical constructs and concluded that "taking several abstract ideas and manipulating them together adds nothing to the basic equation"; hence, the court held that the claims were properly rejected under § 101 . . . Whether one agrees with the court's conclusion on the facts, the holding of the case is a straightforward application of the basic principle that mere laws of nature, natural phenomena, and abstract ideas are not within the categories of inventions or discoveries that may be patented under § 101.[66]

Warmerdam involved a computer data structure that represented basic relationships among objects. To the extent the claims called for a transformation of data, the transformation, according to Judge Plager's analysis, did not involve a useful, concrete, and tangible result. This suggests that computer data structures that represent laws of nature, natural phenomena, and abstract ideas remain off limits to patent protection.

Warmerdam's method claim read as follows:

> 1. A method for generating a data structure which represents the shape of [sic] physical object in a position and/or motion control machine as a hierarchy of bubbles, comprising the steps of:
> first locating the medial axis of the object and then creating a hierarchy of bubbles on the medial axis.

Warmerdam's apparatus claim was of similar scope:

> 5. A machine having a memory which contains data representing a bubble hierarchy generated by the method of any of Claims 1 through 4.

Both of these claims make superficial reference to a position or motion control machine. Nevertheless, the Court found that Warmerdam was actually claiming an abstract relationship.

§ 10.07 Internet Claim Drafting Considerations

Internet inventions lend themselves to a variety of claiming techniques. At a top level, the four classes of statutory subject matter — process, machine, manufacture, or composition of matter — represent a good place to begin the claim drafting analysis. Internet inventions may readily fit into three of these classes. An Internet-enabled system or computer program that communicates with the Internet may be expressed as a machine or apparatus claim. This includes business systems, which represent an important sector of all Internet patents. A process or method may also include one or more steps that engage the

---

[66] *Id.* at 1360.

Internet. Many business methods today employ one or more process steps performed using the Internet. Thus, it is clear that apparatus and method claims may be used to define Internet inventions. The "manufacture" subject matter may not appear as clear cut, but it is a very important type of claim to understand.

The Federal Circuit Court decision in *In re Lowry*[67] and the Patent Office's acceptance of computer program product claims in *In re Beauregard*[68] established that computer software may be patented as an article of manufacture. In such claims, the computer-readable memory containing the computer code served as the manufactured article. A further extension of this claiming concept originated from the Patent Office itself — the propagated signal claim. Under this claiming theory, computer code embodied in a carrier wave (propagating signal) is a statutory "manufacture." This may seem surprising at first blush, for it does not seem quite right that a carrier wave could be an article of manufacture. However, upon more careful reading of the statute, it is clear that the statutory subject matter is not limited to "articles." The statutory language at § 101 states:

> Whoever invents or discovers any new and useful process, machine, manufacture, or composition of matter, or any new and useful improvements thereof, may obtain a patent therefor, subject to the conditions and requirements of this title.[69]

In some decisions, courts have used the term "article of manufacture" to refer to the third-listed class of statutory subject matter. However, as the statutory language plainly shows, the word "article" does not appear in the statute. The following section will further explain the propagated signal claim.

§ 10.08   Propagated Signal "Carrier Wave" Claims

After the Patent Office promulgated its software patent examination guidelines, *Examination Guidelines for Computer-Related Inventions*, in June 1995, officials at the Patent Office determined that additional training materials would be helpful in explaining to its Patent Examiners how to follow the *Guidelines*. The Patent Office thus released a set of training materials that gave specific claim examples, describing how to analyze each in accordance with the *Guidelines*. One of the examples provided was of a computer data signal embodied in a carrier wave. In that example, the Patent Office showed its willingness to treat signals containing computer code as statutory "manufacture" subject matter. This type of claim may be quite important in claiming Internet inventions, as we shall see.

The training example that may well be the genesis of the propagated signal claim is claim 13 from the *Guidelines* training materials. An example of an automated manufacturing plant, reproduced on the next page are excerpted.

---

[67] 32 F.3d 1579 (Fed. Cir. 1994).

[68] 53 F.3d 1583 (Fed. Cir. 1995).

[69] 35 U.S.C. § 101, July 19, 1952, ch 950, § 1.66 Stat. 797.

EXAMINATION GUIDELINES FOR COMPUTER-RELATED
INVENTIONS

Example: AUTOMATED MANUFACTURING PLANT

Disclosure for Claims 1, 3-6 and 8-13

The invention relates to a data compression and encryption system for
monitoring and controlling an automated manufacturing process. The system
translates the outputs of various sensors from an automated plant's
manufacturing process into digital data signals. The system then processes
the digital data signals into a compressed signal of various length codewords,
encrypts the compressed signal, and transmits the compressed and encrypted
signal to a remote supervisory location. At the remote supervisory location,
the signal is decrypted and decompressed. The remote supervisory location
then compares the decrypted and decompressed digital data signals to the
preset ranges for the respective operating parameters of the automated plant's
manufacturing process, generates a digital correction signal on the basis of
the comparison, compresses and encrypts the correction signal, transmits the
correction signal back to the plant location, and applies the correction signal
to the disclosed process controllers, such as valves and motors, to maintain
the automated plant's operation within its design parameters.

The automated plant's manufacturing process is controlled with a
general purpose computer system. In the plant's general purpose computer
system, various memory sections are included to store the plant's operating
parameters and the sensor's outputs. The plant's various sensors and sensing
systems are disclosed.

The remote supervisory location's process is implemented on a general
purpose computer system. The remote supervisory location's general purpose
computer system must have the identical compression and encryption
capabilities of the automated plant's general purpose computer system.

The general purpose computer systems of the automated manufacturing
plant and the remote supervisory location are programmed by a data signal
transmitted from a remote main office location. The data signal includes a
carrier wave and the source code segments for both the compression and
encryption computer programs.

In the preferred embodiment for data compression, the general purpose
computer system at each site is programmed with a computer program to
compress/decompress a digital signal into variable length codewords in
accordance with the Huffman code algorithm. The general purpose computer
system has both an encoder and a decoder on which are stored identical
Huffman code books. The use of compressed signals allows for reduced
transmission time between the sites.

In the preferred embodiment for data encryption, the general purpose
computer system at each site is programmed with a separate computer
program to encrypt/decrypt a digital signal in accordance with the Data
Encryption Standard (DES) algorithm. The DES algorithm uses an encryption key stored in a read-only memory to produce a digital signal whose
content is protected and secured for transmission. In another embodiment for
data encryption, the general purpose computer system has an application
specific integrated circuit (ASIC). The various components of the ASIC are
incorporated by reference from U.S. Patent No. *,***,***.

The disclosure contains both self-documenting source code for the preferred embodiments of the computer programs and high-level written descriptions of the computer programs with flow charts. There is correspondence between the written descriptions, the flow charts, and the specific software. The disclosure states that alternate computer programs based on the high-level written descriptions and flow charts are within the skill of a routineer in the art.[70]

The example presented several claims. Pertinent to this discussion is Claim 13:

A computer data signal embodied in a carrier wave comprising:
   a.   a compression source code segment comprising . . . [recites self-documenting source code]; and
   b.   an encryption source code segment comprising . . . [recites self-documenting source code].[71]

The example then provides analysis, in table form, showing how the examiner should apply the *Guidelines* to the claim. In the table, reproduced on the next page, there are references to several boxes (Box 2, Box 6, Box 8, etc.). These boxes, as well as the associated questions, refer to the examination flowchart comprising part of the *Guidelines* and reproduced here as Figure 10-2.

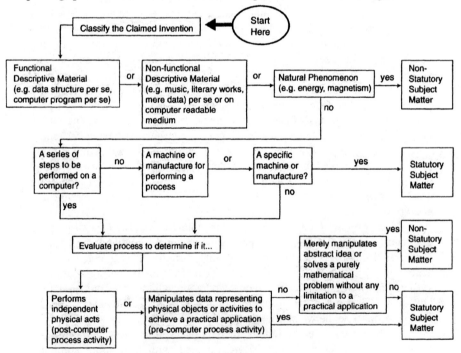

FIGURE 10-2:   PATENT EXAMINATION FLOWCHART

---

[70] Examination Guidelines For Computer-Related Inventions, Example: Automated Manufacturing Plant, page 1, March 28, 1996.
   [71] *Id.* at p. 38.

Chart Showing How to Analyze Example Claim 13[72]

BOX 2	Q.2a. Does disclosed invention have practical application?	YES	Go To: Q.2b	Note 1
	Q.2b. Is disclosed invention in technological arts?	YES	Go To: Q.6a	Note 2
BOX 6	Q.6a. Is claimed invention a computer program per se?	NO	Go To: Q.6b	Note 3
	Q.6b. Is claimed invention a data structure per se?	NO	Go To: Q.6c	
	Q.6c. Is claimed invention non-functional descriptive material?	NO	Go To: Q.6d	
	Q.6d. Is claimed invention a natural phenomenon?	NO	Go To: Q.8	Note 4
BOX 8	Q.8. Is claimed invention a series of steps to be performed on a computer?	NO	Go To: Q.9	
BOX 9	Q.9. Is claimed invention a product for performing a process?	YES	Go To: Q.10	
BOX 10	Q.10. Is claimed invention a specific machine or manufacture?	YES	Go To: END	Note 5
BOX 12	Q.12a. Does process have post-computer process activity?		Go To:	
	Q.12b. Does process have pre-computer process activity?		Go To:	
BOX 13	Q.13a. Does process manipulate abstract idea w/o limitation to a practical application?		Go To:	
	Q.13b. Does process solve math problem w/o limitation to a practical application?		Go To:	

Table Notes for Claim 13

- Note 1: Disclosed invention monitors and controls an automated plant's manufacturing process.

- Note 2: Disclosed invention uses a general purpose computer system.

- Note 3: Claimed invention recites specific software embodied on a computer-readable medium, i.e., specific software embodied in a carrier wave.

---

[72] *Id.* at p. 39. Patent Office at www.uspto.gov/web/offices/pac/dapp/oppd/patco.htm.

- Note 4: Most likely, the "data signal" does not occur as a natural phenomenon. The Examiner bears the burden of establishing that a claimed invention is a natural phenomenon. Therefore, absent object evidence to support the position that the "data signal" is a natural phenomenon, such a position would be untenable.
- Note 5: Claimed invention recites specific software. See *Guidelines*, Section IV.B.2(a)(ii).[73]

The example concluded with a more detailed analysis of each claim. The following excerpt further analyzes claim 13.

> Claim 13 is an article of manufacture claim. It recites a computer program with two claim limitations:
>
> a. Element a. recites a specific source code segment for compression; and
>
> b. Element b. recites a specific source code segment for encryption.
>
> Reviewed as a whole, and given its broadest reasonable interpretation, the claim is limited to a specific article of manufacture. Also, the computer program is embodied on a computer-readable medium — the carrier wave. Thus, claim 13 is a statutory article of manufacture claim.[74]

Note that the example even characterizes the carrier wave as an article of manufacture. That is not to say that all propagating waves are statutory articles of manufacture, however. Keep in mind that a naturally occurring propagating wave, sunlight for example, is a natural phenomenon and is thus off limits to patentability.

The propagated signal claim can be a very important tool in the patent claim drafter's toolkit. The claim can be used, for example, to cover an Internet application where inventive computer code or data is downloaded from server to client. Under the statutory provisions governing infringement, it is an infringement to make, use, offer to sell or sell a patented invention within the United States or to import the patented invention into the United States.[75] Making a patented "carrier wave" manufacture available for download from a Web site could constitute an offer to sell and hence would be covered by the infringement statute. Furthermore the transmission of a patented "carrier wave" manufacture into the United States from an offshore Web site could constitute an infringing importation as well.

§ 10.09   Internet Patent Claim Templates

As discussed above, Internet inventions can present fairly tricky jurisdiction and infringement issues. This is because the Internet permits technological components to be distributed throughout the world and even into space. Therefore, give care when drafting claims to cover the key components in various

---

[73] *Id.* at p. 41.

[74] *Id.* at p. 45.

[75] 35 U.S.C. § 271.

different implementations. Focus on each client, server, and service provider entity in the system alone, as if it will be the sole target of an infringement suit. Draft claims accordingly. While overall system claims may be worth including in a patent, to increase the royalty base in case contributory infringement or inducing infringement may be found, do not limit the patent application to only system claims. Also, be alert to avoid undue use of interactive method claims, where different entities perform different steps of a claimed method. These can be difficult to enforce.

While there are as many different claiming styles as there are different inventions, the following templates may help you solve some recurring Internet patent claim drafting problems.

[A] Claim Drafting Templates for an Internet Server Entity

The server often represents the entity that is most important when seeking effective injunctive relief. If the server can be shut down, often the infringing activity will be terminated. Thus it makes sense to employ a full arsenal of claiming techniques: method claims, apparatus claims, data structure (computer memory) claims, and carrier wave claims directed to a novel server entity. The following claim templates will help in drafting these claims.

1. A method for producing . . . (a useful, concrete and tangible result) . . . comprising:

providing at least one server computer in communication with a computer network;
[Step A] (performed by or in association with the server, e.g., providing input to the server, processing data by the server, generating output from the server);
[Step B] (performed by or in association with the server);
[Step C]
etc.

Note: Preferably, recite each step in relation to the server. That way the claim will be more readily enforced against the party operating the server. Examples include steps that provide input to the server, steps that process data within the server or within components communicating with the server, and steps that provide output from the server.

2. An apparatus, comprising:

at least one server computer having an interface for communicating over a computer network;
[Element A] (associated with the server);
[Element B] (associated with the server);
etc.

Note: Recite the essential components of the apparatus being claimed. If the apparatus performs an information serving function, a database component will often be included.

> 3.   A memory for storing data for access by an application program being executed on a data processing system, comprising:
>
> > a data structure stored in said memory, said data structure including . . .
> > [Element A] (further details of the data structure);
> > [Element B] (further details of the data structure);
> > etc.

Note: The data structure claim is actually one variant of an article of manufacture claim. The claim is modeled after the computer memory claims that were approved by the Federal Circuit Court of Appeals in *In re Lowry*.[76] Other variations include computer program product claims such as the claims at issue in *Ex parte Beauregard*.[77] The *Beauregard* claim is similar to the above template with the following different variations in claim preambles:

> 4.   An article of manufacture comprising:
>
> > a computer usable medium having computer readable program code embodied therein . . .
> > [Element A] (further details of the program code);
> > [Element B] (further details of the program code);
> > etc.

> 5.   A computer program product for use with a computer, said computer program product comprising:
>
> > a computer usable medium having computer readable program code embodied therein . . .
> > [Element A] (further details of the program code);
> > [Element B] (further details of the program code);
> > etc.

> 6.   A computer data signal embodied in a carrier wave, comprising:
>
> > [Element A] (further details of the program code)
> > [Element B] (further details of the program code);
> > etc.

Note: This claim is based on the training examples provided in the Patent Office Examination Guidelines discussed in a preceding section of this chapter.

---

[76] *Supra,* note 67.
[77] Appeal No. 93-0378, slip opinion (Bd. App. 1993).

Another useful technique, particularly with Internet applications, is to claim the user interface as a component. Currently most user interface inventions involve one or more objects displayed on a computer screen. In the future, multi-modal user interfaces may be anticipated, such as natural language speech interfaces. The typical user interface claim includes (a) a user interface "object" or device (which includes displayed objects and speech); (b) a method or functional capability associated with the user interface object or device; and (c) a response or result produced when the user interacts with the object or device. User interfaces may be generated by the server or host computer; alternatively user interfaces may be generated by the client application. A claim to a user interface associated with a server may be modeled according to the following template:

> 1. A system for performing . . . (a useful, concrete, and tangible result), comprising:
>
> a server computer hosting . . . [functionality A] . . . accessible via a computer network by at least one client computer;
> said server providing a user interface . . . [describe user interface object or device] . . . whereby a user [interacts with the object or device to invoke some operation associated with functionality A].

[B] Claim Drafting Templates for an Internet Web Site Entity

A Web site entity will often be covered by the server claims discussed in the preceding section. However, there are situations where a Web site simply provides a link to the server that itself provides additional data processing functionality. For example, a Web site that hosts an e-commerce application may link to a server that provides a catalog lookup function, or a language translation function. To target the claims to the Web site, you may use the templates in the preceding section with appropriate modification. Links to other servers may be handled by reciting them as links associated with the claimed Web site, rather than reciting them as separate server elements in the claims. To illustrate:

> 1. A method for producing . . . (a useful, concrete, and tangible result) . . . comprising:
>
> providing Web site hosted by at least one computer in communication with a computer network;
> [Step A] (performed by or in association with the Web site);
> [Step B] (performed by or in association with the Web site);
> [Step C]
> etc.

If it is necessary to recite functionality that may be performed by a separate server that is not part of the Web site host, this may be done by reciting the component of the Web site that makes the link or communicates with the separate server. For example, in the above template, [Step C] might read:

"communicating with a server based on a link defined by said Web site to verify the buyer's credit history . . ."

[C] Claim Drafting Templates for an Internet Client Entity

The claim drafting templates described above may be readily adapted to cover an Internet client entity. Again, when claiming the client entity the focus should be on aspects that are performed by or in association with the client entity.

1.   A method for producing . . . (a useful, concrete, and tangible result) . . . comprising:

> providing at least one client computer in communication with a computer network;
> [Step A] (performed by or in association with the client);
> [Step B] (performed by or in association with the client);
> [Step C]
> etc.

2.   An apparatus, comprising:

> at least one server computer having an interface for communicating over a computer network;
> [Element A] (associated with the server);
> [Element B] (associated with the server);
> etc.

Note: Preferably, recite each step or element in relation to the client entity. That way the claim will be more readily applied to users of the client application. Examples include steps or elements that provide input to the client, steps or elements that process data within the client or within components communicating with the client, and steps or elements that provide output from the client to other applications.

Another useful technique, particularly with client applications, is to claim the user interface. Currently most user interface inventions involve one or more objects displayed on a computer screen. In the future, multi-modal user interfaces may be anticipated, such as natural language speech interfaces. The typical user interface claim includes (a) a user interface "object" or device (which includes displayed objects and speech); (b) a method or functional capability associated with the user interface object or device; and (c) a response or result produced when the user interacts with the object or device. A claim to a user interface invention may be claimed according to the following template:

1.   A system for performing . . . (a useful, concrete, and tangible result), comprising:

> a client computer running . . . [a client application] . . . and having access to at least one server computer via a computer network

. . . [the client application and/or server computer providing functionality A] . . . ;

said client application providing a user interface . . . [describe
user interface object or device] . . . whereby a user [interacts with the
object or device to invoke some operation associated with functionality A].

Note: The client application could be a simple web browser, or it could be a
custom application designed for a specialized purpose. If a simple web browser is
employed, it is likely that user interaction with the user interface object or device
will invoke functionality associated with the server. Of course, this will depend
on the system being claimed.

§ 10.10   Some Specific Examples of Internet Patents

No chapter on claiming e-commerce patents would be complete without
some examples of actual Internet patents. The following is a sampler of Internet
patents, showing how others have approached the claim drafting exercise. Some
of the patents included have been considered newsworthy by the press or have
been the subject of litigation; others simply represent examples of typical Internet
inventions, employing one or more of the components discussed at the beginning
of this chapter.

[A]   Amazon.com One-click Shopping

United States Patent No. 5,960,411
Hartman, et al.
Issued: September 28, 1999
Method And System For Placing A Purchase Order Via A
Communications Network

Abstract

A method and system for placing an order to purchase an item via the
Internet. The order is placed by a purchaser at a client system and received
by a server system. The server system receives purchaser information
including identification of the purchaser, payment information, and
shipment information from the client system. The server system then
assigns a client identifier to the client system and associates the assigned
client identifier with the received purchaser information. The server system
sends to the client system the assigned client identifier and an HTML
document identifying the item and including an order button. The client
system receives and stores the assigned client identifier and receives and
displays the HTML document. In response to the selection of the order
button, the client system sends to the server system a request to purchase
the identified item. The server system receives the request and combines the

purchaser information associated with the client identifier of the client system to generate an order to purchase the item in accordance with the billing and shipment information whereby the purchaser effects the ordering of the product by selection of the order button.

Representative Claim

1.   A method of placing an order for an item comprising:

under control of a client system, displaying information identifying the item; and

in response to only a single action being performed, sending a request to order the item along with an identifier of a purchaser of the item to a server system;

under control of a single-action ordering component of the server system, receiving the request;

retrieving additional information previously stored for the purchaser identified by the identifier in the received request; and

generating an order to purchase the requested item for the purchaser identified by the identifier in the received request using the retrieved additional information; and

fulfilling the generated order to complete purchase of the item whereby the item is ordered without using a shopping cart ordering model.

[B] Priceline.com Buyer-driven Purchasing

United States Patent No. 6,041,308
Walker, et al.
Issued: March 21, 2000
System And Method For Motivating Submission Of Conditional Purchase Offers

Abstract

A system and method are disclosed for encouraging buyers to submit CPOs to a CPO management system for a desired product. The CPO management system processes each received CPO to determine whether one or more sellers are willing to accept a given CPO. The disclosed CPO management system compensates buyers if the buyer's conditional purchase offer is rejected, or expires before an acceptance is received. If a CPO is rejected by the sellers, or has expired before an acceptance is received, the CPO management system evaluates one or more stored compensation offers to determine if the buyer is eligible for rejection compensation. The compensation offers may optionally require that the conditional purchase offer satisfies one or more additional predefined eligibility criteria. If the predefined criteria is met, the rejection compensation is provided to the buyer. The rejection compensation can include, for example, (i) a cash award, (ii) a prize, or (iii) a coupon or credit that may be redeemed for a discount against future transactions, thereby encouraging future use.

Representative Claim

1. A method of using a computer to compensate buyers who make conditional purchase offers, comprising:

receiving, using a computer, a conditional purchase offer from a buyer for a product, said conditional purchase offer containing at least one buyer-defined condition and a variable condition;

processing said conditional purchase offer to determine if said conditional purchase offer is accepted by a seller; and

compensating said buyer if said conditional purchase offer is not accepted by a seller.

[C] Konrad Client-server Model

United States Patent No. 5,974,444
Konrad
Issued: October 26, 1999
Remote Information Service Access System Based On A Client-server-service Model

Abstract

A local host computing system, a remote host computing system as connected by a network, and service functionalities: a human interface service functionality, a starter service functionality, and a desired utility service functionality, and a Client-server-service (CSS) model is imposed on each service functionality. In one embodiment, this results in nine logical components and three physical components (a local host, a remote host, and an intervening network), where two of the logical components are integrated into one Remote Object Client component, and that Remote Object Client component and the other seven logical components are deployed among the local host and remote host in a manner which eases compatibility and upgrade problems, and provides an illusion to a user that a desired utility service supported on a remote host resides locally on the user's local host, thereby providing ease of use and minimal software maintenance for users of that remote service.

Representative Claim

1. A remote access apparatus for providing end-user access through a human interface to a desired remote utility service on a remote host computer, comprising:

    a)   a local host computer;

    b)   a remote host computer;

    c)   a network connection between said local host computer and said remote host computer allowing data transfer there between;

wherein said local host computer further comprises:

    1)   a human interface service means, for handling input from, and output to, an end-user;

    2)   a human interface server, for mediating requests for human interface services, said requests from human interface clients resident on at least one of said remote host computer and said local host computer, said human interface server operative to process said requests from said human interface clients during normal operation and exception operation; and

    3)   a starter client means, for issuing requests to a starter server means on said remote host computer, said requests for initiating interaction with the desired remote utility service on said remote host computer;

wherein said remote host computer further comprises:

    1)   said starter server means, for responding to requests from said starter client means;

    2)   a desired remote utility service, resident on said remote host computer and platform-independent of said local host computer;

    3)   a remote object client, for issuing requests for human interface services to said human interface server, for issuing requests for said desired remote utility service and for translating a response from said desired remote utility service into a request for human interface services issued to said human interface server; and

    4)   a starter service means, for initiating a remote object client indicated by said starter server means.

[D] AT&T Authentication Using Cookies

United States Patent No. 6,047,268
Bartoli, et al.
Issued: April 4, 2000
Method And Apparatus For Billing For Transactions Conducted Over The Internet

Abstract

A method and apparatus for authenticating transactions accomplished over a data network utilizes a ''cookie'' containing both static information (user-identifying information) and dynamic information (transaction-based information). The transaction-oriented dynamic information portion comprises a random number and a sequence number, the latter tracking the number of billing transactions conducted by the user associated with the account number. The cookie, sent to the user's cookie file upon a previous transaction, is valid for only a single new transaction. A billing server, upon receiving the cookie containing the static and dynamic information portions, identifies the user from the account number in the static portion and accesses from an associated database the expected random number and sequence number that the billing server last sent to that user in the transaction-oriented dynamic portion. If the expected dynamic portion matches the received dynamic portion, the user is authenticated to proceed with the current transaction.

Representative Claim

1. A method of authenticating a user for a transaction on a data network comprising:

sending to a user's client terminal data containing a static information portion and a transaction-oriented dynamic portion, the static information portion identifying an account associated with the user and the transaction-oriented dynamic information portion containing information generated for that user that is valid for a single subsequent transaction;

storing the transaction-oriented dynamic information portion in association with the static information portion;

receiving, from the user's client terminal, the data containing the static information portion and the transaction-oriented dynamic information portion in association with information relating to the single subsequent transaction;

identifying the user's account from the received static information portion;

comparing the transaction-oriented dynamic information portion received from the user's client terminal with the transaction-oriented dynamic information portion stored in association with the static information portion; and

authenticating the user for the single subsequent transaction if the received transaction-oriented dynamic information portion matches the stored transaction-oriented dynamic information for the account associated with the user.

[E] Segal Interactive Analysis for Making Investment Choices

United States Patent No. 6,049,783
Segal, et al.
Issued: April 11, 2000
Interactive Internet Analysis Method

Abstract

A client establishes and/or modifies an interactive account on a server via proprietary sorting and filtering and reporting criteria as a means for timely processing of online financial data and/or other business information to retrieve valuations, sorted lists, etc. The method involves establishing a link with a server preset and programmable with client criteria for investment or decision making, acquiring data, sorting, filtering, etc.

Representative Claim

1.   A method for processing data within a server, with the steps:

(1)   establishing a filtering and sorting system for a client within a separate account on an interactive server;

(2)   setting and modifying upon request filtering and sorting and reporting parameters for each account;

(3)   providing timely access to a source of data and/or data bases as a means for filtering and sorting data and retrieving valuations, or other client information, graph or report; and

(4)   maintaining access to data and data bases for all accounts up to the level of service paid in full at the time of a request for service.

[F]   Amazon.com Merchant Referral System

United States Patent No. 6,029,141
Bezos, et al.
Issued: February 22, 2000
Internet-based Customer Referral System

Abstract

Disclosed is an Internet-based referral system that enables individuals and other business entities (''associates'') to market products, in return for a commission, that are sold from a merchant's Web site. The system includes automated registration software that runs on the merchant's Web site to allow entities to register as associates. Following registration, the associate sets up a Web site (or other information dissemination system) to distribute hypertextual catalog documents that includes marketing information (product reviews, recommendations, etc.) about selected products of the merchant. In association with each such product, the catalog document includes a hypertextual ''referral link'' that allows a user (''customer'') to link to the merchant's site and purchase the product. When a customer selects a referral link, the customer's computer transmits unique IDs of the selected product and of the associate to the merchant's site, allowing the merchant to identify the product and the referring associate. If the customer subsequently purchases the product from the merchant's site, a commission is automatically credited to an account of the referring associate. The merchant site also implements an electronic shopping cart that allows the customer to select products from multiple different Web sites, and then perform a single "check out" from the merchant's site.

Representative Claim

7. A method of selling items with the assistance of associates, the method comprising:

providing a Web site system that includes a browsable catalog of items and provides services for allowing customers to electronically purchase the items;

providing a database which contains information about a plurality of associates that select and recommend items from the catalog within respective areas of expertise, at least some of the associates operating associate Web sites that include item-specific links to the Web site system;

receiving from a computer of a customer a request message which contains an associate identifier and an item identifier and extracting the associate and item identifiers from the message, the request message generated in response to selection by the customer of a link of an associate Web site, the link provided in conjunction with a recommendation of the item by an associate;

transmitting to the customer's computer a Web page which corresponds to the item identifier extracted from the request message;

transacting a sale of the item and/or other items of the catalog with the customer through the Web site system;

using the associate identifier extracted from the request message and the database to identify the associate; and

compensating the associate for the sale.

[G] Scan Technology Factory Automation

United States Patent No. 6,038,486
Saitoh, et al.
Issued: March 14, 2000
Control Method For Factory Automation System

Abstract

A method of operating, controlling, monitoring and analyzing data of control devices used in the manufacturing devices or equipment of a factory automation system is disclosed. According to the method, the system controls itself and executes operations by reading as necessary in real time data in the form of files saved on a memory medium of each type of control device used in manufacturing devices or equipment.

Representative Claim

1.   A control method for a factory automation system that controls itself and effects the operation of control devices used in manufacturing devices or equipment by reading, as necessary while said control devices are in operation, data in the form of files saved on a memory medium for each type of said control devices, wherein centralized control of said control devices is effected by providing a server for each network and transferring and updating as necessary within the server each type of file for all control devices connected to the server.

[H]   Sun Microsystems Article of Manufacture Example

United States Patent No. 6,052,711
Gish
Issued: April 18, 2000
Object-Oriented System, Method And Article Of Manufacture For A Client-server Session Web Access In An Interprise Computing Framework System.

Abstract

An interprise [Internet Enterprise] computing manager in which an application is composed of a client (front end) program which communicates utilizing a network with a server (back end) program. The client and server programs are loosely coupled and exchange information using the network. The client program is composed of a User Interface (UI) and an object-oriented framework (Presentation Engine (PE) framework). The UI exchanges data messages with the framework. The framework is designed to handle two types of messages: (1) from the UI, and (2) from the server (back end) program via the network.

The framework includes a component, the mediator which manages messages coming into and going out of the framework. The system includes software for a client computer, a server computer and a network for connecting the client computer to the server computer which utilize an execution framework code segment configured to couple the server computer and the client computer via the network, by a plurality of client computer code segments resident on the server, each for transmission over the network to a client computer to initiate coupling; and a plurality of server computer code segments resident on the server which execute on the server in response to initiation of coupling via the network with a particular client utilizing the transmitted client computer code segment for communicating via a particular communication protocol. Communication is initiated utilizing the network to acquire characteristics of the client from the network.

Representative Claim

19.   A computer program embodied on a computer-readable medium for enabling a distributed computer system, comprising:

(a)   a client computer code segment for residence on a client computer and including an user interface;
(b)   a server computer code segment for residence on a server computer coupled to the client computer;
(c)   a code segment to enable a connection to a network connecting a plurality of client computers to a plurality of server computers;
(d)   a code segment which executes a client computer application on the client computer which gathers information about the client computer and contacts the server computer using the network;
(e)   a code segment which authenticates the information about the client computer and initiates the server computer code segment responsible for communicating with the client computer; and
(f)   a code segment which transmits the client computer code segment for execution at the client computer to facilitate communication between the client computer and the server computer.

[I]   Sun Microsystems Carrier Wave Claim Example

United States Patent No. 5,850,449
McManis
Issued: December 15, 1998
Secure Network Protocol System And Method

Abstract

A computer network having first and second network entities. The first network entity includes a packet object generator that generates a packet object including an executable source method, an executable destination method, and data associated with each of the methods. The first network entity also includes a communications interface to transmit the packet object. The second network entity includes a communications interface to receive the packet object and an incoming packet object handler to handle the received packet object. The incoming packet object handler includes a source and destination verifier to execute the source and destination methods with their associated data so as to verify the source and destination of the received object packet.

Representative Claim

20.   A computer data signal embodied in a carrier wave, comprising:

instructions for receiving objects transmitted by network entities, wherein at least a subset of the received objects each include source and destination methods and data associated with the source and destination methods; and

an incoming object handler to handle the subset of the received objects, the incoming object handler including a source and destination verifier to execute the source and destination methods of each received object with their associated data so as to verify the source and destination of the received object.

[J]   Lucent Control of Telephone Via Web Browser

United States Patent No. 6,031,836
Haserodt
Issued: February 29, 2000
Web-Page Interface To Telephony Features

Abstract

A method is provided for clients to access server-based telephony features in the Internet or other non-telephony client-server network, in a platform-independent and network-independent fashion and without modification of the clients. A user of a client (101) uses the client's World Wide Web (WWW) browser (113) to download from a WWW server (104) a page (115) that defines a blank feature form that has virtual actuator and associated parameters fields for the telephony features. The user marks up the downloaded page via the WWW browser to indicate feature selection and any feature parameters, and uploads the marked up page to the WWW server. A form-interpreting script (116) executed by the WWW server interprets the marked up page, and the WWW server sends a feature request that corresponds to the user's feature selection and user-specified parameters to a telephony feature server (105). The telephony feature server responds to the request by providing the requested feature to the user's client. If needed, the WWW server also requests the client to establish a TCP/IP connection with the telephony feature server.

Representative Claim

1.   A method of accessing telephony features in a non-telephony client-server network wherein clients and servers communicate with each other via a predefined communications protocol that lacks telephony feature-access commands, comprising the steps of:

the client requesting an individual telephony feature by communicating corresponding data with the server via the predefined protocol;

the server responding to the communicated data by requesting a provider of the telephony features to provide the individual telephony feature to the client;

the server further responding to the communicated data by requesting the client to connect to the provider;

in response to the request to connect to the provider, the client effecting a data and control communications connection with the provider;

in response to the request from the server to provide the individual telephony feature, the provider providing the individual telephony feature to the client, including

the provider instructing the client to redirect a data portion of the data and control communications connection from the provider to a communications entity and to maintain a control portion of the data and control communications connection connected to the provider so that the provider can exert control over the client while the client communicates with the entity, and

in response to the instruction, the client establishing a communications connection to the communications entity.

[K] Internet-based Exercise Device

United States Patent No. 6,053,844
Clem
Issued: April 25, 2000
Interactive Programmable Fitness Interface System

Abstract

An exercise system including an exercise device at a user location includes a controller at the user location for controlling the exercise device. A control location remote from the user location and a communication system for transmitting information between the exercise device and the control location are provided. A sensor at the user location determines user location information and applies the user location information to the communication system for transmission to the control location. Control information is applied to the communication system by the control location in response to the user location information for transmission to the controller to control the exercise device according to the control information. Thus, the present invention is an interactive fitness system for permitting a user of a programmable exercise device to interact with a fitness server device while the user is in a location remote from the fitness server device. For example, the user can interact with the fitness server device from the home of the user. Using the system of the present invention the user can download new fitness equipment programs for controlling the exercise equipment. The user can also interact with fitness experts on-line and provide exercise information and receive control information wherein the received control information can permit interaction between the fitness server device and the user.

Representative Claim

1.   An exercise system including an exercise device at a user location, comprising:

> a controller at the user location for controlling the exercise device;
> an automated control location remote from the user location;
> a communication system for transmitting information between the exercise device and the automated control location;
> a sensor at the user location for determining user identifying information and applying the user identifying information to the communication system for transmission to the automated control location; and
> control information applied to the communication system by the automated control location in response to the user identifying information for transmission to the controller to control the exercise device according to the control information, wherein the control information is automatically derived by the automated control location from a database derived at the automated control location based on the user identifying information.

[L] Go Ahead Software — Smart Web Browser

United States Patent No. 6,055,569
O'Brien, et al.
Issued: April 25, 2000
Accelerating Web Access By Predicting User Action

Abstract

A smart browser working in conjunction with a HTTP server that selectively downloads WWW pages into the browser's memory cache. The determination of which pages to download is a function of a probability weight assigned to each link on a Web page. By evaluating that weight to a predetermined browser criteria, only those pages most probably to be downloaded are stored in the browser's memory cache. The download is done in the background while the browser user is viewing the current Web page on the monitor. This greatly enhances the speed with which the viewer can "cruise" the Web while at the same time conserving system resources by not requiring the system to download all the possible links.

Representative Claim

9.   A system that predicts, in a computer network, what information is to be next downloaded from a server, the system comprising:

a client user's computer, a server computer, a networked link between said client computer and said server computer;

the client computer with the capability to obtain and display information stored on the server computer;

the server computer storing information in the form of pages which in turn contain links to other pages of information;

said links having a probability factor, encoded within, that said links will be downloaded;

said client user's computer enabled to interpret said probability factor;

said client user's computer also enabled to match said probability factor to a predetermined criteria residing in said client user's computer; and

said client user's computer then downloading said information prior to being selected by user, into the browser cache of said client user's computer.

[M]   Cedant Publishing — Simulated Human Merchant

United States Patent No. 6,035,288
Solomon
Issued: March 7, 2000
Interactive Computer-implemented System And Method For Negotiating
Sale Of Goods And/or Services

Abstract

A computer-implemented method and system for negotiating the purchase of goods and/or services by customers utilizes a simulated human merchant having predefined behavioral attributes. An algorithm representing behavioral attributes of a simulated human merchant is used to receive customer input data relating to particular goods and/or services desired to be purchased. The customer input data is processed to generate merchant responses to the customer input data, and the sale of goods and/or services are agreed to at a particular price as a result of processing of customer replies to merchant responses according to the algorithm.

Representative Claim

1.   An interactive, computer-implemented system for negotiating purchases of goods and/or services, comprising:

a database storing merchant character data which simulates a human merchant having predefined behavioral attributes;

interface means for enabling a customer to input data relating to the purchase of particular goods and/or services; and a database engine which utilizes said merchant character data and said data inputted by a customer to generate responses to said data inputted by said customer according to said behavioral attributes;

whereby the occurrence of a sale for said particular goods and/or services at a specific price is determined as a function of customer replies to merchant responses and said merchant behavioral attributes.

[N] Geophonic Networks — Bidding for Energy

United States Patent No. 6,047,274
Johnson, et al.
Issued: April 4, 2000
Bidding For Energy Supply

Abstract

An auction service is provided that stimulates competition between energy suppliers (i.e., electric power or natural gas). A bidding moderator (Moderator) receives bids from the competing suppliers of the rate each is willing to charge to particular end users for estimated quantities of electric power or gas supply (separate auctions). Each supplier receives competing bids from the Moderator and has the opportunity to adjust its own bids down or up, depending on whether it wants to encourage or discourage additional energy delivery commitments in a particular geographic area or to a particular customer group. Each supplier's bids can also be changed to reflect each supplier's capacity utilization. Appropriate billing arrangements are also disclosed.

Representative Claim

1.   A method for creating an automated auction among energy providers and end users in which a moderating computer collects economic incentive data from each provider of a plurality of energy providers, processes the economic incentive data and distributes processed data to a plurality of control computers, each control computer associated with at least one end user, thereby enabling each of the plurality of control computers to select a provider of the plurality of energy providers for the provision of energy to the end users, based on an economic choice, wherein the method comprises:

a.   receiving in the moderating computer, economic incentive data specifying the economic incentive each provider will place on a unit of energy provided to end users associated with at least a portion of the plurality of control computers, processing the economic incentive data to determine which of the economic incentive data correspond to a first control computer and to produce derivative data, and storing the economic incentive data and derivative data in a data base of the moderating computer as first control computer data;

    b.   transmitting at least a portion of the first control computer data to the first control computer; and

    c.   transmitting at least a portion of the first control computer data to at least portion of the plurality of energy providers.

[O] IBM Electronic Catalog with Virtual Sales Agent

United States Patent No. 6,035,283
Rofrano
Issued: March 7, 2000
Virtual Sales Person For Electronic Catalog

Abstract

This invention involves an electronic catalog system which employs the knowledge and experience of a "Sales Agent," which is provided to a computer data base, and used with an inference engine to assist and guide actual customers to products that they will most likely be interested in purchasing. This system employs hypothetical questions and answers, based on the Sales Agent's experience with generic customers, as well as criteria and constraints provided by both the Sales Agent and the electronic catalog content.

Representative Claim

1.   A method of providing electronic catalog shopping comprising the steps of:

    creating a data base containing a Sales Agent's information on generic customer product interests and probable buying habits;

    providing said data base with product questions and anticipated probable answers to said product questions expected from said genetic customer, which are derived from said Sales Agent's prior sales experience, and storing said probable answers to said product questions, in said data base;

    employing a computer logic inference engine to present and advise an actual customer on specific suggested product choices based on said probable answers to said probable product questions provided for said generic customer.

CHAPTER 11

DRAWINGS FOR E-COMMERCE AND BUSINESS MODEL PATENTS

§ 11.01   Drawings in a Patent Application

Why have drawings in a patent application? There are at least five good reasons.

1. In many cases the law requires it.
2. Drawings help explain the invention and make the patent come alive. Trial lawyers, judges, and juries appreciate this.
3. Drawings make the patent easier to comprehend at a glance, and thus easier to find when searching for a particular technology or when studying the state of the art. Patent Examiners appreciate this.
4. Drawings help establish the structure, sequence, and organization of how information is presented in the patent specification. Patent attorneys use drawings as an outline for drafting the written specification.
5. Drawings help extract the key elements of an invention from the complexity of implementation details. Patent attorneys use drawings to sketch out thoughts for drafting claims.

The first reason for having drawings — the law requires it — is discussed further in § 11.02 under Legal Requirements. The remaining reasons for having drawings are practical reasons. These are discussed further in § 11.03 under Practical Considerations.

[A]   Special Considerations for E-Commerce and Business Model Patents

E-commerce patents and business model patents are really no different than any other kind of patent insofar as the drawing requirements are concerned. Because e-commerce patents and many business model patents are implemented using computer software and computer network systems, one would naturally expect the patent drawings to be similar to those found in software patents, and such is the case. Many e-commerce and business model inventions are about information and how that information is packaged, shipped, and used. In this respect, the Internet is just one big information delivery system. User click-throughs from banner ad to retail Web site convey information about the user and his or her interest in the banner ad. Stock prices delivered over the Internet by an on-line brokerage firm convey information about the state of the market economy. The digital content of a streaming video movie delivered to a Web site is information. Indeed, money itself is a form of information. The price of a product conveys information about the product seller's impression of how that product is valued by consumers. In a very real sense, information lies at the heart of virtually every e-commerce and business model invention.

Because information plays such a key role, we need to have ways to represent it, substantively, in patent drawings. That might seem a bit tricky, given its intangible nature and the ephemeral role information often plays. Fortunately, computer software is equally intangible and often ephemeral, as well. Therefore most of the techniques used to represent computer software can be adapted to represent e-commerce and business models.

The following sections provide some thoughts on how business-related inventions may be shown in patent applications. The material draws heavily upon techniques used in drafting software patent applications. Where the business-related invention is implemented using software components, it is quite natural to show those software components in the drawings. Indeed, chances are good you may also be claiming a software patent invention, along with the business-related one. Even where software is not involved, or only marginally involved, these drawing techniques may be helpful because they can help identify the underlying information that flows throughout the business process.

§ 11.02   Legal Requirements for Drawings

The patent statute requires drawings where necessary for the understanding of the subject matter sought to be patented. In the words of the statute:

> The applicant shall furnish a drawing where necessary for the understanding of the subject matter sought to be patented. When the nature of such subject matter admits of illustration by a drawing and the applicant has not furnished such a drawing, the Commissioner may require its submission within a time period of not less than two months from the sending of a notice thereof. Drawings submitted after the filing date of the application may not be used (i) to overcome any insufficiency of the specification due to lack of an enabling disclosure or otherwise inadequate disclosure therein, or (ii) to supplement the original disclosure thereof for the purposes of interpretation of the scope of the claims.[1]

The Code of Federal Regulations further describes the drawings requirement:

> § 1.81 Drawings Required In Patent Applications
>
> (a) The applicant for a patent is required to furnish a drawing of his or her invention where necessary for the understanding of the subject matter sought to be patented; this drawing, or a high quality copy thereof, must be filed with the application. Since corrections are the responsibility of the applicant, the original drawings(s) should be retained by the applicant for any necessary future correction.
>
> (b) Drawings may include illustrations which facilitate an understanding of the invention (for example, flowsheets in cases of processes, and diagrammatic views).
>
> (c) Whenever the nature of the subject matter sought to be patented admits of illustration by a drawing without its being necessary for the

---

[1] 35 U.S.C. § 113.

536

understanding of the subject matter and the applicant has not furnished such a drawing, the examiner will require its submission within a time period of not less than two months from the date of the sending of a notice thereof.

(d) Drawings submitted after the filing date of the application may not be used to overcome any insufficiency of the specification due to lack of an enabling disclosure or otherwise inadequate disclosure therein, or to supplement the original disclosure thereof for the purposes of interpretation of the scope of any claim.[2]

[A] Necessity of Drawings

The statute and regulations do not explain the meaning of the statutory language "when necessary for the understanding." The Federal Circuit Court of Appeals has answered that question in *In re Hayes Microcomputer Products, Inc. Patent Litigation*.[3]

In *In re Hayes* one of the issues the court addresses is whether the patent adequately describes the invention and whether the drawings are adequate to meet the requirements of 35 U.S.C. § 113. The patent-in-suit comprises only 27 lines to describe the heart of the invention and only a single block, identified as microprocessor 55, to represent that heart. Finding both specification and drawing adequate, the Court applies essentially the same rationale: that the specification and drawings are sufficient for one skilled in the art to understand. With regard to the drawing, the Court states:

> According to section 113, "[t]he applicant shall furnish a drawing where necessary for the understanding of the subject matter sought to be patented." Sufficient evidence exists to support the conclusion that, to the extent it was necessary, the drawings were sufficient for a skilled artisan to understand the subject matter of the claimed invention. The microprocessor is identified as element 55 in Figure 1B of the specification. On the facts of this case, no more needed to be included in the drawings to satisfy the description requirement.[4]

Notice that the Court relies on expert testimony evidence to conclude that the drawing requirement of 35 U.S.C. § 113 has been met. The patentee in that case wisely offered expert testimony to prove that the claimed functions were attributed to the microprocessor. Without this evidence, the Court might not have concluded that § 113 had been satisfied, and the patent might have been held invalid.

You should not mistake arguments of counsel for evidence. Therefore, when confronted with a drawing requirements issue, proceed as the patentee in *In*

---

[2] 37 C.F.R. § 1.81.
[3] 982 F.2d 1527, 1536 (Fed. Cir. 1992).
[4] *Id.*

re Hayes did. Place evidence in the record to support the conclusion that a skilled artisan will understand the subject matter of the claimed invention, without drawings of that subject matter.

Admittedly, the *Hayes* patent-in-suit is not a patent that is completely devoid of drawings. Nevertheless, by applying and construing the applicable statute, 35 U.S.C. § 113, the Federal Circuit Court of Appeals shows what the statutory language "necessary for the understanding of the subject matter" must mean. It follows that, in some cases, a patent application requires no drawings. This is the current practice in chemical patent applications.

That some applications may not require drawings implies the converse, that some applications do require drawings. When drawings are required, the Patent Office will not give an application a filing date if the drawings are omitted. In order to receive a filing date, an application must meet all of the basic requirements stated in 35 U.S.C. § 111. Those requirements are:

> Application for patent shall be made, or authorized to be made, by the inventor, except as otherwise provided in this rule, in writing to the Commissioner. Such application shall include (1) a specification as prescribed by section 112 of this title; (2) a drawing as prescribed by section 113 of this title; and (3) an oath by the applicant as prescribed by section 115 of this title.

> *   *   *

> The filing date of an application shall be the date on which the specification and any required drawings are received in the Patent and Trademark Office.[5]

Receiving the earliest possible filing date can be important. An early filing date may make an otherwise key patent or a key publication unavailable as prior art. An early filing date may similarly avoid a public use or sale bar under 35 U.S.C. § 102(b). An early filing date may give the applicant the tactical advantages of the "senior party" in an interference proceeding under 35 U.S.C. § 135, as prescribed by 37 C.F.R. §§ 1.601-1.690.

The kind of problems that can arise when drawings are omitted is shown by *Jack Winter, Inc. v. Koratron Co.*[6] *Jack Winter* was a giant litigation involving 17 separate patent infringement actions consolidated by order of the Judicial Panel on Multidistrict Litigation.[7] The suit revolved around the Koratron patent for permanent press fabric. Accused of infringement were a veritable who's who of the garment industry, including Levi Strauss, Haggar, Deering Milliken, and others. The garment industry "adversaries," as the Court referred to them, raised many issues, ranging from patent invalidity through public use bar, to fraud on the patent office, to patent misuse, to Sherman Act antitrust violations.

---

[5] 35 U.S.C. § 111.
[6] 375 F. Supp. 1 (N.D. Cal. 1974).
[7] 28 U.S.C. § 1407.

During the litigation, Koratron argued that certain public use activity did not operate as a bar under 35 U.S.C. § 102(b) because the patent application was "filed" within one year of the public use activity. True, an application had been filed; however, the Patent Office determined that drawings would be required and that the application could not be given a filing date until a drawing was filed with it. Koratron complied and filed drawings, but not until more than one year after the public use activity.

Koratron argued that the Patent Office erroneously ruled that a drawing was required. The Court rejected the argument out of hand, stating that whether an application required a drawing was a question solely within the discretion of the Patent Office.

On reflection, Koratron's argument is not substantively farfetched. The claimed subject matter deals primarily with chemical treatment of the fabric, whereas the drawings requested by the Patent Office simply show trousers hanging in an oven and various cross sectional views of the trousers' stitching. However, Koratron's argument is procedurally defective. Koratron made no attempt to correct the filing date problem when the application was pending.

Today it is unlikely that Koratron's dilemma will occur. The *Manual of Patent Examining Procedure* (M.P.E.P.) instructs Examiners to carefully distinguish between applications filed without all figures or drawings and applications in which figures are subsequently required.[8] Applications filed without all figures or drawings are treated as *prima facie* incomplete and are not given a filing date. Usually this is detected by the Application Branch before the application ever gets assigned to an Examiner in an Art Unit. On the other hand, acceptance by the Application Branch of an application without a drawing does not preclude the Examiner from requiring one. In such a case, however, the filing date is not lost. On this point, the *Manual of Patent Examining Procedure* provides:

> The acceptance of an application without a drawing does not preclude the examiner from requiring an illustration in the form of a drawing under 37 CFR 1.81(c) or 37 CFR 1.83(c). In requiring such a drawing, the examiner should clearly indicate that the requirement is made under 37 CFR 1.81(c) or 37 CFR 1.83(c) and be careful not to state that he or she is doing so "because it is necessary for the understanding of the invention," as that might give rise to an erroneous impression as to the completeness of the application as filed.[9]

There may be times, however, when the Examiner believes that the application should not have been given a filing date because the application does not contain drawings. The procedure followed in this instance is spelled out in the *Manual of Patent Examining Procedure* as follows:

---

[8] U.S. Dep't of Commerce, Manual of Pat. Examining Proc. § 608.02 (Mar. 1994).
[9] *Id.*

If an examiner feels that a filing date should not have been granted in an application because it does not contain drawings, the matter should be brought to the attention of the supervisory primary examiner (SPE) for review. If the SPE decides that drawings are required to understand the subject matter of the invention, the SPE should return the application to the Application Branch with a typed, signed and dated memorandum requesting cancellation of the filing date and identifying the subject matter required to be illustrated.[10]

The Patent Office thus considers the first sentence of 35 U.S.C. § 113 to address the situation when the application is incomplete and cannot be given a filing date due to lack of drawings. The Patent Office considers the second sentence of 35 U.S.C. § 113 to address the situation when a drawing is not necessary for the understanding of the invention, but nevertheless desired by the examiner. In this latter case, the application is given a filing date.

[B] Content of Drawings

As to what must be shown in the drawings, the Code of Federal Regulations offer some fundamental guidance:

§ 1.83 Content of Drawing

(a) The drawing must show every feature of the invention specified in the claims. However, conventional features disclosed in the description and claims, where their detailed illustration is not essential for a proper understanding of the invention, should be illustrated in the drawing in the form of a graphical drawing symbol or a labeled representation (e.g., a labeled rectangular box).

(b) When the invention consists of an improvement on an old machine the drawing must when possible exhibit, in one or more views, the improved portion itself, disconnected from the old structure, and also in another view, so much only of the old structure as will suffice to show the connection of the invention therewith.

(c) Where the drawings do not comply with the requirements of paragraph (a) and (b) of this section, the examiner shall require such additional illustration within a time period of not less than two months from the date of the sending of notice thereof. Such corrections are subject to the requirements of § 1.81(d).[11]

The regulations make it clear that you are not required to show conventional features in great detail. "[C]onventional features disclosed in the description and claims, where their detailed illustration is not essential for a proper understanding

---

[10] *Id.*
[11] 37 C.F.R. § 1.83.

of the invention, should be illustrated in the drawings in the form of a graphical drawing symbol or a labeled representation (e.g., a labeled rectangular box).''[12] On the other hand, the Patent Office will sometimes require structural details that are of sufficient importance to be described in the specification.[13] The reason for this is expressed in the 1911 Commissioner's decision, *Ex parte Good*:

> This detail (a flange) is not found in the claims, and the only question is, therefore, whether the added illustration is necessary to make the described structure fairly clear and intelligible from the drawings.

> * * *

> The question is not whether one skilled in the art can decipher the invention, but whether the drawing is so clearly and artistically executed as to facilitate a ready understanding of the invention both at the time of examination and in searches afterward in which reference to the patent must be made.[14]

§ 11.03   Practical Considerations Prior to Litigation

Even if drawings are not required, there are still many good reasons to include them. Drawings help explain the invention and make the patent come alive. When drafting a patent application, it is easy to lose sight of the many purposes the patent may serve. Certainly it is possible that the patent will be involved in litigation. If so, it is a pretty safe bet that at least one of the patent drawings will be blown up to poster size for presentation to the judge or jury. Often the patent owner will call the inventor to the witness stand to explain the invention. For the patent owner, this testimony is a great opportunity to sell the importance of the invention and the merits of the patentee's case to the judge or jury. Do not waste this opportunity by having poor drawings. Rather, use the drawings, presented clearly and arranged logically, so they tell the story.

[A] Selecting Drawings

How do you structure the patent drawings so they tell a story? Begin by picturing yourself asking the inventor to show what he or she had been working on when the invention was conceived. Perhaps there are drawings that help to show this. Next, imagine asking the inventor to explain what problem he or she was trying to solve. There may well be drawings that show this. Finally, imagine asking the inventor how the problem was solved, what makes the invention work. Certainly you will want drawings of this.

Collect all of the drawings possible. It may be that you will choose not to use all of them, but the exercise of collecting them is valuable in two respects.

---

[12] *Id.* § 1.83(a).
[13] *Supra,* note 8 at § 608.02(d).
[14] *Ex parte* Good, 1911 C.D. 43, 164 O.G. 739 (1911).

First, collecting these drawings helps bring the invention into focus and helps put the invention in proper context with the prior art. Second, building this collection of drawings may prove immensely valuable if the application is ever involved in an interference.

[B] Preserving Drawings for Interference Evidence

The rules of practice in interference proceedings require each party, in its preliminary statement, to state the date that drawings of the invention were first made, or first introduced into the United States (for inventions made abroad).[15] Copies of these first drawings are required to be submitted with the preliminary statement in a sealed envelope.[16]

In an interference, when the first party to conceive the invention is normally awarded the patent, submitting dated drawings can be critical. Without the drawings, the invention is treated as if conceived when the application was filed. 37 C.F.R. § 1.629(d) states this:

> If a party files a preliminary statement which contains an allegation of a date of first drawing or first written description and the party does not file a copy of the first drawing or written description with the preliminary statement as required by § 1.623(c), § 1.624(c), or § 1.625(c), the party will be restricted to the earlier of the party's filing date or effective filing date as to the allegation unless the party complies with § 1.628(b).[17]

Thus every effort should be made to locate all drawings made by or under supervision of the inventor, as they may become critical interference proceeding evidence.

In some cases the inventor may have discarded the first sketches or drawings; or the inventor may have failed to date the drawings when made. This is not a desirable position. If an interference does arise, you will only be able to rely on corroborated, dated drawings and other physical evidence. Therefore, get in the habit of reminding inventors to sign and date their drawings and have them witnessed, to support later corroboration.

[C] Organizing Drawings

Once you have gathered all existing drawings, mentally sketch out any additional drawings that may be needed, and start organizing the drawings in groups. Put drawings that show what the inventor had been working on when the invention was made in one group. These may represent the prior art, or the environment, or a field of use for the invention. When making this prior art and environment group, be alert to early drawings that may contain a glimmer of the

---

[15] 37 C.F.R. §§ 1.622-1.625.
[16] 37 C.F.R. §§ 1.623-1.626.
[17] 37 C.F.R. § 1.629(d).

invention in its formative stages. Such drawings should not be placed in the prior art group.

Put drawings that show the problem the inventor was trying to solve in another group. These may help show why the invention is important. Drawings that show why the invention is important are extremely valuable because they can later become the icon for the invention itself. Few laypersons, judges and juries included, are able to understand the technical intricacies of how an invention works; yet most are able to understand that the invention solves a certain problem. Quite naturally, laypersons equate invention with a solution to that certain problem. The patentee wants the invention viewed in this way because it allows the patent to cover equivalents more easily.

Litigation issues aside, good drawings also make the patent easier to comprehend at a glance. Not everyone needs to study a patent in detail. Searching for prior art or conducting research of a particular technology, a researcher needs to be able to quickly spot the feature the researcher is looking for. Similarly, a manager, managing a large patent portfolio, needs to be able to quickly distinguish the subject matter of one patent from another. Titles, abstracts, inventor names, and patent numbers do not convey the essence of the patent subject matter as well as drawings.

The Patent Office has long required applicants to provide drawings for quick comprehension. In *Ex parte Sturtevant*, decided by the Patent Commissioner in 1904, this nugget appears:

> It is a great desideratum of Patent Office drawings that they should tell their story to the eye without making it necessary to go into the specification for explanation, which should be apparent upon inspection.

> * * *

> The suggestion by the applicants that the Examiner's objection should not be insisted upon, simply because one skilled in the art could make the invention from the present disclosure, does not overcome this necessity for such representation of the invention in the drawings as will make it intelligible for the purposes of search.[18]

[D] Drawing for *Patent Office Gazette*

Finally, there is the Official Gazette, published every Tuesday, reporting each patent issued that day.[19] The *Patent Office Gazette* includes, when applicable, one view taken from the patent drawings. Regarding this view, the regulations state that "[o]ne of the views should be suitable for publication in the Official Gazette as the illustration of the invention."[20]

---

[18] 108 O.G. 563 (1904).
[19] *Supra*, note 8 at § 1703.
[20] 37 C.F.R. § 1.84(j).

With software inventions, it is not always easy to pick one drawing that suitably represents the whole invention. Many software inventions are illustrated through flowcharts and flowcharts do not always show the big picture. At a certain level of detail, flowcharts tend to flow on and on, from page to page, with no single page representing the whole invention. To rectify this, consider adding an additional drawing that has been carefully crafted to be suitable for publication in the Official Gazette. This drawing may be a diagrammatic representation of a claim of broad or intermediate scope, for example. This drawing may otherwise be an object diagram or data flow diagram, showing the software components from which the invention is made.

[E] Use of Drawings to Outline Specification

Not only do drawings help the reader, they also help in drafting the application. Drawings can establish the structure, sequence, and organization of how to present information in the patent specification. Drawings can serve as an outline for drafting the written specification and for constructing the claims.

Usually an invention is best explained using a top-down, general-to-specific approach. Explain the logical and physical structure first. Follow that with a description of how that structure operates. Properly selected drawings help orchestrate this explanation. In other words, start with drawings that show the big picture. Put the invention in context by showing where the invention fits in the environment. Next, show the invention broken down into major subcomponents. Some of these subcomponents may be old. For instance, some subcomponents may exist simply to connect the invention to the environment. Other subcomponents may be new and may form part of the claimed invention.

Finally, break the subcomponents that form part of the claimed invention into even smaller subcomponents, using additional views. You now have a series of drawings to use, as an outline, to draft the specification in a well-structured, top-down manner.

[F] Use of Drawings to Develop Claims

Drawings can also help you extract the key elements of an invention from the complexity of implementation details. Use drawings to sketch out your thoughts for drafting claims. At first this may seem obvious, given the requirement of 37 C.F.R. § 1.83(a) that the drawings must show every feature of the invention specified in the claims. There is more here than meets the eye.

Claim drafting involves finding the key abstract concepts that comprise the essence of the invention. Complexity often complicates the task. One successful technique is to decompose the complex invention into more manageable parts. Break the invention into its essential components; examine those components; study how the components interact; identify which components or interactions are new and which are not. Then make drawings to capture your mental analysis to later draft claims at leisure. For further details regarding the claim drafting process, see Chapter 10.

Interestingly, software developers often work this way. Today, many software developers use computer-aided software engineering tools (CASE tools) to build software. Starting by drawing graphical symbols to represent the key abstractions of the software problem, programmers continue adding details of their design to these drawings until the drawings are so complete that the CASE tool can take over and write all the computer program source code needed. Patent attorneys do not yet have computer-aided claim drafting tools comparable to software developers' CASE tools, but that day is not far off. In the meantime, get ready for the patent application drafting tools of the future by making drawings to represent your main claims.

§ 11.04   Flowcharts

Most programmers today do not use them, yet flowcharts are currently the most used diagram in software patents. Why is this? To some extent the answer is simple inertia. Practically everyone who drafts his or her first software patent application will look at a sample patent first. Likely the sample patent contains a flowchart. Believing a flowchart to be necessary, one is included. Thus the flowchart tends to perpetuate itself.

Business method patents also seem to teem with flowcharts. This is quite natural. Where a method of any sort is involved, business methods included, the flowchart is probably what most application drafters turn to first.

That is not to say that flowcharts are not important. When software is viewed as a computer-implemented process, it is quite natural to resort to flowcharts. Likewise, when one desires to highlight the underlying process in a business model, the flowchart is probably called for. In fact, 37 C.F.R. § 1.81(b) specifically recognizes the propriety of flowcharts or "flowsheets." "Drawings may include illustrations that facilitate an understanding of the invention (for example, flowsheets in cases of processes, and diagrammatic views)."[21] Also, historically, computer programs were structured *procedurally*, as a set of step-by-step instructions for the computer to perform. Procedurally, structured programs and flowcharts are a natural fit.

Flowcharts have the *imprimatur* of the Code of Federal Regulations; why use anything else? One reason is that flowcharts do not always give a good overview of the big picture, particularly if many separate sheets are involved. If application preparation cost is a concern, lengthy flowcharts, paid for on a per-sheet basis, can get expensive.

Also, some nonprocedural business models and e-commerce systems simply do not fit the flowchart mold. Interactive web-based inventions or event-driven software programs, for example, do not fit easily into flowcharts. To illustrate, whereas you can easily construct a flowchart to explain how to mail a letter; consider how difficult it is to construct a flowchart to explain how the post office

---

[21] 37 C.F.R. § 1.81(b).

functions. For the post office, an organizational chart, or some other object-oriented notation seems more appropriate.

The bottom line is not to abandon flowcharts. Flowcharts are excellent for many business-related inventions. However, do not, by habit, turn to flowcharts in every case. Consider the other possibilities. Several possibilities are described in this chapter. Also, ask the inventor for suggestions on what notation best suits the invention. You may learn of a new notation that you can use in other applications.

[A] ISO Flowchart Standards

There are international standards on flowchart symbols, known as ISO 5807. The standards describe several different kinds of flowcharts, including program flowcharts, data flowcharts, system flowcharts, program network charts, and system resource charts. The standards also describe a number of graphical symbols used to make these charts. Program flowcharts are most frequently used in patents to describe processes, including business and software processes. However, you may want to be familiar with the other kinds of flowcharts, as well.

All four kinds of ISO 5807 standard charts use basically two classes of symbols: process symbols and data symbols. The process symbols graphically represent process steps or method steps. The most commonly used process symbols are the rectangle, which represents any basic process, and the diamond, which represents a decision. The most commonly used data symbols are the slanted rectangle or parallelogram, which represents any basic data, and the cylinder or drum, which represents direct access storage, such as a file stored on a hard disk. Figure 11-1 shows all of the ISO 5807 process symbols and Figure 11-2 shows all of the ISO 5807 data symbols.

[B] Program Flowcharts

Program flowcharts use process symbols, such as the familiar rectangular box and the diamond, to show processes or method steps. Lines interconnecting the process symbols show the flow of control. The emphasis is on showing the flow of control, that is, how one process leads to the next, or how one process passes control to the next. Therefore, use the program flowchart to explain or emphasize the sequence of procedures performed. An example of a program flowchart is shown in Figure 11-3.

[C] Data Flowcharts

Data flowcharts use data symbols, such as the parallelogram, to represent data and data storage media, and use process symbols, such as the rectangular box and the diamond, to show processes performed on the data. Many business processes involve storing and processing of business data. You can use these symbols to represent such data. Lines interconnecting these symbols show the

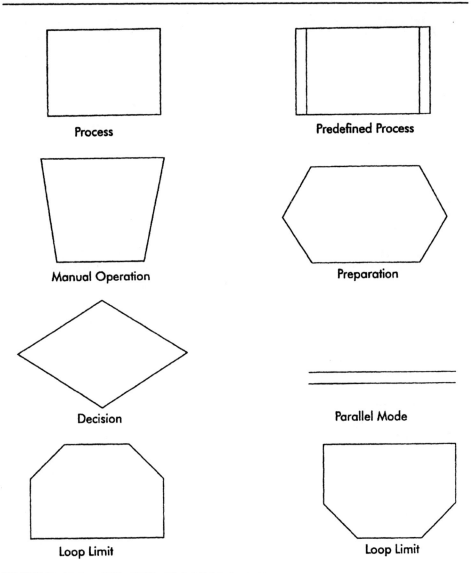

Process

Predefined Process

Manual Operation

Preparation

Decision

Parallel Mode

Loop Limit

Loop Limit

FIGURE 11-1: ISO 5807 PROCESS SYMBOLS

flow of data between processes and data storage media. The emphasis is on the data and how it flows from process to process. Therefore, use the data flowchart to explain or emphasize how data is changed through a series of manipulating steps. Data may be anything from electrical signals received from an Internet router or credit card processing terminal, to numerical data representing physical parameters (for example, from turnstile counters), to statistical data (for example, from a model of historical stock prices), to computer data. Likewise, the processes may be anything from physical, chemical, and electrical processes, to computational processes. Of course, many business-related patents will involve computational processes. Yet, recognize that these processes nevertheless usually involve data representing physical parameters. Therefore, even in business model

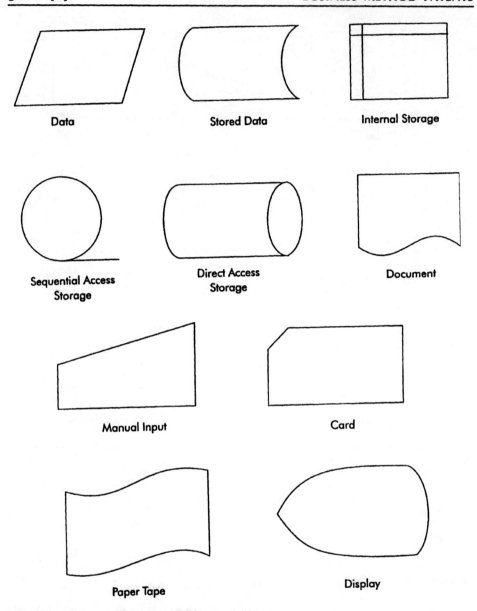

FIGURE 11-2: ISO 5807 DATA SYMBOLS

and e-commerce patents, data flowcharts focus on the problem solved by the invention and not on how the computational processes are performed within the computer. An example of a data flowchart is given in Figure 11-4.

[D] System Flowcharts

System flowcharts use data symbols to show the existence of data and process symbols to indicate the operations to be executed on data. Lines interconnecting the symbols show data flow between processes as well as control

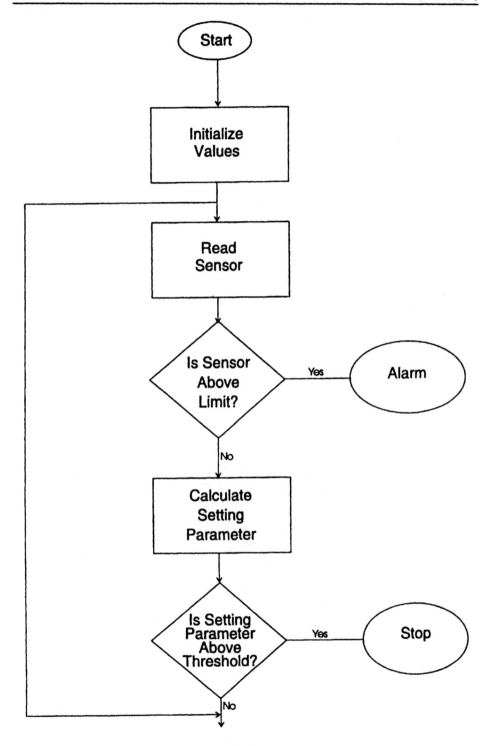

FIGURE 11-3:   PROGRAM FLOWCHART

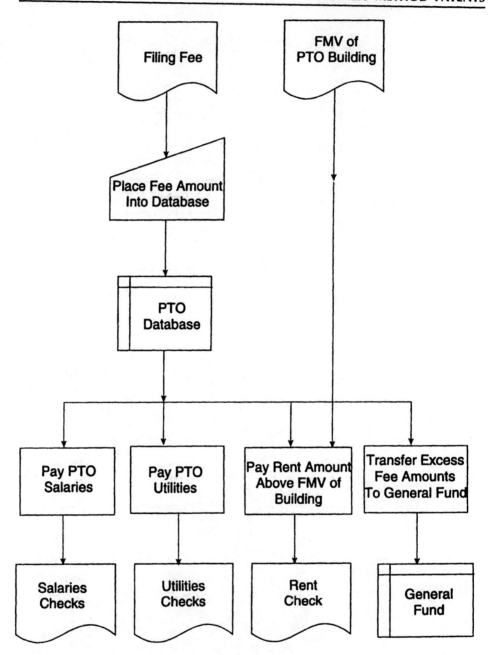

FIGURE 11-4:   DATA FLOWCHART

flow between processes. The emphasis is on the functional components of a system. Thus system flowcharts may frequently show databases or keyboard entry feeding processors with data, and further show processors feeding other processors with data. An example of a system flowchart is given in Figure 11-5.

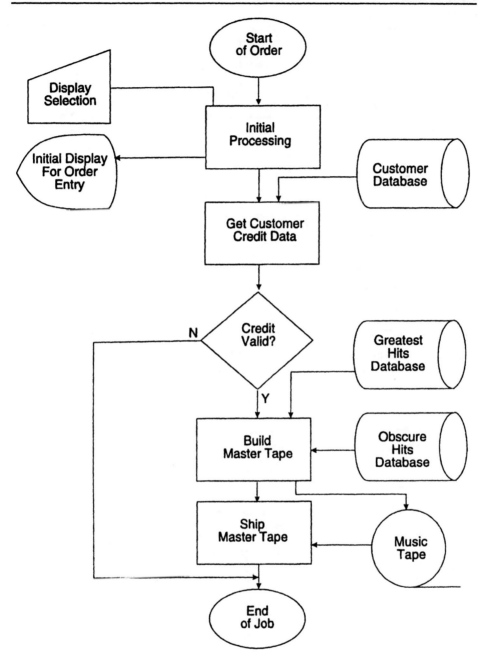

FIGURE 11-5: SYSTEM FLOWCHART

[E] Program Network Charts

Program network charts use data symbols and process symbols to show the existence of data and to show what operations are performed on that data. Lines show the relationship between processes and data, including the activation of processes and the flow of data to and from processes. The emphasis is on the

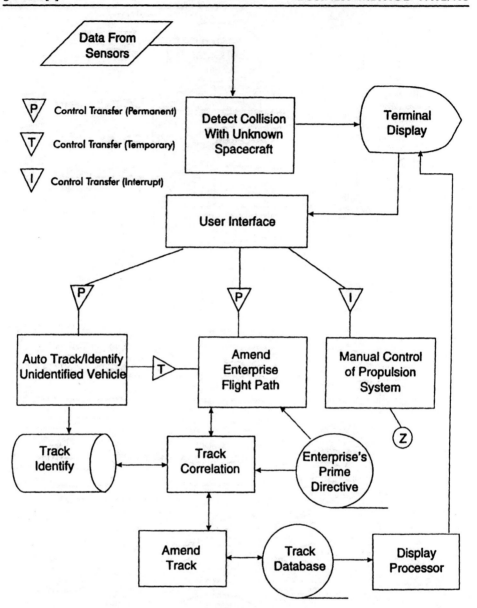

FIGURE 11-6: PROGRAM NETWORK CHART

relationships or interactions between data and processing elements. The program network chart is similar to the system flowchart, a principal difference being that the program network chart shows each data and processing element only once, whereas the system flowchart shows data devices and processors possibly more than once, as needed to describe the process flow. An example of a program network chart is given in Figure 11-6.

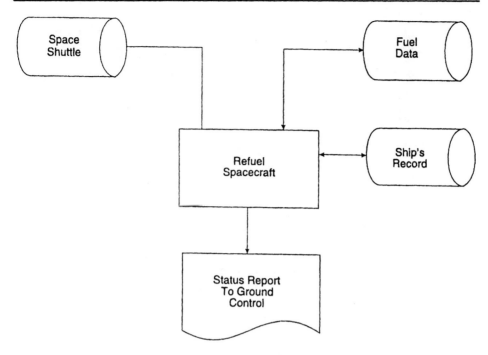

FIGURE 11-7: SYSTEM RESOURCES CHART

[F] System Resources Charts

System resources charts use data symbols to show input, output, and storage devices and use process symbols to show processors, such as central processing units. Lines interconnecting these symbols represent data transfer between data devices and processors. Lines also show control transfer between processors. The emphasis is on how the system resources (data units and process units) are interrelated. Thus system flowcharts may frequently show databases connected to processing functions. An example of a system resources chart is given in Figure 11-7.

§ 11.05 Pseudocode

Pseudocode is one notation to consider if the process you are describing is long and involved. Pseudocode is simply source code, where the syntax rules are relaxed and where unimportant details are left out. That is not to say that the unimportant details must be left out if the code is relatively short. An HTML or XML listing of a key web page could be included in its entirety, for example.

Compared to flowcharts, pseudocode is quite compact. Because there are no boxes and diamonds to draw, you can get the content of many pages of flowcharts onto a single page of pseudocode. Also, where appropriate, you can prepare pseudocode in the word processor and include it as an appendix to the application. This process can save considerable patent drawing cost.

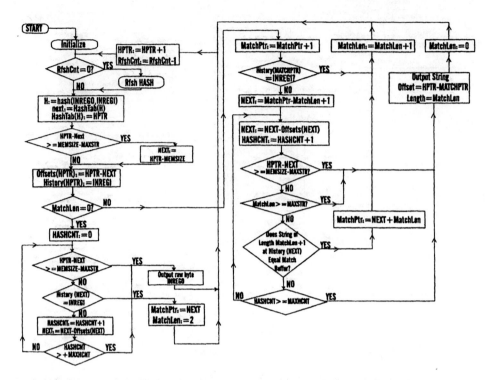

FIGURE 11-8:   FLOWCHART FROM U.S. PATENT NO. 5,016,009

To demonstrate, U.S. Patent No. 5,016,009[22] contains a flowchart (Figure 5A). Figure 11-8 shows the original flowchart as contained in the patent and how that flowchart might appear as pseudocode.

All these advantages are not without a price. Pseudocode is boring to look at and does work well at showing multiple layers of routines nested within other routines. Note how in Figure 11-8 the nested routines, demarked by indentation, and the terms START and NEXT are not easy to spot without drawing brackets or using a ruler. You are therefore better off using pseudocode only when concentrating on implementation details of a preferred embodiment of the invention, or when you need to get an application on file quickly to meet a statutory deadline.

[A]   Pseudocode in the Application

If using pseudocode, determine where it should go — in the specification or as a drawing. You have a choice. The Code of Federal Regulations states that computer program listings of ten printout pages or less may be submitted in the

---

[22] Whiting, "Data Compression Apparatus & Method," U.S. Patent No. 5,016,009, issued May 14, 1991. This is the patent-in-suit in Stac Elec. v. Microsoft Corp., No. 93-00413 (C.D. Cal. Jan. 23, 1994).

patent application either as part of the specification or as drawings.[23] There seems to be little reason not to treat pseudocode as a form of computer program listing. The regulation defines a computer program listing as "a print-out that lists in appropriate sequence the instructions, routines, and other contents of a program for a computer," a definition broad enough to cover pseudocode.[24]

The regulations provide that if submitted as part of the specification, the listing should comply with 37 C.F.R. § 1.52, and should be placed at the end of the specification but before the claims. If submitted as a drawing, the listing should comply with 37 C.F.R. § 1.84, and should include at least one figure numeral per sheet.[25]

If the pseudocode is particularly short, in the nature of a table, consider placing the pseudocode in the body of the specification as a table. Tables, but not drawings or flowcharts, are permitted in the specification.[26] If placing pseudocode in the body of the specification as a table, keep in mind that the table should be limited to 5 inches (12.7 cm) width, so that it may appear as a single column in the printed patent.[27]

On the other hand, if your pseudocode is long, that is, more than ten pages, you have the option of submitting it on either paper or microfiche.[28] The Patent Office prefers microfiche, as it saves government printing costs.[29]

§ 11.06 Entity-relationship Diagram

You may encounter a business-related invention that relies heavily on databases. In such cases, the entity-relationship diagram may be helpful. The entity-relationship diagram (or ERD as it is sometimes called) concentrates on the data model of a system, that is, it concentrates on what data entities exist and how they relate to one another. Data entities can be objects, which endure over time, or they can be events, which last only a moment.

To illustrate, the business-related invention might be a new kind of patent printing service that pulls patent data from the U.S. Patent and Trademark Office Web site and prints it. An embodiment of the invention might employ several different data entities. One data entity might be a patent database, containing information about the identity of each patent to be downloaded. Another data entity might be a formating database containing information about paper sizes and printing layouts. In this same example, a data entity event might communicate between entities. For example a "format" event might take the

---

[23] 37 C.F.R § 1.96; *see also* U.S. Dep't of Commerce, Manual of Pat. Examining Proc. § 608.05 (Mar. 1994).

[24] 37 C.F.R § 1.96; *see also* Manual of Pat. Examining Proc., *id.*

[25] 37 C.F.R. § 1.96; *see also* Manual of Pat. Examining Proc., *id.*

[26] 37 C.F.R. § 1.58.

[27] *Id.*; *see also* Manual of Pat. Examining Proc., *id.* at § 608.01.

[28] 37 C.F.R. § 1.96; *see also* Manual of Pat. Examining Proc., *id.* at § 608.05.

[29] Manual of Pat. Examining Proc., *id.*

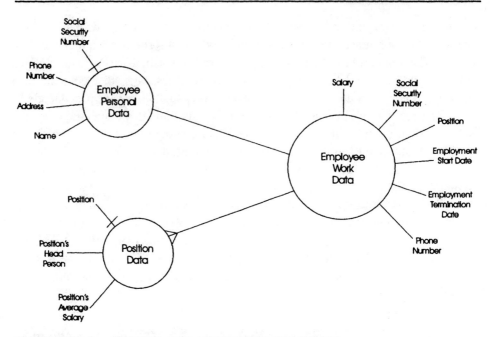

FIGURE 11-9: ENTITY RELATIONSHIP DIAGRAM

patent information from the Web site and place it into a selected form for printing.

Relationships between two data entity objects have a property called "cardinality." Cardinality denotes the number of entities that can exist on each side of a relationship. For example, there is a one-to-one relationship between a person and that person's physical location. Simply put, a person cannot be in two places at the same time. In the patent printing system described above, there may be a one-to-many relationship between downloaded patent content and the number of possible printing formats the system provides. Across a relationship there are basically three common kinds of cardinality: one-to-one, one-to-many, many-to-many.

In the entity-relationship diagram, data entities are represented by rectangular boxes containing the name of the data object or event. Relationships between data entities are represented by connecting lines, whose ends can be adorned to show cardinality. On each side of a relationship, if there is only one entity on a given side, the connecting line on that side is noted with a bar. If there are many entities on a given side, the connecting line on that side is noted with a crows-foot. To include the possibility that two entities may be related, but not in all instances, the connecting line is additionally marked with a small circle. Figure 11-9 shows examples of each of these notations.

§ 11.07   Booch Notation

Grady Booch is one of the pioneers of object-oriented software design. He has developed a notation, called the Booch notation, to aid in designing object-oriented software. It can be readily adapted to represent business models as well. Not all object-oriented programmers use the Booch notation, although many use a notation similar to it. Today, many use the Unified Modeling Language (UML) notation described in § 11.12. Nevertheless, the Booch notation is quite expressive. Used properly, Booch devotees claim it can even help write object-oriented programs with fewer logical errors or oversights.

The Booch notation starts with the premise of an architectural model of a software system, or the need to design an architectural model of a software system. As in the physical world, there can be different ways of looking at the model, that is, different views of the model. No one view tells the whole story. For instance, it may be helpful to view a software architectural model in terms of its logical abstractions. In object-oriented terms, what classes exist and how are those classes related. It may also be helpful to view the model in terms of its physical components — what software and hardware components or modules must be assembled and in what manner. Further, it may be helpful to see how the assembled system functions, dynamically as opposed to how it is constructed statically.

The Booch notation provides diagram notation to express four different views of a software model. The four view concept is illustrated in Figure 11-10. Booch notation facilitates four views of a software model, using the following diagrams:

- Class diagrams — indicating what classes (software or business abstractions) exist and how those classes are related

- Object diagrams — indicating what mechanisms are used to regulate how objects collaborate

- Module diagrams — indicating where each class and object should be declared in the program

- Process diagrams — indicating how multiple processes are scheduled and which processor is assigned which task.

In the above four views, class diagrams and object diagrams concentrate on the logical structure of the software or business model. Often the logical structure relates closely to the problem being solved by the software system or business system. In contrast, module diagrams and process diagrams concentrate on the physical structure of the software model or business model. Often the physical structure relates closely to code written to solve the problem in software.

Describing a business model or e-commerce invention using Booch notation, it is not necessary to use all four types of diagrams in every application. Just as a golfer carries more clubs in the golf bag than may be needed in a given round, it is helpful to carry these four Booch diagrams in mind, to call upon as

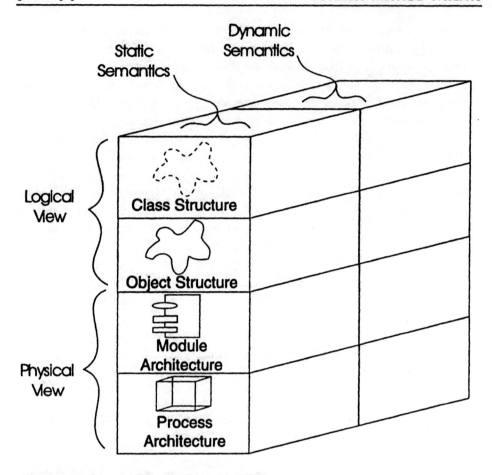

FIGURE 11-10: FOUR VIEW CONCEPT

needed. The following describes the basics of the four Booch diagram notations. If you find the notation helpful, consult Booch's book for the details.[30]

[A] Class Diagrams

The class diagram is used to show what classes exist in the software or business model and how these classes are related. The class diagram provides a logical view of the model. That is, it shows the key abstractions and mechanisms at work in the problem space. The logical view is distinguished from a physical view, which concentrates on the concrete software and hardware components that comprise the implementation space.

A class is an object-oriented programming concept that may be foreign to procedural programmers. If the whole notion of classes seems foreign, you are not alone. Many very fine programmers today still do not think this way, and consequently do not program this way. It takes time to become class conscious,

---

[30] Grady Booch, *Object-Oriented Analysis and Design with Applications*, 2d ed. (1994).

particularly if you are used to considering software as a series of procedural steps. Do not conclude that if the programmer did not use object-oriented programming, there is no need to worry about classes. You may not need classes to describe the preferred embodiment, but why be limited? If you have the ability to explain an invention in object-oriented terms, why not do so? It may someday help broaden the interpretation of the claims. Also, because identifying classes forces identification of the key abstractions, drawing class diagrams may help ferret out important elements of the patent claims.

A class is a group of things with common powers and attributes. In the Booch notation, a class is represented by a dotted-line cloud. Booch explains that the cloud is intended to suggest the boundaries of an abstraction or a concept, which may not have concrete boundaries.[31] For our purposes, there is no requirement that class diagrams comprise dotted-line clouds. Any shape will work, including ovals or rectangles, if you are not comfortable with clouds.

Viewing the software model or business model as classes, you often find that classes interact with each other in a variety of ways. The Booch notation identifies four different relationships through which classes can interact with each other. Booch calls these four relationships:

1. ''association''
2. ''inheritance''
3. ''has''
4. ''using''

An *association* between classes simply implies that there is a general relationship between the two, which Booch calls a semantic connection. The association relationship is thus the most general. It may be used if unsure or uncommitted as to the specific nature of the relationship between classes.

An *inheritance* relationship between classes is a more specific association relationship. It denotes a relationship from general to specific. For example, we can describe an inheritance relationship between the class ''rowboat'' and the class ''watercraft.'' A rowboat *is a* form of watercraft. Thus, sometimes the inheritance relationship is referred to as the ''is a'' relationship.

The *has* relationship between classes is another, more specific association relationship, sometimes called an ''aggregation'' relationship. It denotes a relationship between the whole and its parts. For example, we can describe a has relationship between the class ''Congress'' and the class ''Senate'' or the class ''House of Representatives.''

Finally, the *using* relationship between classes is yet another more specific association relationship. The using relationship denotes a client-server relationship, in which the client is using services provided by the server. For example, the student-teacher relationship can be described as a using relationship, in the sense that the class ''student'' is *using* the teaching services of the class

---

[31] *Id.* at p. 177.

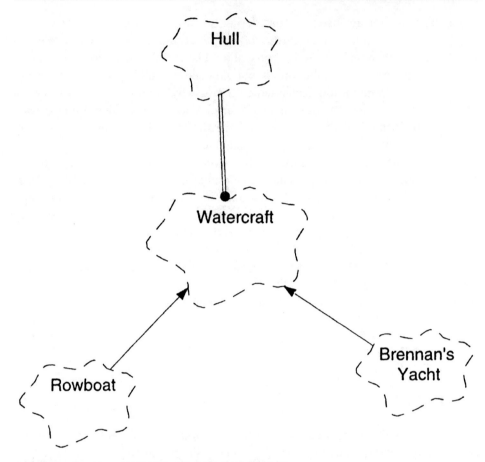

FIGURE 11-11: BOOCH CLASS DIAGRAM

"teacher." Using relationships are also found in more abstract situations. For example, Booch gives an example of a hydroponic gardening system in which there is a using relationship between the class "temperature" and the class "actuator."[32] Figure 11-11 gives an example of a Booch class diagram.

[B] Object Diagrams

Like class diagrams, object diagrams give a logical (as opposed to physical) view of the software model. Whereas class diagrams concentrate on the key abstractions, object diagrams concentrate on the objects based on those abstractions. Object diagrams show prototypes or examples of a system based on the software model or business model. Object diagrams are tied closely to class diagrams because objects are treated as tangible embodiments of or instances of abstract states. For example, object Senator Edward Kennedy is an instance of the class Senator.

---

[32] *Id.* at 181.

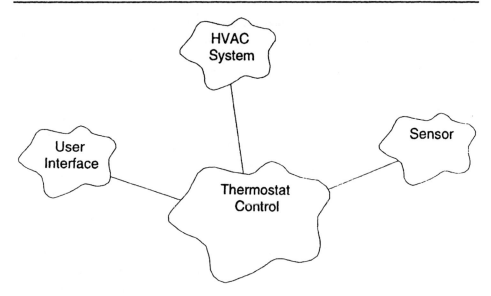

FIGURE 11-12: BOOCH OBJECT DIAGRAM

Like class diagrams, object diagrams also show relationships. Typically these are relationships by which objects communicate or interact. Objects interact with one another through messages. For example, object A may ask object B (via a message) to perform a service or to invoke an operation. This is the most common type of message. Object A's message may ask object B to invoke one of the operations defined by B's class. Thus object "Popeye's Remorse" of the class "rowboat" may ask object "boat" of the class "watercraft" to invoke the operation "float" on its behalf.

The Booch notation represents objects as solid line clouds, making them appear more tangible than the more abstract concept of classes. Messages are drawn as simple lines between linked objects. Customarily, the message lines are labeled to further describe the nature of the message. Figure 11-12 shows an example of an object diagram with several message lines.

§ 11.08 Use of Object-oriented Notation to Illustrate Business Systems

You now know more about object-oriented system design than many procedural language programmers. The question is why. Since one of the purposes of including drawings in a patent application is to explain how to make and use the invention, you may wonder why you would ever need to give a logical view of the problem space. After all, it would seem that a physical view of the implementation should be more than adequate.

The answer is quite simple. 35 U.S.C. § 112 has been interpreted to require the specification to teach, in sufficient detail, to avoid undue experimentation. The drawings help with this teaching. Thus by showing the logical structure of a business system, in object-oriented terms, you are teaching important principles of the invention, namely the logical architectural design of the system. In that way, you are permitting a broader reading of the scope of the patent, while at the same time helping avoid undue experimentation. Object-oriented thinking can be a good fit when describing business models and business.

In fact, a fair amount of study has gone into the use of object-oriented technology to represent business systems. A growing number of business system designers swear by it. In essence, the approach involves representing business concepts in terms of software objects. The software objects can then be used directly to construct software modules to support the business system, whether it is a modern e-commerce system or a conventional business system. Proponents of object technology maintain that it produces software modules which are more understandable, and easier to maintain and modify. These advantages allow the system to grow and adapt more readily to changing business conditions.

David A. Taylor, author of *Business Engineering with Object Technology*,[33] presents, in his book, a systematic way to analyze business systems using object-oriented principles. His primary focus is on the design and reengineering of business systems, but his methodology works equally well for analyzing and describing existing systems. Thus, it can be a useful tool in illustrating business systems in patent specifications.

Taylor calls his methodology *convergent engineering*. It seeks to engineer the business and its supporting software as a single, integrated system. That is quite a departure from the way most companies do it. Conventional practice typically adopts one of two approaches: (1) evolve the business processes first and then try to develop software that fits, or (2) purchase the software first and then try to force fit the business processes to conform to what the software requires. Taylor's approach provides a sensible breath of fresh air — engineer business and software processes together so they are sure to fit. To achieve his goal Taylor proposes object-oriented business models that closely follow object-oriented software thinking. This results in business object models that are easy to convert into software objects when it is time to construct the supporting software systems that will be used by the business.

This may sound fine in theory, but exactly how do you apply object-oriented software thinking to business models? Fortunately, Taylor lays out a framework that does exactly that. Taylor arrived at his object-oriented, business model framework by analyzing the philosophies underlying the evolution of business over the last 50 years. This evolution is an interesting study in its own right. Each generation of management has chanted its own mantra for success. Taylor collects these and develops an object-oriented business framework that gives equal prominence to each.

Being based on history, one would hardly expect Taylor's business model framework to hold for the next 50 years. No doubt as new generations of management develop new mantras for success, Taylor would advocate updating

---

[33] D. Taylor, *Business Engineering with Object Technology* (John Wiley & Sons, 1995).

his business framework to include them. To give you an idea of where Taylor is coming from, consider how business has evolved in the last 50 years.

In the 1950s and 1960s, most businesses focused on *resources*. The primary goal was to optimize the use of these resources. If two cents could be saved by reusing the scraps on the shop room floor, then, by all means, reuse the scraps. In the 1970s and 1980s, businesses gradually switched to a new way of thinking, one that focused on the *organization*. Reusing scrap became passé. Business managers of that era began tinkering with their business organizations, seeking higher productivity. Traditional departments and complex org-charts were abolished in favor of ad hoc teams. "Downsizing" the organization was in vogue. Middle management layers were cast out as companies put themselves on crash weight-loss programs. In the 1990s, *business process reengineering* became the new mantra. Managers directed their companies to simplify operations, reduce business cycle times, increase value added, cut costs, improve reliability, and focus on core competencies.

It is easy to get caught up in the management style of the moment. Focusing on core competency simply sounds more impressive than reusing scrap. However, one should not discount the importance of the business mantras from earlier eras. There is a reason General Motors grew to be the largest automotive manufacturer in the world. What worked then is still powerful — its now ingrained in the fiber and no longer needs to be exalted in mantra. Accordingly, Taylor proposes a business object framework that includes three major business subclasses — devoted to each of the three business mantras of past:

- Resource (1950s and 60s)

- Organization (1970s and 80s)

- Process (1990s)

Taylor breaks each of these three major business subclasses into their respective component parts. Like this:

- Resource
  - Human
  - Facility
  - Financial
  - Information
- Organization
  - Company
  - Division
  - Department
  - Team

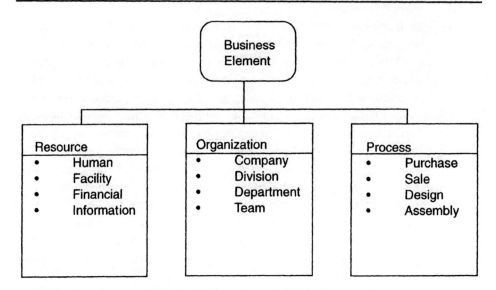

FIGURE 11-13: TAYLOR'S BUSINESS ELEMENTS

- Process

  - Purchase

  - Sale

  - Design

  - Assembly

There are, of course, other ways to break down resources, organizations, and processes into component parts. You will probably discover others as you analyze different business models. To unify even the three major business subclasses, Taylor defines something larger that he calls the *business element*. Taylor's business element subsumes the major business subclasses as constituent parts, as shown in Figure 11-13.

So far, the business element framework simply shows what different types of things are needed to define the business enterprise. One hardly needs an object-oriented framework to do that. The real power of the business element framework lies in what the members of each subclass can do. Some of this functionality can be ascribed to all objects of a subclass. Any such capabilities that are common to all subclasses (Resource, Organization, Process) are made a part of the Business Element object. Taylor proposes several capabilities that should be common to all subclasses (and hence attributed to the Business Element object). They are:

1. Supporting simultaneous business activity
2. Scheduling and recording its activities
3. Interacting intelligently with users of a system
4. Managing component objects of the same class

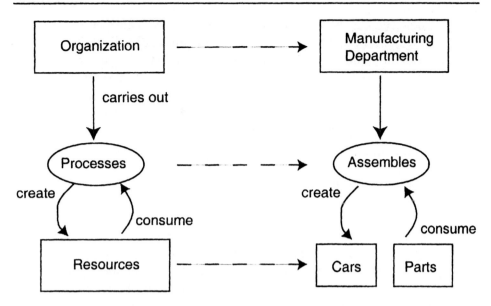

FIGURE 11-14: TAYLOR RELATIONSHIP — RESOURCES SUBCLASS VS. ORGANIZATION SUBCLASS

Other functions may be unique to objects of a particular subclass. Taylor's framework allows these functions to be ascribed to only objects of the subclass that exhibit those capabilities.

In describing a business model or e-commerce invention using Taylor's framework, you are certainly not limited to following the framework in its strictest sense. The purpose of using the framework is to help you organize the business system you are working with, and to help you discover logical structure you otherwise might have overlooked. Do not be surprised if you discover subclasses other than those identified by Taylor, which may have capabilities other than those illustrated here. Because the business system you are working with is probably new (patententably new, you hope) you should expect to discover subclasses and associated functionality not found elsewhere.

After thinking about the constituent parts of your business element object, the next step is to lay out how those objects work together. The capabilities of your business element object, and of the objects of its individual subclasses, may be shown in the drawings in terms of how they relate to one another. In all likelihood, objects of one subclass will make use of or communicate with objects of the other subclasses. An object (e.g. department) within the Organization subclass will probably take selected objects from the Resource subclass and apply certain objects from the Processes subclass. This can be shown diagrammatically in the your drawings. Similarly, objects within the Process subclass will undoubtedly have links with objects in the Resources subclass; and objects within the Resources subclass will have links with objects in the Organization subclass. Figure 11-14 shows how these relationships will probably exist in many business systems you will analyze.

By now you may be thinking that the Taylor business element framework is fine as a theoretical construct, but not terribly helpful in describing the particular business system you are working on. Don't give up yet. The power of Taylor's framework is about to be revealed. So far, the framework has concentrated on what elements make up a business system. If nothing else, the framework serves as a useful checklist, so you won't forget to include an important element in the patent drawings. The elements within the framework also have capabilities — they interact with each other and perform functions. You want patent drawings that will show these capabilities. You want drawings that show which entities communicate with which others, which communicate with which others and what messages are passed. Taylor's framework will help you discover these capabilities so that they can be captured in the drawings.

Because the Business Element object holds the highest position within the framework, and many subclasses inherit its capabilities, we will start by examining what the Business Element can do. Keep in mind that the framework will identify what the business element in your system *probably* can do. Not every system will fully exploit every capability Taylor has envisioned, and some may have features Taylor did not anticipate. Generally, however, to the extent every business element has certain capabilities, listed above, every business element will need certain structure to carry out those capabilities. For example, to support simultaneous business activity, the business element will need some way to keep track of what it has already done — it must know its history. To interact intelligently with users of the system, the business element will need the ability to handle requests for services and to make requests of others. It will also need a scheduling mechanism to keep things flowing smoothly. Taylor addresses these needs of the business element by embedding scheduling, active service handling, and history components in every business element. A graphic illustration of such components is found in Figure 11-15. You may find that the business system you are describing also has these basic components.

When a business element interacts with the outside world, it does so through messages in the form of requests and responses. An example of such interaction is provided in Figure 11-16, showing how a business element asks for help. The business element sends a request for help [1] to a Person object that serves as the electronic agent for the human user. The Person object places the incoming request in a request storage area and then posts the request [2] to the computer screen so the human user can read it. The human user selects [3] the request for viewing and then provides a response [4] to the business element.

The organization objects, process objects, and resource objects that make up the business element each inherit this interactive ability from the business element. They may individually also have other capabilities that are not necessarily common to their siblings. Organization objects, for example, will often have a predefined structure. Often an organization will have a hierarchical structure that defines who reports to whom, who has access to which resources and who manages which processes. Figures 11-17 and 11-18 show two different business structures, one a more traditional hierarchical structure and the other a

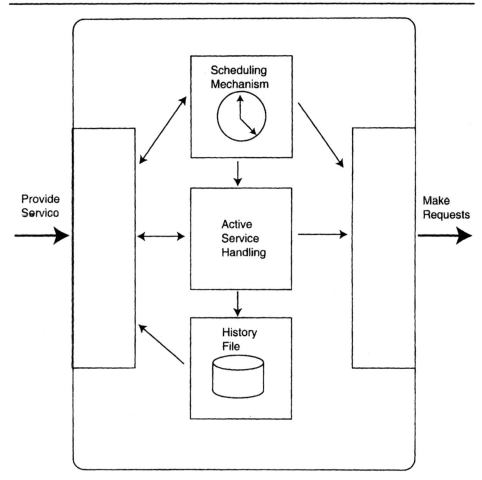

FIGURE 11-15:   TAYLOR BUSINESS ELEMENT COMPONENTS

team-based structure. Figure 11-19 illustrates yet another example. This one is adapted from the "hub and spoke" structure described in the patent at issue in *State Street Bank v. Signature Financial.*

The organization objects will typically have their own internal processes. These may differ from the more generic processes performed by all business elements. Such internal processes may be shown inside the organizational object, as shown in Figure 11-20. In the illustrated example, the Sales object performs a series of processes (using higher level functionality such as scheduling, activating and recording, of its parent business element). These Sales object-specific processes are performed on Resources made available by the Resource component of the parent business element.

In some cases it may also be useful to illustrate individual Process objects in fuller detail. This may be done functionally, as illustrated in Figure 11-21 which shows an example of a Purchase object. The illustration shows, in rectangular boxes, the individual documents generated by the Purchase object. These documents should not be confused with the processes that create them.

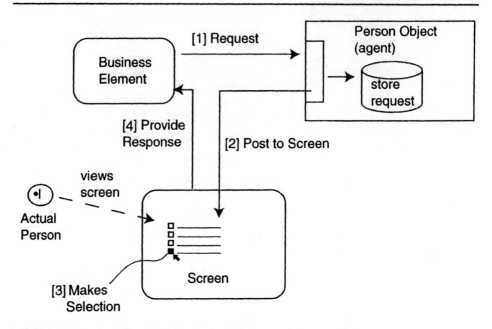

FIGURE 11-16: BUSINESS ELEMENT INTERACTION WITH OUTSIDE WORLD

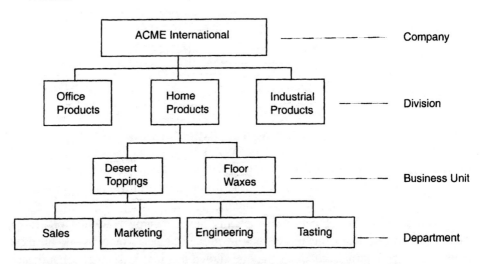

FIGURE 11-17: TRADITIONAL HIERARCHICAL STRUCTURE

Like Process objects, Resource objects may be more fully illustrated to show their internal detail. Often this internal detail is structural, like the internal detail of the Organization object. Figure 11-22 shows an example of a Resource object, breaking it down into constituent parts. Resource objects can have internal processes, just like the Organization object did. Under Taylor's business object framework, resources are responsible for scheduling their utilization. Figure 11-23 shows a Resource object performing this scheduling function.

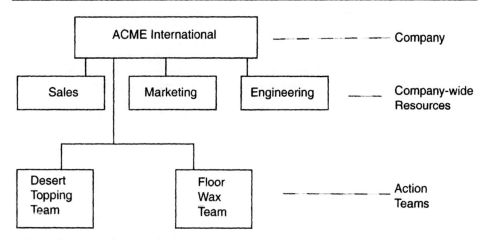

FIGURE 11-18: TEAM-BASED STRUCTURE

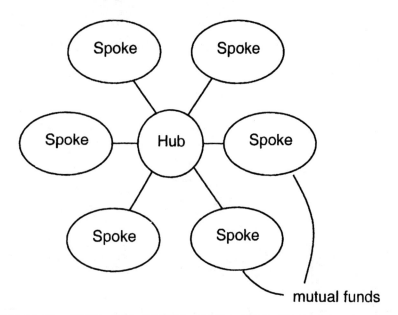

FIGURE 11-19: HUB AND SPOKE FROM *SIGNATURE FINANCIAL*

§ 11.09 Data Flow Diagram

The data flow diagram is a potentially important diagram for business method patents because many business methods involve information and its flow. Money is a form of information and many business systems will need to track its flow. The data flow diagram graphically models data and processes that operate on data. A form of data flow diagram, the ISO 5807 data flowchart, was described at § 11.04[C]. Presented here are two other data flow diagrams, the Gane/Sarson data flow diagram[34] and the Yourdon/DeMarco data flow dia-

---

[34] C. Gane & T. Sarson, *Structured Systems Analysis: Tools and Techniques* (1979).

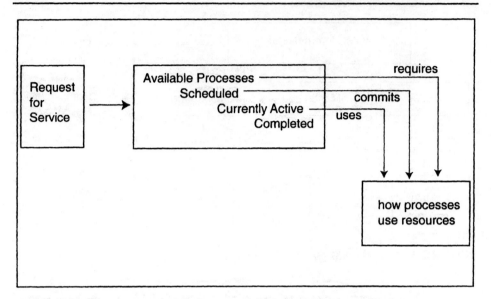

**FIGURE 11-20: SALES OBJECT PERFORMING PROCESSES —
SCHEDULING**

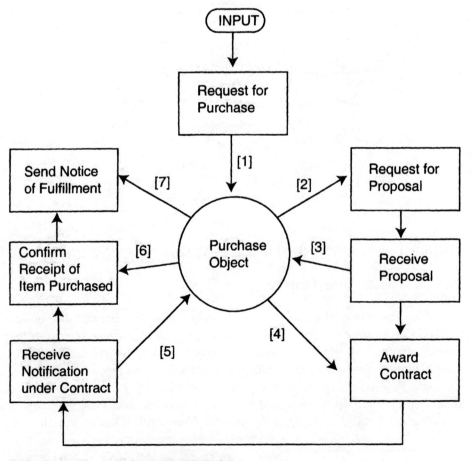

FIGURE 11-21: PURCHASE OBJECT

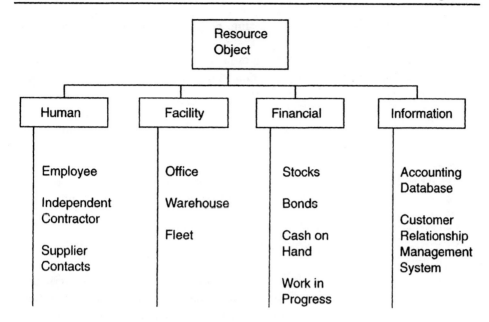

FIGURE 11-22: RESOURCE OBJECT

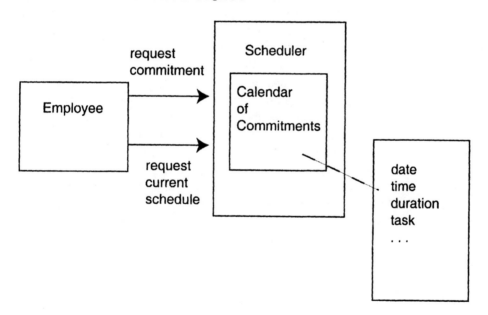

**FIGURE 11-23: RESOURCE OBJECT PERFORMING SCHEDULING
FUNCTION**

gram.[35] Both are widely used to describe systems, including software systems. Both also embody a system design philosophy of the respective authors. While the Gane/Sarson data flow diagram and the Yourdon/DeMarco data flow diagram

---

[35] T. DeMarco, *Structured Analysis and System Specification* (1979).

use similar symbols, the system design philosophies have some substantive differences.

If you are devising data flow diagrams for a patent application from scratch, you probably do not need to worry about the philosophical differences between Gane/Sarson and Yourdon/DeMarco. In this case, simply use one notation or the other to represent a system that is already designed. On the other hand, if you receive Gane/Sarson or Yourdon/DeMarco data flow diagrams from an inventor, it is helpful to know a little about the design philosophies of each. The design philosophy greatly affects the layout and content of the diagram. Sections 11.09[A] and [B] describe the Gane/Sarson and the Yourdon/DeMarco data flow diagrams, the notation of each, and some of the underlying design philosophy of each.

[A] Gane/Sarson and Yourdon/DeMarco Notations

Data flow diagrams, whether Gane/Sarson or Yourdon/DeMarco, use a collection of symbols to represent the data-related objects of a system. These objects include the "external entity," a generic entity representing any source of data flow into the system or any destination of data flow out of the system. These objects also include the "process," any process or function that transforms data in some way, and the "data store," a place where data is stored in the system. In addition, the data flow diagram uses an arrow symbol to represent data flow. Data flow is different from control flow. In a flowchart, connecting lines show the transfer of control from one process or method step to the next. In a data flow diagram, data flow arrows represent the path over which data are passed.

The respective Gane/Sarson and Yourdon/DeMarco object symbols are shown in Figure 11-24. Note that both Gane/Sarson and Yourdon/DeMarco use a square to represent the external entity. Both use an arrow to represent data flow. Gane/Sarson uses a rectangle with rounded corners to represent a process, whereas Yourdon/DeMarco uses a circle. The data store symbols are similar. Yourdon/DeMarco uses two parallel horizontal lines, whereas Gane/Sarson uses two parallel horizontal lines, closed at one end, to form an open-ended rectangle.

To compare the look of the Gane/Sarson data flow diagram with the look of the Yourdon/DeMarco data flow diagram, the same example system is illustrated using each in Figures 11-25 and 11-26.

[B] Gane/Sarson and Yourdon/DeMarco Design Philosophies

Although each has its own look, the substantive difference between Gane/Sarson and Yourdon/DeMarco lies in design philosophy. In designing a system, Gane/Sarson emphasizes focus on the whole, and recommends that the data flow diagram should include as many processes and data stores as you can fit on a page. Yourdon/DeMarco, on the other hand, emphasizes focus on the primitive processes, and recommends that the data flow diagram should include no more than about seven processes on it. When the number of processes rises much above seven, Yourdon/DeMarco recommends a layered approach in which

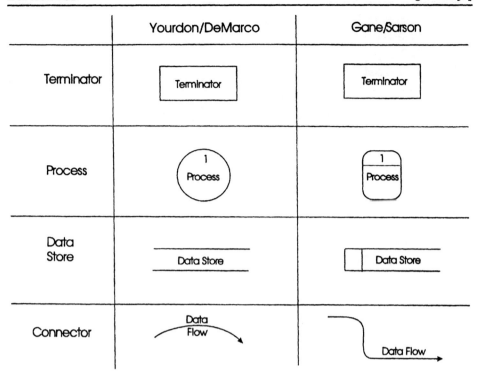

	Yourdon/DeMarco	Gane/Sarson
Terminator	Terminator	Terminator
Process	1 Process	1 Process
Data Store	Data Store	Data Store
Connector	Data Flow	Data Flow

FIGURE 11-24: YOURDON/DEMARCO AND GANE/SARSON OBJECT SYMBOLS

elements on an overview diagram are decomposed into component parts through a series of exploded views.

The number seven is not arbitrary. Psychologists have determined that most humans can only comprehend seven separate things at once, plus or minus two.[36] (Think about that next time you are cleaning off the top of your desk.) Given that patent drawings are intended to help the reader understand the invention, there is good reason to adopt the Yourdon/DeMarco philosophy when laying out patent drawings. Although there is certainly nothing wrong with showing an entire data flow design on a single sheet of drawings, you may wish, in addition, to include a series of subdiagrams that concentrate on key parts of the system.

Another difference between Gane/Sarson and Yourdon/DeMarco is how the prior art is treated. Often, when a new software system is designed, it is to replace an existing system. The existing system may be computer implemented, or it may be manually implemented. Whether the existing system is used as the starting point for the new system, or whether the new system is developed from scratch, and not based on the prior system, can make a big difference. Yourdon/DeMarco recommends always using the prior system as the starting point. Gane/Sarson recommends always having the option of starting from scratch. Thus, Your-

---

[36] G. Miller, *The Magical Number Seven, Plus or Minus Two: Some Limits on our Capacity for Processing Information*, 63 The Psychol. Rev. 86 (1956).

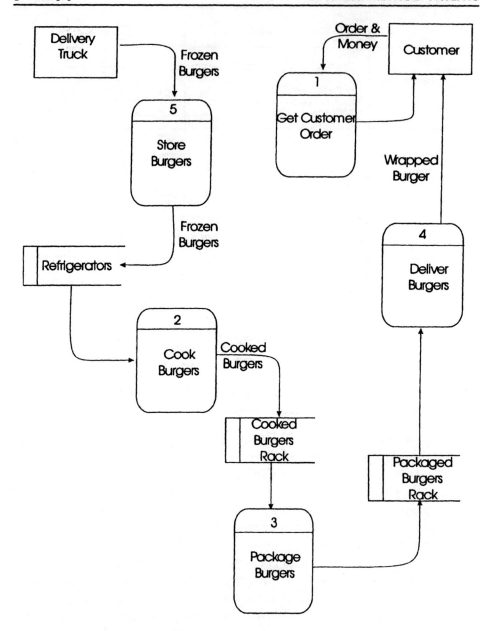

FIGURE 11-25:   GANE/SARSON DATA FLOW

don/DeMarco recommends a series of steps in which the existing system is analyzed and modeled; that model is then adapted to meet the needs of the new system. In the Yourdon/DeMarco approach, first model the existing system, then ask how this model needs to be changed to arrive at a proposed new model, which is then built into the new system. In contrast, Gane/Sarson recommends first building a logical model of the new system and then deciding whether it is

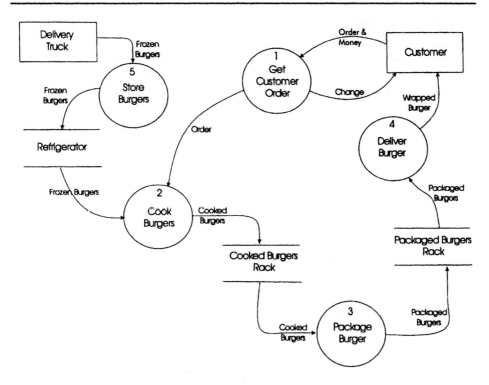

FIGURE 11-26: YOURDON/DEMARCO DATA FLOW

more productive to modify the logical model of the existing system or whether it is more productive to start from scratch.

Because the Yourdon/DeMarco approach models the prior art system first, then identifies how that model must be changed, it is easy to spot the new and potentially patentable subject matter when the Yourdon/DeMarco approach has been used. If the inventor is a strict follower of the Yourdon/DeMarco approach, you may be able to get precise design documents showing how the new system differs from the prior art from the inventor. On the other hand, when the Gane/Sarson technique has been used, it may not be clear from the design documents how the new system differs from the prior art.

A final difference between Gane/Sarson and Yourdon/DeMarco is how the data flow diagram actually relates to the underlying data model. Each data flow arrow on the diagram is a path over which data travels; each data store on the diagram is a place where data are held when not traveling. However, the diagram gives no details about the data structures themselves. For example, a data flow diagram may indicate that ''sales'' data may flow from process A to data store B, but that diagram does not indicate that ''sales'' data comprises customer name, invoice number, item number, dollar amount, date, and so forth. Thus both Gane/Sarson and Yourdon/DeMarco recommend providing detailed data structure documentation, so it is clear what data structures are involved in the data flow diagram.

§ 11.10   Representing Data Structures

Gane/Sarson represent the data structure in "outline" form in which hierarchy is shown by indentation. Gane/Sarson might write a "Sales" data structure in this way:

Sales
>   Customer-name
>   Invoice-number
>   Date
>   Item
>   Item-number
>   Quantity
>   Item-dollar-amount
>   Total-sale-dollar-amount

Yourdon/DeMarco represents the data structure in "in-line" equation form. Yourdon/DeMarco might write the same "Sales" data structure:

Sales = Customer-name + Invoice-number + Date + [Item-number
+ Quantity + Item-dollar-amount] + Total-sale-dollar-amount

Clearly, either data structure notation is suitable for patent application purposes. If a particular data structure comprises a key part of the invention, you may wish to show the data structure as its own block diagram figure. See Figure 11-27 as an example. This is especially true if the data structure happens to be an element of the claimed invention. The patent rules specifically require, "[T]he drawing must show every feature of the invention specified in the claims."[37]

§ 11.11   Patent Drawing Checklist

Here is a simple checklist to help you think about what type of drawings to use in your next business model or e-commerce application.

PATENT DRAWING CHECKLIST

<u>Why Prepare Drawings</u>:

___ to analyze what to claim

___ to organize before drafting the specification

___ as future litigation exhibits

---

[37] 37 C.F.R. § 1.83.

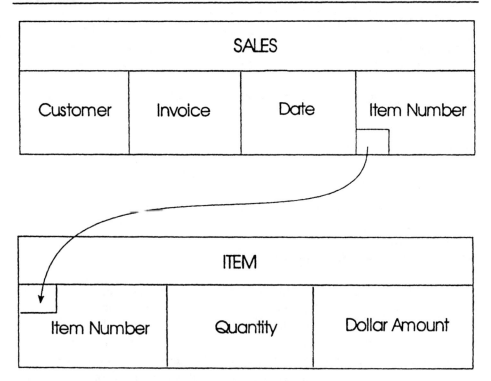

FIGURE 11-27:   DATA STRUCTURE DIAGRAM

Common Pitfalls:

___ Failure to retain dated, corroborated drawings. Without dated, corroborated (witnessed) drawings you may not be able to sustain your position in a patent interference

___ Failure to submit all necessary drawings in the application. The Patent Office will refuse to grant an application filing date if the written specification refers to a drawing (for example, in the Brief Description of the Drawings), but the drawing is inadvertently not submitted.

Types of Drawings Useful in E-commerce and Business Model Applications:

___ Flowcharts (there are numerous styles to choose from)

___ Program Flow — illustrates flow of program control

___ Data Flow — illustrates flow of data from process to process

___ System Flow — illustrates functional components of system

___ Entity Relationship Diagrams

___ Pseudocode

___ Class/Object Diagrams

___ Data Structure Diagrams

___ Program Network Diagrams — illustrates relationships between processes and data

___ System Resource Diagrams — illustrates how system resources (process and data) are interrelated

Some Pseudocode Guidelines:

___ Insert specification as appendix or as drawings

___ If listing is short, pseudocode can be presented as a table, for reproduction in body of specification — must fit inside a single patent column when printed

§ 11.12   Unified Modeling Language

The universe, from the mind-boggling galactic cluster to each infinitesimal subatomic particle, dances and swirls to a music written upon a single sheet. Physics texts speak of four fundamental forces: electromagnetism, gravity, the strong nuclear force, and the weak nuclear force. These forces, unseen but through their effect, mediate everything that happens in the physical universe. These forces, some physicists believe, are actually different forms of the same thing, related in a way we cannot yet perceive. Some day, physicists hope, a visionary will discover the secret of nature and will explain to the rest of us how these forces are one. In a single theory, a "grand unified theory," the visionary shall explain everything, or so the story goes.

The software universe — within which e-commerce and business models abound — seeks a similar messiah, someone to unify and explain this ever-expanding discipline. Unlike the physical universe, the software universe began its expansion before our very eyes. Recorded history traces the origin of the software universe back to its very beginning in 1969, when IBM unbundled software from hardware so that the two could be sold separately. However, this does not mean that the grand unified theory of software is going to be any easier to discover than the grand unified theory of the physical universe. Software being a product of human ingenuity, the grand unified theory of software must, in the end, be a grand unified theory of human nature.

Considering their radically different beginnings, the physical universe and the software universe are alike in one rather remarkable way. Both seem to be expanding. Nature abhors a vacuum; software abhors an empty hard disk.

The latest effort to unify software comes from the collaborative effort of three gurus of object-oriented software development: Grady Booch, James Rumbaugh, and Ivar Jacobson. They call their latest effort the Unified Modeling Language, or UML. Previously working independently, each of these three men had developed his own modeling techniques to drive software development and each had amassed a substantial following. Now working together, these three represent a dominant force in the science of software development.

Will the Unified Modeling Language affect software patents? Very likely. The UML supports a graphical diagramming technique that software developers use prior to actual program code writing. UML diagrams can rigorously describe, and yet simplify, software designs, and can help developers crystallize their ideas so that the program code they write later is solid. UML diagrams are both the architectural blueprints of a software project and the test bed in which developers can try out new ideas. Because UML diagrams have the potential to rigorously describe a software project, and thereby teach one of ordinary skill how to practice a software invention, you may want to use UML diagrams to help meet the patent application disclosure requirements of 35 U.S.C. § 112.

Although UML diagrams can be drawn with pencil and paper, many software developers use computer-aided software engineering (CASE) tools to lay out their UML diagrams. The CASE tools allow developers to quickly try different "what-if" scenarios, and the more sophisticated versions can actually write program code based on the diagrams. In preparing a software patent application, you may wish to ask the inventor for copies of all diagrams generated by the CASE tool the inventor is using. If the CASE tool was able to generate working program code from the diagrams, then you have a strong argument that those diagrams meet the enablement requirement of 35 U.S.C. § 112.

Sections 11.12[A]-[I] describe the various different types of UML diagrams. Each type serves a different purpose. Some show how the software components are arranged; some show the sequence of how the components operate; and some show how the preferred implementation may be built.

[A] Use Case Diagrams

Figure 11-28 shows a use case diagram. The use case diagram conveys a typical interaction between computer system and a user. A *use case* is an example or scenario showing how the system operates. The use case diagram thus shows a feature of the invention in operation by illustrating, from the user's standpoint, how the software would operate under certain conditions or to perform certain tasks. The use case might show, for example, how the user interacts with the software to reconfigure operating parameters upon start-up, or to instruct a software agent on how to carry out a user-defined task.

Functionally, use case diagrams are nothing new. Chemical patent practitioners have used use case diagrams for years, although they have not called them by that name. Chemical practitioners commonly employ "examples" to disclose how to practice the invention. Often there is little choice but to disclose the invention through examples because the chemical formula or process may be susceptible to too many variations to describe all possibilities in a single, rigorous way. One could, for example, describe how to bake a loaf of nut bread by giving an example of one recipe or formula, knowing that many other variations are also possible. By giving one example, the chemical practitioner expects the reader to

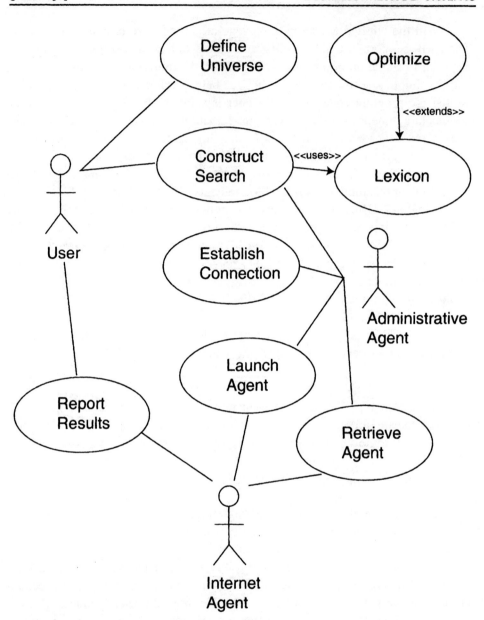

FIGURE 11-28: UML USER CASE DIAGRAM

be able to fill in the details necessary to adapt the recipe to bake other culinary equivalent nut breads.

The use case diagram is normally used to describe a user-visible function. Software developers will employ use case diagrams during the initial design phase, to help understand the primary functions the system must perform. Software patent practitioners can employ use case diagrams to give an overview of the system, before embarking on a detailed description of the system's component parts, or to give an example of the system in operation after the component parts have been described.

The use case diagram introduces a graphical symbol not normally found in conventional flowcharts. It is the "actor" symbol that resembles a stick-figure person. Because it resembles a person, the actor symbol connotes a *human* actor; however, the UML documentation describes this actor as being either a human or *software* actor. Recognizing that the UML actor notation may be misunderstood as symbolizing only a human end user, the software patent attorney may wish to describe in the written specification that the actor may be either a human or a software agent. Indeed, even if the currently preferred embodiment is designed for interaction with human users, the invention may be capable of modification to interact with software agents (as substitutes for human users). Use of use case diagrams and a suitably drafted specification may help extend the coverage of the patent through the doctrine of equivalents.

[B] Class Diagrams

The class diagram shows objects in a system and the static relationships of those objects. Static relationships are those permanent relationships that do not vary as the system operates. The class diagram is the key to unlocking many object-oriented software systems.

Class diagrams are not new. However, the UML puts a new spin on the subject by combining the collective wisdom of several prior diagramming notations. UML class diagrams define three types of static relationships: associations, generalizations, and aggregations.

Associations define conceptual relationships between an instance of one class and an instance of another class. For example, a Rowboat instance of the class Boat might use the Oar instance of the class Propulsion.

Generalizations define inheritance relationships between two classes, describing how the classes may be conceptually arranged into types and subtypes. For example, a boat is a kind of conveyance; thus there is a generalization relationship between the class Boat and the class Conveyance.

Aggregations define part-of relationships among classes. They describe how a given class is made up of component parts. For example, the class Sailboat may comprise the following component part classes: class Hull, class Mast, class Deck, and class Sail.

The UML class diagram uses a rectangular box to represent a class of objects. The rectangular box is subdivided into three regions. In the top region goes the name of the class; in the middle region go any attributes specific to that class; in the bottom region go any methods performed by that class.

Associations between classes are drawn as simple, solid lines. Optional arrowheads can be added to the lines to show which class sends messages to the other or causes the other to take action. Generalizations and aggregations are also shown by solid lines. To differentiate the generalization, place a triangle symbol at the end of the line that points to the hierarchically superior class. To

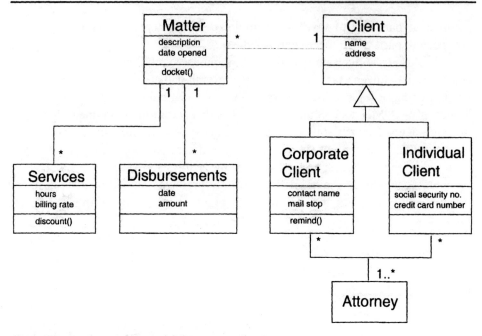

FIGURE 11-29: UML CLASS DIAGRAM

differentiate the aggregation, place a diamond at the end of the line that points to the hierarchically superior class.

If desired, you can annotate association, generalization, and aggregation lines with numbers at the ends to show cardinality, such as a one-to-one or a one-to-many relationship. The UML syntax uses a two-dot ellipsis to denote a range and the asterisk to denote "many"; thus, the range 1 to 8 is written 1..8 and one-or-more is written 1..*.

Figure 11-29 shows an example of a class diagram for a hypothetical client billing system. Note that the Client class has two defined attributes, name and address. There is a one-to-many relationship between the Client class and the Matter class; thus one client may have many pending legal matters. The Client class is a generalization of two subclasses, Corporate Client and Individual Client. Note the triangle symbol used to denote this generalization relationship. Several of the classes include defined methods. For example, the Services class includes the method "discount()," which the Services class may invoke to apply a discount to the charges for legal services. Similarly, the Matter class includes the method "docket()," which might invoke a docket recordkeeping function.

[C] Interaction Diagrams

Interaction diagrams show the behavior of a single use case, giving example objects and the messages passed between them. Interaction diagrams show the dynamic relationships among classes. Combined with class diagrams (which show the static relationships among classes), these diagrams can show how a working embodiment of a software system is implemented.

The UML provides a choice of two styles of interaction diagrams, the sequence diagram and the collaboration diagram. Use the sequence diagram when you want to emphasize the sequence of messages passed between objects. Use the collaboration diagram when you want to show the messages passed, but want to place more emphasis on the objects themselves. Both styles of interaction diagrams are described in §§ 11.12[D] and [E] below.

[D] Sequence Diagrams

The sequence diagram shows individual objects as they come into being, as they communicate with other objects, and (on occasion) as they depart or cease to exist. In the sequence diagram, the vertical axis represents the passage of time (from top to bottom) and the horizontal axis represents the passing of messages between objects. Each object appears as a labeled rectangle where it first comes into being. A dashed line, called the object's *lifeline*, extends downward from the rectangle, demarking the time during which that object exists.

In a typical sequence diagram, some objects are assumed to exist at the outset. These are arranged horizontally across the top of the page. Other objects come into being after the software program has started. These are shown at the appropriate vertical location (time) where that object comes into being. Often a message from an existing object causes the later object to enter the scene.

Messages between objects extend as horizontal arrows from the lifeline of the sending object to the lifeline of the receiving object. Labels above each message line tell the specifics of the message sent. Because time flows vertically, from top to bottom, messages above occur before messages below. The first message to occur is the one closest to the top of the diagram.

Figure 11-30 shows a sequence diagram for a hypothetical client intake system. All illustrated objects, except the Billing Record object, exist at start-up. The Matter Entry Window object begins the sequence by sending a message, "generate( )," to the Client object. In a software system, this might involve creating, or instantiating, an object to represent particulars about a client. Next, the Client object sends a message, "generate( )," to the New Matter object, thereby creating an object to represent particulars about the legal matter of interest to that client.

The New Matter object sends a message, "check( )," to the Existing Client object, to determine if the proposed legal work conflicts with that of an existing client. Existing Client sends a message, "Return," to New Matter, to advise if there is a conflict or not. Thereafter, New Matter issues a message, "create( )," causing a new object, Billing Record, to come into being. Finally, New Matter issues a message, "update( )," to the Existing Client object, causing the Existing Client records to be updated.

Note that the "update( )" message issued by New Matter has a special instruction indicated in brackets above the message. This is an example of a *condition* placed upon the message "update( )." When a condition is present, the associated message issues only if the specified condition is true.

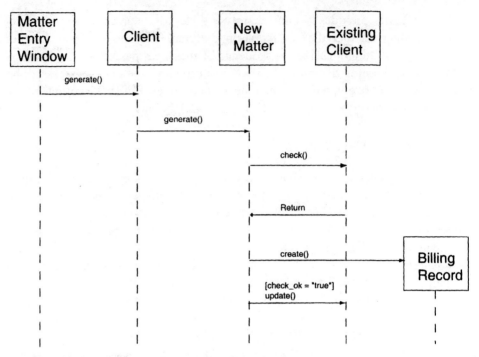

FIGURE 11-30: UML SEQUENCE DIAGRAM

[E] Collaboration Diagrams

Whereas the sequence diagram concentrates on messages and their timing, the collaboration diagram concentrates on objects and their communicative relationships with one another. The collaboration diagram represents objects as labeled rectangles with interconnecting lines to show the communication pathways. Objects that communicate with each other are connected by lines; objects that do not communicate with each other are not connected.

The collaboration diagram shows the flow of messages by labeled arrows beside a communication pathway. To show sequence of messages, the labels are numbered.

Figure 11-31 shows a collaboration diagram of the system also shown in the preceding sequence diagram example (Figure 11-30). Note how the collaboration diagram lends itself well to showing communication pathways between objects and showing which objects collaborate. The collaboration diagram downplays the actual sequence of message passing and does not readily convey an object's lifeline.

[F] Package Diagrams

In analyzing a software invention, the software attorney often confronts the puzzle of how to unfold the software system into meaningful component parts. Like an origami peacock, a finished software product can be seen in different ways. It can be seen through its functional parts — head, beak, wings, and

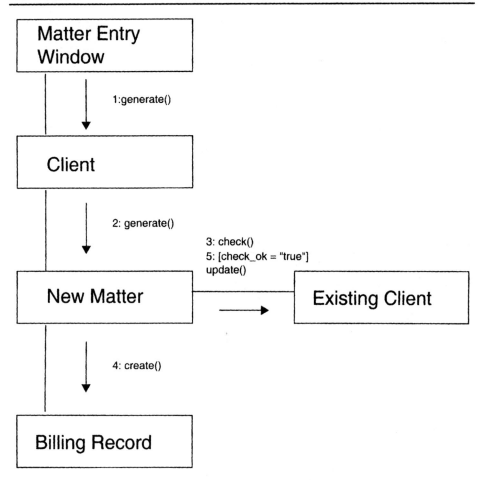

FIGURE 11-31: UML COLLABORATION DIAGRAM

body — or through its physical construction — tucks, corners, folds, and paper. Neither view is incorrect and yet neither view tells the same story.

For patent applications, often it is more helpful to describe software in terms of its functional components. The UML class diagram serves this purpose well. Given a good functional description, most skilled programmers can readily understand how to build an embodiment of the system described. However, with very large or complex systems it may be helpful to provide an implementation view, showing how the software components are actually packaged in the preferred embodiment. Use the UML package diagram for this purpose.

The package diagram is an offshoot of the class diagram, in which classes are grouped together into higher level units that show the dependency among the units. The UML specification calls these higher level units *packages*. There is a dependency between two packages when any two classes within the packages are dependent upon one another. The package diagram shows how the different functional units are grouped together in the working embodiment. For example, Figure 11-32 illustrates a database for tracking client and matter information. A

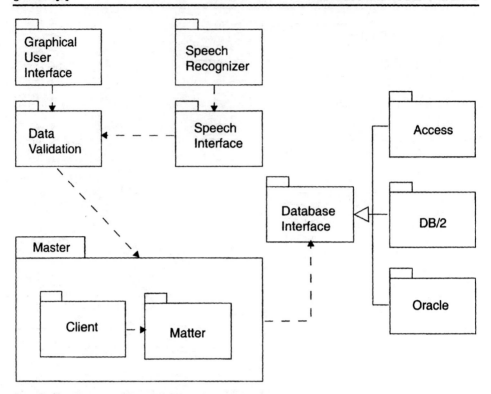

FIGURE 11-32:   UML PACKAGE DIAGRAM

user may input information either through a graphical user interface or through speech. The database itself stores information in three different database systems: Microsoft Access, IBM DB/2, and Oracle.

File-folder-like symbols represent the packages, which in turn may be made up of one or more related classes (not shown). If desired, a metapackage can be used to group together several related packages, as shown by the master package that encloses the client and matter packages. The metapackage, grouping packages of like domain, saves the need to draw individual class dependencies where those are relatively self-explanatory.

Aside from showing how the preferred embodiment of the invention is laid out, the package diagram can also help break the ice during the early phases of analyzing and understanding a software system prior to drafting a patent application. Although the software patent attorney normally wants to develop a functional description of the software components, this is sometimes hard to extract from the inventor, who is used to thinking of the invention in more implementation-related terms. Constructing a package diagram during the early application drafting phase helps to encourage the inventor to disclose all component parts (packages) that constitute the working embodiment. Armed with the package diagram, the attorney can then inquire as to the constituent classes or software modules that make up each package and ultimately determine the function of each.

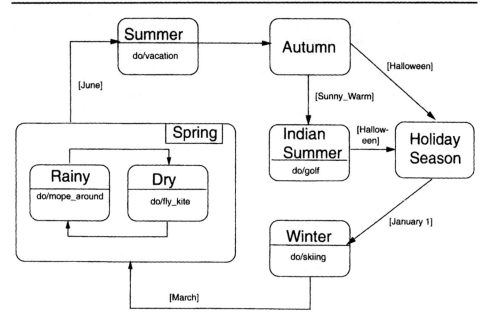

FIGURE 11-33: UML STATE DIAGRAM

[G]   State Diagrams

Many software systems change from state-to-state in a predictable way, like the red-green-yellow-red state change sequence of the familiar traffic light. Software engineers call a system with such predictable behavior a *state machine*, and they frequently use a diagramming technique known as the state diagram to show the sequence. The UML extends the state diagram somewhat, adding some useful navigational cues and a few other object-oriented concepts.

Figure 11-33 shows a state diagram for a sequence of states familiar to many, the changing of the four seasons. The UML state diagram uses a labeled box with rounded corners to represent a state. Nothing unusual about that. However, inside the box below the state name or label, an optional region lists any activity associated with that state. For example, the Indian Summer state has a listed activity, "do/golf," and the Summer state includes the activity, "do/vacation." In both cases, the "do/ . . ." is UML syntax.

Lines with arrows connect the state boxes, showing what possible state transitions can occur. These state transition lines can be annotated to give additional information. The UML syntax for such annotation has these three parts, all optional: Event, [Guard], /Action. An *event* can be any activity that occurs either upon entry into or exit from a given state. An *action* can be any activity that occurs while within, and without leaving, a given state. A *guard* performs state transition control logic. When a guard is indicated next to a transition line, the state transition cannot occur unless the guard expression is true.

In the sample figure, the guard [Halloween] prevents transition into the Holiday Season state. In other words, according to this state diagram, the holiday season shall not begin until after Halloween.

Another innovative feature of the UML state diagram is the superstate. The superstate notation allows several related states to be grouped together, thereby simplifying transitions into and out of those states. In the sample figure, the superstate Spring comprises two substates, the Rainy state and the Dry state. Note that either substate may be entered from the preceding Winter state and that either state may exit (in June) to the Summer state. While within the Spring superstate, numerous transitions between the Rainy and Dry states are possible.

[H] Activity Diagrams

No doubt many patent attorneys still use flowcharts more than any other type of drawing to describe inventive methods, processes, and procedures. Conventional flowcharts are good at showing sequential process, but not good at showing parallel processes. To illustrate, the simple sequential process of sharpening a pencil can be rendered by flowchart because the steps sequentially follow one another like pearls on a string. The insert-pencil-in sharpener step *must* precede the turn-sharpener-handle step. The order in which the steps are performed can be precisely stated in advance.

In contrast, the simple parallel or concurrent process cannot be so easily rendered by flowchart because the steps do not have to be performed in one and only one order. Consider the process of inviting friends to a party at the Grand Water Buffalo Lodge. Invite-Joe, invite-Mary, . . . , invite-Murphy, reserve-Lodge: these commands can be processed in any order, by yourself or concurrently with the help of a friend. A simple flowchart of this invitation process would likely be inaccurate, by implying that a certain invitation order is required.

The activity diagram solves how to render parallel processes through a clever notational trick, called the *synchronization bar*. The synchronization bar shows where two or more concurrent processes need to stop and wait for each other. Perhaps one process always reaches the synchronization bar first. Thus it always must wait for the other process to catch up. More generally, however, any process may be first to arrive, or last to arrive, depending on conditions that cannot be predicted.

In the party invitation example, the activity diagram might look like the one in Figure 11-34. Note how invitations and securing the lodge are performed concurrently, whereas the food is not ordered until after the invitations are made and the lodge is reserved.

The basic building block of the activity diagram is the activity symbol, drawn as a rounded box. Activities can be essentially any step in the process or task that must be accomplished. If desired, decision activities may be drawn as diamonds, similar to those used in conventional flowcharts but without being labeled inside. The label inside a decision activity diamond turns out to be

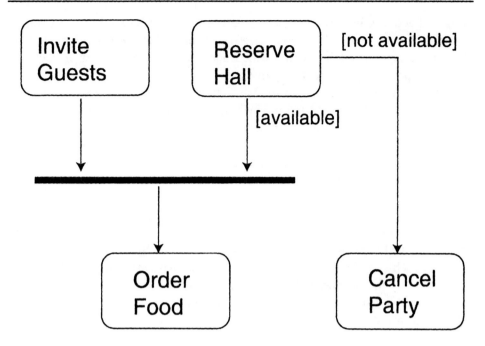

FIGURE 11-34:   UML ACTIVITY DIAGRAM

unnecessary because the lines issuing from the decision diamond are instead labeled to show what states may result from that decision activity.

So how do you label the lines to show that a decision has been made? The UML defines something called a *guard*, used to denote that a given activity can result in one of a set of mutually exclusive options. In official UML notation, a *guard* is a logical expression, placed in brackets, that will be either true or false. To use the notation, simply identify an activity that results in one of a set of mutually exclusive options. Then label each line issuing from that activity symbol with a different logical expression [in brackets], thereby accounting for all mutually exclusive possibilities.

For example, in Figure 11-34, the activity "Reserve Hall" can result in two mutually exclusive outcomes: [available] and [not available]. The guard notations on the respective lines show what activity next occurs, depending on which guard expression happens to be true. The guard notation thus allows you to dispense with the need to internally label decision diamonds as is done in conventional flowcharts.

Dispensing with the need to label decision diamonds is a significant improvement. In conventional flowcharts, decision diamonds must sometimes be made unduly large to contain the text description of the decision the diamond represents. Patent Office drawing regulations dictate that font size should not be less than 0.32 cm (1/8 inch) in height. That equates to 12-point type.

In general, an activity may result from a triggering event. Often such triggering events originate from other activities. You can designate the important triggering events by labeling the line between the triggering activity and the triggered activity. In UML notation, an event produced by an illustrated activity is simply labeled with a brief text narrative. Actions that precede an activity, but that are not produced by an illustrated activity, are labeled with a backslash character preceding the text narrative.

Labeling of transitions from one activity to another follows a standard labeling syntax that is also used in UML state diagrams. The transition label can have up to three parts, all optional: Event, [Guard], /Action. Figure 11-33 illustrates examples of these.

As noted earlier, the activity diagram permits you to draw several parallel processes, using the synchronization bar to show where those parallel processes must stop and wait on each other. Another useful trick is to arrange all activities of like classes into vertical columns called "swimlanes." Divide the page up to resemble an aerial view of an Olympic-style swimming pool. Draw a series of dashed vertical lines to define the swimlanes, assign a different class or software entity to each swimlane, and then arrange each activity symbol in the lane of the class or entity that performs that activity. It may take several tries to get a good layout, but the result richly communicates not only what the process is, but also who performs it.

[I] Deployment Diagrams

The deployment diagram shows physical relation between hardware and software components. Hardware components, such as computers, workstations, and even simple devices like sensors, are shown as boxes. Inside the boxes are the software components (physical modules of code) that actually reside on those hardware components. Dashed lines connecting these software components show how the software components depend upon or communicate with one another. Note that a software component can communicate with another software component that is not in its same box. In other words, a software component can communicate with a software component that resides on a different hardware component. Thus the dashed lines show logical or communicative relationships among software components, not necessarily physical connections among such components.

Physical connections are shown for the hardware components. Outside the hardware boxes, lines designating physical connections between hardware components show the communication paths over which the system interacts.

Figure 11-35 shows an example of a patent docketing system using the deployment diagram. Note that the physical connections have been labeled to show the protocol used to implement the physical connection, in this case TCP/IP protocol. Also note that several software modules communicate with one another, though not through direct physical connection. This communication is shown with dashed lines.

In the illustration, the docket database physically resides on the docketing server, a hardware component. A user can access docket information using a Windows PC, another hardware component. How the information gets from the

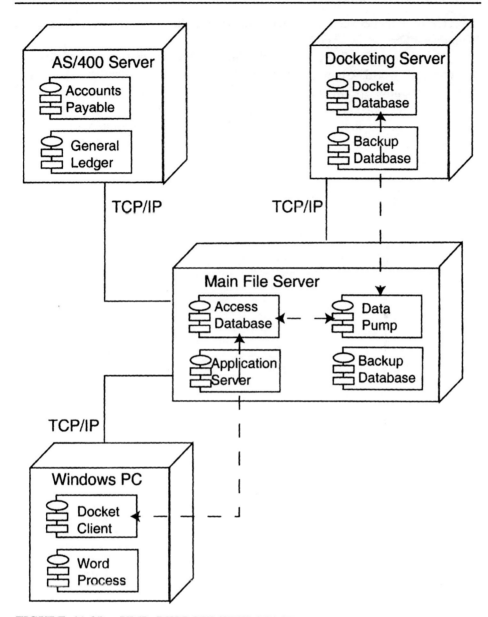

FIGURE 11-35: UML DEPLOYMENT DIAGRAM

docketing server to the Windows PC is a bit circuitous. As illustrated, the data is extracted by a data pump software component residing on the main file server, yet another hardware component. From there the data is supplied to an Access database, a software component residing on the main file server. Finally, using a docket client software component residing on the Windows PC, the user extracts the data from the Access database. If desired, the user could then copy and paste the extracted docket information into a word processor application.

Note that although the data flows logically from docket database to data pump, to Access database, to docket client, the actual physical data flows through the TCP/IP link between the docketing server, the main file server, and the Windows PC.

The deployment diagram can help show how a software invention relates to the physical world. Showing this can be important. Remember, the Patent Examiner is now specifically instructed by the software patent guidelines to "review the detailed disclosure and specific embodiments of the invention to determine what the applicant has invented."[38] Thus, the Examiner will examine your specification and drawings in determining whether the applicant's invention is statutory subject matter under 35 U.S.C. § 101. Showing how the software invention is embedded in the physical world can help demonstrate that the invention is statutory.

---

[38] *Examination Guidelines for Computer-Related Inventions* (March 29, 1996); *see also* *M.P.E.P.* § 2106.

CHAPTER 12

THE PATENT SPECIFICATION

§ 12.01   Purpose of Specification

The fundamental purpose behind the patent system is to promote science and the useful arts. It has been written that a patent is a contract between the government and the inventor. In exchange for a government grant of a patent monopoly, the inventor is required by this contract to disclose the invention, so that others may learn from it and thereby promote science and the useful arts. This contract between the government and the inventor becomes the guiding principle that dictates how the specification is to be drafted. Simply stated, the specification must disclose what the inventor claims the invention to be and what the inventor may later wish to claim the invention to be.

§ 12.02   Statutory Requirements of Specification

While it is the United States Constitution at Article I, Section 8, Clause 8, that creates the requirement that a patent application must include a specification, it is Title 35, United States Code, Section 112 that sets forth the actual legal standards that the specification must meet. 35 U.S.C. § 112 states:

> The specification shall contain a written description of the invention, and of the manner and process of making and using it, in such full, clear, concise, and exact terms as to enable any person skilled in the art to which it pertains, or with which it is most nearly connected, to make and use the same, and shall set forth the best mode contemplated by the inventor of carrying out his invention.

Dissecting the above statute, the written specification must meet the following three requirements:

1. It must contain a written description of the invention. This is sometimes called the written description requirement.
2. It must be sufficiently descriptive to enable one skilled in the art to practice the invention. This is sometimes called the enablement requirement.
3. It must disclose the best mode of practicing the invention contemplated by the inventor. This is sometimes called the best mode requirement.

Each of these requirements must be met for the specification to meet the terms of 35 U.S.C. § 112.

The price for failing to meet the terms of 35 U.S.C. § 112 is a high one. Since the patent grant is premised upon the full disclosure of an invention in exchange for the grant of a patent monopoly, the failure to provide a proper specification results in an invalid patent. It is for this reason that a great deal of care is given to drafting the specification.

§ 12.03   *Regulations Governing* Specification Content

It would be helpful to have more than one terse statutory sentence to guide in drafting the specification. Fortunately, there is more. The patent statute at 35 U.S.C. § 6(a) empowers the Commissioner of Patents and Trademarks to make regulations defining how the patent system shall work. These regulations are collected together or codified in the Code of Federal Regulations under Title 37 — Patents, Trademarks, and Copyrights. Regulation 37 C.F.R. § 1.71, sometimes referred to by its shorthand name, Rule 71, tells more about what the specification should contain. This regulation states at § 1.71:

DETAILED DESCRIPTION AND SPECIFICATION OF THE IN-VENTION.

(a) The specification must include a written description of the invention or discovery and of the manner and process of making and using the same, and is required to be in such full, clear, concise, and exact terms as to enable any person skilled in the art or science to which the invention or discovery appertains, or with which it is most nearly connected, to make and use the same.

(b) The specification must set forth the precise invention for which a patent is solicited, in such manner as to distinguish it from other inventions and from what is old. It must describe completely a specific embodiment of the process, machine, manufacture, composition of matter or improvement invented, and must explain the mode of operation or principle whenever applicable. The best mode contemplated by the inventor of carrying out his invention must be set forth.

(c) In the case of an improvement, the specification must particularly point out the part or parts of the process, machine, manufacture, or composition of matter to which the improvement relates, and the description should be confined to the specific improvement and to such parts as necessarily cooperate with it or as may be necessary to a complete understanding or description of it.

The above regulation, to some extent, simply reiterates the language of the statute, 35 U.S.C. § 112. However, the regulation goes further than the statute, adding a bit more detail. Comparing the regulation with the statute, details added by Rule 71 include:

1. The specification must set forth the precise invention claimed.
2. The specification must distinguish the invention from the prior art.
3. The specification must describe completely a specific embodiment of the invention. Frequently, since the specification must describe the best mode contemplated by the inventor, the specific embodiment described completely in the specification is the best mode embodiment. However, under certain circumstances, a patent applicant may also describe other

specific embodiments, in addition to the best mode embodiment. This is sometimes done when the invention can be practiced in alternate ways. By including a description of such alternate ways, the patent applicant may improve the chances of having the claims expansively read under the doctrine of equivalents.

4. The specification must explain the mode of operation or principle, where applicable.

5. If the invention represents an improvement, the specification must particularly point out the part or parts of the process, machine, manufacture, or composition of matter to which the improvement relates. Arguably, nearly every invention can be viewed as an improvement over the prior art. Therefore, this requirement needs to be considered together with the above requirements when the specification is drafted.

6. Although not expressed as a requirement, when the invention is an improvement, the description should be confined to the specific improvement and to such parts as necessarily cooperate with it, or as may be necessary to a complete understanding or description of it. By way of example, if the invention is an improved swivel head for a video camera tripod, a detailed description of the legs of the tripod, or of the standard threaded attachment to the camera would probably not be necessary for an understanding of the invention and probably should not be described in detail in the specification. Similarly, if a software invention resides in an application program, which in turn runs on one or more existing operating system platforms, it should not be necessary to include a detailed description of the operating system or systems, unless the inventive application program interacts with the operating system in a nonstandard or inventive way.

§ 12.04   Case Law on Specification Content

Statutes and regulations often need court interpretation to come alive. Such is the case with 35 U.S.C. § 112 and its associated 37 C.F.R. § 1.71 regulations. While the statute and regulations are clear that the specification must contain a written description, must enable, and must disclose the best mode contemplated, what does this really mean? Every invention is different and therefore there can be no rigid rules or simple formulas.

Understand that courts do not meddle in scrutinizing patent specifications or give advisory opinions. Courts will only review the sufficiency of a patent specification if necessary to resolve an actual litigated dispute. Then, the sufficiency of that specification hinges on the facts of that case. Thus, the case law is more like clay than concrete. It provides guidelines and judicial reasoning, which help attorneys understand how to comply with the statutes and regulations.

§ 12.05   Written Description Requirement

The written description requirement ensures that the patent applicant is truly in possession of the claimed invention as of the application filing date. The written description requirement most often comes into play when claims, not presented in the application when filed, are presented thereafter.[1] This happens frequently in interference proceedings when one party copies the claims of another party to provoke interference.

It may seem anomalous that 35 U.S.C. § 112 has been interpreted as requiring a separate written description requirement, when the invention is, necessarily, the subject matter defined in the claims. Why have this requirement when 35 U.S.C. § 112, second paragraph, requires "one or more claims particularly pointing out and distinctly claiming the invention"?

One explanation is historical. The written description requirement was a part of the patent statutes before claims were required. Back then, the specification served not only to teach or enable one to make and use the invention, but also to serve as a notice or warning of what not to infringe.[2]

Today, the continued existence of the written description requirement involves policy. Even though the second paragraph of 35 U.S.C. § 112 requires that claims particularly point out and distinctly claim the invention, that requirement provides the notice or warning of what not to infringe. The written description requirement guards against the inventor's overreaching, by insisting that the inventor recount the invention in such detail that the claims can be determined to be encompassed within the original creation.[3]

At first glance, it would also seem that the written description requirement and the enablement requirement are redundant. The enablement requirement, discussed at § 12.06, involves providing sufficient teaching to enable one of skill to make and use the invention. In contrast, the written description requirement involves providing proof in the specification that the applicant actually invented the thing claimed. In a case involving chemical subject matter, the Court of Customs and Patent Appeals expressly stated that "it is possible for a specification to *enable* the practice of an invention as broadly as it is claimed, and still not *describe* that invention."[4]

Whether the written description requirement has been met is a question of fact.[5]

§ 12.06   Enablement Requirement

The enablement requirement comes from the first paragraph of 35 U.S.C. § 112, which reads:

---

[1] Vas-Cath. Inc. v. Mahurkar, 935 F.2d 1555 (Fed. Cir. 1991).

[2] *Id.*

[3] *Id.*

[4] *In re* DiLeone, 436 F.2d 1404, 1405 (C.C.P.A. 1971).

[5] *In re* Hayes Microcomputer Prods., Inc., 982 F.2d 1527 (Fed. Cir. 1992).

The specification shall contain a written description of the invention, and of the manner and process of making and using it, in such full, clear, concise, and exact terms as to enable any person skilled in the art to which it pertains, or with which it is most nearly connected to make and use the same . . .

[A] Nature of Enablement Requirement

Enablement is a question of law with underlying factual issues to be resolved.[6] It is the invention claimed which must be enabled.[7] The specification must enable at the time the application is filed. Thus, if a software invention needs a specific component to work, such as a specific operating system, that component must exist at the time of filing. That does not mean that the specification may later become nonenabling if the manufacturer of the component changes its design.[8]

The test for enablement is whether the specification teaches one skilled in the art to make and use the invention without undue experimentation. When the challenged subject matter is a computer program, enablement is determined from the viewpoint of a skilled programmer, using programming knowledge and skill. The amount of disclosure that will enable may vary according to the nature of the invention, the role of the program in carrying out the invention, and the complexity of the contemplated programming, all from the viewpoint of the skilled programmer.[9]

Every detail need not appear for the specification to enable. Block diagrams and functional descriptions are permissible, provided they represent conventional structure and can be determined without undue experimentation. A minimum amount of experimentation is not fatal.[10]

[B] Scope of Enablement Based on Predictability

The scope of the claims must bear a reasonable correlation to the scope of enablement provided by the specification. If the invention pertains to an art where the results are predictable, then a broad claim can be enabled by disclosure of a single embodiment.[11] In cases involving unpredictable factors, such as most chemical reactions and physiological activity, the scope of enablement varies inversely with the degree of unpredictability of the factors involved.[12] Many software inventions can be described with mathematical precision and thus arguably fall into the predictable category. However, this may not be true for all

---

[6] Spectra-Physics, Inc. v. Coherent, Inc., 827 F.2d 1524 (Fed. Cir. 1987).

[7] *In re* Knowlton, 481 F.2d 1357 (C.C.P.A. 1973).

[8] *In re* Comstock, 481 F.2d 905 (C.C.P.A. 1973).

[9] Northern Telecom, Inc. v. Datapoint Corp., 908 F.2d 931 (Fed. Cir. 1990).

[10] Hirschfeld v. Banner, 462 F. Supp. 135 (D.D.C. 1978).

[11] Spectra-Physics, Inc. v. Coherent, Inc., 827 F.2d 1524 (Fed. Cir. 1987).

[12] *In re* Fisher, 427 F.2d 833 (C.C.P.A. 1970).

software inventions. For example, consider a software invention that mutates its behavior based on a genetic algorithm, or consider a software invention that predicts the weather. It may be that the results produced by these inventions are not predictable at all.

[C] Undue Experimentation

The principal measuring stick for enablement is undue experimentation. The focus here is on the word undue, not on experimentation. Undue experimentation involves a standard of reasonableness, with due regard for the nature of the invention and the state of the art.[13] Shedding light on the issue of undue experimentation, there is a line of cases that states eight factors to be considered. The Court of Appeals for the Federal Circuit, following this line of cases, has adopted these factors in at least one reported decision:

1. quantity of experimentation necessary
2. amount of direction or guidance presented
3. presence or absence of working examples
4. nature of the invention
5. state of the prior art
6. relative skill of those in the art
7. predictability or unpredictability of the art
8. breadth of the claims.[14]

This list is most often cited in cases involving unpredictable art, such as chemical or biological art. However, with the possible exception of the third factor (the presence or absence of working examples), the above factors apply equally well to inventions involving more predictable subject matter.

It is tempting to search for a rule that a certain length of experimentation time is per se undue experimentation. Certainly there have been cases where a certain length of experimentation time was found to be unreasonable. A case in point is *White Consolidated Industries v. Vega Servo-Control, Inc.*[15] This was a 1983 case involving a computer language translator needed to enable a numerical control system for machine tools. The Federal Circuit found the one-and-a-half to two person-years to develop a translator to be "clearly unreasonable."

However, to generalize that a certain experimentation time is, by definition, unreasonable, disregards the other factors recognized by the Federal Circuit. Thus, *White Consolidated Industries* should not be taken as a quantitative rule of "per se" unreasonableness. The test is "not merely quantitative," as the Federal Circuit has noted, "since a considerable amount of experimentation is permissible, if it is merely routine, or if the specification in question provides a reasonable

---

[13] *In re* Wands, 858 F.2d 731 (Fed. Cir. 1988).
[14] *Id.* at 737.
[15] 713 F.2d 788 (Fed. Cir. 1983).

amount of guidance with respect to the direction in which the experimentation should proceed."[16]

In *The Mythical Man-Month*, Frederick Brooks, Jr., describes his rule of thumb for scheduling software tasks:

    1/3 planning

    1/6 coding

    1/4 component test and early system test

    1/4 system test, all components in hand[17]

Fred Brooks managed IBM's development of OS/360, so he ought to know.

Note that half of Brooks's estimate is devoted to debugging and that only one-sixth is devoted to actual coding. In evaluating undue experimentation, does debugging time count? At least one court has ruled that it does not. In *Hirschfeld v. Banner*,[18] Hirschfeld sued Commissioner Banner under 35 U.S.C. § 145, seeking a patent for a "Digitally Controlled Electro-Optical Imaging System." Hirschfeld's application described a technique to enhance the image of a television camera tube using a computer. The application described the computer application in words, but provided no source code listing and no flow chart.

To prove the specification enabling, Hirschfeld's attorneys gave the specification to witness Grey, who was skilled in the art of computer-controlled optical systems. Using only the specification (and knowledge of programming), Grey wrote a program within four hours. This evidence convinced the court that the specification was enabling. Interestingly, here is how the court treated debugging and the fact that the program did not run:

> The program contained certain routine programming errors of the type customarily expected and eliminated during a routine "debugging" operation. Such errors would quickly and easily be eliminated by one skilled in the art. Grey's program was not written for any specific computer and therefore contained some general portions written in English rather than the Fortran IV computer language used elsewhere throughout the program. The details for completing the program for a specific computer installation would require no undue experimentation on the part of one skilled in the art.[19]

§ 12.07   Best Mode Requirement

The best mode requirement also comes from the first paragraph of 35 U.S.C. § 112:

---

[16] *In re* Wands, 858 F.2d 731, 737 (Fed. Cir. 1988).
[17] Frederick Brooks, Jr., The Mythical Man-Month 20 (2d ed. 1982).
[18] 462 F. Supp. 135 (D.D.C. 1978).
[19] *Id.* at 140.

> The specification . . . shall set forth the best mode contemplated by the inventor of carrying out the invention.

Although from the same statutory section as the enablement requirement, the best mode requirement has a different purpose. Its purpose is to ensure that the public fully benefits from the patent grant. When the patent expires, it is the best mode taught by the patent that the public inherits.

[A] Nature of Best Mode Requirement

The purpose of the best mode requirement is to restrain inventors from applying for patents while at the same time concealing from the public preferred embodiments of their inventions that they have, in fact, conceived.[20] It is the best mode *contemplated* by the inventor that counts, not some superior mode which may have existed, unknown to the inventor. In other words, there is no statutory requirement that the disclosed mode be the optimum one.[21]

The best mode requirement is different from the enablement requirement in its focus. As the court in *In re Glass* put it, "[t]he question of whether an inventor has or has not disclosed what he feels is his best mode is, however, a question separate and distinct from the question of the *sufficiency* of his disclosure to satisfy the [enablement] requirements."[22] Best mode is a question of fact.[23] It focuses on the state of mind of the inventor, whereas enablement focuses on the ability of the specification to teach an unidentified "one of skill in the art."

[B] Concealment

For best mode, concealment is the issue. The case law has interpreted the best mode requirement to mean that there must be no concealment of a mode known by the inventor to be better than that which is disclosed.[24] The relevant time for the concealment inquiry is the time of filing the application. Best mode is determined by the knowledge of facts within the inventor's possession at the time of filing the application.[25] There is no objective standard for best mode; only evidence of concealment, whether accidental or intentional, is considered.[26]

---

[20] *In re* Gay, 309 F.2d 769 (C.C.P.A. 1962).

[21] *Id.*

[22] *In re* Glass, 492 F.2d 1228, 1223 (C.C.P.A. 1974).

[23] Amgen, Inc. v. Chugai Pharmaceutical Co., 927 F.2d 1200 (Fed. Cir. 1991) (citing DeGeorge v. Bernier, 768 F.2d 1318 (Fed. Cir. 1985)). *See also In re* Hayes Microcomputer Prods., Inc., 982 F.2d 1527 (Fed. Cir. 1992).

[24] Amgen, Inc. v. Chugai Pharmaceutical Co., 927 F.2d 1200 (Fed. Cir. 1991).

[25] Spectra-Physics, Inc. v. Coherent, Inc., 827 F.2d 1524 (Fed. Cir. 1987).

[26] *Id.*; DeGeorge v. Bernier, 768 F.2d 1318 (Fed. Cir. 1985).

Whether best mode is met depends on the scope of the invention, the skill in the art, evidence of the inventor's belief and other surrounding circumstances.[27]

At the specification drafting stage, most best mode problems can be avoided by asking the inventor a few simple questions prior to filing. Has the inventor thought about the best way to practice the invention? If so, is that best way described in the application about to be filed? These questions may elicit information about a best mode that might otherwise be missed. This may seem obvious advice; yet consider how many applications are filed each year based on disclosure documents that are more than a year old. The older the disclosure, the more likely it is that the inventor has further perfected the invention, possibly contemplating a best mode not originally disclosed.

[C] Accidental or Unintentional Concealment

While intentional concealment of the best mode is certainly fatal, the cases leave open the possibility that accidental concealment may also be fatal. Accidental concealment might apply to the case where the inventor improves the invention, after the initial disclosure, but before the application is filed, and does not disclose the improvements. Accidental or unintentional concealment can also occur if the disclosure is poor. Even though there may be a general reference to the best mode, the quality of the disclosure may be so poor as to effectively result in concealment.[28]

Whether the disclosure is so poor that it results in concealment may involve the same factors as are used to judge enablement (for example, whether only routine skill or undue experimentation is required to fill in the gaps). This was the view taken by the Court of Customs and Patent Appeals in *In re Sherwood*, where the Court found "the specification in our view delineates the best mode in a manner sufficient to require only the *application of routine skill* to produce a workable digital computer program. Therefore, the quality of appellant's disclosure is not so poor as to result in the concealment of the best mode."[29]

The invention in *In re Sherwood* was an apparatus and method for producing cross-sectional seismic maps depicting the position and shape of subterranean geological formations. These maps are used to locate valuable oil and mineral deposits. To produce high quality maps, the invention used a computer to mathematically manipulate seismic data. The Patent Office rejected the application on best mode grounds because the affidavit evidence showed the inventor had a working software program at the time the application was filed. A listing of the program was not disclosed. Reviewing the Patent Office rejection, the Court did not agree that a program listing was required. Appreciating the wide spectrum onto which a programming task may fall, the Court ruled:

---

[27] *In re* Hayes Microcomputer Prods., Inc., 982 F.2d 1527 (Fed. Cir. 1992).

[28] Spectra-Physics, Inc. v. Coherent, Inc., 827 F.2d 1524 (Fed. Cir. 1987); *In re* Sherwood, 613 F.2d 809 (C.C.P.A. 1980).

[29] 613 F.2d 809, 817 (C.C.P.A. 1980).

In general, writing a computer program may be a task requiring the most sublime of the inventive faculty or it may require only the droning use of a clerical skill. The difference between the two extremes lies in the creation of mathematical methodology to bridge the gap between the information one starts with (the input) and the information that is desired (the output). If these bridge-gapping tools are disclosed, there would seem to be no cogent reason to require disclosure of the menial tools known to all who practice this art.[30]

Having now reviewed the legal requirements of the specification, we turn to the mechanics. Sections 12.08 to 12.13 will explore what the business model or e-commerce patent specification actually looks like, who its readers are, and how to draft the specification, starting either from the top and working down, or starting from the bottom and working up. Most importantly, how to apply the written description, enablement, and best mode requirements and satisfy 35 U.S.C. § 112 will be discussed.

§ 12.08   Form and Style of Specification

Different audiences read the patent specification: attorneys, judges, patent examiners, juries, business people, engineers, potential licensees, and venture capitalists, to name a few. These audiences all have different reasons for reading and they do so at different times.

For example, patent examiners read the specification to find out what the invention is and to check that the specification meets legal requirements. They do this while the application is pending, while the technology is still (in most cases) state-of-the-art.

Business people read the specification when considering spending money, for the right to make, use, or sell the invention. They may have engineers who also read the specification to help decide whether the technology is a good investment. Such business people and engineers may read the specification when the patent first issues, while the invention is still state-of-the-art, or they may not read it until years after the patent issues.

Judges and juries read the specification to determine whether someone has infringed the patent claims. Infringement issues can arise anytime during the life of the patent, and, depending on when a suit is filed, the technology disclosed in the specification may still be state-of-the-art, or it may be outdated.

The specification is, then, a multipurpose document. It is a snapshot in time, capturing and recording what the inventor has discovered. It is a technical description, preserving how to make and use the invention, so it will not be forgotten. It is a marketing vehicle, bringing the advantages of the invention to investors. It is a legal instrument, amplifying the meaning of the patent claims and serving as the *quid pro quo* for the patent grant. The specification should contain something for everyone.

---

[30] *Id.* at 816-817.

[A] Patent Examiner

What are patent examiners looking for in the specification? Primarily, examiners scan the specification to verify that it supports the claims. Of course, every examiner is different, but consider what you would do if you had a day and a half to review the specification and drawings, study the claims, conduct a search, and issue an office action. A day and a half is not far from the norm for experienced examiners. In 1993 Group 2300, the patent examining group that handles many software patent applications, received 8,661 new applications.

Confronted with this workload, no doubt many examiners scan the specification, look at the drawings and the description of the drawings, and then quickly turn to the claims. The claims define what prior art must be searched and the search for prior art takes the most time. Therefore examiners use the specification as a quick introduction to the technology of the claims.

Since examiners do not have a lot of time to study the specification, you should make their job easier. Make the specification a road map for understanding the claims. Remember, once the examiner approves the specification and allows the patent, the specification carries a presumption that it meets the requirements of 35 U.S.C. § 112. That presumption can only be overcome by clear and convincing evidence. Therefore, you want the examiner to be impressed. Of course, the specification is more than just a road map for the claims. Nevertheless, there is considerable benefit in helping the examiner do the job.

How can the specification be made into a road map? Start with the claims the examiner is being asked to search and do what the examiner does. Examiners are trained to diagram at least the independent claims before conducting a search. Examiners draw claim diagrams to refer to while searching. Claim diagrams remind examiners, when searching, to focus on the claims and not merely on the disclosed embodiment. Draw a diagram of your claims.

There is nothing special that your claim diagram should resemble. You could draw boxes or ovals to represent each claim element or limitation, with connecting lines to show the relationship or association between elements or limitations. The claim diagram does several things. It shows what claim elements are essential to the invention. More importantly, it shows the relationships between those elements.

Later when considering what the basic building blocks that make up the invention and its preferred embodiment are, you will explore various interfaces, which are the connections or relationships between entities. A claim diagram will help you identify these building blocks and interfaces, and will remind you to adequately disclose them.

Next, look at the drawings you intend to submit with the application. At least one of the drawings should suggest or resemble the claim diagrams already sketched. Perhaps none of the drawings show precisely the claim elements and connections you have sketched. This is probably because the claims are more broadly drafted than the embodiment shown in the drawings, and because a single claim element may be composed of several components shown in the drawings. If you encounter this situation, see if it is possible to put dotted lines around several components that comprise a single claim element. In most cases you should be able to do that.

Consider leaving the dotted lines there. The examiner may find them helpful when preparing a diagram of your claims. Consider adding a new drawing, showing the claim diagram. Introduce the reader to the invention by referring to it. Do not misunderstand. There is no requirement to provide a diagram of the claimed structure.[31] You are providing this diagram simply to help the examiner. You do not need to identify this diagram as a claim diagram, and probably should not, since the diagram may not fully correspond to the claims that are ultimately allowed. Instead, you can identify this diagram as representing some of the principles of the invention, or as an overview of some of the important concepts of the invention, or something similar.

In July 1994, the Court of Appeals for the Federal Circuit ruled *en banc* that the Patent Examiner is required by 35 U.S.C. § 112 to look to the specification in determining whether the invention claims patentable subject matter under 35 U.S.C. § 101.[32] If you use means-plus-function claims, be aware that the examiner is going to inspect the specification to see if you disclose structure for each of the means-plus-function elements.

Do not disappoint the examiner in this regard, or you may receive a § 101 rejection in addition to a § 112 rejection. These rejections can be difficult to overcome, without adding new matter. Adding new matter requires filing a new application, which may not be possible if the one year statutory bar has expired.

[B] Business Reader

The next reader of the specification may be a business person, potential licensee, or venture capitalist. These readers probably do not completely understand patents. They may not know that the patent claims define the scope of the patent monopoly, and they are therefore likely to think the patent covers whatever is described in the part they do understand, or have found time to read. Often, business people, potential licensees, and venture capitalists read only the title, abstract, and the first few paragraphs of the specification. Then, if they are interested in pursuing the patent further they will ask for a patent attorney's second opinion. Having reached this point and having committed to spend money for legal assistance, they are interested.

Therefore, the patent specification is a good place to sell the invention, particularly in the first few paragraphs. Put the features and benefits of the invention in the first few paragraphs, where you still have the reader's attention. Although the specification must enable one of skill in the art, there is no requirement that it must begin doing so in the first few paragraphs. Save the

---

[31] U.S. Dep't of Commerce *Manual of Pat. Examining Proc.* § 904.01(a) (Mar. 1994).

[32] *In re* Alappat, 31 U.S.P.Q.2d (BNA) 1545 (Fed. Cir. 1994).

enablement for the description of the preferred embodiment. By that time, you have lost most of the business people, potential licensees, and venture capitalists, anyway.

It is best to refrain from such salesmanship in the abstract, even though this is one thing everyone reads, due to its brevity. The abstract is strictly a searching aid. Load the abstract with keywords and keep needless verbiage and claim language legalese to a minimum. The abstract's purpose is to make one specification jump out when searching by hand or by keyword. Again, anyone who has searched patents will understand how the well-written abstract makes one patent get noticed, while a poorly written one glazes over the reader's eyes and is ignored.

[C] Courts and Specification

Courts read patents because they have to. Federal judges are busy and reading patents is rarely considered a delight. Patent cases frequently take years to litigate, with countless discovery disputes, motions, protective orders, and requests for preliminary injunctions and motions for summary judgment. Federal judges are aware of this and no doubt many dread receiving a patent case, because it represents a truckload of papers they will be asked to read.

In a patent case, one document the judge is going to try to read is the patent. The patent, therefore, is the plaintiff's vehicle to victory. If the patent specification can convey the importance of the invention, how lacking the art would be without it, and why the patentee deserves a patent monopoly, then the specification has done its job. If the patent specification can explain, in layman's terms (not to one skilled in the art, but to the judge) what the invention is about, the patent claims will have meaning and the judge is far more apt to construe them in the patentee's favor.

§ 12.09   Mechanics of Specification

The Code of Federal Regulations gives basic guidelines for the elements of a patent application and the arrangement of those elements. These guidelines illustrate what form the specification should take. As set forth in 37 C.F.R. § 1.77, the elements of a patent application are and should appear in the following order:

1. Title of invention
2. Cross-reference to related applications, if any
3. Cross-reference to microfiche appendix, if any
4. Brief summary of the invention
5. Brief description of the drawings, if any
6. Detailed description
7. Claims
8. Abstract of the disclosure
9. Oath or declaration
10. Drawings, if any.

Consult the *Manual of Patent Examining Procedure* § 608 for additional details regarding the desired form and style of the specification.

§ 12.10 Defining the E-Commerce or Business Model Invention

The first step in describing the e-commerce or business model invention is to define what the invention is. Inventors usually have a pretty good idea of what they *think* the invention is. Inventors, however, often do not understand what it means to be an invention and they may have difficulty differentiating between the invention and their creation. Many inventors view their entire creation or product as the invention, even though the true invention may be in a specific improvement or discovery that makes that creation or product better than the competition. Thus, start by ferreting out the invention.

Experienced patent attorneys often draft claims before writing the specification. This is good practice as it forces you to find the invention. Having found the invention, you know better what must be disclosed in the specification and what is superfluous. Some may argue, "It helps me to understand the invention if I write the specification first." That may be so. Many people write to understand, and there is nothing wrong with sketching out thoughts in writing, as an aid to understanding the invention. If you do write the specification first, be aware that it may be necessary to go back and extensively revise the specification after the claims are written to ensure that the specification supports the claims.

[A] Collaboration Between Attorney and Inventor

To ferret out the invention and to draft a good specification, patent attorneys and inventors need to collaborate. Patent attorneys have the legal knowledge of what it takes to be a patentable invention; inventors have the technical knowledge of their development or discovery. It is only natural that both should be actively involved in drafting the document that will define the inventor's patent rights for the next 20 years and record the inventor's discovery for civilization.

This collaboration is not easy — patent attorneys and inventors speak different languages. Both must strive to communicate with the other, if an understanding of the invention is to be reached. Experienced patent attorneys are aware that many inventors find it frustrating to explain their invention to a patent attorney. Inventors live and breathe their inventions and they are accustom to discussing their inventions with peers who understand the technicalities of what they are doing. Patent attorneys may often start out knowing very little about an inventor's particular invention, but having been placed in a position of ignorance so often, are usually quite adept at learning the technology quickly.

Begin perhaps by explaining to the inventor what the legal objectives are. Explain how the patent grant will help the inventor or the inventor's company keep the competition in check, or how a patent will help attract investors. Also explain the importance of accurately describing the invention and of providing an enabling disclosure of the best mode. Explain that once issued the patent will be preserved forever as a part of the collective knowledge of humanity. That way you will get the inventor interested in collaborating with you. Make your inventor understand that this is not simply an explanation of the invention to you; the inventor is collaborating with you to draft an important document that neither could do alone.

Having communicated to the inventor your respective roles, the next step is to ferret out the invention.

[B] Checklist of Specifics to Discuss with the Inventor

The following checklist of things to discuss with the inventor may be helpful.

CHECKLIST 12-1: ITEMS TO DISCUSS WITH INVENTOR

___ 1. What does the inventor think the invention is?

___ 2. What was the objective in making the invention?

___ 3. Component Parts of the Invention

    ___ a) What component parts make up the invention?

    ___ b) What component parts, if any, are new?

    ___ c) How are the component parts connected? How do they work together?

    ___ d) Are any of the connections new or nonstandard?

    ___ e) What variations are possible?

    ___ f) What parts took longest to develop? Why?

___ 4. Purpose and Function of the Invention

    ___ a) What does the invention do; what is it used for?

    ___ b) Are there other ways to do what the invention does?

___ 5. Problems Solved by the Invention

    ___ a) What problems does the invention solve?

    ___ b) In solving those problems, what was the most difficult part?

    ___ c) Before the invention, how were those problems addressed?

___ 6. Competition

    ___ a) What competes with the invention?

    ___ b) What are the advantages of the invention over the competition?

    ___ c) How are these advantages achieved?

    ___ d) What features will the competition want to copy?

    ___ e) How are these features achieved? How may the competition add them?

___ 7. Commercialization of the Invention

    ___ a) Is the invention used in a commercial product?

___ b) Has the product been introduced? If so, when, and under what circumstances?

___ c) What advantages and features help sell the product?

___ d) How are these advantages and features achieved?

___ e) Does the invention improve upon a prior product? If so, what products and in what ways?

___ f) Provide a full description of prior product to cite as prior art.

___ 8. What is the skill in this art?

___ a) Who else works in this field? What is the general level of training and experience required?

___ b) What journals are published about this field?

___ 9. Best Mode Contemplated by the Inventor (This question should be asked or repeated just prior to filing.)

___ a) Has the inventor thought about a best mode of practicing the invention?

___ b) If a best mode has been contemplated, is it fully disclosed?

[C] Attorney-client Privilege

Inventor-attorney communication is essential in finding and describing the invention. This raises a delicate subject, the attorney-client privilege. Attorneys are taught that communications between attorney and client are privileged. The privilege belongs to the client. The concept of privilege is deeply rooted in common law. The attorney-client privilege is designed to allow the client to freely communicate with the attorney, without fear that the attorney will later testify as a witness against the client. The attorney is also duty-bound to keep communications with the client confidential.

To qualify as attorney-client privilege, the following criteria must be met:

1. Asserted holder of the privilege is or sought to become a client
2. Person to whom the communication was made is a member of the bar of a court and is acting as a lawyer in connection with the communication
3. Communication relates to a fact of which the attorney was informed:
   (a) by attorney's client
   (b) without the presence of strangers
   (c) for the purpose of securing primarily either an opinion of law, or legal services, or assistance in some legal proceeding
   (d) not for the purpose of committing a crime or tort
4. Privilege has been claimed, and not waived by the client.[33]

---

[33] United States v. United Shoe Mach. Corp., 89 F. Supp. 357, 358 (D. Mass. 1950).

Thus you might assume that the communications between inventor and patent attorney fall squarely into the realm of the attorney-client privilege. After all, the client is clearly seeking legal services, specifically assistance in drafting, and prosecuting a patent application.

The majority view is that the attorney-client privilege does apply to communications between patent attorney and client. See, for example, *Natta v. Zeltz*[34] (disclosure policies of the patent laws do not preclude the proper application of the attorney-client privilege); *Chubb Integrated Systems, Inc. v. National Bank of Washington*[35] (attorney-client privilege from discovery is not lost in a patent case merely because the communication contains technical data); *Knogo Corp. v. United States*[36] (accompaniment of a draft of an application to be submitted to the patent office with a letter from the client to his attorney does not destroy the attorney-client privilege).

There is, unfortunately, a minority view in which the attorney-client privilege does not apply. The minority view treats the patent attorney as a mere conduit between the inventor and the Patent Office.[37] Under this view, information given to the patent attorney to be submitted to the Patent Office is not protected. See for example, *W.R. Grace & Co. v. Viskase Corp.*[38] (information received solely for the purpose of conveying that information to the Patent Office is not protected by the attorney-client privilege).

The following lists demonstrate the types of things that are protected and that are not protected. Recognize, however, that courts have considerable latitude in deciding whether to apply the attorney-client privilege or not. Often the decision is made by a magistrate. As these lists demonstrate, there are numerous inconsistencies. The following examples illustrate when privilege has held:

1.   Purely technical exposition of invention is privileged[39]
2.   Technical explanation of general field of technology and technical comparison of the invention and other devices are privileged[40]
3.   Letter accompanied by a draft of specification and claims from inventor to attorney is privileged[41]
4.   Communications that contain opinion of the attorney regarding patentability and scope of the patent claims are privileged[42]
5.   Inventor's memo discussing prior relevant art is privileged[43]

---

[34] 418 F.2d 633, 636 (7th Cir. 1969).

[35] 103 F.R.D. 52, 56 (D.D.C. 1984).

[36] 213 U.S.P.Q.2d (BNA) 936, 942 (N.D. Ill. 1991).

[37] Jack Winter, Inc. v. Koratron Co., 54 F.R.D. 44 (N.D. Cal. 1971); Duplan Corp. v. Deering Milliken, Inc., 397 F. Supp. 1146 (D. S.C. 1974).

[38] 21 U.S.P.Q.2d (BNA) 1121, 1122 (N.D. Ill. 1991).

[39] Knogo Corp. v. United States, 213 U.S.P.Q. (BNA) 936 (Ct. Cl. 1980).

[40] *Id.*

[41] *Id.*

[42] *Id.*

[43] F.M.C. Corp. v. Old Dominion Brush Co., 229 U.S.P.Q. (BNA) 150 (W.D. Mo. 1985).

6. Invention Disclosure Forms are privileged[44]
7. Interoffice memos with handwritten comments recounting legal opinion of counsel are privileged notwithstanding that an attorney did not prepare them[45]
8. Testing and analysis of accused product is privileged[46]
9. Infringement opinion by plaintiff's counsel relating to third party patent is privileged.[47]

The following examples illustrate when privilege did not hold:

1. Cover letters accompanying a Patent Office communication are not privileged[48]
2. Documents relating to foreign application filing are not privileged, when the documents are of public record and constitute information necessary to complete foreign filing[49]
3. Information Disclosure Statement is not privileged[50]
4. Factual information communicated in order that attorney could disclose it in a patent application is not privileged[51]
5. Client authorization to file and prosecute application is not privileged[52]
6. Papers submitted to Patent Office are not privileged[53]
7. Technical information communicated to attorney but not calling for legal opinion or interpretation and meant primarily for aid in completing patent applications is not privileged.[54]

Since the conduit theory exists, be careful to avoid making a written record that could later be contrary to the inventor's interests. This caution does not prevent collaboration with the inventor. It simply means to use care when taking notes or when writing letters. This applies to letters written by the inventor to you, as well as letters from you to the inventor. A useful practice is to include a statement in the opening paragraph of letters that the client has requested specific legal advice, or that you are responding to a request for legal advice.

If there ever is an attempt to pierce the attorney-client privilege, it will probably be in the context of litigation discovery, when an accused defendant is trying to limit the patent coverage to the way the inventor originally explained

---

[44] Illinois Tool Works, Inc. v. K.L. Spring & Stamping Corp., 207 U.S.P.Q. (BNA) 806 (N.D. Ill. 1980).
[45] *Id.*
[46] W.R. Grace & Co. v. Viskase Corp., 21 U.S.P.Q.2d (BNA) 1121 (N.D. Ill. 1991).
[47] *Id.*
[48] *Knogo Corp.*, 213 U.S.P.Q. (BNA) 936 (Ct. Cl. 1980).
[49] *Illinois Tool Works*, 207 U.S.P.Q. 806.
[50] *W.R. Grace & Co.*, 21 U.S.P.Q.2d 1121.
[51] Hercules, Inc. v. Exxon Corp., 434 F. Supp. 136 (D. Del. 1977).
[52] Jack Winter, Inc. v. Koratron Co., 54 F.R.D. 44 (N.D. Cal. 1971).
[53] *Id.*
[54] *Id.*

the invention to the attorney when the application was being drafted. This could be disastrous, since, as noted, sometimes an inventor does not fully understand the true scope of the invention, particularly during the initial inventor-attorney collaboration.

§ 12.11   Describing E-Commerce Invention

A proper e-commerce specification lies somewhere on a spectrum between a black box and full source code. A black box is a descriptive concept engineers use when they want to simplify by hiding the details. The concept is simple. The black box is described simply as performing a given function, so that a given input, subjected to that function, produces a given output. The details of how the function is performed are not disclosed. E-commerce systems and software can be described like this. For example, an Internet-based customer resource management (CRM) system can be described as a black box that takes information about customers' buying habits, stores them in a database and produces a series of on-screen displays to present the information in a convenient way. How the CRM system does this is not revealed. If the CRM system is the invention, then this description is not very useful, since it tells nothing but the end result and leaves everything to experimentation.

At the other end of the spectrum lies full source code, which arguably contains every detail one skilled in the art needs to make and use the invention. However, full source code may not provide a very good understanding of what the invention does.

The principal function of source code is to supply a rigorously complete description that the compiler processes to produce an executable computer program. Good programmers place comments in their source code to allow themselves and others to more quickly comprehend what the source code is about.

Even with well-documented comments, one person's source code can be difficult for another person to read and comprehend. Therefore, it is best not to rely entirely on source code in describing a software invention in the specification. In patent application practice, source code provides too much information and may not adequately identify what is new and inventive from what is old and commonplace.

There are many different ways to adequately describe an e-commerce or software invention. Almost certainly, the description of an e-commerce or software invention will fall somewhere between the above two ends of the spectrum. It is therefore possible to arrive at a fully enabling disclosure by starting from either end of the spectrum. We will consider both, using top-down and bottom-up techniques, which software professionals will recognize as techniques that they also use in developing software. To the extent that most e-commerce inventions involve software, it is helpful to have an understanding of these two programming techniques.

To a software professional, top-down programming starts with broad concepts and overall program structure and then iteratively refines those concepts and structures into subparts, sub-subparts, and so forth, until a complete, compilable source code has been created.

Bottom-up programming is the reverse process. The sub-subparts are written first and are then connected together and grouped into larger and larger parts, until the complete, compilable source code has been created.

The patent specification can be built in much the same way. Both the software professional and the patent practitioner need to understand this, since this similarity between software program writing and patent application writing can be a valuable common ground when the software professional and patent practitioner meet to discuss the invention.

[A] Functional Components of E-Commerce Software

A computer program, whether intended for e-commerce or for any other purpose, is made up of building block components, just like any other machine. These components are data structures, data processors, and interfaces. These components are functional elements. Data structures are places to organize and store data. Data processors manipulate data by performing processes or algorithms upon the data. Interfaces connect data structures and data processors to the outside world, or to other data structures and data processors. You can use these three building block components to describe virtually any computer program.

Do not mistake these building block components for the source code files that may contain these components. Source code is really something different. Source code is a list of instructions, written in a selected computer language, and then converted into computer machine language, which the computer uses to build the software machine described by the instructions. The software machine is made up of the building block components; the source code is simply a detailed blueprint telling the computer how to assemble those components into the software machine.

Source code is often organized into separate files; files are organized into separate modules; modules are organized into separate functions or routines, and so forth. This organizational structure is largely for the benefit of the programmer. A program is easier to understand and to debug if it is well organized in this way.

If given source code as part of a disclosure, you can break it down fairly easily into files, modules, functions, and the like. Most programmers place file name and module description headers in the title block of each section of the source code that give the file, module, and function breakdown. Most programmers also record who authored the code, who modified it, and when and why in the title block. This can be a helpful indication of who may be a joint inventor of subject matter sought to be patented.

Finding the basic building blocks takes a little more work. The best approach is to have the inventor identify the basic building blocks for you. Whether you are working alone, with only the source code, or with the inventor's help, here is what to do. Keeping in mind what you are trying to patent, first identify what data is fed to the system as inputs. Similarly, identify what data is produced by the system as outputs. Then identify what data structures are required to build the preferred embodiment. Often these data structures will receive input data or supply output data. Be aware that there may also be additional data structures, used somewhere between input and output, but not necessarily directly involved with input and output.

Having identified the data structures, next identify the key data processors that manipulate the data in these data structures. The objective is to identify the processors responsible for making the claimed invention work.

In locating the key data processors, it is often helpful to use the process of elimination. Essentially all programs have some interface to a computer operating system, and some interface to the user. Often the claimed invention does not reside in either the operating system interface or the user interface. In other words, often the operating system interface and the user interface are conventional, or obvious variations of the conventional. If you know that these interfaces are conventional, locate the portions of the source code that implement these interfaces, and eliminate them from further detailed consideration. This will often eliminate huge sections of code because a great deal of most computer programs involves the necessary, but strictly conventional, handling of operating systems and user interface details. This is especially true where graphical user interfaces (for example, Windows or Web browsers) are involved.

Once you have identified the data structures and key data processors, you are now ready to describe the preferred embodiment as a software machine. Show and describe how the data structures and processors are connected and how they work together to produce the claimed result. Use of one or more flow charts to illustrate how the key processors operate may be helpful.

Although the user interface may be conventional, you may also want to include examples of the user interface screens, if that will help give the reader an overview understanding of what the invention does.

Ordinarily, the way the source code has been organized into files, modules, and functions is a matter of design choice. Different programmers may structure their source code differently, and yet achieve the same overall result. Thus, providing a description of how the source code of the preferred embodiment has been structured is usually not necessary. Nevertheless, if the source code is particularly complex, you may want to show how it is organized and arranged, using an organizational chart (similar to a company organizational chart).

In drafting the specification in this fashion, do not lose sight of the claims. Be absolutely certain each claim element is fully supported by a building block component or process step in the preferred embodiment described. Also, be sure to include a full description of any necessary initialization variables or parameters. Courts often look to what is taught in the specification when deciding whether the subject matter is patentable under 35 U.S.C. § 101. The *en banc* decision in *In re Alappat* is a good example.[55]

Data structures and interfaces are described in greater detail in subsections [1] through [3] below.

[1] Data Structures

Data are simply organized information; data structure simply describes how the data are organized. In a spreadsheet program the data may be organized into cells identified by unique column and row numbers. In a word processing program the data may be organized into a single long chain, or "string" of letters, the letters forming words, the words forming sentences, the sentences forming paragraphs, and so forth.

These are examples of fairly specialized data structures. Software professionals have many different textbook data structures to choose from and when these do not quite fit, they make up their own. The textbook data structures include the array, the linked list, the stack, the queue, and the deque. The list goes on and on. In a fully developed computer program there may be many individual data structures.

Why worry about data structures? Part of the software professional's craft is choosing just the right data structures for the job. If the data structure is chosen wisely, the program becomes much easier to write. Chosen wisely, the data almost seems to process itself. Chosen poorly, the data resists. The practical implications of this phenomenon can be astounding, as AT&T learned in the early 1980s.

The local telephone operating companies needed a way to keep track of its cables and equipment in the field. Bell Labs developed a database to do this. The database was written using conventional CODASYL technology. The program comprised over 155,000 lines of COBOL code amounting to a printout four inches thick. The CODASYL database took 50 people five years to develop. The program received many complaints. It was extremely slow and difficult to maintain; it rigidly resisted modification every time new telephone equipment was deployed. It required a maintenance staff of 50 people. The data structure did not fit the problem.

In the early 1980s Bell Labs took a fresh look at the problem. This time, instead of trying to force the problem to match an existing data structure, the programmers developed a new data structure, specifically to model this problem. They named the new data structure "a directed hypergraph." With the new data structure in place, a new database program was quickly written, in half the time, by half the number of programmers. The new program when printed out was less than one-half an inch thick. It was fast, easy to maintain, and easy to modify

---

[55] *In re* Alappat, No. 92-1381 (Fed. Cir. 1994).

when new equipment was deployed. Now only five people are needed to maintain it. The data structure fit the problem.[56]

[2] Common Data Structures

Data structures are important. When drafting a patent specification, how do you know if a data structure needs detailed explanation or if a data structure needs no explanation to one of ordinary skill in the art? Like so many questions about what is known in the art, it is possible to find the answer in books. The authority on data structures is Donald E. Knuth's *The Art of Computer Programming*.[57] Another well-organized and useful book on data structures is Robert L. Kruse's *Data Structures & Program Design*.[58] Consult these sources to help identify what is conventional and what is not. For quick reference, [a] through [e] compare some common data structures you may encounter.

[a] Array

Anyone who has taken a beginning course on computer programming has encountered the array. An array is simply a table of data values, all values of the same data type. Being of the same data type means that each of the data values is stored in the same way, using the same amount of memory space. Values of the same data type are computationally handled in the same way. For example, an array could comprise a table of integers, or a table of floating point numbers, or a table of alphanumeric characters, or even a table of five letter words. It makes no difference, as long as all data values are of the same type.

The array is not just any table, it is a look-up table in which each data value is assigned to its own indexed storage cell. These cells are sometimes called array elements. To look up the data value stored in a particular cell or array element, you simply refer to that cell or array element by index number. For example, consider the following ten element array of five letter words, A[n], where n=10.

$$A[1] = alpha$$
$$A[2] = spock$$
$$A[3] = james$$
$$A[4] = mccoy$$
$$A[5] = earth$$

---

[56] For a complete account of how the directed hypergraph data structure solved AT&T's database problem, *see* A. Jay Goldstein, *A Directed Hypergraph Database: A Model for the Local Loop Telephone Plant*, 61 Bell Sys. Tech. J. (Nov. 1982). See also Jon Bently, *Programming Pearls 30* (1986).

[57] Donald E. Knuth, *The Art of Computer Programming* (1973); this is a multivolume work. Two other volumes have been published to date: Donald E. Knuth, *Seminumerical Algorithms*, 2d ed. (1980); Donald E. Knuth, *Sorting and Searching* (1973).

[58] Robert L. Kruse, *Data Structures & Program Design*, 2d ed. (1987).

A[6]   =   venus
A[7]   =   pluto
A[8]   =   quark
A[9]   =   light
A[10] =   omega

By specifying array element A[7] the computer looks up the data value stored at that data structure location and returns the value "venus." Due to the way arrays are stored in memory, all elements must be the same size (same data type). Here the array contains only five letter words. Each letter is, in turn, of the data type "character." Depending on the particular computer language implementation, the character data type may comprise a single byte, the smallest addressable unit of data. This is the case in ANSI C, where the *char* data type consists of one byte of storage.[59]

An array can be a simple, one-dimensional list, or it can be multi-dimensional. A two-dimensional array might appear as a spreadsheet-like table having horizontal and vertical indices. Both horizontal and vertical indices must be specified to address a given array element or data cell. The two-dimensional array is sometimes called a rectangular array.

Another important characteristic of the array is that, by definition, the array has a fixed size, that is, a fixed number of array elements or data cells. This can be quite a limitation, because you cannot, for example, stuff 11 data values into a 10 element array. The array is a "static" data structure. Once declared, or "bound" to specific memory locations, the size of the array cannot be increased. This limitation carries a hidden advantage, however. Because the array elements are each bound to a specific memory location, the computer can access the elements quite quickly. The array is therefore a good data structure to achieve speed.

[b] List and Linked List

Sometimes a static data structure, such as an array, is too limiting. In some programming problems you simply do not know, in advance, how many data values you will have. In such cases, a better data structure is one that can vary in size as the program runs. Such a data structure is considered a dynamic data structure. The most generalized dynamic data structure is the list, or linked list. A *list* is simply a finite, ordered collection of elements that can vary in number. Lists can take many forms and arrangements, some of which are separately discussed at § 6.29 and § 6.30. The key difference between the list and the array is that the list is not a predefined size, whereas the array is a predefined size.

An important feature of a list is that you can insert an element between two existing elements, and all the existing elements automatically make room for the

---

[59] Mark Williams Company, *ANSI C: a Lexical Guide 95* (1988).

new element. Precisely how this is done depends on the type of list. Imagine a family of penguins lined up, single file, on the ice. A new penguin joins the family, squeezing in, between two penguins already in the line. All the penguins slide sideways on the ice, automatically making room for the new penguin. New element insertion may be thought of like this. The reverse process is also possible. With a list you can remove an element and the remaining elements automatically move in to fill the void.

For what kinds of things are lists practical? Grocery lists come to mind, of course. Yet there are uses for lists you might not consider. The words in this sentence, indeed the words in this entire book, may be stored as a list by a word processor. This is how a word can be so easily inserted in or deleted from a sentence.

Of course, computer data are not penguins. Data do not actually slip and slide around in memory, the way penguins might on the ice. To understand the way lists actually work, you need to know about the pointer. The pointer is a data type with a very special purpose. The pointer stores a value that points to a location in computer memory. Locations in computer memory are referred to by address number. To point to a given address, simply store the address number in the pointer. Having done so, the pointer now points to the given memory address. By comparison, the integer is a data type that stores a value representing a number. You can add +1 to the integer 10 and get 11. Similarly you can add +1 to memory address 65389 and get memory address 65390.

Understanding pointers, now envision the list composed of individual elements, each having a data cell to hold a data value, and each having at least one pointer to hold the memory address or addresses of that element's neighbor elements. To insert a new element into the list between two neighbor elements, simply change the neighbor elements' pointers to point to the new element and add the neighbor elements' addresses to the pointers of the new element. Now the new element is integrated into the list, just as if it had been present there in the first place. To remove the element, simply reverse the process. Unlike the penguins, which must physically slide sideways to make room for a new penguin, there is no need for the data to physically move. Although *logically* lined up in the list, the new element may physically occupy a memory address far removed from its adjacent neighbors. The computer does not care. It references different memory addresses with equal ease.

[c] Stack

The *stack* is a special kind of list in which all insertions and deletions are made at one end, not in the middle. Called the "top" of the stack, this one end is where all data is "pushed" onto or "popped" from the stack. The analogy frequently given is that the stack works like the stack of lunchroom trays in the lunch room, or like the stack of salad plates at the salad bar.

The stack exhibits last-in, first-out (LIFO) behavior, making it an ideal scratch-pad to temporarily store sequential information that needs to be popped off, in the reverse order it was pushed in. A program needs information fed to it in this way to be able to return where it left off when it executes procedures, which call other procedures (which may call still other procedures). Without the stack to guide it, the program is unable to retrace its steps and gets hopelessly lost. MS-DOS uses a stack data structure to keep track of its place in much the same way.

[d] Queue

Another variation on the basic list is the queue. The *queue* is a first-in, first-out (FIFO) structure in which all insertions to the list are made at one end, and all deletions to the list are made from the other end. The queue is used quite heavily, as it mimics many real world problems, such as standing in line at the ATM machine or waiting to take a ticket to gain entry to the parking garage.

A special queue worth knowing about is the deque, pronounced "D-Q" or "deck." The term *deque* is actually a contraction for "double ended queue." Being double-ended, insertions or deletions can be made from either end, but never from the middle.

[e] Binary Tree

The *binary tree* is a linked data structure with a split personality. The lists discussed so far are linked in a sequential fashion. To get from one end of the list to the other, you must visit each element sequentially, like counting a string of pearls. The binary tree is different. Starting from the top you have two choices: either branch left or right. Having made your selection, you again have two choices: either branch left or right. The elements that make up the binary tree are sometimes called *nodes*. The node at the top is called the *root*. The two binary trees branching left and right from the root are called the *left subtree* and the *right subtree*. You sometimes see the outermost nodes referred to as *leaves*, in keeping with the tree metaphor. Actually, the way most binary trees are drawn, it would have been more accurate to call it the inverted binary tree, but nobody does that.

The term binary refers to the fact that there are two (and only two) subtrees branching from each node. Thus the number of nodes at every level increases as binary numbers increase with each placeholder digit: 1 . . 2 . . 4 . . 8 . . 16 . . 32, and so on.

The binary tree is an extremely useful data structure, which makes possible a very quick searching technique called binary search. In binary search, you make a comparison at a given node of the tree and then traverse right or left, depending upon the outcome of the comparison. As it turns out, on average, the binary search technique is considerably faster than a linear search technique where you must make a comparison with every element of the list.

To demonstrate this, consider the guessing game in which you must guess the number I am thinking. The number may be from one to ten. Using a linear search technique you must ask: Is the number one? Is the number two? Is the number three? and so on, until I answer, Yes. If you are lucky, I was thinking of the number one, and you have the answer after only one guess. If you are unlucky, I was thinking of the number ten, and you have the answer after ten guesses.

Now change to a binary search technique. Start by picking a number in the middle and ask, "Is the number five?" I answer either "Higher," "Lower," or "Yes." Based on the answer, if the number is lower, pick again, selecting a number in the middle of the left branch of the tree. If the number is higher, pick a number in the middle of the higher branch of the tree. You will have the answer "Yes" after at least four questions.

Computers are blindingly fast at doing such comparisons. Therefore, why the big fuss over reducing the number of guesses from ten to four? Clearly if all you have is ten numbers, there is little computational difference between linear search and binary search. However, increase that number to one million and the point spread opens wide. Even taking into account that the linear search, on average, will find a match before reaching the end of the list, it will still take the linear search $N/2$ guesses, that is 500,000 guesses. The binary search will do it in 20 guesses, at most.

[3] Interfaces

Software professionals use the term *interface* to refer to a working connection between two entities. Sometimes the connection is physical. Two software entities can be so integrally joined together that they simply merge into a new, larger software entity. More often, the connection is metaphysical — there is no physical connection. The two software entities retain their separate identities and simply communicate with one another.

Strictly speaking, the term interface describes the abstract concept of a working connection. In practice, the term interface is more often used to refer to the means or mechanism by which the working connection is made. Thus, the interface between two software entities will probably describe precisely how the entities exchange information and under what circumstances.

For example, software entity A may request software entity B to sort a list of numbers by supplying B with:

1. Precise information about when the list may be found
2. Size of the list
3. What type of sorting is requested
4. How B should notify A when finished, and so forth.

The interface would need to specify in what order this information must be passed from A to B, how the information must be expressed, and even how A is to notify B that this request is intended for you. Likewise the interface would also need to specify how B informs A when the sorting task is complete, how B informs A that an error has occurred, and what that error was. In a properly designed interface, nothing is left to chance. Every possible contingency is provided for with mathematical precision.

Interface is a very powerful concept. It makes division of labor possible. Programmers use division of labor to great advantage. Why write the same information sorting routine ten times, just because information needs to be sorted for ten different purposes? Instead, why not write one, general purpose sorting entity and allow other parts of the program to ask it for sorting help when needed? To do this, programmers simply design the individual entities they need and then design an interface to connect them. In this way, each individual entity specializes in only one part of the problem. The interface connects the entities together, allowing them to work in concert while still retaining their individual specialties.

[a] Operating System Interface

Virtually every computer program has an interface. The operating system interface is the mechanism that connects the computer program to the operating system. The operating system interface may, for example, give a program access to the disk drive, so that information contained in a data structure can be stored on disk. It may also give a program access to a printer, for example. Practically every software program will have an operating system interface.

Rarely does the invention reside in the operating system interface, although it can happen. In those rare cases where the invention does modify the operating system, describe the operating system in detail. Usually, however, the operating system is not part of the invention and you should not need to include an extensive description of it.

[b] User Interface

The user interface is one of the most interesting interfaces. It supplies the connection between the software program and the human user. Although admittedly dehumanizing, the user interface concept treats the software program as one entity and the user as another entity. As far as the software program is concerned, the two entities are equals. Today, the mention of user interface evokes notions of graphical screen displays, menus, button bars, function keys, point and click, and the ubiquitous mouse. These represent the user interface stone age, as far as what is possible.

In his science fiction book *Neuromancer,*[60] William Gibson describes a future world where software programs are connected directly to the brain by inserting chips into medically implanted sockets behind the ear. The programs give the wearer instant knowledge of a selected subject. If your automatic transmission needs an overhaul, simply select the How to Repair Your Automatic Transmission chip, insert it behind the ear, and repair the transmission yourself, with complete confidence and expert knowledge of how to do the job. Gibson's vision is not farfetched. No doubt there will be many future biomedical

---

[60] William Gibson, *Neuromancer* (1984).

inventions of great significance that will involve this very type of user interface. Although the socket behind the ear may not be cosmetically appealing, it is no less appealing than a hearing aid inserted in the ear, and there are many other places to put the socket.

The user interface of the future need not involve implants at all. Spoken language may be used. Many will remember the dialogues between Dave Bowman and the HAL 9000 computer, in *2001, A Space Odyssey*. Through HAL's advanced technology, the user interface between HAL and Dave was all but invisible to us. HAL and Dave simply talked with one another, as two people do, using Dave's natural language, English. HAL understood more than English, however. HAL also understood how to hold a conversation with a human. This is undoubtedly a desirable goal for the user interface of the future.

§ 12.12   Describing the Business Model Invention

In contrast with the e-commerce patent specification, which will often comprise a description of the system's computer hardware or software components, the business model specification has the potential to be far more diverse. Unlike e-commerce inventions, which typically hinge on computer technology, business model inventions can encompass virtually any human endeavor and may involve very little technology. Thus in drafting business model specifications you can expect to have a fair amount of latitude as to what to include.

One useful technique in piecing together a business model specification is to first identify what general business domain the invention falls into. It may be banking, insurance, shipping, sales, marketing, customer support, you name it. Identifying the applicable business domain will help you later in searching the prior art to learn the terminology used by business people in that domain. When drafting the specification it is helpful to have this terminology at hand, to allow you to draft a specification that is more readable and that uses applicable business terminology correctly.

After identifying the business domain and learning the terminology, next identify the key players in your business model. In a sales model, the key players may be buyers and sellers. In an insurance model, the key players may be insurer, insured, sales agents, claims adjustors, and government officials. Next identify how these players interact with each other. Typically the players interact by exchanging something — information, goods, services, money. Identify which are the important media of exchange (often money) and track how those media flow as the business model operates. In the kanban system described in § 8.02[3], the one medium of exchange was information which traveled on a written card, the kanban card, attached to the goods being manufactured. In a banking system, money is typically the medium of exchange, which flows among automatic teller machines, account ledgers, the federal reserve system, and the like.

You may also want to consider both static and dynamic views of the business model. The static view shows you how the model is set up. The dynamic view shows you how the model behaves as it operates. You may want to use a series of time-lapse snapshots of the business model in operation as a way of showing the dynamic behavior of the model.

The following subsections introduce some basic business model terminology and outline some of the more common business domains, key players, and media of exchange.

[A] Basic Business Model Terminology

Sometimes the most basic terms can be the most elusive. Trying to define what is a business method patent, or a business model patent, one quickly discovers the need to understand the term "business." Here are some definitions of that term and several other closely related terms that we probably use every day but give little thought to.

> *Business* — of or relating to or characteristic of trade or traders; the activity of providing goods and services involving financial and commercial and industrial aspects; a commercial activity engaged in as a means of livelihood or profit.

> *Commerce* — the movement of money, the buying and selling of products and services.

> *Banking* — engaging in the business of banking; transacting business with a bank.

> *Bank* — an organization, usually a corporation, chartered by a state or federal government, which does most or all of the following: receives demand deposits and time deposits, honors instruments drawn on them, pays interest on them; discounts notes, makes loans, and invests in securities; collects checks, drafts, and notes; certifies depositor's checks, and issues drafts and cashier's checks.

> *Finance* — the commercial activity of providing funds and capital; the management of money and other assets; the management of money and credit and banking and investments; to raise money through the issuance and sale of debt and/or equity; asset-based lending.

[B] Basic Business Domains

Typically a business model patent will lie within a specific business domain. A new type of electronic cash might best be described as being within the financial domain. Thus, the language of finance is a good one to adopt when describing electronic cash in the patent specification. The following list contains some of the basic business domains. The list is not exhaustive, but rather is provided to spur your thinking when drafting a business model specification.

- financial
- production

- the supply chain
- the distribution chain
- transportation — the movement of goods
- communication — the movement of services
- billing and collection
- internal operating procedures
  - distributing information
- distribution and sales procedures
- labor and resource allocation
- supplier and customer service
- product (service) creation and testing
- business customized software and information systems
- education
- market intelligence management
- data management
- return handling
- order processing
- e-commerce
- insurance
  - calculating premiums and commissions
  - calculating insurance vesting
  - health insurance eligibility verification
- business to customer
- advertising
- order processing
- credit check
- distribution
- support
- warranty claims
- repair
- follow-on sales
- business-to-business (and intra-business)
- supplier-manufacturer
- manufacturer-merchant
- warranty claims processing
- inventory check

- shipping schedule

- accounting

- product/service information dissemination

- consider product (purchase approval)

- price determination

- product availability

- order placement

- process order

- check inventory

- schedule delivery

- invoice generation

- ship product

- receive product

- confirm receipt

- invoice processing

- invoice payment

- payment processing

- follow-on sales

- warranty claims processing

- product support

- maintenance

§ 12.13   A Section 112 Checklist

35 U.S.C. § 112 governs how the patent specification is written and interpreted. The attorney who needs a quick answer to often-asked questions about 35 U.S.C. § 112 can use the following checklist. The cases cited in this checklist are good introductions to further research.

CHECKLIST 12-2:   THE ATTORNEY'S § 112 CHECKLIST

Written Description Requirement[61]

___ Purpose: To show applicant was in possession of the invention.

___ Test: Whether the specification reasonably conveys to the artisan that the inventor had possession of the invention on the filing date.[62]

---

[61] Vas-Cath Inc. v. Mahurkar, 935 F.2d 1555 (Fed. Cir. 1991); *In re* Hayes Microcomputer Prods., Inc. Patent Litigation, 982 F.2d 1527 (Fed. Cir. 1992).

[62] *Id.*

___ Relevant Time: Application filing date[63]

___ Question of: Fact[64]

___ Burden of Proof: Clear and convincing evidence[65]

Enablement Requirement

___ Purpose: To supply the *quid pro quo* for the patent monopoly by teaching how to make and use the invention.

___ Test: Whether the specification teaches one skilled in the art to make and use the invention without undue experimentation.[66]

___ Relevant Time: Application filing date[67]

___ Question of: Law[68]

___ Burden of Proof: Thoroughly convincing evidence[69]

Best Mode Requirement

___ Purpose: To restrain inventors from concealing the preferred embodiments actually conceived.[70]

___ Test: There is no objective standard for best mode; only evidence of concealment, whether accidental or intentional, is considered.[71] Whether best mode is met depends on the scope of the invention, the skill in the art, evidence of the inventor's belief and other surrounding circumstances.[72]

___ Relevant Time: Application filing date[73]

___ Question of: Fact[74]

---

[63] Vas-Cath Inc. v. Mahurkar, 935 F.2d 1555 (Fed. Cir. 1991).

[64] *Id.*

[65] DeGeorge v. Bernier, 768 F.2d 1318 (Fed. Cir. 1985).

[66] Northern Telecom, Inc. v. Datapoint Corp., 908 F.2d 931 (Fed. Cir. 1990).

[67] *In re* Comstock, 481 F.2d 905 (C.C.P.A. 1973).

[68] Spectra-Physics, Inc. v. Coherent, Inc., 827 F.2d 1524 (Fed. Cir. 1987).

[69] Hirschfeld v. Banner, 462 F. Supp. 135 (D.D.C. 1978).

[70] *In re* Gay, 309 F.2d 769, (C.C.P.A. 1962).

[71] Spectra-Physics, Inc. v. Coherent, Inc., 827 F.2d 1524 (Fed. Cir. 1987); DeGeorge v. Bernier, 768 F.2d 1318 (Fed. Cir. 1985).

[72] *In re* Hayes Microcomputer Prods., Inc. Patent Litigation, 982 F.2d 1527 (Fed. Cir. 1992).

[73] Spectra-Physics, Inc. v. Coherent, Inc., 827 F.2d 1524 (Fed. Cir. 1987).

[74] *In re* Hayes Microcomputer Prods., Inc. Patent Litigation, 982 F.2d 1527 (Fed. Cir. 1992); Amgen, Inc. v. Chugai Pharmaceutical Co., 927 F.2d 1200 (Fed. Cir. 1991); DeGeorge v. Bernier, 768 F.2d 1318 (Fed. Cir. 1985).

CHAPTER 13

EXPLOITING THE BUSINESS MODEL AND E-COMMERCE PATENT PORTFOLIO

§ 13.01 Introduction

This chapter focuses on exploiting business model and e-commerce patents — a multifaceted topic. The entrepreneur or small start-up company may have only one or two patents in the portfolio. Thus exploiting the portfolio boils down to exploiting those one or two patents. Whether the entrepreneur or start-up company is successful in profitably exploiting the portfolio will usually turn on the relative strength of the patents, and whether they can be enforced within the jurisdiction.

Companies with large patent portfolios have other considerations. They typically have many patents which, considered alone, are unlikely to yield licensing revenue. However, when considered as part of an entire portfolio, they can add value by blocking design around options and by simply increasing the size of the licensed estate. In a cross-licensing arrangement between two companies, the relative size of the respective portfolios can be a factor.

To illustrate, imagine a cross-licensing negotiation between two competing companies, each holding 250 patents on their respective product lines. Simply playing the odds, it is likely that each company infringes at least a handful of patents owned by the other. Rather than painstakingly identifying which require licenses by each party, it is more expedient to cross-license each company's entire portfolio. Not only does that avoid the complexities of identifying and valuing royalties under specific patents, it gives each company the flexibility to develop future products without concern that additional patent licenses will be required.

On the other hand, imagine the same cross-licensing negotiation between two competing companies, where one holds 250 patents and the other holds six. Unless those six contain something extremely valuable, chances are the company holding 250 patents will be unwilling to grant a cross-license to its entire portfolio. In this case, numbers matter.

This chapter comprises three parts. The first part focuses on the jurisdictional and infringement issues that may arise with business model and e-commerce patents. The focus is more microscopic — how is the single patent within the portfolio enforced, particularly where the Internet has enabled the contemplated licensee to operate outside the United States. The second part focuses on large scale portfolio development. The focus is more macroscopic — how does a company develop a strategic business model or e-commerce patent portfolio. The third part is related to the second. It focuses on mapping techniques useful in managing a large portfolio to ensure that the portfolio meets the strategic goals of the company.

§ 13.02 Part I — Jurisdiction Issues Involving Business Model and E-Commerce Patents

Business model and e-commerce patents abide by the same rules as other patents. In many respects, exploiting a business patent is no different from other patents. However, because of the specific technology involved, and because many business inventions involve the Internet, there are some special legal issues to consider when drafting an Internet patent application. The issues revolve around how the Internet patent will be enforced. Here is a list of issues to consider:

- How difficult will it be to get personal jurisdiction over an infringer — what if the infringer is an offshore Web site?

- Who is the direct infringer — is it necessary to sue the end user to snare the manufacturer as a contributory infringer?

- Will the statutory defenses available for business methods weaken the impact of the claims?

- Do the claims raise statutory subject matter questions?

The jurisdiction and infringement issues arise because the Internet detaches structure from function and distributes that function all over the world. Nicholas Negroponte described in his book, *Being Digital*,[1] the paradoxical way in which legal systems still fail to adequately differentiate between information and tangible things — to differentiate, as Negroponte puts it, between "bits and atoms." Negroponte describes how he was precluded from carrying across the Canadian border a CD-ROM containing a presentation he was giving in Vancouver. Negroponte was permitted to cross, but import/export laws interceded, and his CD-ROM got held up in customs. Fortunately, Negroponte had a copy of the same presentation on his Web site at home. He simply logged onto the site from Vancouver and downloaded the information. In short, the "atoms" got tangled in red tape at the border, while the "bits" flowed across without raising even a customs agent's eyebrow. Patent claims to Internet inventions raise many of these same "bits versus atoms" enforcement issues.

The patent statute defines what acts constitute infringement at 35 U.S.C. § 271. Most of these statutory provisions were crafted well before the Internet was conceived and well before the bits versus atoms issue was apparent. What does and does not constitute infringement of a patent claim may well depend on where the component parts of the Internet-based system are distributed and on how the claim has been drafted.

Thus in exploiting a business model or e-commerce patent that involves the Internet, it is well to consider jurisdiction first.

[A] The Nature of Jurisdiction

Jurisdiction is tied to sovereignty. Traditionally, every nation was thought to have sovereign power within its boundaries, and complete authority to control persons and things within those boundaries. Thus each nation has traditionally had the absolute power to regulate conduct that occurred within its territorial

---

[1] Negroponte, Nicholas. *Being Digital* (New York: Alfred Knopf, 1995).

boundaries. This territorial notion of jurisdiction is still quite strong, but it is beginning to soften the edges. Why is this? For one thing, territorial notions of jurisdiction often break down when cross-border disputes arise. When conduct involved in the dispute occurs on both sides of a territorial border, it is not often clear which law should be applied. This has forced all nations to rework their notions of jurisdiction. The Internet has accelerated the process. The Internet has made it possible to have a virtual "presence" within territorial boundaries, without ever having had a physical presence there.

If one has a virtual presence within a country by virtue of overt actions taken to "push" content there via the Internet, many believe the content supplying party should be subject to the laws of the receiving country. For example, if a Web site in the Chinese language offers to ship products to buyers in China, that Web site would likely be subject to Chinese law, even though the Web site is hosted by a computer sitting in Bermuda.

On the other hand, if a party's content has been "pulled" into the receiving country by residents there, with no overt act directing content there by the supplying party, many believe the supplying party should not be subject to the laws of the receiving country. Thus, if a Chinese resident orders a subscription to a French newspaper from a Web site in France, written in the French language, it is unlikely the French Web site would be subject to Chinese law. Of course, these are the two easy examples. The Internet has the potential to host a myriad of far more difficult jurisdictional puzzles.

In exploiting a business model or e-commerce patent, it is helpful to understand that jurisdiction actually encompasses at least three different aspects. First, there is personal jurisdiction — will the law of a particular nation apply to a particular person or company. Second, there is prescriptive jurisdiction — which nation will have the power to regulate which actions. Third, there is enforcement jurisdiction — will the jurisdiction where the defendant resides or owns property exercise its power to assist the plaintiff in satisfying a judgment from a foreign court or tribunal.

[1] Personal Jurisdiction

In the United States, personal jurisdiction is largely a creature of common law; whereas in Europe and Japan, personal jurisdiction is a creature of statute. Thus, U.S. personal jurisdictional disputes are often resolved on a case by case basis by applying broad jurisdictional principles. Whether a person has had *minimum contact* within a forum to exerting personal jurisdiction over that person is the overriding principle under U.S. law. Minimum contact can arise through physical presence, from actions that produce in the jurisdiction even if done from afar. Placing goods into the *stream of commerce* can subject the manufactured goods to the jurisdiction where the goods are shipped and used. Under U.S. law, the courts have the final say on whether to exercise personal jurisdiction, and that decision often involves a fact-specific judgment call.

For example, in *Asahi Metal Industry Co. v. Superior Court*,[2] the Supreme Court unanimously ruled that the California court did not have personal jurisdiction over a Japanese company being sued as a third party by a Taiwanese component parts manufacturer seeking indemnification. The Court ruled that it would be unreasonable for the California court to exert personal jurisdiction under the facts of that case. Asahi Metal had sold valve assemblies to the Taiwanese company, which in turn manufactured the tube used in a tire sold to Honda to manufacture motorcycles. The tire burst on a California highway. The plaintiff sued there and settled with all parties, leaving only the claim between Asahi and the Taiwanese company pending in the California court.

Although the Court ruled unanimously that personal jurisdiction was lacking under the facts, there was disagreement over whether the minimum contacts standard had been met. Some members of the Court felt that Asahi's use of the stream of commerce was sufficient to constitute "purposeful availing" to satisfy the minimum contacts requirement. The "regular and anticipated flow of products from manufacturer to distribution to retail sale" was enough to assert jurisdiction. Justice O'Connor argued that it was not, urging that "mere awareness" was insufficient to satisfy the minimum contacts requirement. She wrote that for jurisdiction to hold, the defendant must in some fashion purposefully direct its action toward the forum state.

Persons unfamiliar with U.S. law need to be aware that the makeup of the U.S. Supreme Court does change over time. Thus, while the Court in *Asahi* ruled unanimously against asserting jurisdiction under the recited facts, the Court could rule differently in the future, as new Justices with different views toward jurisdiction are appointed.

In Europe or Japan, it is doubtful that a jurisdictional issue, such as the one presented in *Asahi* would have required an appeal to the highest court of the land, as it did in the United States. Both Europe and Japan have explicit jurisdictional statutes that describe under what circumstances jurisdiction will and will not attach. While these statutes do give a court some latitude, where exceptional circumstances warrant, the rules of personal jurisdiction are far more fixed.

What impact has the Internet had on the law of personal jurisdiction? The basic rules still apply, but the issue certainly has not gotten any simpler. Just as a manufacturer of valve assemblies could place products in the international stream of commerce and find itself before the U.S. Supreme Court arguing jurisdictional *esoterica*, so too does the Web page publisher that places Web content on the World Wide Web. In *Bensusan Restaurant Corp. v. King*,[3] for example, a district judge for the Southern District of New York held that defendant's mere operation of a Web site, without more, was insufficient to confer New York jurisdiction over the Missouri defendant. The dispute was over the name of a blues music club. Plaintiff's and defendant's clubs were physically located 1500 miles apart, although web browser search engines placed them virtually side-by-side in

---

[2] 444 U.S. 286 (1980).
[3] 937 F. Supp. 295 (S.D.N.Y. 1996), *aff'd*, 126 F.3d 25 (2d Cir. 1997).

cyberspace. Citing dicta from the *Asahi* case, the district judge analogized that the operation of a Web site was like Asahi's placing a valve assembly into the stream of commerce, and that without some act purposefully directed to the forum state there was no proper basis for personal jurisdiction.

Bensusan was a trademark or trade name case. Personal jurisdiction issues in patent cases can be even more difficult to unsort because the patent infringement statute, 35 U.S.C. § 271, prescribes several different types of acts that constitute direct infringement, contributory infringement and inducing infringement. This issue is explored more fully in section § 13.03 below.

[2] Enforcement Jurisdiction

Getting a judgment is one thing. Enforcing it can be quite another. In many domestic patent infringement actions, the court having jurisdiction over the defendant in the infringement action will also have jurisdiction to enforce its judgment if liability is found. Thus if a court finds infringement and awards both injunctive relief and money damages, that court will usually retain jurisdiction to enforce the injunction through contempt proceedings. Levying execution on the defendant's property to satisfy the money judgment can be another matter. If the defendant has assets within the trial court's jurisdiction, then the trial court will exercise jurisdiction to order seizure and liquidation of the assets. However, if the assets are located in another jurisdiction, the plaintiff must bring a collection action in the other jurisdiction to proceed against those assets.

The Internet has complicated the picture. Because many business model and e-commerce patents have an Internet component, those patents present a far greater likelihood of being infringed by a defendant whose major assets are located outside the jurisdiction of the trial court, and indeed outside the territorial limits of the United States. Thus, even if a U.S. district court finds that it has jurisdiction over an alien defendant in a patent infringement action, this is no guarantee that the plaintiff will be able to collect on any damage award rendered.

Collection of a damage award would require a collection action in the jurisdiction where the alien defendant's assets are located. If that is another country, then the collection action becomes a matter of international law. Usually collection actions of this nature are based on comity, and require that both countries exercise reciprocal recognition of each other's judgments. However, even where reciprocal enforcement provisions exist, you should expect to encounter significant procedural differences, and significant expense.

Where the expense and procedural red tape make levy of execution impractical, the plaintiff's only recourse may be to use the U.S. Customs Service to seize infringing goods as they enter U.S. soil. The National Intellectual Property Rights Coordination Center would be your first stop in exploring this option.[4]

---

[4] National Intellectual Property Rights Coordination Center, 1300 Pennsylvania Avenue, NW, Rm. 3.5A, Washington, DC 20229.

[3] Proceeding Against Intermediaries

Where the defendant in a patent infringement action resides outside the jurisdiction of any U.S. district court, the plaintiff may have no recourse other than to consider proceeding against an intermediary supplier. Where Internet components are used in the business model or e-commerce invention, this intermediary may be an Internet service provider. Indeed, even where there is a basis for personal jurisdiction over a non-resident defendant — because the defendant's actions purposefully placed infringing goods into the stream of commerce — the patent owner may still want to consider suit against intermediary suppliers who are more likely to be subject to and respect an injunctive order of the U.S. district court.

This being said, however, the trend in both United States and Europe is to limit liability for Internet intermediaries, where the intermediary merely provides access and electronic meeting facilities, as opposed to taking part in the substantive commercial transactions and information exchange.[5]

Proceeding against an Internet service provider is most likely to be successful if one or more of the patent claims are directly infringed by the service provider. This is one reason to consider, where possible, having claims covering the information packages being delivered by the service provider.[6] Nevertheless, regardless of the scope of the claims, when proceeding against a service provider in an infringement action you should always be mindful that liability may be limited by statute. For example, in the Digital Millennium Copyright Act,[7] Congress had modified the liability of Internet service providers with respect to third party copyright violations on their sites. Section 230 of the Telecommunications Act of 1996 also limits intermediary liability.[8] It is well worth considering this act before filing suit against a service provider. If you don't, you may find your suit being dismissed on the pleadings, as unfortunate Ken Zeran did in his suit against America Online (AOL).

In *Zeran v. America Online*,[9] an unknown party had posted a message on an AOL bulletin board advertising "Naughty Okalahoma T-Shirts," featuring tasteless slogans related to the April 19, 1995, bombing of the Federal Building in Oklahoma City. The advertisement, presumably perpetrated as a prank, instructed interested parties to contact Ken Zeran at his home phone number in Seattle, Washington. As a result of the advertisement, Zeran received a high volume of calls, mostly from irate people who had been offended by the advertisement. Zeran could not change his phone number because he ran a business out of his home. He therefore called America Online to request removal

---

[5] "Achieving Legal and Business Order in Cyberspace: A Report on Global Jurisdiction Issues Created by the Internet," *The Business Lawyer*, Vol. 55, Aug. 2000, pp. 1895-1896.

[6] See Chapter 10, § 10.08 on propagated signal claims.

[7] Pub. L. No. 105-304, 1998 U.S.C.C.A.N. (112 Stat.) 2860.

[8] 47 U.S.C. § 230 (1997).

[9] 129 F.3d 327 (4th Cir. 1997), *cert. den.*, 524 U.S. 937 (1998).

of the advertisement. AOL complied with his request, but refused to post a retraction for reasons of company policy.

The next day, another similar message appeared on the AOL bulletin board, urging persons to call back if the number was busy. Over the next four days, Zeran received an average of one phone call every two minutes. Eventually, AOL advised Zeran that it would close the account from which the anonymous advertisements had been posted. However, by this time Zeran's problems had escalated. A radio station, KRXO, in Oklahoma City learned of the AOL posting, relating the contents of the message over the air. After the radio broadcast, Zeran was again inundated with calls, including death threats, from Oklahoma City residents.

Zeran filed suit against the radio station KRXO and later filed a second suit against AOL. Zeran did not file suit against the party who posted the offensive message. AOL moved for dismissal on the pleadings pursuant to Fed. R. Civ. P. 12(c), asserting Section 230 of the Telecommunications Act[10] as a complete defense. The district court granted AOL's motion and dismissed the suit against AOL. Zeran appealed and the Fourth Circuit affirmed with these words:

> The relevant portion of § 230 states: "No provider or user of an interactive computer service shall be treated as the publisher or speaker of any information provided by another information content provider." 47 U.S.C. § 230(c)(1). By its plain language, § 230 creates a federal immunity to any cause of action that would make service providers liable for information originating with a third-party user of the service. Specifically, § 230 precludes courts from entertaining claims that would place a computer service provider in a publisher's role. Thus, lawsuits seeking to hold a service provider liable for its exercise of a publisher's traditional editorial functions — such as deciding whether to publish, withdraw, postpone or alter content — are barred.[11]

Section 230(e)(2) defines "interactive computer service" as "any information service, system, or access software provider that provides or enables computer access by multiple users to a computer server, including specifically a service or system that provides access to the Internet and such systems operated or services offered by libraries or educational institutions." The term "information content provider" is defined at § 230(e)(3) as "any person or entity that is responsible, in whole or in part, for the creation or development of information provided through the Internet or any other interactive computer service." The parties did not dispute that AOL falls within the act's "interactive computer service" definition and that the unidentified third party who posted the offensive messages fit the definition of an "information content provider."

The *Zeran* case illustrates that the courts are prepared to strictly enforce the Telecommunications Act's federal immunity for service providers, even where

---

[10] 47 U.S.C. § 230.
[11] 129 F.3d at 330.

the facts in a plaintiff's favor are otherwise compelling. You should therefore consider whether the Telecommunications Act is implicated where the publishing of information content is an operative fact leading to liability. Some e-commerce inventions may involve publishing of information content.

[B] American Bar Association Proposes Default Jurisdictional Rules for Cyberspace

The rules by which personal jurisdiction will be extended to cyberspace are still being forged. In an effort to promote uniformity and to aid courts and lawyers who confront this issue, in 1998 the American Bar Association sponsored a two year Cyberspace Jurisdiction Project. The mission was to study how personal jurisdiction rules should be interpreted or extended to promote a fair system of jurisdictional jurisprudence for commercial transactions in cyberspace. A steering committee and working groups were formed and those groups produced a published report, the highlights of which are summarized here.

[1] Proposed Default Jurisdictional Rules

- Every Internet party should be subject to personal jurisdiction somewhere.

- Personal jurisdiction should not be based solely on a passive Web site that does not target a particular state.

- Personal jurisdiction over a web content provider should be proper where the content provider:
  - habitually resides in or has a principal place of business in the jurisdiction, or
  - targets the web content to that jurisdiction and the claim arises out of that content of the Web site, or
  - engages in a transaction generated through the Web site, where the content provider can be fairly considered to knowingly engage in business transactions within the jurisdiction

- Good faith efforts to prevent access to a site by users outside the jurisdiction should insulate the content provider. Examples of such efforts include, disclosures, disclaimers, software and other technological blocking or screening mechanisms to prevent or discourage access.

§ 13.03   Infringement and Territorial Issues Involving Internet Patents

The patent statute confers upon its owner the right to exclude others from making, using, selling, or offering for sale the invention within the United States or importing the invention into the United States. These rights are set forth in 35 U.S.C. § 271, which reads:

§ 271 Infringement of patent

(a) Except as otherwise provided in this title, whoever without authority makes, uses, offers to sell, or sells any patented invention, within the United States or imports into the United States any patented invention during the term of the patent therefor, infringes the patent.

(b) Whoever actively induces infringement of a patent shall be liable as an infringer.

(c) Whoever offers to sell or sells within the United States or imports into the United States a component of a patented machine, manufacture, combination or composition, or a material or apparatus for use in practicing a patented process, constituting a material part of the invention, knowing the same to be especially made or especially adapted for use in an infringement of such patent, and not a staple article or commodity of commerce suitable for substantial non-infringing use, shall be liable as a contributory infringer.

(d) No patent owner otherwise entitled to relief for infringement or contributory infringement of a patent shall be denied relief or deemed guilty of misuse or illegal extension of the patent right by reason of his having done one or more of the following: (1) derived revenue from acts which if performed by another without his consent would constitute contributory infringement of the patent; (2) licensed or authorized another to perform acts which if performed without his consent would constitute contributory infringement of the patent; (3) sought to enforce his patent rights against infringement or contributory infringement; (4) refused to license or use any rights to the patent; or (5) conditioned the license of any rights to the patent or the sale of the patented product on the acquisition of a license to rights in another patent or purchase of a separate product, unless, in view of the circumstances, the patent owner has market power in the relevant market for the patent or patented product on which the license or sale is conditioned.

(e)

(1) It shall not be an act of infringement to make, use, offer to sell, or sell within the United States or import into the United States a patented invention (other than a new animal drug or veterinary biological product (as those terms are used in the Federal Food, Drug, and Cosmetic Act and the Act of March 4, 1913) which is primarily manufactured using recombinant DNA, recombinant RNA, hybridoma technology, or other processes involving site specific genetic manipulation techniques) solely for uses reasonably related to the development and submission of information under a Federal law which regulates the manufacture, use, or sale of drugs or veterinary biological products.

(2) It shall be an act of infringement to submit —

(A) an application under section 505(j) of the Federal Food, Drug, and Cosmetic Act or described in section 505(b)(2) of such Act for a drug claimed in a patent or the use of which is claimed in a patent, or

(B) an application under section 512 of such Act or under the Act of March 4, 1913 (21 U.S.C. 151-158) for a drug or veterinary biological product which is not primarily manufactured using recombinant DNA, recombinant RNA, hybridoma technology, or other processes involving site specific genetic manipulation techniques and which is claimed in a patent or the use of which is claimed in a patent, if the purpose of such submission is to obtain approval under such Act to engage in the commercial manufacture, use, or sale of a drug or veterinary biological product claimed in a patent or the use of which is claimed in a patent before the expiration of such patent.

(3) In any action for patent infringement brought under this section, no injunctive or other relief may be granted which would prohibit the making, using, offering to sell, or selling within the United States or importing into the United States of a patented invention under paragraph (1).

(4) For an act of infringement described in paragraph (2) —

(A) the court shall order the effective date of any approval of the drug or veterinary biological product involved in the infringement to be a date which is not earlier than the date of the expiration of the patent which has been infringed,

(B) injunctive relief may be granted against an infringer to prevent the commercial manufacture, use, offer to sell, or sale within the United States or importation into the United States of an approved drug or veterinary biological product, and

(C) damages or other monetary relief may be awarded against an infringer only if there has been commercial manufacture, use, offer to sell, or sale within the United States or importation into the United States of an approved drug or veterinary biological product. The remedies prescribed by subparagraphs (A), (B), and (C) are the only remedies which may be granted by a court for an act of infringement described in paragraph (2), except that a court may award attorney fees under section 285.

(f)

(1) Whoever without authority supplies or causes to be supplied in or from the United States all or a substantial portion of the

components of a patented invention, where such components are uncombined in whole or in part, in such manner as to actively induce the combination of such components outside of the United States in a manner that would infringe the patent if such combination occurred within the United States, shall be liable as an infringer.

(2) Whoever without authority supplies or causes to be supplied in or from the United States any component of a patented invention that is especially made or especially adapted for use in the invention and not a staple article or commodity of commerce suitable for substantial non-infringing use, where such component is uncombined in whole or in part, knowing that such component is so made or adapted and intending that such component will be combined outside of the United States in a manner that would infringe the patent if such combination occurred within the United States, shall be liable as an infringer.

(g) Whoever without authority imports into the United States or offers to sell, sells, or uses within the United States a product which is made by a process patented in the United States shall be liable as an infringer, if the importation, offer to sell, sale, or use of the product occurs during the term of such process patent. In an action for infringement of a process patent, no remedy may be granted for infringement on account of the noncommercial use or retail sale of a product unless there is no adequate remedy under this title for infringement on account of the importation or other use, offer to sell, or sale of that product. A product which is made by a patented process will, for purposes of this title, not be considered to be so made after—

(1) it is materially changed by subsequent processes; or

(2) it becomes a trivial and nonessential component of another product.

(h) As used in this section, the term "whoever" includes any State, any instrumentality of a State, and any officer or employee of a State or instrumentality of a State acting in his official capacity. Any State, and any such instrumentality, officer, or employee, shall be subject to the provisions of this title in the same manner and to the same extent as any non-governmental entity.

(i) As used in this section, an "offer for sale" or an "offer to sell" by a person other than the patentee, or any designee of the patentee, is that in which the sale will occur before the expiration of the term of the patent.[12]

---

[12] 35 U.S.C. § 271 (July 19, 1952, ch. 950, § 1, 66 Stat. 811), (*as amended* Sept. 24, 1984, Pub. L. No. 98-417, Title II, § 202 (98 Stat.), 1603; Nov. 8, 1984, Pub. L. No. 98-622, Title I, § 101(a) (98 Stat.), 3383; Aug. 23, 1988, Pub. L. No. 100-418, Title IX, Subtitle A, § 9003 (102 Stat.), 1564; Nov. 16, 1988, Pub. L. No. 100-670, Title II, § 201(i) (102 Stat.), 3988; Nov. 19, 1988, Pub. L. No.

The above statutory language can be a bit daunting when read top to bottom. Sections 271(a) sets forth the basic definition of what acts constitute direct infringement. Section 271(b) makes it clear that actively inducing infringement is actionable under the patent laws; and section 271(c) sets forth the basic definition of what constitutes contributory infringement. Section 271(d) addresses the topic of patent misuse by setting forth certain acts which *do not* constitute misuse. For our purposes, § 271(e) need not be considered here, as it relates to biotechnology, drugs and veterinary biological products. Section 271(f) extends the infringement statute to cover the exportation of components that are combined outside the United States. Section 271(g) extends the infringement statute to cover the importation into the United States of products made by a patented process.

[A] The Internet as an Information Pipeline

It is well settled that business inventions, if properly claimed, will constitute statutory subject matter. Nevertheless, courts still have difficulty with the statutory subject matter issue, particularly where the data gathering steps are simple and where the post solution activity is trivial. At one level, the Internet is simply a new form of information packaging. It is the pipeline through which valuable products and services flow to customers and end users. Having a patent on the pipeline would offer significant commercial advantage. Hence many Internet patents, if distilled to their essence, attempt to claim the pipeline, or at least an important piece of it.

As demonstrated by the district court and Federal Circuit Court decisions in *AT&T Corp. v. Excel Communications, Inc.*,[13] courts still differ in opinion on where to draw the statutory subject matter line. The *AT&T* case, although not Internet technology, per se, precisely involved a question of information packing, specifically the automatic routing of interexchange telephone calls based on an indicator embedded in the message record associated with each call.

Internet inventions offer a variety of interesting scenarios to which one or more of the sections of 35 U.S.C. § 271 may be applied. Being a global network, the Internet has all but removed the natural geographic barriers that have traditionally delimited the respective jurisdictions of sovereign states. It is now technologically easy to construct an Internet system, such as an information system or business method system, that spans several countries. Computer process steps or business process steps may be performed outside the United States, with the final information or business product being offered to persons within the United States. Internet data may be harvested in the United States, shipped abroad for processing and returned to the United States for consumption.

100-703, Title II, § 201 (102 Stat.), 4676; Oct. 28, 1992, Pub. L. No. 102-560, § 2(a)(1) (106 Stat.), 4230; Dec. 8, 1994, Pub. L. No. 103-465, Title V, Subtitle C, § 533(a) (108 Stat.), 4988.

[13] AT&T Corp. v. Excel Communications, Inc., 172 F.3d 1352 (Fed. Cir. 1999).

Sections 271(f) and (g) may provide fertile statutory language to address such extra-territorial issues.

As if these extra-territorial issues were not enough, one other potential issue remains: what to do about inventions in outer space. With much of the international Internet traffic being transmitted via satellite, one must anticipate this issue. Fortunately, Congress foresaw the issue and, in 1990, enacted special statutory provisions that begin to address issues involving inventions made, used, or sold in outer space. The statute, reproduced below, treats a space object much the same as the law treats sailing ships.

35 U.S.C. § 105 Inventions in outer space

(a) Any invention made, used or sold in outer space on a space object or component thereof under the jurisdiction or control of the United States shall be considered to be made, used or sold within the United States for the purposes of this title, except with respect to any space object or component thereof that is specifically identified and otherwise provided for by an international agreement to which the United States is a party, or with respect to any space object or component thereof that is carried on the registry of a foreign state in accordance with the Convention on Registration of Objects Launched into Outer Space.

(b) Any invention made, used or sold in outer space on a space object or component thereof that is carried on the registry of a foreign state in accordance with the Convention on Registration of Objects Launched into Outer Space, shall be considered to be made, used or sold within the United States for the purposes of this title if specifically so agreed in an international agreement between the United States and the state of registry.[14]

The statutory provisions of §§ 271(a) to (i) and § 105 yield endless possibilities. The following sections will explore some different scenarios and show how the provisions of 35 U.S.C. §§ 271(a) to (i) and § 105 may be applied. These statutory sections should be considered when drafting claims to Internet inventions.

The scenarios presented below apply a hypothetical U.S. patent with four independent claims: (a) method claim, (b) apparatus or system claim, (c) article of manufacture computer memory claim, and (d) article of manufacture carrier wave claim.

The subject matter of the claimed inventions involves a system for producing and selling recorded music over the Internet. Musicians working at home or in a local studio make a multi-track recording that is stored in a proprietary format. The proprietary format maintains a separate record of each

---

[14] 35 U.S.C. § 105 (added Nov. 15, 1990, Pub. L. 101-580, § 1(a), 104 Stat. 2863).

track and each track maintains a pointer to a master absolute time code record that may be used to "fast-forward" and "fast-rewind" the multi-track recording.

The multi-track recording, stored in the proprietary format, is then uploaded via the Internet to a mix-down server where the multi-track data is accessed and the recorded audio is mixed down into a stereo master recording. A compressed audio version of the stereo master recording is produced and stored on the music distribution server. The server engages in Internet-based electronic commerce transactions with purchasers and delivers copies of the compressed music product to purchasers after receiving electronic payment.

Here are hypothetical claims (a) through (d) for the scenarios presented in the following sections:

(a) A method for producing and selling a recorded music product, comprising the steps of:

recording multiple discrete audio content segments and storing said segments as separate digital tracks on a storage medium;

providing said digital tracks via a computer network to a mix-down host computer and using said mix-down host computer to mix said digital tracks into a stereo digital master recording;

compressing said stereo digital master recording according to a predefined compression scheme to generate said recorded music product;

placing said recorded music product on a server for delivery to a purchaser via a computer network in connection with an electronic commerce transaction.

(b) A system for producing and selling a recorded music product, comprising:

a digital audio recorder that records audio content as discrete digital tracks on a storage medium;

a communication processor communicating with said storage medium and coupled to a computer network for providing said digital tracks to another node on said computer network;

a mix-down host computer coupled as a node on said network and thereby being receptive of said digital tracks, the mix-down host computer being operative to mix said digital tracks into a stereo digital master recording;

a compressor receptive of said stereo digital master recording that alters said stereo digital master recording according to a predetermined compression scheme to generate said recorded music product; and

a commerce transacting and music delivery server coupled to a computer network and having memory for storing said recorded music product, said server being operative to conduct an electronic commerce transaction with a purchaser and deliver said recorded music product to said purchaser.

(c) A computer program embodied on a computer-readable medium for configuring a multi-track digital audio storage system, comprising:

a data structure instantiating code segment that establishes a multi-track storage record in memory having:
(1) an absolute time code record;
(2) a plurality of separate digital audio segment storage records, each maintaining a pointer to said absolute time code record.

(d) A computer-readable multi-track digital audio data signal embodied in a carrier wave, comprising:

(1) an absolute time code record;
(2) a plurality of separate digital audio segment storage records, each maintaining a pointer to said absolute time code record.

The above claims will be tested in a variety of scenarios presented in the following sections. The scenarios are intended not as a definitive statement of this complex area, but rather as illustrations of the types of enforcement issues that you may encounter when applying 35 U.S.C. § 271.

[B] Entire Internet Process Performed Inside the United States

Perhaps the easiest scenario to analyze is where the entire Internet-based system is deployed within the United States. However, even here, the analysis gets tricky because of the inherently distributed nature of the Internet. While it is possible that a single U.S.-based recording studio would set up operations to perform each of the steps recited in method claim (a), it is more likely that several entities might perform their respective tasks independent of the other.

To illustrate, assume that the musicians working in the United States purchase a CD-ROM product containing the multi-track recording software that records according to the proprietary format. The product is intended to allow the musicians to choose from a variety of different mix-down service providers. Mix-down studios wishing to become service providers purchase a different mix-down CD-ROM product. That product implements the basic mix-down functions and offers additional signal processing modules. The mix-down component includes computer code that establishes a buffer in memory, configured according to the proprietary format, for storing material while performing mix-down. It is assumed here that all mix-down studios are located within the United States.

For purposes of this example, further assume that the electronic commerce and delivery server functions are split. Electronic commerce transactions are processed by a credit card clearinghouse server in Minnesota which notifies a music delivery server in Nashville to deliver the recorded music product to the purchaser.

Under the above factual scenario, arguably no one party will directly infringe the method claim (a) or the system claim (b) under § 271(a), assuming the musicians and their recording operation are separate from the mix-down and subsequent processing operations. The article of manufacture computer program claim (c) would be infringed under § 271(a) by a competitor who has designed a product that stores data in the proprietary format or that creates a buffer designed to process data stored in the proprietary format. The carrier wave claim (d) would likewise be infringed under § 271(a) by a party who sends computer data over the Internet in the proprietary format.

If method claim (a) and system claim (b) are not infringed, why have them? One reason is to increase the royalty base and scope of injunctive relief, in the event a party is found guilty of contributory infringement or of actively inducing infringement. The value of the manufacture claims (c) and (d) should be clear. These claims make it possible to go after the software manufacturer or service provider without suing the end user.

[C] Substantially All of Internet Process Performed Outside United States

Assume for this scenario the entire operation from recording to final mix-down and compression is carried out by a non-U.S. company in Norway. The only portion arguably carried out in the United States is the purchaser's side of the electronic commerce transaction and the ultimate delivery of the product to a purchaser in the United States. Putting aside the issue of whether the Norwegian company is subject to U.S. jurisdiction, does this scenario fit any of the statutory criteria for finding infringement?

Under § 271(g) it is an infringement to import into the United States a product which is made by a process patented in the United States. The process steps of claim (a) are performed by the Norwegian company and might therefore arguably constitute infringement under § 271(g). However, there is some question as to what is the *product* of the patented process. In this case, the process recited in that claim has elements of a business method transaction (the final step involves payment and delivery). Is the *product* of the patented process the recorded music product or is it the service of selling the recorded music? The claim preamble recites a basis for either interpretation:

> A method for producing *and selling* a recorded music product, comprising the steps of: . . .

Thus it might be argued that the "product" of the claimed process is not a recorded music product, but rather the "service" of delivering recorded music. Whether a service may constitute a "product" under § 271(g) could be subject to debate. Webster's Third New International Dictionary defines "product" as follows:

> product — **2a:** something produced by physical labor or intellectual effort: the result of work or thought [use for hammocks and other ~s — P. E. James] [even the simplest poem is the ~ of much . . . work — Gilbert Highet] **2b:** a result of the operation of involuntary causes or an ensuing set of conditions:

consequence, manifestation [a ~ of liberal arts education — B. W. Hayward] [he was the ~ of his time — Allan Nevins] . . . **3:** the amount, total, or quantity produced: the output of an industry or firm [our national ~ . . . has quickly risen to an enormous volume — George Soulle] **4:** a substance produced from one or more other substances as a result of chemical change.

The above definition shows that the ordinary meaning of the word "product" encompasses both tangible (for example, substances produced by chemical change) and intangible (for example, product of liberal arts education) meanings. Elsewhere in the patent statute, Congress included "process" as one of the forms of statutory subject matter.[15] The term process in that context is not limited to products that produce a physical transformation. Indeed, any process that produces a useful, concrete and tangible result (for example, a business method) will constitute statutory subject matter.[16] A physical transformation is not required.[17] Thus, it is consistent with other parts of the patent statute to construe "product" to encompass services that result from patented business method processes.

To complicate matters further, consider a slightly different scenario where the method steps are not all performed by one Norwegian company. Instead, different individuals throughout Europe record, mix, and compress the musical content, with only the final distribution server being located in Norway. In this case, it is unlikely that method claim (a) will be directly infringed and thus § 271(g) is not helpful. What about the manufacture claims (c) and (d)? Assuming that only the final recorded music product is shipped via Internet into the United States, claims (c) and (d) are probably not helpful either. Note that these claims cover the unique format used in the recording and mix-down processes. The final recorded music product would not use this format. To cover this situation it would be necessary to have claims that would infringe under the "imports into the United States" clause of § 271(a).

In this example, because the recorded music product itself is of conventional structure, there is probably little that can be done to prevent importation of the music product. However, if the product were structurally new to meet the novelty requirements of 35 U.S.C. § 102, then a product-by-process claim might be used. For example, if the compression algorithm were new and thus imparted new structure to the recorded music product, then a product-by-process claim may be constructed and thereby invoke the importation clause of § 271(a).

[D] Internet Process Partially Performed in United States — Components Exported

In this example, assume that the musicians record the audio content in the United States and upload the recorded digital tracks to a mix-down host computer

---

[15] 35 U.S.C. § 101.

[16] State Street Bank & Trust Co. v. Signature Financial Group, 149 F.3d 1368 (Fed. Cir. 1998).

[17] *AT&T Corp.*, 172 F.3d 1352.

in Taiwan. The Taiwanese host mixes and compresses the tracks into a final recorded music product. The final product is then transferred to a number of Internet sites throughout the world, including a server in the United States, which also processes electronic commerce transactions as a prerequisite to downloading the recorded music to a purchaser in the United States.

Under these facts, § 271(f) merits consideration. Congress enacted that section as part of the Patent Law Amendments Act of 1984 to close a loophole whereby copiers could avoid U.S. patents by supplying components of a patented product from the United States for assembly abroad. The statutory language, however, went beyond merely closing this loophole. Section 271(f)(1) applies where all or a substantial portion of the components of a patented invention are exported for assembly abroad. The exported components may even be staple articles or commodities which are also suitable for substantial non-infringing use. Section 271(f)(2) applies to the exportation of even a single component of a patented invention. However, that component must be specially adapted for use in the invention and may not be a staple article or commodity of commerce suitable for substantial non-infringing use. In addition, the exporter must have knowledge of and intent to commit the infringing activity as set forth in the statute.

At first blush, it would seem that § 271(f) might apply to this scenario. The application is not clear cut, however. At least one court has held that § 271(f) does not apply to the carrying out of patented processes abroad.[18]

In *Standard Havens Products v. Gencor Industries*, the plaintiff, Standard Havens Products, owned a patent with claims to a method for continuously producing an asphalt composition. Gencor Industries made plants for producing the asphalt composition according to the patented method. Gencor sold plants to both United States and to one foreign customer in England. The Court found that the sale of plants to customers within the United States constituted contributory and inducing infringement. However, as to the sale to the customer in England, the Court ruled that sale of the plant did not result in infringement.

> As to the sale to a foreign customer, Standard Havens asserts that Gencor made the sale in the United States. The '938 patent claims a method for producing asphalt, not the apparatus for implementing that process. Thus, the sale in the United States of an unclaimed apparatus alone does not make Gencor a contributory infringer of the patented method. Moreover, infringement by the foreign customer has not been shown because there is no evidence of the plant's use in the United States. Further, there can be no inducement of infringement or contributory infringement under 35 U.S.C. § 271(b), (c) in the absence of direct infringement . . . Similarly, there is no evidence in the record showing that the foreign purchaser shipped products back to the United States made abroad by the patented process, *see* 35 U.S.C.

---

[18] Standard Havens Products, Inc. v. Gencor Industries, Inc., 21 U.S.P.Q.2d 1321 (Fed. Cir. 1991).

§ 271(g). Finally, we do not find the provisions of 35 U.S.C. § 271(f) implicated.[19]

The Court found that § 271(f) was not implicated where only method claims are present. Why is this so? Arguably this conclusion follows from the wording of the statute. Note that § 271(f) applies to exporting components of a "patented invention." The statute dealing with importing components, § 271(c), is worded differently. It defines infringing importation to apply to a component of "a patented machine, manufacture, combination and compositions, or a material or apparatus for use in practicing a patented process."

Therefore, applying the *Standard Havens Products* ruling to the present scenario, § 217(c) does not apply to our process claim (a). Claim (b) likewise fails to fall under § 271(c) because none of the components of that apparatus claim are being exported. Data is being exported, but the components used to generate that data are not. Thus, claims (a) and (b) would probably be infringed only if the facts warrant a finding of contributory infringement or active inducement of infringement. Claims (c) and (d) present another issue. On the facts presented, no computer readable medium (for example, CD-ROM, or computer memory) is being exported. A computer data signal embodied in a carrier wave is being exported. Thus, claim (d) appears to present the better case for application of § 271(c). However, because the claimed combination is already fully assembled (into the claimed data structure) before it is exported, no one is combining the components abroad. Thus § 271(c) would not seem to apply.

As this example demonstrates, it can be extremely tricky to draft claims to catch potential infringers where the allegedly infringing activity is distributed throughout the world. This example has focused on exporting. The flip side is importing, which the next section will explore.

[E] Internet Process Partially Performed in United States — Components Imported

If we flip the previous scenario, such that the musicians are in Europe and the unlicensed mix-down server is in Nashville, the infringement analysis comes out differently, particularly for the Nashville mix-down company. Section 271(a) makes it an infringement to import a patented invention into the United States. Section 271(c) makes it contributory infringement to import into the United States a (non-staple article) component of a patented machine, manufacture, combination or composition, or material. That section also makes it contributory infringement to import a non-staple material or apparatus for use in practicing a patented process. Because the proprietary data structure is patented under claim (c) as a computer program in memory and under claim (d) as a data signal embodied in a carrier wave, the Nashville server used with these structures may be found to infringe. The Nashville server employs a computer program to

---

[19] *Id.* at 1332.

generate the proprietary data structures. Thus, it directly infringes claim (c) under § 271(a). By advertising its mix-down services on the Internet, it also actively induces infringement under § 271(b). The importation of the carrier wave data signal directly infringes claim (d).

Also, depending on its knowledge and involvement in the subsequent commerce transactions and distribution of the music product, the Nashville server may also contributorily infringe claims (a) and (b) under § 271(c) by supplying the mix-down and compression components of the patented machine, which are also used in practicing the patented process.

[F] Internet Process Partially Performed in United States — Orbiting Satellite Employed

For this final example, consider two scenarios. In the first scenario, the commerce transacting and music delivery server is located on a distributed network of geosynchronous satellites launched by a U.S. corporation under the jurisdiction of the U.S. government. In the second scenario, the server is located on a space station launched by the Russian government and licensed to a United States corporation for use as a music distribution server.

Under 35 U.S.C. § 105, the satellite in the first scenario is under the jurisdiction or control of the United States. Thus, the components located on that satellite are treated as if they are planted on U.S. soil. The infringement analysis proceeds as if the space-born components are located in the United States.

The second scenario is different. Absent treaty to the contrary, the Russian space station is a space object that would be carried on the Russian registry in accordance with the Convention on Registration of Objects Launched into Outer Space. Thus the server would be treated as if it lies outside the United States.

§ 13.04   The Special Defense to Business Method Infringement

In late 1999, Congress enacted special defense provisions to insulate a party from charges of business method patent infringement, if it is shown that the accused party was practicing the method before the patent was applied for. The enacted provisions did not define what is meant by a business method, leaving open the question whether an apparatus claim involving a business system would be subject to this statutory defense. In any event, the availability of this defense should be factored into the patent exploitation strategy. The full text of the statute is reproduced in the following subsection.

[A]   35 U.S.C. § 273 — Special Business Method Defense

35 U.S.C. § 273 Defense to infringement based on earlier inventor (Effective on the date of enactment, but only to actions occurring thereafter.)

(a) DEFINITIONS. For purposes of this section

(1) the terms "commercially used" and "commercial use" mean use of a method in the United States, so long as such use is in connection with an internal commercial use or an actual arm's-length sale or other arm's-length commercial transfer of a useful end result, whether or not the subject matter at issue is accessible to or otherwise known to the public, except that the subject matter for which commercial marketing or use is subject to a premarketing regulatory review period during which the safety or efficacy of the subject matter is established, including any period specified in section 156(g), shall be deemed "commercially used" and in "commercial use" during such regulatory review period;

(2) in the case of activities performed by a nonprofit research laboratory, or nonprofit entity such as a university, research center, or hospital, a use for which the public is the intended beneficiary shall be considered to be a use described in paragraph (1), except that the use

(A) may be asserted as a defense under this section only for continued use by and in the laboratory or nonprofit entity; and

(B) may not be asserted as a defense with respect to any subsequent commercialization or use outside such laboratory or nonprofit entity;

(3) the term "method" means a method of doing or conducting business; and

(4) the "effective filing date" of a patent is the earlier of the actual filing date of the application for the patent or the filing date of any earlier United States, foreign, on international application to which the subject matter at issue is entitled under section 119, 120, or 365 of this title.

(b) DEFENSE TO INFRINGEMENT

(1) IN GENERAL. It shall be a defense to an action for infringement under section 271 of this title with respect to any subject matter that would otherwise infringe one or more claims for a method in the patent being asserted against a person, if such person had, acting in good faith, actually reduced the subject matter to practice at least 1 year before the effective filing date of such patent, and commercially used the subject matter before the effective filing date of such patent.

(2) EXHAUSTION OF RIGHT. The sale or other disposition of a useful end product produced by a patented method, by a person entitled to assert a defense under this section with respect to that

useful end result shall exhaust the patent owner's rights under the patent to the extent such rights would have been exhausted had such sale or other disposition been made by the patent owner.

(3) LIMITATIONS AND QUALIFICATION OF DEFENSE. The defense to infringement under this section is subject to the following:

(A) PATENT. A person may not assert the defense under this section unless the invention for which the defense asserted is for a method.

(B) DERIVATION. A person may not assert the defense under this section if the subject matter on which the defense is based was derived from the patentee or persons in privity with the patentee.

(C) NOT A GENERAL LICENSEE. The defense asserted by a person under this section is not a general license under all claims of the patent at issue, but extends only to the specific subject matter claimed in the patent with respect to which the person can assert a defense under this chapter, except that the defense shall also extend to variations in the quantity or volume of use of the claimed subject matter, and to improvements in the claimed subject matter that do not infringe additional specifically claimed subject matter of the patent.

(4) BURDEN OF PROOF. A person asserting the defense under this section shall have the burden of establishing the defense by clear and convincing evidence.

(5) ABANDONMENT OF USE. A person who has abandoned commercial use of subject matter may not rely on activities performed before the date of such abandonment in establishing a defense under this section with respect to actions taken after the date of such abandonment.

(6) PERSONAL DEFENSE. The defense under this section may be asserted only by the person who performed the acts necessary to establish the defense and except for any transfer to the patent owner, the right to assert the defense shall not be licensed or assigned or transferred to another person except as an ancillary and subordinate part of a good faith assignment or transfer for other reasons of the entire enterprise or line of business to which the defense relates.

(7) LIMITATION ON SITES. A defense under this section, when acquired as part of a good faith assignment or transfer of an entire enterprise or line of business to which the defense relates, may only be asserted for uses at sites where the subject matter that would

otherwise infringe one or more of the claims is in use before the later of the effective filing date of the patent or the date of the assignment or transfer of such enterprise or line of business.

(8) UNSUCCESSFUL ASSERTION OF DEFENSE. If the defense under this section is pleaded by a person who is found to infringe the patent and who subsequently fails to demonstrate a reasonable basis for asserting the defense, the court shall find the case exceptional for the purpose of awarding attorney fees under section 285 of this title.

(9) INVALIDITY. A patent shall not be deemed to be invalid under section 102 or 103 of this title solely because a defense is raised or established under this section.

§ 13.05   Part II — Developing Strategic Portfolios for the Future

Compared with more conventional machines, processes, articles of manufacture, and compositions of matter, business method and business model inventions are more difficult to identify and claim. Companies may sense they need more business method patents in their portfolio, but they do not know where to look for this kind of innovation. The following sections will help you discover where to look for innovative business models and business methods, and how to develop portfolios around them.

§ 13.06   The Strategic Pyramid

In his book, *Value-Driven Intellectual Capital*,[20] Patrick Sullivan classifies all corporate intellectual property (IP) strategies into a hierarchy of five levels, which he graphically represents as a pyramid. Sullivan attributes the five level pyramid to Julie L. Davis, an Arthur Andersen intellectual asset management consultant. Arthur Andersen refers to the pyramid as the IP Value Hierarchy. The pyramid is intended to encompass all forms of intellectual property, including copyrights and trademarks, but it certainly applies well to patents.

At the base of the pyramid, lowest on the strategic scale lies the popular *defensive* IP strategy. At the top of the pyramid, Sullivan places the *visionary* IP strategy. Corporate patent counsel will no doubt be interested in assessing where in the hierarchy their portfolio strategy lies. In a nutshell, here are the five levels, listed in order of *decreasing* IP value.

1. Visionary
2. Integrated
3. Profit Center
4. Cost Control
5. Defensive

---

[20] Sullivan, Patrick. *Value-Driven Intellectual Capital* (John Wiley & Sons, 2000).

[A] Defensive Patent Strategy

Taking the list from bottom up, a *defensive* strategy is one in which patents are used to provide limited product protection, but mostly as a shield to protect against litigation. Many companies have adopted this defensive strategy. Their portfolios are filled with scores of patents of doubtful licensing potential. Companies with a defensive strategy maintain these large portfolios to have ample fodder for cross-licensing as a means of avoiding litigation.

[B] Cost Control Strategy

Having generated large portfolios under a defensive strategy, many companies then graduate to a *cost control* strategy, where they attempt to minimize the cost of maintaining the large portfolio by judiciously pruning the less valuable patents from the portfolio. Typically patent portfolio managers will prune at several different levels. Through a well-managed patent disclosure assessment process, some managers will prune away inventions that do not warrant patent filing at the outset. Although rarely done, some managers will also prune patent applications during prosecution, by abandoning those applications that are not faring well in prosecution. More frequently, however, managers will prune their less desirable patents at the maintenance fee stage, by simply withholding fee payment.

International patents are typically pruned separately. A global patent portfolio can be astronomically expensive. Thus, experienced portfolio managers will spend considerable thought and effort in determining which foreign patents and patent applications to prune from the portfolio.

Although it makes the portfolio more valuable, by reducing the cost to maintain, the cost control strategy does virtually nothing to improve the inherent value of the patents in the portfolio.

[C] Profit Center Strategy

After costs of maintaining the portfolio are under control, many companies graduate to the next level where a *profit center* strategy is implemented. When this occurs, portfolio managers find themselves trying to justify how their portfolios add to the bottom line of the company. A primarily defensive patent protection strategy is hard to bring to the bottom line. Thus many portfolio managers find themselves organizing their portfolios as IP profit centers. The managers focus changes from a passive, defensive role to a proactive, licensing role. A common practice involves moving all patent assets to a centralized business unit which then engages in patent mining operations, seeking out and licensing the best patents and otherwise making deals to extract value from the rest of the portfolio.

Once a company has graduated to a profit center strategy it becomes important to assess what the portfolio contains. Many consultants recommend conducting an intellectual property audit at this stage. The audit, which can be quite extensive, involves assessing all intellectual assets, not just those considered as intellectual property. Intellectual assets encompass such items as supplier relationships and corporate best practices, which are not classically thought of as intellectual property, per se. Database and graphical tools are frequently used to collect, organize and assess the assets uncovered during the audit. Some database and graphical tools are discussed at the end of this chapter.

[D] Integration Strategy

The profit center strategy, described above, tends to develop an inward focus. All intellectual property assets are collected under one department and that department must bring value to the bottom line, mostly through licensing. However, some companies evolve beyond this and achieve an *integration* strategy in which intellectual property assets are assessed when making decisions within other departments of the company. It is a very sophisticated company that operates at this level. Instead of pigeonholing intellectual property as mere assets to be licensed, companies with an integration strategy link their IP assets to all of their day-to-day operations. Thus, when the director of research and development constructs the roadmap for the next quarter, he or she has ready access to what the IP portfolio contains, and factors that into the decision-making process. Likewise, when the human resources director considers hiring technicians from a related industry, or when top management considers making a corporate acquisition, the benefits to the IP portfolio and potential value to be extracted from the IP portfolio are considered.

[E] Visionary Strategy

The zen-like final stage, which very few companies reach, is adoption of a *visionary* strategy. A visionary strategy entails identifying the direction of future technological trends, future customer preferences and future political and social changes that will impact not only the company but also all of its competitors. A visionary strategy can no longer be managed as simply an IP function. It requires leadership from the company's top management. Those visionaries within the company — and every company has them — must be identified and their ideas about the future need to be systematically harvested and incorporated into a body of strategically placed patents and other intellectual capital assets.

Perhaps the reason few companies achieve this highest level of sophistication is that the visionary strategy requires a deep understanding by top management of how to guide the company along a visionary path. The larger a company is, the more difficult it becomes to keep everyone working towards a common visionary goal.

There is considerable reason today to have at least a few visionary business model patents in the portfolio, even if the company has not otherwise achieved a visionary strategy in all respects. The reason is simple. A well-placed business model patent can neutralize a competitor's patent portfolio, as the next section will explain.

§ 13.07   Neutralizing a Competitor's Portfolio with Business Model Patents

Incremental improvement patents should not be cast out of the portfolio, but they are not where to spend one's strategic effort. Often it is one breakthrough patent that will dominate an industry. Consider this example.

For years two competitors, *Alpha* and *Beta*, have engaged in a grueling tug-of-war over market share. Both companies make conveying equipment used for factory automation on assembly lines. *Alpha* is the market leader by 15 percentage points. Until recently both companies have used price and well-focused sales forces to fight the market share battle. Neither has made much progress. Market share of the respective companies has remained steady within a few percentage points. While both companies occasionally develop new technology, the industry is mature and the improvements are incremental.

Because both companies are so focused on gaining market share, neither has devoted much effort to new product innovation. Management regards engineering more as an expense than as an asset. Neither company has a significant patent portfolio. In an effort to improve profitability, *Alpha*, the market leader, downsizes its engineering department. The remaining engineering staff is directed to support product sales. What little new technology innovation *Alpha* had been working on is reduced to a trickle. The most innovative engineers, facing new job assignments in sales support, leave the company to seek more challenging careers. With guillotine precision, *Alpha*'s management decapitates its innovative head.

Why would *Alpha*'s management do this? It seems a foolish move. Yet many well-run companies will do just that to improve stock prices and the bottom line. Such moves are particularly prevalent where the company sells staple articles in a mature industry and customer service is perceived as more important than new product innovation.

At first *Alpha*'s strategy has the desired effect. The engineering money saved goes to the bottom line. With increased profitability, stock prices rise. The investors are happy; management is happy. *Alpha*'s strategy also has an initial positive effect on market share. With engineering forces redirected to support product sales, the company slices modestly into *Beta*'s market share.

Beta reacts, not by copying *Alpha*, a battle it deems it cannot win, but by increasing engineering and innovation. It is a risky move because hiring more engineers pulls money from the bottom line, making *Beta*'s stock look less attractive. Fortunately for *Beta*, the investment works. *Beta*'s engineers design a new, more robust product line with features that captivate the market. *Beta* applies for patents on its new innovations and about the time its patents are beginning to issue, *Beta* begins taking significant market share away from *Alpha*.

By the time *Alpha*'s management realizes its decision to cut engineering was a mistake, *Alpha* and *Beta* have switched places as market leaders and *Beta* continues to innovate. *Beta*'s product line is now a full two generations ahead of *Alpha* and *Alpha*'s gutted engineering forces have no chance of catching up.

Having squandered its market share lead, what can *Alpha* do now? *Beta*'s patents block *Alpha* from coming out with the products its customers now want.

Alpha is boxed in by *Beta*'s patent portfolio. It needs a way to neutralize *Beta*'s portfolio. One way would be to undertake a major redesign of all of its products, avoiding *Beta*'s patents while trying to introduce new features that will win back customers. Unfortunately, that approach takes time, and *Alpha*'s crippled engineering staff now lacks the talent and direction to pull it off. *Alpha*'s other choice is to focus its creative efforts on new business models and exploit the business model patent to develop a portfolio that might neutralize *Beta*'s technology-based patent portfolio.

With no viable engineering solution on the short-term horizon, *Alpha* elects the business model solution. It establishes an innovation team, staffed by *Alpha*'s most visionary and creative members selected from all major departments: sales, marketing, engineering, upper management, and human resources. Using a moderator hired from an outside consulting firm, the innovation team holds a series of off-site brainstorming meetings at which they try to predict how their industry will evolve and how they should position their company.

The team focuses much effort on the customer, and on how *Alpha* can provide greater value to the customer. Instead of the classic product-centric focus, the team adopts a service-centric focus. *Alpha*'s and *Beta*'s customers purchase their respective products because they provide services through the products' features and functionality. The question becomes — How can *Alpha* structure its business and its products to increase the value of the services provided?

After several months and many hours of brainstorming, *Alpha*'s innovation team has an entirely new business model mapped out. Instead of simply selling products with a focus on new features — as *Beta* had successfully done — *Alpha* will sell service. By taking a hard look at the products of both *Alpha* and *Beta*, the innovation team enumerated all of the services those products provided to customers. The team listed all functional features and linked those to the underlying processes their customers were trying to perform by using the products. With this list in front of them, the team came up with new, more innovative ways for their customers to perform their underlying processes more efficiently and cost effectively. They discovered that both *Alpha*'s and *Beta*'s products were simply the wrong tools to meet their customers' objectives in several respects.

Alpha's innovation team mapped out new solutions to their customers' business problems. The team had discovered that customers were losing substantial money each time the production line had to be shut down due to equipment failure. Both *Alpha*'s and *Beta*'s products, although well-designed, did periodically fail when certain consumable components became depleted. Their customers knew this and employed special technicians to monitor the consumable components and to replace them prior to failure. Replacing the components was easy, but predicting failure was difficult. The monitoring job was monotonous. Research revealed that most technicians quit after six to twelve months, so that their customers were constantly expending money and effort hiring and retraining new technicians.

Alpha's innovation team realized that the consumable component was not likely to be eliminated in the near future. Thus, replacement of that component was a given. In the industry as it currently existed, companies such as *Alpha* and *Beta* were forcing the component replacement cost onto their customers. Customers didn't like it, but they had no choice. *Alpha*'s innovation team decided they could help eliminate this cost and thereby deliver better value to customers. It proposed using a series of simple, inexpensive runtime monitoring devices that would feed data to an Internet data collection site where predictive failure analysis would determine when component depletion was likely.

According to the business model developed by the team, *Alpha* would install the monitoring devices on a customer's equipment for free and would undertake to replace the consumable components as a service, not only to customers using *Alpha*'s equipment, but also to customers using *Beta*'s equipment. *Alpha* would offer a range of fees reflecting the quality of service desired by the customer. Customers that could tolerate a four hour down time were charged one price; customers that could tolerate a 30 minute down time were charged a higher price.

Using this business model as a base, *Alpha* commissioned several patent applications to be written. Although the details of the monitoring devices would have to be worked out, *Alpha* filed its initial set of patent applications on the business models that would be used regardless of what monitoring device would ultimately be chosen to implement the working embodiment. It took six months to structure the business organization to carry out the new replacement service and several years of diligent sales and marketing effort before the idea caught on. Compelled by the undeniable cost savings, in two years most of *Alpha* and *Beta*'s customers had signed on. *Alpha*'s replacement service is very popular.

Now with its foot in the doors of *Beta*'s customers, *Alpha* expands its business, supplying not only consumable components but entire systems on a service basis. Customers had traditionally bought *Alpha*'s and *Beta*'s products to convey articles being manufactured from one assembly line workstation to the next. With its demonstrated success in the consumable replacement service business, *Alpha* introduced a more far-reaching service, a conveying service. For a subscription fee, *Alpha* would take over the customer's conveying operation, running it directly from within the customer's plant. If conveyance equipment needed replacing, *Alpha* would replace it, at no cost to the customer.

Alpha was free to use equipment from any manufacturer, so long as it got the job done. Sometimes *Alpha*'s service team would use *Beta*'s products, where those were the more cost effective solution. However, more often the service team would use *Alpha*'s products because the profit on those products went to *Alpha*'s bottom line. The strategy put *Alpha* in control of the respective companies' market shares. *Alpha*'s service team was sensitive to finding the most cost effective solution; and that could entail purchasing some product from *Beta*.

However, *Alpha*'s service team was far less likely to be swayed by cosmetic feature improvements designed to attract sales to traditional end user customers.

Not to be caught off guard as it had been during the technology era, *Alpha* applied for and obtained several business model patents on both its initial consumable replacement concept and on various additional aspects of its service business. The patents were designed to block *Beta*'s entry into the service business. Some of its patents on the consumable replacement concept focused on the use of remote sensing and statistical analysis to support a replacement business model. Others focused on the mechanisms by which *Alpha* was able to guarantee different levels of quality of service, at different subscription rates. Although these patents were not an absolute barrier to entry, they did significantly cripple that effort. If *Beta* wanted to enter the consumable replacement market, it would have to do so without remote sensing. Without that ability, it would have to replace consumables far more often to meet quality of service guarantees. That would significantly cut into *Beta*'s profit margin.

Alpha's patents on its more ambitious conveying service model were even more clever. *Alpha*'s attorneys recognized that the end customers had been operating conveying "services" for years. Admittedly these "services" were being performed by plant employees using company-owned equipment. Nevertheless, the basic conveying operations were clearly prior art. Therefore, *Alpha*'s attorneys focused on a different aspect of the new conveying service, namely how *Alpha* was able to manage many numerous, often varied, conveyance systems for a plurality of customers as a business. Operating one conveyor was easy — plant employees did that all the time. Operating a sea of conveyors of different make, model, and function scattered throughout the country, each with different quality of service contract requirements — that was a far more difficult problem. *Alpha* used computer tools to handle the management task. *Alpha*'s attorneys applied for patents on these computer tools.

When *Beta* saw its market share gains evaporating, and realized why, it found itself blocked from entering the service field by *Alpha*'s business model patents. Unable to compete in the newly created service marketplace, *Beta* was forced to refocus its business as a supplier to *Alpha*. *Alpha* had neutralized *Beta*'s technological patent lead using business model patents.

Why did *Alpha*'s strategy work? It is not that the business model patent is superior to the technology patent — quite the contrary. A breakthrough technology patent has tremendous power to dominate an industry. However, *Beta*'s patents did not fall into the breakthrough category. They were, for the most part, incremental improvement patents, covering innovative improvements, but improvements nevertheless. *Alpha*'s strategy worked because *Alpha* was able to reinvent the marketplace and was then able to obtain broad blocking patents that barred entry into that new marketplace.

Before *Alpha*'s innovation team discovered how to exploit a service market, industry focus was on satisfying the end user customer with technological features. Both *Alpha* and *Beta* were selling products to factory automation *engineers* within the customer companies. Knowing they and their technicians would have to operate the equipment, these engineers quite understandably purchased products based on technological features. When *Alpha* invented the service market, it began selling to a different audience. It was selling to *company management*, who were more interested in guaranteed quality of service and overall product throughput than in technological features. *Alpha* created the new market and was thus able to obtain blocking patents before *Beta* had time to react.

So how do you identify visionary business model patents within your client's company? Essentially, you will need to predict the future. Whether your client's company is a leader or a follower, where is the industry headed. Although no book can answer that question, the business section of most major bookstores are filled with books describing what other visionaries believe the future will bring. One popular view today is that traditional brick and mortar manufacturing companies are evolving into service businesses. If this is so, it indeed suggests that business model patents may become a valuable component of every company's patent portfolio.

The following section explains how the evolution from product to service is currently taking place. It may help crystallize your thinking on how your client's business is evolving and on how business model patents may be strategically placed.

§ 13.08   Service and Flow Economy

Paul Hawken, Amory Lovins, and L. Hunter Lovins published a book in 1999 entitled *Natural Capitalism*.[21] In it, they predict the coming of a new industrial revolution during which companies will begin to discover opportunities for saving both money and resources through innovative new business practices. *Natural Capitalism* is not one of those gee whiz books that promises a futuristic world where technology solves every problem. Rather, more circumspectly, it shows how companies have already begun to adopt business methods that improve their bottom line while eliminating waste and reducing damage to the environment. Many of the ideas expressed are not new; but the authors integrate them well into a mural that could depict our future.

Books like *Natural Capitalism* help us see the present, in context of what the future may bring. Such vision can be extremely important when drafting patent claims because claims written today will be tested in light of what the future may bring. A skilled patent attorney drafts claims to predict the future.

One of the most powerful concepts expressed by the authors of *Natural Capitalism* is that of *service and flow*. It is a concept that traces its roots to a Japanese man named Taiichi Ohno (1912-90). Ohno-sensei engineered Toyota's world respected production system based on a simple concept: drive out waste. Ohno-sensei's word says it best: *muda*. In the Japanese language, *muda* means waste, futility, or purposelessness. Ohno-sensei built an infrastructure at Toyota

---

[21] Hawken, et al., *Natural Capitalism: Creating the Next Industrial Revolution* (1999).

designed to identify and eliminate *muda* wherever it could be found. What qualified as *muda*? Everything that did not contribute to customer value was *muda*. Excess scrap material was *muda*. Unnecessary paperwork was *muda*. Even poor productivity from Toyota's suppliers was *muda*.

Under Taiichi Ohno's guidance, Toyota squeezed out *muda* and leaped ahead of the rest of the world in manufacturing prowess. The word spread quickly. Soon companies throughout the world were sending emissaries to Japan to learn Ohno-sensei's techniques. Shortly thereafter, many companies had established *kaizen* programs to spread the word and to explore how Ohno-sensei's techniques could be put into practice in their companies. *Kaizen* means improvement, continuous improvement. Many *kaizen* programs focus on continuous improvement through statistical controls. However, at the core, most can be summed up simply as eliminating *muda*.

James Womack and Daniel Jones captured and expounded upon the essence of Taiichi Ohno's ideas in their best selling book, *Lean Thinking*.[22] They give several excellent accounts of where hidden waste can be identified and cast out for significant gains in product quality and customer value.

To get a sense of the insight provided by the *muda* concept, consider, as Womack and Jones do, the simple beverage can. The beverage can begins as bauxite ore buried in the side of some mountain. Hundreds of hours of labor and many kilowatts of energy later the ore becomes an aluminum container that someone casually pops open, consumes its contents and tosses it away. Womack and Jones calculate that it takes 319 days for that beverage can to be manufactured, filled with beverage and stored in your refrigerator, so that it will be available for you to consume its contents in a matter of minutes. When you consider the can as a value stream for cola delivery, it's truly wasteful nature is instantly revealed:

- mine ore

- reduce ore in mill

- smelt ore

- hot rolling process

- cold rolling process

- make can

- store can in warehouse

- ship can to bottler

- fill can

- store filled can in bottler warehouse

---

[22] Womack and Jones. *Lean Thinking: Banish Waste and Create Wealth in Your Corporation* (1996).

- ship to intermediate store warehouse

- ship to store

- sell to customer

- place in customer's refrigerator

To the frequent flyer, an even more compelling argument in favor of "lean thinking" may be found in our airports. Air travel is packed to overflowing with waste. To make a simple trip from Omaha to Washington D.C., for example, one must first go to the Omaha airport and stand in the check-in line. From the traveler's point of view this check-in line is pure *muda*. Unless one happens to enjoy standing in lines, the check-in line provides zero customer value. It is designed simply to assist the airline personnel in processing the self-sorting cargo (people) it will ship from Omaha to Washington.

But that is not all. Having checked in, received a boarding pass, and deposited checked luggage with the clerk, one proceeds through the metal detectors, which usually involves standing in a second line. Depending on the amount of residual metal in one's belt buckle, ballpoint pen, and steel arched shoes, one may be further routed from the metal detector line to the hand search line, where security guards use hand held metal detectors, chemical sniffers, and simple hand searches to determine if one is carrying anything *verboten*.

After passing inspection, one may proceed to the gate, where yet another line is encountered. This line is ostensibly designed to load passengers on the plane in an orderly fashion. Yet, notwithstanding that everyone with a boarding pass knows his or her seat assignment, only a crude attempt is made to sort the passengers prior to boarding the plane. Boarding is from the rear of the aircraft, but no effort is made to presort passengers so that the person assigned to seat 20E does not block persons further back from taking their seats as he struggles to jam a three cubic foot parcel into a two-point-five cubic foot space.

Well over an hour has passed, and the plane hasn't even left the ground. Before it can leave, it must taxi to another line, where it queues up behind other planes bound for other destinations. Counting the taxi line, one has been forced to wait in as many as five separate lines — all *muda*, all waste. When the plane finally takes off it is usually free sailing until landing — unless there happens to be congestion at its destination, in which case the plane enters a holding pattern, in line behind other aircraft also waiting to land.

The airlines have determined that it is more efficient and profitable for them to use large planes, filled with many passengers. Because statistically few passengers a week travel between Omaha and Washington, D.C., the airline routes passengers wishing to make this journey through Chicago. By funneling passengers from all over the Midwest to the Chicago hub, the airline can fill a jumbo airliner with passengers destined for Washington, D.C. While arguably efficient from the airline's standpoint, one traveling this route must layover in Chicago, while baggage is transferred and more lines are negotiated. Again, to the customer, this layover is pure waste. Weather permitting, eventually, the flight leaves Chicago bound for Washington, hopefully with luggage and passenger on board.

Once the D.C. bound plane finally lands, one must proceed to a carrousel and wait again, this time to retrieve the checked luggage. A further line is encountered at the car rental desk, followed by a further line to wait for the rental car courtesy bus and a still further line waiting to have papers processed so that one can finally drive the rental car beneath the final anti-theft gate.

If you break down the Omaha to Washington, D.C. trip into its component parts, all of the steps make sense — *to the service provider*, but not to the customer. The check-in line saves the airline money because it allows one check-in attendant to process hundreds of customers a day. The metal detector line makes sense, again, because it allows a few security guards to check thousands of passengers a day. Likewise, routing traffic from small cities through larger hub cities seems like an efficient use of resources. A large plane would seem to be more cost effective, benefiting from economies of scale. More cost effective, that is, until one factors in the customer time wasted waiting in all of those lines.

The airline infrastructure is so ingrained in our society that it seems like heresy to suggest that bigger, faster planes should be replaced with smaller, slower ones to provide better customer value. However, from the standpoint of eliminating waste and delivering more convenient customer service, smaller, slower planes may indeed be the way to go. The next time you make a business trip, calculate all of the hours spent checking in, waiting in line, waiting for luggage, waiting for ground transportation. Then consider how much more pleasant, and possibly even quicker, the trip would have been if you could have simply driven to the small airport next to your local shopping mall and walked from your car to the waiting mini-jet. Even if the mini-jet is 50 percent slower than the big jumbo airliners, you would probably get to your destination in about the same time, given all the time you save standing in lines and participating in the airline's mass self-sorting process.

§13.09 A Systematic Approach to Portfolio Development

In any single software patent, the claims define the metes and bounds of the patent grant. Chapter 10 explores the internal workings of the business patent claim and describes various e-commerce and business model claim drafting strategies. Claim drafting is a fundamental skill, like that of the stone mason. Claim elements fitted together by a skilled patent attorney present a formidable barrier. But to build a strategic patent portfolio requires a further skill, with a higher vision, like that of the architect who designs the castle walls.

A strategically designed patent portfolio should protect a company's market share and the technological advantage of its products. Individual patents become carefully chosen stones within a well-designed castle wall. The strategic patent portfolio is no accident. It is the result of a properly designed intellectual property strategy that takes into account the company's business climate, the relative maturity of the company's technologies, and the aggressiveness of the company's competitors.

The most effective intellectual property strategy is designed to flag the commercially important innovations for patent protection while continuously fine tuning its parameters to remain on course as management steers around commercial obstacles.

This section provides a methodology for building the strategic patent portfolio. The next few subsections describe techniques for analyzing the topographical coverage of an existing patent portfolio. From this topographical view, individual claims disappear, revealing the overall portfolio landscape. Subsequent sections describe how to design an intellectual property strategy that will flag important innovations and that can be readily fine-tuned. This chapter then ends with a series of architectural patterns specifically for crafting the software patent portfolio.

[A] Analyzing the Topography of a Patent Portfolio

Patent claims define the metes and bounds of every patent in a portfolio. Whereas a single patent claim can usually be understood after reasonable study, the patent claims of an entire portfolio present far too much information for the human mind to grasp and retain. If you doubt this, select any 30 patents from a portfolio. Study the claims of each, and then try, on 30 consecutive days, to recall a claim from each of the patents and what it covers. The human mind is a marvelous instrument, but it simply isn't tuned to perform this task. Alternatively, study the claims of all 30 patent claims and then try to immediately recall what each one is about. The average human mind can't perform that feat either.

To properly analyze a patent portfolio of even moderate size there needs to be a way to concisely represent the essence of a diverse collection of patent claims so that the human mind can understand and retain the information. Many patent portfolio managers employ computer database technology to attack this problem. Computers excel at precisely the kind of mundane recordkeeping that the human mind repels.

Using a computer database to track patents in a portfolio makes sense. Patent numbers, inventors' names, assignees, filing dates, issue dates, number of claims, full text of the abstract and claims: these are all obvious candidates to include in a patent database. However, these data do not adequately capture the essence of a body of patent claims. What is needed is a way to see all claims at once, not singly, but one at a time.

Conceptually, patent claim language is like music. It must be read serially, one word phrase at a time, to be appreciated, just as music is perceived serially, one musical phrase at a time. With patent claims and with music, there is no way to perceive the entire work at once. In contrast, observe any good landscape painting or other work in the visual arts and the effect is immediate. You can perceive the entire scene in an instant and yet remain free to study the details at your leisure.

This suggests that the visual arts provide a better way to perceive a body of claims, if only claims could be transformed into a visual domain. One technique for doing this involves *claim clustering*. Claim clustering is the grouping together of patents that claim similar subject matter. Once grouped, you can graphically display the clusters, like cities on a topographical map, and readily determine from the sizes of the clusters how different subject matters are covered in the portfolio.

This may all sound rather esoteric, but fortunately there are several easy ways to cluster patent claims. One way is to form patent clusters using the U.S. Patent Classification System. Patents are classified by the U.S. Patent Office into patent classes and subclasses. Patents are so classified based on the subject matter of the patent claims. Thus you can use patent class/subclass information to group or cluster patents that have common claimed subject matter.

The Patent Office prints patent class/subclass numbers on the face of every patent. Patent classification data are also readily available electronically from most commercial databases that supply patent information. It is relatively straightforward to include patent class/subclass numbers in your portfolio tracking database. As an alternative to using the U.S. Patent Classification you may wish to consider using computer linguistic technology to analyze patent claim language.

Both U.S. Patent Classification and linguistic analytic techniques have their advantages. The U.S. Patent Classification System reflects the Patent Examiner's view of what he or she perceived as the main thrust of the patented invention. Linguistic technology reflects more the actual words used by the patent applicant in defining his or her invention. You may find that a combination of both techniques yields the best results because the two techniques combined can give a stereoscopic view of the patent claim data that neither alone can give.

Regardless of what techniques you use to categorize claims in the portfolio, the objective is to cluster the patents in your portfolio into groups of related subject matter. Doing this may not allow you to recall the details of any given claim, but it will allow you to see the entire portfolio as a body of claims. From this view you can spot patterns in your portfolio coverage and assess where the portfolio is strong or weak.

[B] Improving Patent Clusters

Clustering patents according to common subject matter often yields too many clusters to be manageable. Clustering a portfolio of 500 patents into 150 clusters does not give the easily grasped view you seek. Ten to 20 clusters would be better. If you get too many clusters when you cluster patents in your portfolio, you may need to group selected clusters into superclusters or categories that represent concepts that are more manageable. Ideally, the categories you select should represent familiar subject matter, such as recognized technologies, or product features. For example, a software company that manufactures compilers, word processors, database systems, operating systems, and spreadsheets may want to map its patent portfolio clusters onto these five categories.

Keep in mind that clustering techniques used to analyze the portfolio of a single company can also be used to compare the portfolios of two or more companies. Often such analysis can give valuable insight into the posture of a given company, vis-à-vis its competitors.

Once you have clustered the patents in your client's portfolio, you can construct different visual presentations, depending on your interest. Pie charts and bar charts present the data quite effectively. Use a pie chart to show the percentage breakdown of the client's top ten patent portfolio categories, for instance. Include patent filing dates in your cluster data and you can readily show patent filing trends graphically, as well. You may use bar charts to illustrate annual portfolio growth by category, for example.

The visual presentation and analysis you elect to use is up to you. Your goal is to understand what the portfolio covers, so that you can design a strategy to systematically improve it.

Analyzing the topography of your existing patent portfolio comes first, but that is just the first step in developing a strategic portfolio. The next step is to design a process by which your future portfolio will be built. Many don't give the latter a second thought, but they should.

[C] Drawbacks of Classic Invention Disclosure Programs

Examine the patent portfolio of any company with more than 50 patents and see if you can discover that company's intellectual property strategy. To be sure, most patents in the portfolio relate to the core business of the company, but beyond that, can you find any rhyme or reason why the innovations reflected in these patents were chosen for protection, while other innovations were excluded? In many cases, the answer lies, not in the omniscient wisdom of patent counsel, but in the day-to-day operation of the company's invention disclosure program — its invention harvesting mechanism.

Many companies harvest inventions using an invention disclosure program that rewards inventors with a small to moderate bonus for each invention disclosure submitted. A company patent review committee then determines which submitted disclosures warrant patent protection and which do not. Many companies give bonuses to only those disclosures that are accepted for filing, or actually filed as patent applications. Some give an additional award if the patent is granted or if the invention is used in a commercial product.

This classic invention harvesting mechanism can yield a sizeable patent portfolio, but the results often lack focus. This happens for several reasons. First, some innovators like to write; others do not. Thus some find it easy to compose and submit invention disclosures regardless of how busy their schedules. Others, perhaps bowing to product-release pressures, never find the time to disclose. This basic difference among people lies at the heart of a basic flaw in classic invention harvesting. Potentially important innovations can slip into commercial products, and ultimately into the public domain, without ever being considered for patent protection.

A second reason the classically harvested portfolio can lack focus originates in simple office politics. Patent review committees often approve disclosed inventions for patent filing simply to encourage the submitters, or the submitters' department. They do this as a means of perpetuating a steady flow of disclosures. Patent review committees know that once rejected, a contributor may be unlikely to participate in the disclosure program again. Entire departments can lapse into a non-disclosure malaise if not periodically rewarded. Unfortunately, keeping the disclosure pumps primed can become a means unto its own end. The original reason for building the patent portfolio to enhance the company's strategic position is all but ignored.

Another flaw in classic invention harvesting lies in recordkeeping. The typical patent review committee will maintain statistics on the number of disclosures received, per department, the number of applications filed and the number of patents granted. Rather than focusing on quality of the portfolio, these statistics focus on the size of the portfolio. Some patent review committees, in the interest of controlling the process, even maintain statistics on how quickly the disclosure harvesting process draws in raw disclosures and pumps out issued patents. Such statistics miss the point. Although it is important to keep careful tabs on timing, to avoid unexpected statutory bars, it by no means follows that hastily prosecuted applications yield high quality results.

If classic invention disclosure programs are so fraught with shortcomings, does this mean they should be discarded? In most cases the answer is no. Classic disclosure programs capture some important innovations, although not necessarily all of them. Thus you can use the classic invention disclosure program as one part of the overall intellectual property strategy. To improve the strategy, supplement it with an auxiliary innovation development system that is based on the company's business model as described in the next section.

[D] Developing Innovation with a Business Model

Most companies undergo exhaustive budget reviews at least annually. During such budget reviews, project leaders from each department meet with management to vie for their share of next year's project funding. Project leaders spend the weeks leading up to annual budget review carefully mapping out next year's business strategies. Then, during budget review, management scrutinizes each strategy, funding those with promise and pruning back those that lack it.

The annual budget review represents an excellent opportunity for you to formulate the company's intellectual property strategy, for at this unique time, engineering plans and management plans are sharply in focus. To take advantage of this opportunity, you should engage management and engineering in innovation review, during or shortly after budget review.

The idea is quite simple. Assess where the research and development dollars have been allocated and why. Then use this information to plan a proactive innovation development program for the following year. In this way you can keep tabs on different projects throughout the year, making sure that important patent applications get filed regardless of whether formal invention disclosures have been submitted or not.

Periodic innovation review replaces office politics with cold, hard numbers. Unless there are specific reasons to the contrary, the intellectual property protection given to different projects should roughly match the proportions allocated to fund those projects. In other words, if management allocates 50 percent of a company's research and development budget to one project, then you should plan to allocate roughly 50 percent of your proactive innovation development efforts on this project as well.

[E] Putting It All Together

Harvesting inventions through classic disclosure programs or through periodic innovation review is an important first piece, but it is not the entire jigsaw puzzle. You still need a strategy for assessing which innovations to protect and which to defer.

To construct your intellectual property strategy, it may be helpful to have a model, such as the one shown in Figure 13-1. The model describes a system where all of the company's intellectual and technological assets enter as raw materials and at least a portion of them exit as finished patents. From the raw material side, you want to ensure that nothing potentially valuable is left out. Therefore, you should consider all forms of intellectual and technological assets, including know-how, unpatented technology, best practices, knowledge databases, technological contracts, partnership agreements, teaming agreements, license agreements, and the like.

Figure 13-1 represents these raw material intellectual and technological assets as a pool from which your invention protection systems will draw. If you are using both a classic invention disclosure program and periodic innovation review, then these two mechanisms draw from this pool. At this stage, treat both mechanisms as equally likely to deliver important innovations. Do not assume, for example, that the inventions developed through periodic innovation review are more important than inventions disclosed through the classic invention disclosure program. Assessing which innovations are the most important comes in the next step.

Once your harvesting mechanisms have done their work and all innovations have been identified, you will next rank the results. Think of this ranking as a bubble-sort operation, where the most important innovations rise to the top, while the less important ones sink to the bottom. There are many ways to rank innovation. The next section on metrics describes a simple "good-better-best" ranking scheme that may help.

After ranking comes the cut. Few clients have an unlimited budget for protecting intellectual property assets. Thus you can expect somewhere in your ranked list there will be a line, above which innovations are protected and below

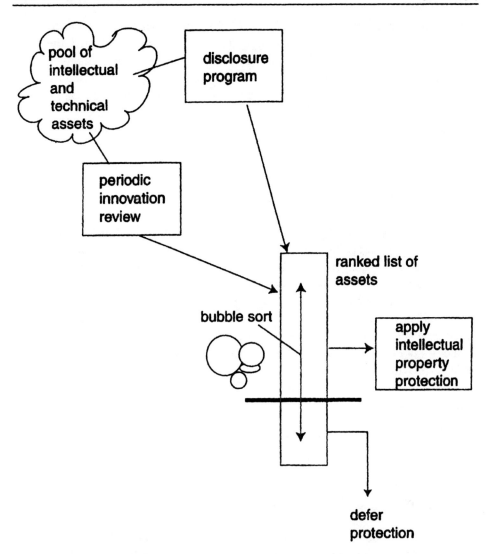

FIGURE 13-1:   INTELLECTUAL PROPERTY PROTECTION SYSTEM

which protection is deferred. While different innovations require different forms and levels of protection, and no two innovations are exactly alike, you can probably predict in advance what it will cost to secure protection for each innovation in the list.

Based on the available budget, adjust the line up or down until an optimum point is reached. Your client should be satisfied that all items above the line warrant protection and all items below the line can safely be deferred. If not, encourage your client to rethink the intellectual property protection budget.

By sorting all intellectual and technological assets in a systematic way, you can help your client avoid needless or haphazard spending on intellectual property protection, while ensuring that the most important innovations are protected first. With a sorted list, you and your client know which innovations made the cut and which did not. You and your client are now in control of the intellectual property strategy. Through this strategy you have tied the purse strings directly to the reins of portfolio control. To keep the strategy in tune, periodically meet with your client to determine if the cut line remains in the correct place. If not, simply adjust the budget accordingly.

[F] Metrics for Measuring and Improving the Effectiveness of Your Intellectual Property Strategy

No intellectual property strategy can be expected to run forever without fine-tuning. Thus, you should design your intellectual property strategy to be flexible and self-tuning. Achieving continuous improvement is a key objective. Meeting this objective implies measurement. Relevant measurable aspects of the strategy and processes must be identified, measured, and understood. Only then can the strategy and processes be improved, based on the newly gained understanding. Although measurables can be used to shed light on "how well" the company and its business partners are performing, the reason for employing measurables in the intellectual property strategy is to allow the strategy to be intelligently fine-tuned. In a systems engineering sense, the intellectual property strategy should be a stable "closed-loop" system, with appropriate "measurables" providing the feedback needed to keep the system on course.

While there are many measurable aspects to an intellectual property strategy, you may want to concentrate on these fundamental business ones:

• What intellectual property assets does the company have?

• What is the value of each asset to the company?

• How is the asset being protected?

• How effective is that protection?

Clearly, these questions invite broad-brush answers and will not serve as the quantifiable measurables needed to fine-tune the intellectual property strategy. Nevertheless, these broad questions can be broken down into more focused questions that can provide useful measurables for evaluating the effectiveness of the intellectual property strategy. In the following sections, each of these broad-brush questions is more fully explored.

[G] Identifying the Intellectual Property Asset

Before you can measure an intellectual property asset's value, you must first identify the asset. That may seem self-evident, but it is very important that this be done correctly. Otherwise the wrong system will be measured and your intellectual property strategy may not be fine-tuned correctly.

It is common to think of intellectual property "assets" as being patents, trademarks, and copyrights. However, for our purposes here, the term intellectual property *asset* is used more broadly to include know-how, unpatented technol-

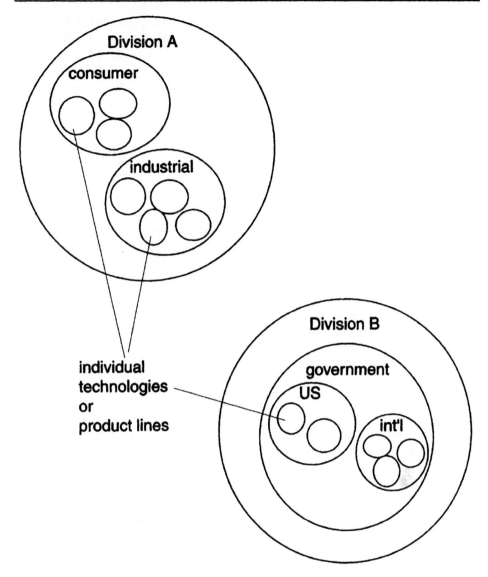

FIGURE 13-2: NESTED BUBBLE MODEL

ogy, best practices, knowledge databases, and the like. In this broader context, patents, trademark registrations, and copyright registrations represent forms of *legal protection* of certain kinds of intellectual property assets. This distinction is made to ensure that all segments of the company's business are considered (including the service-providing segments).

[H]   Nested Bubble Model

To aid in identifying the company's intellectual property assets, a hierarchical list, or the nested bubble model, may be useful. As illustrated in Figure 13-2, the sum of the company's current intellectual property assets is shown in the outermost bubble. Within this bubble are separate bubbles, each representing one of the company's operating groups. Each of these bubbles is in turn subdivided into operating units or divisions (where applicable), and those subdivisions are further subdivided into individual technologies or product lines.

The bubbles may be subdivided as finely as needed to expose all meaningful, homogeneous bodies of technology. For example, at the most fundamental or elemental level, these bodies of technology may represent individual products or services; or they may represent customers or groups of customers; or they may represent collections of knowledge about a technology that may be exploited in the future. The choice is up to you. Use whatever makes sense.

The objective in constructing the nested bubble model is to identify all important intellectual property assets. Later in the analysis, each of the lowest level (most elemental) bubbles will represent one intellectual property asset to which measurables are applied to arrive at different figures-of-merit. The higher level bubbles are important, too, because they provide a nested framework to help ensure that nothing is overlooked. At the conclusion of the analysis, each of these higher level bubbles will inherit an aggregate figure-of-merit from its lower level members. The aggregate figure-of-merit provides a high level topographical overview of the effectiveness of the strategy.

[I] Linking of Assets and Protection

Once all relevant bubbles have been identified, the company's current intellectual property protection (patents, trademark and copyright registrations, license agreements, and intellectual property-related contracts) can be linked to one or more technology bubbles as may be relevant. Some technology bubbles will have many links, indicating that these may be better protected than those with fewer links.

The presence or absence of such links is a gross measurable of the effectiveness of the intellectual property strategy. It is only a gross measurable because the presence of a link indicates simply that *some* protection has been provided. How effective that protection is will depend on other factors. (Measurables for these other factors are considered below.)

After all assets are identified, an attempt should be made to measure each asset's value. Similarly, after all forms of protection for these assets have been identified and linked to the relevant asset or assets, the quality or effectiveness of protection should also be measured. These aspects are discussed in the following two sections.

[J] Measuring the Asset's Value

What is the value of the intellectual property asset to the company? That question can be broken down into these smaller questions:

- To what extent does the asset enhance profitability?

- To what extent does having the asset protected protect marketshare?

- What licensing revenue does the asset generate (if any) apart from enhancing profitability (in item 1)?

There are several ways to provide quantifiable answers to these questions.

One way is with dollars, by using asset valuation accounting techniques, described in the literature. Formal asset valuation is probably not workable for our purposes. Although formal asset valuation techniques could be used to provide quantifiable (dollar amount) answers to the above three questions, the human resource cost to do so is probably prohibitive.

Another, less costly way to provide quantifiable answers is by a simple "good-better-best" rating scheme. The following is proposed:

Rate 0: Negligible. The asset does little to enhance profitability and its protection does little to protect market share. Royalty revenue is negligible.

Rate 1: Moderate. The asset enhances profitability some, but not substantially; or protection of the asset partially blocks competitors, but not fully; or royalty revenue exceeds the cost of maintaining IP protection, but is not substantial.

Rate 2: Substantial. The asset substantially enhances profitability; or protection of the asset fully or significantly blocks competitors; or royalty revenue is substantial.

Rate 3: Supreme. The asset is of supreme value, representing a significant component of the company's bottom line, the crown jewels of the company. Note: it is expected that few assets will achieve this highest rating.

Using the above questions, try to rate each asset in the portfolio. By all means, keep the rating scheme simple. Fine distinctions between the value of one asset and another are not needed and will only complicate your task. Here you simply want to separate the wheat from the chaff. Also, keep in mind that here you are rating the importance of the assets themselves, not the quality of the assets' legal protection. It is important to keep these concepts separate. A heavily protected asset of little value is quite different, indeed, from a poorly protected asset of great value.

[K] Measuring the Effectiveness of Protection

How well an intellectual property asset is protected is a difficult question to answer. Effective legal protection is a multifaceted issue that is very difficult to measure because there are many variables that are simply unknown. Predicting how a competitor will respond in the face of a patent is like predicting how the stock market will respond in the face of news that droughts are predicted next summer. Measuring a legal protection system is much like measuring an economic system. No set of measurables will work in every situation.

Nevertheless, if the intellectual property strategy is to be fine-tuned, some reasonable measurables will need to be identified. Proposed here is a business-centered approach that measures effectiveness by how well the legal protection blocks competition. Blocking competition is how premium prices are charged, market share is protected, and royalty revenue is generated.

Before ranking how well each of the intellectual property assets is protected, a little information gathering is in order. Return to the bubble model described in § 13.09[H]. For each asset identified (for each of the smallest bubbles) identify what forms of legal protection are currently applied to that asset. Then try to identify the key elements that are required to compete in this technology. Try to identify those essential parameters, features, and selling points that a competitive product or service must have in the current marketplace. Answers to these questions should be provided by the company's most experienced people (those with the best vision of the big picture).

Then list these identified parameters for each technology bubble, associating each list with the corresponding bubble, so that the list can be compared with the previously identified legal protection that is applied to that bubble.

Armed with the above information, ask the following question for each asset: How well does the current protection block competition? Then use the following "good-better-best" rating system to rate your answers:

Rate 0: Negligible. Little or no legal protection is provided.

Rate 1: Moderate. The company's current products (or services) are at least partially protected against direct knock-offs and product copying; however, it is possible to compete without significant price penalty using different technology.

Rate 2: Substantial. The company's current products (or services) are substantially protected and a competitor would not be able to compete using different technology without a significant price penalty.

Rate 3: Supreme. The protection forecloses all conceivable competition. There are no known ways to compete using a different technology. Note: it is expected that this highest rating will be rare — most likely associated with breakthrough discoveries that are far ahead of the rest of industry — and will undoubtedly be eroded as newer technologies supplant the protected asset.

[1] The Final Effectiveness Tally

The foregoing analysis produces a two-dimensional "good-better-best" matrix. The matrix includes an entry for every intellectual property asset you have identified. Associated with each asset is a first figure-of-merit, describing the commercial value of the asset, and a second figure-of-merit, describing how effectively the asset is currently protected. The matrix readily shows whether important assets are well protected, and whether unnecessary protection is applied to unimportant assets.

Higher level matrices should be constructed for each of the higher level (outwardly nested) bubbles. These higher level matrices may simply contain entries for each lower level matrix that the higher level bubble contains. The individual figure-of-merit scores of the lower level matrices may be aggregated or averaged to serve as figures-of-merit in the higher level matrices. These higher level matrices provide a good high level topographical view of how effectively intellectual property legal protection is being applied to any given subdivision within the company, or to the company as a whole.

Technology	Asset Figure-of-merit	Protection Figure-of-merit	Link to Lower Level
technology A	-1- (moderate)	-1- (moderate)	technology X
technology B	-2- (substantial)	-1- (moderate)	technology Y

FIGURE 13-3: EXAMPLE OF MATRIX

[L] Fine-tuning the Intellectual Property Strategy

The figures-of-merit provided by the matrices take a snapshot of the company's intellectual property strategy. To fine-tune the strategy, the matrices should be periodically (for example, annually) recomputed and compared with prior period figures. It is expected that the figures will change (both for the better and for the worse). The value of a given technology may increase or decrease, due to market growth, introduction of new technology, change in management focus, and so forth. The effectiveness of legal protection may also increase or decrease, due to acquisition of new patents, introduction of new technology that legally circumvents existing protection, abandonment of patents no longer deemed important due to changes in marketing strategies, and the like.

To begin the fine-tuning process, the above measurables may now be applied to the company's current position. All current patents in the company's portfolio can be factored into the equation. Then, at a suitable time next year the same measurables should be computed and compared with this year's "tally." Comparison of the individual matrices and their figures-of-merit will allow you to fine-tune your intellectual property strategy more systematically.

[M] Summary of How to Measure Effectiveness of Intellectual Property Strategy

Measuring effectiveness of the intellectual property strategy involves the following steps. These steps may be performed now, as part of the portfolio analysis, and periodically (annually) thereafter. The figures-of-merit produced by

675

these steps allow the value of your client's intellectual property assets to be assessed as well as the effectiveness of the legal protection applied to those assets.

Step 1. Identify all intellectual property assets (using bubble model as an aid).

Step 2. For each asset, list all current legal protection applied to that asset (e.g., patents, trademark registrations, copyright registrations, license agreements, IP-related contracts).

Step 3. Using the "good-better-best" ratings described in § 13.09[K] above, rate the value of each asset. (The rating system provides an asset figure-of-merit designed to reflect how the asset enhances profitability.)

Step 4. Using the "good-better-best" ratings described in § 13.09[K] above, rate how well each asset is protected. (The rating system provides a protection figure-of-merit to reflect how well competition is blocked by the legal protection.)

Step 5. Record the figures-of-merit for each asset in a matrix for comparison in subsequent evaluations. (Note: the matrices may be hierarchically linked to mirror the nested nature of the bubble model. This will allow a quick, high level overview on a group, division, or company-wide basis.)

[N] Architectural Patterns for Building an Information-based Patent Portfolio

Form follows function. This rule of good design, espoused by the mid-20th-century Bauhaus movement, applies to software design, just as it does to industrial product design and architecture. In constructing a strategic software patent portfolio, it is often helpful to look through the form of a software design to see its underlying function. Function is often the more important aspect to protect.

How do you expose the underlying function in the form of a software design? Often you can spot recurring architectural patterns, design patterns that software developers use again and again. These will lead you to the underlying function of a software design.

The following sections discuss several common architectural software patterns. Once you understand them, you can use them to build software patent portfolios. The trick is to spot the pattern, understand the underlying function, and then figure out how to apply software patent protection to protect the underlying function. Sometimes a single patent will suffice; sometimes several patents will be required.

Keep in mind that these architectural patterns are intended to reveal the function behind broad technological movements, entire industries or entire product lines, and not merely the functional underpinning of a single invention.

For example, you will use these architectural patterns when your client says, "We have developed a new paradigm that makes the desktop metaphor obsolete."

[O] Information Flow and Information Packaging

In searching for strategic architectural components with which to build a software patent portfolio, look first at the information flow. Information is a basic commodity, a raw material, that software engines process, pump, and package for consumption by the end user. Internet technology, digital video, electronic commerce — these are all examples of specialized information flow processed, pumped, or packaged by software.

If information is a commodity, like oil, then it pays to own the pipeline, and if not the pipeline, at least the shipping containers used in a major downstream distribution channel. Thus, software patent protection on the packaging and delivery system for important information can be extremely valuable. For example, a software patent protecting a cost effective satellite data processing technique could prove extremely valuable in delivering video-on-demand digital television. Likewise, a patented software encoding scheme that renders cellular telephones more reliable in fringe reception areas could also prove extremely valuable. In these examples, software defines a key component in getting the information from point A to point B.

To build a strategic portfolio that capitalizes on information flow, first identify the source of the information and learn why the information is valuable and to whom. In doing this, try to identify all important users or consumers of the information. Then map the information flow from its source to the ultimate users or consumers. In effect, you are simply identifying the information flow from point A to point B.

Once you understand the information flow, then determine where in the flow your client's technology sits. The closer to the information source the technology resides, the more powerful your client's position is. Determine how your client's technology affects the flow of information. Does it correct a bottleneck or improve the reliability of the information flow? Does it add value to the information flow? Think about these things as you plan your strategic patent portfolio.

[P] Time Versus Space

Transporting information from point A to point B involves the same basic laws of physics that NASA grappled with in landing humans on the moon or in charting the grand tour of Voyager to the outlying planets of our solar system. Transportation, whether of bits of information or atoms of matter, comes down to a fundamental problem of time versus space. The larger something is, the longer it takes to deliver. Given a thruster rocket of predetermined size, the larger mass accelerates more slowly than the smaller mass. Given a Central Processing Unit

(CPU) of a predetermined clock speed, the larger file takes longer to deliver than the smaller one.

Time versus space imposes significant metaphysical consequences upon software involved in transporting information from point A to point B. With rare exception, the software developer always strives to deliver the largest payload in the shortest amount of time. Ingenuity appears endless in this endeavor. Data are compressed, stored in cache buffers, shipped via alternate routes, spontaneously generated all in a never-ending battle to beat the clock.

If you discover this pattern in your client's technology, try to determine how your client's technology addresses the time versus space problem. Are the data being compacted or compressed so they take up less space? Is the information pipeline being widened, or split into several parallel paths? Are the information pumps being made larger by increasing clock speed, or are they being made more efficient by eliminating unnecessary processing steps? These are the kinds of questions to ask. Once you have a good grasp of the time versus space problem, then determine if alternate ways of addressing the problem are as good. Try to discover the essence behind why your client's technology solves the problem the best.

§ 13.10   Part III — Portfolio Mapping Techniques

Not all business model patents are created equal. Some may represent mere incremental improvement in a current business practice; others may represent a strategic bet on how future commerce will unfold. Today as you sit reading this, your company or client sits in the center of its particular piece of commerce, with varying degrees of certainty about what the future will bring. It is relatively easy to predict what tomorrow will bring (most likely the same as yesterday). It is more difficult to predict what six months will bring, and far more difficult to predict what six years will bring.

Picture a bull's eye evolution diagram, Figure 13-4, where now rests directly in the center and time radiates outwardly in concentric circles about the center. You sit in the here and now, precisely in the center of the bull's eye. The first concentric circle represents next week, the next circle represents six months from now, the next circle a year from now, and so forth. You can be certain that business practices will evolve. In what direction, no one knows. For example, business evolution may follow the meandering path A (see diagram), or it may follow the meandering path B. Because the circumference of next week's circle is still relatively small, meandering paths A and B cross that circle only a short distance apart. Thus if you predict the future will evolve along path A, when in fact the future follows path B, you won't be terribly wrong a week from now. However, as time unfolds in ever expanding circles, your chosen path A begins to diverge radically from the one the future has chosen.

The diagram illustrates how wrong, or right, a choice can be. Of course, in the business world no sensible business person would stubbornly continue along a path making buggy whips and bridles when the rest of the world was making

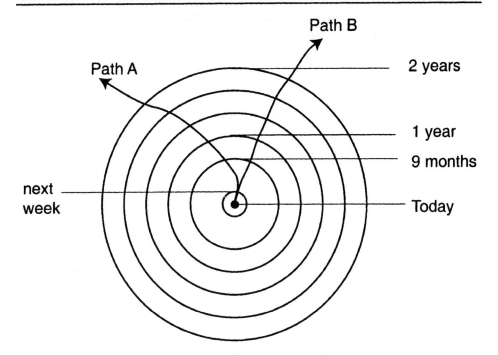

FIGURE 13-4: BULL'S EYE EVOLUTION DIAGRAM

seat belts and wiring harnesses. Businesses change paths all the time. Thus, you might criticize our bull's eye diagram for failing to account for management's propensity for jumping paths whenever judgment dictates. That is not the point of the diagram. The diagram is not about jumping paths, it is about placing bets. If you were to place five strategic patent bets on the bull's eye timeline, where would you place them? Bets placed close to the center are likely to represent incremental improvements. Bets placed far from the center represent dreams and vision.

The bull's eye timeline holds one immutable fact. The farther you move from the center, the more likely your bet is to be wrong. That is because the circumferences of the concentric circles grow as you move out from the center. Of course this geometric dispersion in bet-placing accuracy holds true for everyone placing bets. Everyone's risk of betting wrong grows as the circles increase. With the increased risk comes increased reward. Imagine being the first to bet (and patent) that some day wireless Internet commerce will beam like sunshine on every part of the planet. The potential rewards from licensing such a patent would be huge.

Understanding the bull's eye timeline and its implications is one thing. Building a strategic patent portfolio in a systematic fashion is another. It is the difference between the day trader, who haphazardly buys stocks on intuition, whims, and tips from others, and the systematic investor, who develops and sticks with an investment strategy, buying and selling in market sectors he or she understands. The former day trader may get lucky if he or she times the market perfectly. Hopefully the lucky hit will not be eaten up by losses sustained on all

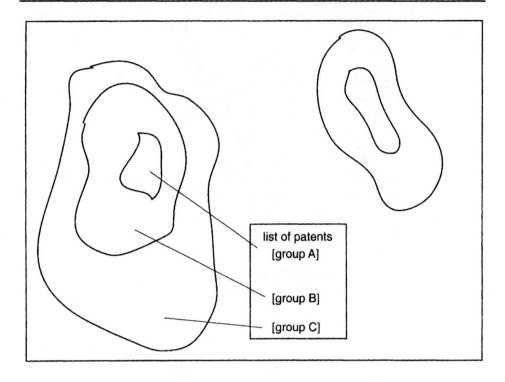

FIGURE 13-5: PORTFOLIO MAP

of the misplaced bets. In time, however, the systematic investor stands a far better chance of retiring with a substantial nest egg.

[A] Portfolio Maps for Systematic Portfolio Development

So how do you develop a strategic business patent portfolio? Start by constructing a model of the territory you expect your portfolio to occupy. The model should reflect the patents your client currently owns, as well as the patents currently owned by others. If these patents were laid out graphically, they might resemble geologic structures on a topographical map. Large portfolios might appear as large oceans; small portfolios might appear as mountain peaks. See Figure 13-5. The map of existing portfolios shows the regions already covered by your client and its competitors. All of the geologic structures that appear on this map represent patent bets already placed. Thus, if transposed onto the bull's eye timeline, all of these geologic structures would lie at the center.

In developing a systematic approach to portfolio development, the challenge is to predict where future geologic structures will develop before anyone else figures it out. Powerful commercial forces drive global business. These forces mold the commercial landscape, just as tectonic forces deep within the earth mold the geologic landscape. If the topographical map of Figure 13-5 could be animated, using the bull's eye timeline to predict where future commercial tectonic forces will be brought to bear, new mountain ranges and

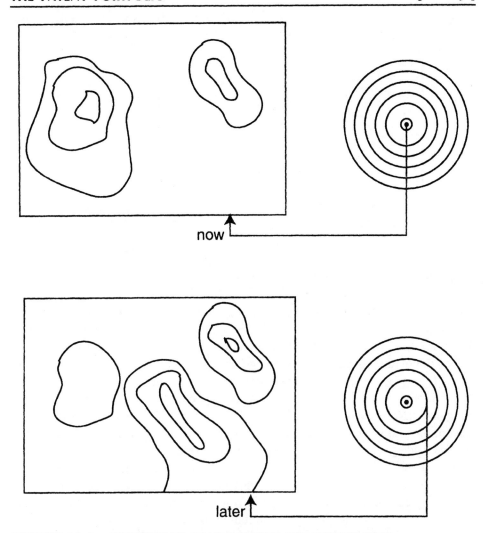

FIGURE 13-6: PORTFOLIO MAP LINKED TO BULL'S EYE

oceans of technology would emerge before the eye, as old technologies erode away and evaporate.

The modeling technique you select to represent your portfolio can be a simple list of current patents and anticipated future developments; or it can be a more complex graphical model. Figure 13-6 shows a graphical model in which the portfolio is represented topographically. The East-West and North-South dimensions of the map may correspond to patent class and subclass designations. The height of the topographical feature may correspond to the number of patents under that particular class and subclass.

To explain, U.S. patents are assigned to a particular class and subclass when they are issued. The Patent Examiner makes the assignment based on what subject matter is claimed. All U.S. patents are assigned to one primary class and subclass. Sometimes, examiners will assign a patent to additional classes and subclasses for cross-reference purposes. For example, a patent primarily assigned

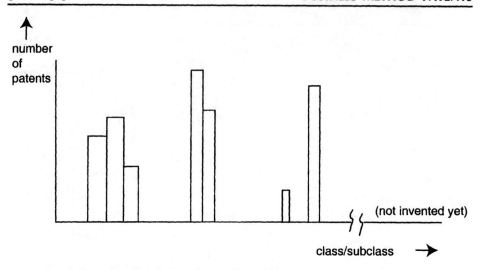

FIGURE 13-7:   SIMPLIFIED SKYLINE VIEW

to a class and subclass dealing with e-commerce business methods may also be secondarily assigned to an additional class and subclass dealing with data models for Internet transactions. These classifications may be used as the East-West and North-South dimensions of the topographical map.

Represent the primary class/subclass assignment along the x-axis (East-West dimension) and any additional secondary class/subclass assignments along the y-axis (North-South dimension). To ensure that each patent has at least some data in the y-axis dimension, you can include the primary classification as well as any secondary classifications when plotting along the y-axis dimension. Thus, every patent will be represented by at least one point at the x position and y position each corresponding to its primary classification. Use of both primary and secondary assignments gives a multidimensional quality to the patents within a portfolio. In a simpler model, one could omit the secondary classification information. In which case, the map might appear as shown in Figure 13-7, a view that resembles the Manhattan skyline as viewed from the shores of New Jersey.

In both single- and multidimensional maps, there is space that lies outside the current patent classification system. This space represents technology that has not been invented yet and, thus, not yet classified by the U.S. Patent Office. Into this space your client's future portfolio may expand. Additionally, from time to time the Patent Office may create new subclasses within an existing class, to subdivide a growing subclass into more manageable pieces. When this happens the topographical map may need to expand a bit, like the bellows of an outwardly spreading accordion.

As patents are granted, add them to the topographical map; as patents expire, remove them. By making a series of topographical maps to represent different timeframes, one can begin to see the direction a portfolio is growing. To better understand the commercial tectonic forces that are driving the portfolio growth, you can associate a bull's eye timeline diagram with each map. Start with the map that captures the earliest timeframe and associate that map with the center of the bull's eye. Plot points corresponding to each successive map on concentric circles radiating outwardly from the bull's eye center.

When plotting each point, look at the topographical map as a whole and try to think of what forces were primarily responsible for the direction the portfolio has taken. The forces may be technological (transistors replace vacuum tubes), sociological (aging baby boomers retire), regulatory (FCC releases new wireless communication frequencies), financial (banks merge to compete globally), judicial (business method patents upheld by the courts), or any other force driving your client's industry. These forces can be shown as directions or points on the bull's eye compass. In Figure 13-6 note that the primary force driving the portfolio is Internet e-commerce. In this figure, there is no portfolio growth attributable to use of horse-drawn vehicles (although in a business that caters to the horse fancier, this would probably not be the case).

The topographical map is not a crystal ball. It cannot show you the future. It can, however, help you analyze where your client's portfolio lies today, and the direction it has been growing (or shrinking). By developing similar maps of the portfolios of key competitors, you and your client can be better prepared to steer future portfolio growth in a systematic fashion. The next subsection presents some suggestions on how to architect a systematic program for invention harvesting and portfolio development.

The topographical mapping technique offers a good way to see the big picture. There are other mapping techniques that allow you to see the portfolio in a more granular way — allowing you to see what the claims of an individual patent actually cover, and how they may interrelate with other claims in the same patent or in other patents. This granular technique draws inspiration from the mathematics of fractals.

[1] Portfolio Mapping Using Fractal Thinking

In his groundbreaking book, *The Fractal Geometry of Nature*,[23] Benoit Mandlebrot described a new way to see and understand the world. Since ancient times mathematicians had sought to understand the world through simple geometric shapes. In the world of classical geometry everything is made up of simple shapes — circles, squares, triangles, polygons. Classical geometry describes some things well, particularly things made by the human hand. Classical geometry fails miserably at other things that have a more chaotic appearance, such as clouds, trees, and coastlines. Nature paints from a rich pallet, unconstrained to simple geometric shapes.

Mandlebrot's insight was that nature is inherently fractal — the small parts of a thing are the same as the big parts. If you examine a rocky coastline through a telescope you may see an intricate, jagged pattern made up of many thousand jagged rocks. If you examine one of those rocks under a microscope, you will see

---

[23] Mandlebrot, Benoit B. *The Fractal Geometry of Nature* (1988).

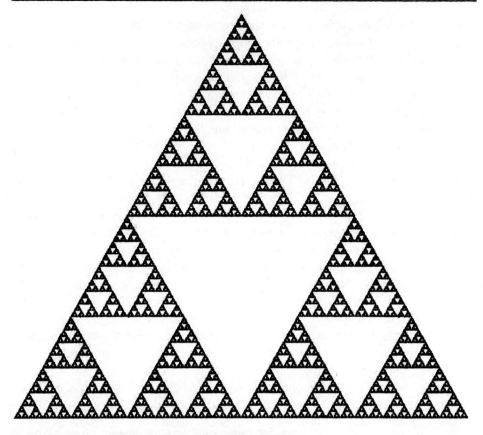

FIGURE 13-8:   THE SIERPINSKI TRIANGLE

the same intricate, jagged pattern made up of many thousand chunks of smaller jagged rocks. The same is true of clouds. If you could zoom in on a portion of a fluffy cloud, say at a magnification factor of 100, you would likely find that the magnified portion looks no different than the original cloud. Computer graphic artists now widely exploit Mandlebrot's insight to generate realistic images of countless complex shapes, including clouds, foliage, rocks, and even snowcapped mountain ranges.

The property of fractals that small parts are the same as big parts has been termed *self-similarity*. The concept is illustrated in the fractal shape known as the Sierpinski triangle, shown in Figure 13-8. Note how the Sierpinski triangle can be decomposed into three pieces, each a Sierpinski triangle in its own right, and that each piece can, in turn, be decomposed into three pieces, and so forth.

[2]   The Fractal Patent Portfolio

Patent portfolios also have fractal properties. Just as the Sierpinski triangle can be described as a collection of self-similar parts, combined to make larger self-similar parts, so too can the patent portfolio be described. Claim elements are the parts of which claims are made. Claims are the parts of which patents are made. Patents are the parts of which portfolios are made. There is self-similarity among these elements, claims, patents, and portfolios in this important respect. All are designed to describe intellectual property, for the purpose of excluding others. One would expect fractal thinking to apply to patents because inventions and discoveries are all the product of the human mind. The human mind thus imposes the self-similar property — There's nothing you can do that can't be done (John Lennon).

The Court of Appeals for the Federal Circuit identified long ago that patent claims, the basic construct that defines the metes and bounds of a patent invention, are in turn made up of claim elements or claim limitations. The Federal Circuit Court considers each claim element to material an essential when construing a patent claim. As Judge Nies stated it in *Pennwalt v. Durand-Wayland*:

> An infringement analysis, thus, requires that the courts look at *each* element of the claim, that is, proceed through the claim element-by-element, and look for correspondence in the allegedly infringing device. If an accused device does not contain at least an equivalent for each limitation of the claim, there is no infringement because a required part of the claimed invention is *missing*.[24]

Thus, individual claim elements are the building blocks of which patent claims are made. At one time perhaps each element was, or could have been, an invention in its own right. To illustrate, consider the following e-commerce claim:

> A subway turnstile system for facilitating commerce over a computer network, comprising:
> a turnstile;
> a remote sensing system associated with said turnstile and employing responder carried by a customer that provides the identity of said customer when said customer is in a predetermined proximity to said turnstile;
> an e-commerce server coupled to said remote sensing system that supplies information over a computer network for presentation by a client application;
> an electronic funds transfer system communicating with said server for debiting an account associated with said customer each time the turnstile is used by said customer; and
> a database associated with said server that stores customer turnstile usage data.

In the above claim, each element, even the turnstile, at one time represented a new invention or discovery. Presumably, however, at the time the above claim was drafted, the individual elements — the turnstile, the e-commerce server, the electronic funds transfer system, the database — were old in the art. Thus, the

---

[24] Pennwalt Corp. v. Durand-Wayland, Inc., 833 F.2d 931, 949 (Fed. Cir. 1987).

drafting attorney presumably needed to include all elements, in combination, to recite a novel invention.

Under the Federal Circuit decisions, such as *Penwalt*, all elements in the above claim must be present before the terms of the claim are met. One would not infringe the above claim by merely furnishing a turnstile, or by connecting a turnstile to an e-commerce server, for example. However, at one time, perhaps a thousand years ago, even the turnstile was a new invention and could have been patented, assuming a patent system existed then. Quite a few years newer than the turnstile, the remote sensing element might still be the subject of patent today. Thus the elements of our subway turnstile system each represent nuggets of intellectual property that may or may not have been covered by patent in the past. These claim element nuggets, taken individually, do not exclude others, but when combined they represent a property right that the owner may use to exclude others. The right to exclude others is statutory:

> Except as otherwise provided in this title, whoever without authority makes, uses, offers to sell, or sells any patented invention, within the United States, or imports into the United States any patented invention during the term of the patent therefor, infringes the patent.[25]

The patent system is inherently granular. It focuses on intellectual property in claim-size chunks. Under the U.S. patent laws, if any one claim of a patent is infringed, the patent is infringed. Boolean mathematicians would say there is an "OR" relationship among claims. If claim 1 OR claim 2 OR claim 3 is infringed, the patent is infringed. That is because each claim represents a stand alone property right to exclude others. The patent law treats these claim-sized granular chunks as fully protectible property rights. In contrast, Boolean mathematicians would say there is an "AND" relationship among elements of a single claim. Elements A AND B AND C . . . must all be present before the claim is infringed. Thus A patent is infringed if any one or more of its claims are infringed. The patent law treats these element-sized granular chunks as expired or public domain property. As to this patent, the individual claim elements are devoid of any exclusionary right. (The individual claim element may, itself, be the subject of a patent and may in that way support an exclusionary right, however.)

[B]   Ranking Patent Claims in a Portfolio Map

Patent claims and the granular elements that make them up have one important aspect in common. They both describe some nugget of knowledge. Whether that nugget has an exclusionary right associated with it depends on whether that nugget is novel, nonobvious, and meets all other requirements of the patent law. We may thus consider the exclusionary right as an attribute or property of these nuggets of knowledge. If the nugget meets all patentability

---

[25] 35 U.S.C. § 271.

requirements, the nugget may be used to exclude. If the nugget does not meet all patentability requirements it may not be used to exclude.

If the exclusionary right is one attribute of each nugget of knowledge, another equally important attribute is strength or immunity from validity attack. Immunity from attack is rarely black and white. Like fractal structures in nature, whose dimensionality falls somewhere between two and three dimensions, patent structures have a validity that falls somewhere between valid and invalid. This is not to say that a patent's validity cannot be precisely determined as either valid or invalid. Courts do that all the time. However, the vast majority of patents are untested by the courts. Statistically speaking, some patents, if tested, would be found valid and others invalid. This leads one to adopt a fractional validity attribute to express that in any portfolio some of the patents would be valid if tested and some would not.

The exclusionary right attribute and the immunity from validity attack attribute provide us with two metrics that may be used in graphing or mapping patent property. You can express or diagram a single patent claim using an entity-relation diagram that shows each claim element and its relationship to other claim elements. The entity-relationship diagram is an excellent way to map out claim drafting strategies prior to the actual writing of claims. Figure 13-9 shows an example of an entity-relationship diagram of a single claim. See § 10.03[A] for a discussion of this technique.

You can also diagram a single patent using an entity-relationship diagram that shows each claim and its relationship to other claims. Some claims are directly dependent upon each other. For example, in a typical patent, claim 1 is an independent claim and claim 2 may be drafted in dependent form to expressly incorporate all of the terms of claim 1.

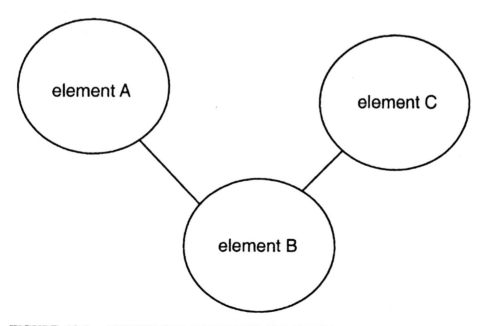

FIGURE 13-9:   ENTITY-RELATIONSHIP DIAGRAM

1. An apparatus comprising,
   element A;
   element B; and
   element C.
2. The apparatus of claim 1 further comprising,
   element D.

You can diagram the above dependent relationship as parent-child relationships. See Figure 13-10.

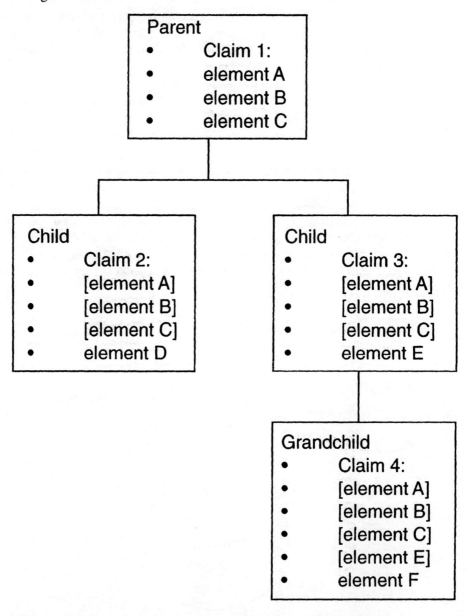

FIGURE 13-10:　PARENT-CHILD CLAIMS RELATIONSHIPS

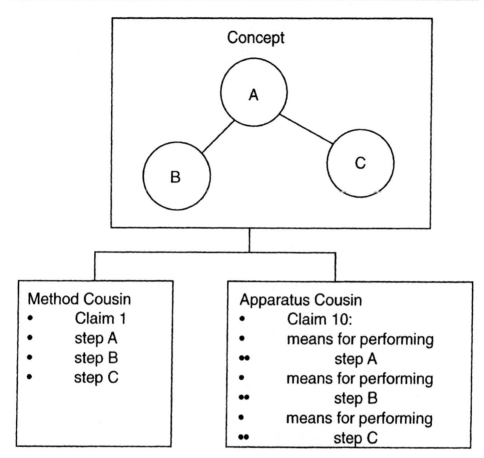

FIGURE 13-11:   COUSIN CLAIMS RELATIONSHIPS

Other claims in a patent may be related, but not necessarily in independent-dependent, or parent-child relationship. For example, method claim 1 and apparatus claim 10 might express similar subject matter without reciting dependency on one another.

1. A method comprising:
    step A;
    step B; and
    step C.

10. An apparatus comprising:
    means for performing step A;
    means for performing step B; and
    means for performing step C.

You can diagram above relationship as a cousin relationship. See Figure 13-11.

For each claim you can assign two figures-of-merit to reflect the strength of that claim's exclusionary right and its immunity from validity attack. These figures-of-merit are purely subjective, but if uniformly applied they can help assess the overall value of a patent and the portfolio to which it belongs.

How do you determine what figures-of-merit to apply? Good advice would be to keep it simple. Adopt a "good, better, best" metric that expresses a patent claim's exclusionary value as a 1, 2 or 3, depending on the following simple test above.

Exclusionary Value	Test
3	The claim is very broad; it covers all practical implementations of what the industry is doing or will be doing.
2	The claim is of average scope; it covers several important implementations of what the industry is doing or will be doing.
1	The claim is narrow; it is a "picture claim" covering one implementation of what the industry is doing or will be doing.

FIGURE 13-12:   EXCLUSIONARY VALUE TEST

You can derive a numeric metric reflecting the claim's immunity to validity attack, although the methodology for doing this is different than for deriving the exclusionary value metric. One simple technique is to count the number of claim elements and to use this to count the metric. There are exceptions to every rule. That said, often the fewer number of claim elements or limitations, the more vulnerable a claim is to validity attack. A three element claim is usually easier to attack than a seven element claim. This is because a three element claim is usually considerably broader than a seven element claim and thus more likely to read on the prior art. When a claim reads on the prior art it can be rendered invalid. Dependent claims include more elements than their parent claim and thus they are generally more immune to validity attack.

Of course, some claims may include gratuitous elements that restrict the claim scope very little, if at all. These elements need to be discounted in arriving at a figure-of-merit for the claim's immunity to validity attack. To do this, ask yourself whether two or more elements are so strongly related that one would typically not find one without the other. If so, then these elements should be counted as a single element when generating a figure-of-merit for immunity to validity attack.

By way of example, consider the following claim:

1. A universal electronic transaction card for storing, transmitting, and receiving information, including personal information for a user of the universal electronic transaction card, account information for accounts with service institutions in which the user has an account, and transactional information for accounts with service institutions in which the user has an account, for a plurality of service institutions, comprising:

a. housing means for housing inputting means, memory means, communications means, display means, and processing means, the housing means adapted to fit in a pocket or purse;

b. inputting means for inputting information, including personal information for the user, account information for a plurality of service institutions in which the user has an account, and transactional information for each service institution for which account information exists, into the memory means;

c. memory means for storing information, including personal information for the user, account information for a plurality of service institutions in which the user has an account, and transactional information for each service institution for which account information exists;

d. communications means for electronically communicating information, including personal information, account information, and transactional information, with service institutions;

e. display means for displaying information for a plurality of service institution accounts, including personal information, account information, and transactional information;

f. processing means for processing information, including personal information, account information, and transactional information;

g. means for providing and storing electric power, and,

h. security means for preventing unauthorized use of the universal electronic transaction card and for preventing unauthorized access to the information stored in the memory means of the universal electronic transaction card.

In the above claim, one might, at first blush, assign it a score of eight because there are eight claim elements listed, a-h. However, many of these elements add little to the overall immunity to validity attack in light of personal organizers, such as the Palm Pilot, which are assumed in this example to be part of the prior art. The chart in Figure 13-13 shows how the immunity to validity attack metric might be assigned to this claim when the Palm Pilot is considered as prior art.

Using the above analysis, the example claim gets an immunity score of two. The score is somewhat subjective, because individual judgment is used in grouping elements that add little to distinguish the claim over the prior art. Although subjective, the metric is useful when comparing a large number of claims, such as may be found in a large patent portfolio. Whether a single claim got a score of 2 or 3 or 4 makes little difference in the overall picture. The power of this ranking technique lies in the power of statistics and the law of averages. If a hundred or a thousand claims are ranked using the claim element counting metric to assess ability to withstand a validity attack, the cumulative score, or average score, of an entire portfolio is little affected by any one subjective value judgment.

Once every claim in the portfolio has been scored, both for exclusionary figure-of-merit and for immunity from attack, the scores may be associated with

Claim Element	Comment	Cumulative Score
1. A universal electronic transaction card . . .	The Palm Pilot is not a "card" per se — add one to score	1
a. housing means . . .	The Palm Pilot has this — add nothing to score	0
b. inputting means . . .	The Palm Pilot has this, except arguably it is not programmed to input "account information" and "transactional information" — add one to score	1
c. memory means . . .	The Palm Pilot has this. To the extent the memory is specifically configured to handle "account information" and "transactional information" that limitation has already been accounted for in b — add nothing to score	0
d. communication means . . .	The Palm Pilot has this. To the extent the application communicates "account information" and "transactional information" this has already been accounted for in b — add nothing to score	0
e. display means . . .	The Palm Pilot has this. To the extent the application displays "account information" and "transactional information" this has already been accounted for in b — add nothing to score	0
f. processing means . . .	The Palm Pilot has this. To the extent the application processes "account information" and "transactional information" this has already been accounted for in b — add nothing to score	0
g. means for providing and storing electric power	The Palm Pilot has this. Add nothing to score	0
h. security means . . .	The Palm Pilot has this in its password feature. Add nothing to score	0

FIGURE 13-13: THE IMMUNITY TO VALIDITY ATTACK METRIC

the points on the topographical map of Figure 13-14. These metrics add another dimension to the map, allowing you to see which domains within a portfolio represent the potentially most valuable property. With these added metrics, the portfolio map shows you at a glance which patents represent incremental improvements and which represent potential breakthrough technology.

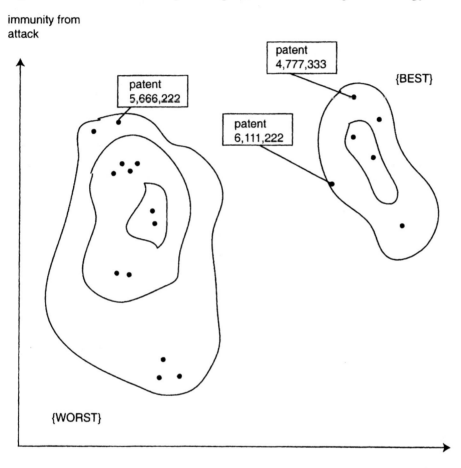

FIGURE 13-14: TOPOGRAPHICAL MAP

UNCITRAL MODEL LAW ON ELECTRONIC COMMERCE

UNCITRAL MODEL LAW ON ELECTRONIC COMMERCE WITH GUIDE TO ENACTMENT 1996

WITH ADDITIONAL ARTICLE 5 BIS AS ADOPTED IN 1998
UNITED NATIONS

CONTENTS

GUIDE TO ENACTMENT OF THE UNCITRAL MODEL LAW ON ELECTRONIC COMMERCE

Resolution adopted by the General Assembly
[on the report of the Sixth Committee (A/51/628)]
51/162 Model Law on Electronic Commerce adopted by the United
Nations Commission on International Trade Law

The General Assembly,

Recalling its resolution 2205 (XXI) of 17 December 1966, by which it created the United Nations Commission on International Trade Law, with a mandate to further the progressive harmonization and unification of the law of international trade and in that respect to bear in mind the interests of all peoples, in particular those of developing countries, in the extensive development of international trade,

Noting that an increasing number of transactions in international trade are carried out by means of electronic data interchange and other means of communication, commonly referred to as "electronic commerce," which involve the use of alternatives to paper-based methods of communication and storage of information,

Recalling the recommendation on the legal value of computer records adopted by the Commission at its eighteenth session, in 1985,[1] and paragraph 5(b) of General Assembly resolution 40/71 of 11 December 1985, in which the Assembly called upon Governments and international organizations to take action, where appropriate, in conformity with the recommendation of the Commission,[1] so as to ensure legal security in the context of the widest possible use of automated data processing in international trade,

Convinced that the establishment of a model law facilitating the use of electronic commerce that is acceptable to States with different legal, social and economic systems, could contribute significantly to the development of harmonious international economic relations,

Noting that the Model Law on Electronic Commerce was adopted by the Commission at its twenty-ninth session after consideration of the observations of Governments and interested organizations,

Believing that the adoption of the Model Law on Electronic Commerce by the Commission will assist all States significantly in enhancing their legislation governing the use of alternatives to paper-based methods of communication and storage of information and in formulating such legislation where none currently exists,

1. *Expresses* its appreciation to the United Nations Commission on International Trade Law for completing and adopting the Model Law on Electronic Commerce

---

[1] See *Official Records of the General Assembly, Fortieth Session, Supplement No. 17* (A/40/17), chap. VI, sect. B.

contained in the annex to the present resolution and for preparing the Guide to Enactment of the Model Law;

2. *Recommends* that all States give favourable consideration to the Model Law when they enact or revise their laws, in view of the need for uniformity of the law applicable to alternatives to paper-based methods of communication and storage of information;

3. *Recommends* also that all efforts be made to ensure that the Model Law, together with the Guide, become generally known and available.

85th plenary meeting
16 December 1996

UNCITRAL Model Law on Electronic Commerce
[Original: Arabic, Chinese, English, French, Russian, Spanish]

Part one.   Electronic commerce in general

Chapter I.   General provisions

Article 1.   Sphere of application*

This Law** applies to any kind of information in the form of a data message used in the context*** of commercial**** activities.

---

*The Commission suggests the following text for States that might wish to limit the applicability of this Law to international data messages: "This Law applies to a data message as defined in paragraph (1) of article 2 where the data message relates to international commerce."
**This Law does not override any rule of law intended for the protection of consumers.
***The Commission suggests the following text for States that might wish to extend the applicability of this Law: "This Law applies to any kind of information in the form of a data message, except in the following situations: [. . .]."
****The term "commercial" should be given a wide interpretation so as to cover matters arising from all relationships of a commercial nature, whether contractual or not. Relationships of a commercial nature include, but are not limited to, the following transactions: any trade transaction for the supply or exchange of goods or services; distribution agreement; commercial representation or agency; factoring; leasing; construction of works; consulting; engineering; licensing; investment; financing; banking; insurance; exploitation agreement or concession; joint venture and other forms of industrial or business cooperation; carriage of goods or passengers by air, sea, rail or road

---

Article 2.   Definitions

For the purposes of this Law:

(a) "Data message" means information generated, sent, received or stored by electronic, optical or similar means including, but not limited to, electronic data interchange (EDI), electronic mail, telegram, telex or telecopy;

(b) ''Electronic data interchange (EDI)'' means the electronic transfer from computer to computer of information using an agreed standard to structure the information;

(c) ''Originator'' of a data message means a person by whom, or on whose behalf, the data message purports to have been sent or generated prior to storage, if any, but it does not include a person acting as an intermediary with respect to that data message;

(d) ''Addressee'' of a data message means a person who is intended by the originator to receive the data message, but does not include a person acting as an intermediary with respect to that data message;

(e) ''Intermediary,'' with respect to a particular data message, means a person who, on behalf of another person, sends, receives or stores that data message or provides other services with respect to that data message;

(f) ''Information system'' means a system for generating, sending, receiving, storing or otherwise processing data messages.

Article 3. Interpretation

(1) In the interpretation of this Law, regard is to be had to its international origin and to the need to promote uniformity in its application and the observance of good faith.

(2) Questions concerning matters governed by this Law which are not expressly settled in it are to be settled in conformity with the general principles on which this Law is based.

Article 4. Variation by agreement

(1) As between parties involved in generating, sending, receiving, storing or otherwise processing data messages, and except as otherwise provided, the provisions of chapter III may be varied by agreement.

(2) Paragraph (1) does not affect any right that may exist to modify by agreement any rule of law referred to in chapter II.

Chapter II. Application of legal requirements to data messages

Article 5. Legal recognition of data messages

Information shall not be denied legal effect, validity or enforceability solely on the grounds that it is in the form of a data message.

Article 5 bis. Incorporation by reference

(as adopted by the Commission at its thirty-first session, in June 1998)

Information shall not be denied legal effect, validity or enforceability solely on the grounds that it is not contained in the data message purporting to give rise to such legal effect, but is merely referred to in that data message.

Article 6. Writing

(1) Where the law requires information to be in writing, that requirement is met by a data message if the information contained therein is accessible so as to be usable for subsequent reference.

(2) Paragraph (1) applies whether the requirement therein is in the form of an obligation or whether the law simply provides consequences for the information not being in writing.

(3) The provisions of this article do not apply to the following: [. . .].

Article 7. Signature

(1) Where the law requires a signature of a person, that requirement is met in relation to a data message if:

(a) a method is used to identify that person and to indicate that person's approval of the information contained in the data message; and

(b) that method is as reliable as was appropriate for the purpose for which the data message was generated or communicated, in the light of all the circumstances, including any relevant agreement.

(2) Paragraph (1) applies whether the requirement therein is in the form of an obligation or whether the law simply provides consequences for the absence of a signature.

(3) The provisions of this article do not apply to the following: [. . .].

Article 8. Original

(1) Where the law requires information to be presented or retained in its original form, that requirement is met by a data message if:

(a) there exists a reliable assurance as to the integrity of the information from the time when it was first generated in its final form, as a data message or otherwise; and

(b) where it is required that information be presented, that information is capable of being displayed to the person to whom it is to be presented.

(2) Paragraph (1) applies whether the requirement therein is in the form of an obligation or whether the law simply provides consequences for the information not being presented or retained in its original form.

(3) For the purposes of subparagraph (a) of paragraph (1):

(a) the criteria for assessing integrity shall be whether the information has remained complete and unaltered, apart from the addition of any endorsement and any change which arises in the normal course of communication, storage and display; and

(b) the standard of reliability required shall be assessed in the light of the purpose for which the information was generated and in the light of all the relevant circumstances.

(4) The provisions of this article do not apply to the following: [. . .].

Article 9. Admissibility and evidential weight of data messages

(1) In any legal proceedings, nothing in the application of the rules of evidence shall apply so as to deny the admissibility of a data message in evidence:

(a) on the sole ground that it is a data message; or,

(b) if it is the best evidence that the person adducing it could reasonably be expected to obtain, on the grounds that it is not in its original form.

(2) Information in the form of a data message shall be given due evidential weight. In assessing the evidential weight of a data message, regard shall be had to the reliability of the manner in which the data message was generated, stored or communicated, to the reliability of the manner in which the integrity of the information was maintained, to the manner in which its originator was identified, and to any other relevant factor.

Article 10. Retention of data messages

(1) Where the law requires that certain documents, records or information be retained, that requirement is met by retaining data messages, provided that the following conditions are satisfied:

(a) the information contained therein is accessible so as to be usable for subsequent reference; and

(b) the data message is retained in the format in which it was generated, sent or received, or in a format which can be demonstrated to represent accurately the information generated, sent or received; and

(c) such information, if any, is retained as enables the identification of the origin and destination of a data message and the date and time when it was sent or received.

(2) An obligation to retain documents, records or information in accordance with paragraph (1) does not extend to any information the sole purpose of which is to enable the message to be sent or received.

(3) A person may satisfy the requirement referred to in paragraph (1) by using the services of any other person, provided that the conditions set forth in subparagraphs (a), (b) and (c) of paragraph (1) are met.

Chapter III. Communication of data messages

Article 11. Formation and validity of contracts

(1) In the context of contract formation, unless otherwise agreed by the parties, an offer and the acceptance of an offer may be expressed by means of data messages. Where a data message is used in the formation of a contract, that contract shall not be denied validity or enforceability on the sole ground that a data message was used for that purpose.

(2) The provisions of this article do not apply to the following: [. . .].

Article 12. Recognition by parties of data messages

(1) As between the originator and the addressee of a data message, a declaration of will or other statement shall not be denied legal effect, validity or enforceability solely on the grounds that it is in the form of a data message.

(2) The provisions of this article do not apply to the following: [. . .].

Article 13. Attribution of data messages

(1) A data message is that of the originator if it was sent by the originator itself.

(2) As between the originator and the addressee, a data message is deemed to be that of the originator if it was sent:

(a) by a person who had the authority to act on behalf of the originator in respect of that data message; or

(b) by an information system programmed by, or on behalf of, the originator to operate automatically.

(3) As between the originator and the addressee, an addressee is entitled to regard a data message as being that of the originator, and to act on that assumption, if:

(a) in order to ascertain whether the data message was that of the originator, the addressee properly applied a procedure previously agreed to by the originator for that purpose; or

(b) the data message as received by the addressee resulted from the actions of a person whose relationship with the originator or with any agent of the originator enabled that person to gain access to a method used by the originator to identify data messages as its own.

(4) Paragraph (3) does not apply:

(a) as of the time when the addressee has both received notice from the originator that the data message is not that of the originator, and had reasonable time to act accordingly; or

(b) in a case within paragraph (3)(b), at any time when the addressee knew or should have known, had it exercised reasonable care or used any agreed procedure, that the data message was not that of the originator.

(5) Where a data message is that of the originator or is deemed to be that of the originator, or the addressee is entitled to act on that assumption, then, as between the originator and the addressee, the addressee is entitled to regard the data message as received as being what the originator intended to send, and to act on that assumption. The addressee is not so entitled when it knew or should have known, had it exercised reasonable care or used any agreed procedure, that the transmission resulted in any error in the data message as received.

(6) The addressee is entitled to regard each data message received as a separate data message and to act on that assumption, except to the extent that it duplicates another data message and the addressee knew or should have known, had it exercised reasonable care or used any agreed procedure, that the data message was a duplicate.

Article 14. Acknowledgement of receipt

(1) Paragraphs (2) to (4) of this article apply where, on or before sending a data message, or by means of that data message, the originator has requested or has agreed with the addressee that receipt of the data message be acknowledged.

(2) Where the originator has not agreed with the addressee that the acknowledgement be given in a particular form or by a particular method, an acknowledgement may be given by

(a) any communication by the addressee, automated or otherwise, or

(b) any conduct of the addressee,

sufficient to indicate to the originator that the data message has been received.

(3) Where the originator has stated that the data message is conditional on receipt of the acknowledgement, the data message is treated as though it has never been sent, until the acknowledgement is received.

(4) Where the originator has not stated that the data message is conditional on receipt of the acknowledgement, and the acknowledgement has not been received by the originator within the time specified or agreed or, if no time has been specified or agreed, within a reasonable time, the originator:

(a) may give notice to the addressee stating that no acknowledgement has been received and specifying a reasonable time by which the acknowledgement must be received; and

(b) if the acknowledgement is not received within the time specified in subparagraph (a), may, upon notice to the addressee, treat the data message as though it had never been sent, or exercise any other rights it may have.

(5) Where the originator receives the addressee's acknowledgement of receipt, it is presumed that the related data message was received by the addressee. That presumption does not imply that the data message corresponds to the message received.

(6) Where the received acknowledgement states that the related data message met technical requirements, either agreed upon or set forth in applicable standards, it is presumed that those requirements have been met.

(7) Except in so far as it relates to the sending or receipt of the data message, this article is not intended to deal with the legal consequences that may flow either from that data message or from the acknowledgement of its receipt.

Article 15. Time and place of dispatch and receipt of data messages

(1) Unless otherwise agreed between the originator and the addressee, the dispatch of a data message occurs when it enters an information system outside the control of the originator or of the person who sent the data message on behalf of the originator.

(2) Unless otherwise agreed between the originator and the addressee, the time of receipt of a data message is determined as follows:

(a) if the addressee has designated an information system for the purpose of receiving data messages, receipt occurs:

(i) at the time when the data message enters the designated information system; or

(ii) if the data message is sent to an information system of the addressee that is not the designated information system, at the time when the data message is retrieved by the addressee;

(b) if the addressee has not designated an information system, receipt occurs when the data message enters an information system of the addressee.

(3) Paragraph (2) applies notwithstanding that the place where the information system is located may be different from the place where the data message is deemed to be received under paragraph (4).

(4) Unless otherwise agreed between the originator and the addressee, a data message is deemed to be dispatched at the place where the originator has its

place of business, and is deemed to be received at the place where the addressee has its place of business. For the purposes of this paragraph:

(a) if the originator or the addressee has more than one place of business, the place of business is that which has the closest relationship to the underlying transaction or, where there is no underlying transaction, the principal place of business;

(b) if the originator or the addressee does not have a place of business, reference is to be made to its habitual residence.

(5) The provisions of this article do not apply to the following: [. . .].

Part two.   Electronic commerce in specific areas

Chapter I. Carriage of goods

Article 16. Actions related to contracts of carriage of goods

Without derogating from the provisions of part one of this Law, this chapter applies to any action in connection with, or in pursuance of, a contract of carriage of goods, including but not limited to:

(a)   (i) furnishing the marks, number, quantity or weight of goods;

(ii) stating or declaring the nature or value of goods;

(iii) issuing a receipt for goods;

(iv) confirming that goods have been loaded;

(b)   (i) notifying a person of terms and conditions of the contract;

(ii) giving instructions to a carrier;

(c)   (i) claiming delivery of goods;

(ii) authorizing release of goods;

(iii) giving notice of loss of, or damage to, goods;

(d)   giving any other notice or statement in connection with the performance of the contract;

(e)   undertaking to deliver goods to a named person or a person authorized to claim delivery;

(f)   granting, acquiring, renouncing, surrendering, transferring or negotiating rights in goods;

(g)   acquiring or transferring rights and obligations under the contract.

Article 17. Transport documents

(1) Subject to paragraph (3), where the law requires that any action referred to in article 16 be carried out in writing or by using a paper document, that requirement is met if the action is carried out by using one or more data messages.

(2) Paragraph (1) applies whether the requirement therein is in the form of an obligation or whether the law simply provides consequences for failing either to carry out the action in writing or to use a paper document.

(3) If a right is to be granted to, or an obligation is to be acquired by, one person and no other person, and if the law requires that, in order to effect this, the right or obligation must be conveyed to that person by the transfer, or use of, a paper document, that requirement is met if the right or obligation is conveyed by using one or more data messages, provided that a reliable method is used to render such data message or messages unique.

(4) For the purposes of paragraph (3), the standard of reliability required shall be assessed in the light of the purpose for which the right or obligation was conveyed and in the light of all the circumstances, including any relevant agreement.

(5) Where one or more data messages are used to effect any action in subparagraphs (f) and (g) of article 16, no paper document used to effect any such action is valid unless the use of data messages has been terminated and replaced by the use of paper documents. A paper document issued in these circumstances shall contain a statement of such termination. The replacement of data messages by paper documents shall not affect the rights or obligations of the parties involved.

(6) If a rule of law is compulsorily applicable to a contract of carriage of goods which is in, or is evidenced by, a paper document, that rule shall not be inapplicable to such a contract of carriage of goods which is evidenced by one or more data messages by reason of the fact that the contract is evidenced by such data message or messages instead of by a paper document.

(7) The provisions of this article do not apply to the following: [. . .].

Guide to Enactment of the UNCITRAL Model Law on Electronic Commerce (1996)

Purpose of this guide

1. In preparing and adopting the UNCITRAL Model Law on Electronic Commerce (hereinafter referred to as "the Model Law"), the United Nations Commission on International Trade Law (UNCITRAL) was mindful that the Model Law would be a more effective tool for States modernizing their legislation if background and explanatory information would be provided to

executive branches of Governments and legislators to assist them in using the Model Law. The Commission was also aware of the likelihood that the Model Law would be used in a number of States with limited familiarity with the type of communication techniques considered in the Model Law. This Guide, much of which is drawn from the *travaux préparatoires* of the Model Law, is also intended to be helpful to users of electronic means of communication as well as to scholars in that area. In the preparation of the Model Law, it was assumed that the draft Model Law would be accompanied by such a guide. For example, it was decided in respect of a number of issues not to settle them in the draft Model Law but to address them in the Guide so as to provide guidance to States enacting the draft Model Law. The information presented in this Guide is intended to explain why the provisions in the Model Law have been included as essential basic features of a statutory device designed to achieve the objectives of the Model Law. Such information might assist States also in considering which, if any, of the provisions of the Model Law might have to be varied to take into account particular national circumstances.

I. INTRODUCTION TO THE MODEL LAW

A. Objectives

2. The use of modern means of communication such as electronic mail and electronic data interchange (EDI) for the conduct of international trade transactions has been increasing rapidly and is expected to develop further as technical supports such as information highways and the INTERNET become more widely accessible. However, the communication of legally significant information in the form of paperless messages may be hindered by legal obstacles to the use of such messages, or by uncertainty as to their legal effect or validity. The purpose of the Model Law is to offer national legislators a set of internationally acceptable rules as to how a number of such legal obstacles may be removed, and how a more secure legal environment may be created for what has become known as "electronic commerce." The principles expressed in the Model Law are also intended to be of use to individual users of electronic commerce in the drafting of some of the contractual solutions that might be needed to overcome the legal obstacles to the increased use of electronic commerce.

3. The decision by UNCITRAL to formulate model legislation on electronic commerce was taken in response to the fact that in a number of countries the existing legislation governing communication and storage of information is inadequate or outdated because it does not contemplate the use of electronic commerce. In certain cases, existing legislation imposes or implies restrictions on the use of modern means of communication, for example by prescribing the use of "written," "signed" or "original" documents. While a few countries have adopted specific provisions to deal with certain aspects of electronic commerce, there exists no legislation dealing with electronic commerce as a whole. This may

result in uncertainty as to the legal nature and validity of information presented in a form other than a traditional paper document. Moreover, while sound laws and practices are necessary in all countries where the use of EDI and electronic mail is becoming widespread, this need is also felt in many countries with respect to such communication techniques as telecopy and telex.

4. The Model Law may also help to remedy disadvantages that stem from the fact that inadequate legislation at the national level creates obstacles to international trade, a significant amount of which is linked to the use of modern communication techniques. Disparities among, and uncertainty about, national legal regimes governing the use of such communication techniques may contribute to limiting the extent to which businesses may access international markets.

5. Furthermore, at an international level, the Model Law may be useful in certain cases as a tool for interpreting existing international conventions and other international instruments that create legal obstacles to the use of electronic commerce, for example by prescribing that certain documents or contractual clauses be made in written form. As between those States parties to such international instruments, the adoption of the Model Law as a rule of interpretation might provide the means to recognize the use of electronic commerce and obviate the need to negotiate a protocol to the international instrument involved.

6. The objectives of the Model Law, which include enabling or facilitating the use of electronic commerce and providing equal treatment to users of paper-based documentation and to users of computer-based information, are essential for fostering economy and efficiency in international trade. By incorporating the procedures prescribed in the Model Law in its national legislation for those situations where parties opt to use electronic means of communication, an enacting State would create a media-neutral environment.

B. Scope

7. The title of the Model Law refers to "electronic commerce." While a definition of "electronic data interchange (EDI)" is provided in article 2, the Model Law does not specify the meaning of "electronic commerce." In preparing the Model Law, the Commission decided that, in addressing the subject matter before it, it would have in mind a broad notion of EDI, covering a variety of trade-related uses of EDI that might be referred to broadly under the rubric of "electronic commerce" (see A/CN.9/360, paras. 28-29), although other descriptive terms could also be used. Among the means of communication encompassed in the notion of "electronic commerce" are the following modes of transmission based on the use of electronic techniques: communication by means of EDI defined narrowly as the computer-to-computer transmission of data in a standardized format; transmission of electronic messages involving the use of either publicly available standards or proprietary standards; transmission of free-

formatted text by electronic means, for example through the INTERNET. It was also noted that, in certain circumstances, the notion of "electronic commerce" might cover the use of techniques such as telex and telecopy.

8. It should be noted that, while the Model Law was drafted with constant reference to the more modern communication techniques, e.g., EDI and electronic mail, the principles on which the Model Law is based, as well as its provisions, are intended to apply also in the context of less advanced communication techniques, such as telecopy. There may exist situations where digitalized information initially dispatched in the form of a standardized EDI message might, at some point in the communication chain between the sender and the recipient, be forwarded in the form of a computer-generated telex or in the form of a telecopy of a computer print-out. A data message may be initiated as an oral communication and end up in the form of a telecopy, or it may start as a telecopy and end up as an EDI message. A characteristic of electronic commerce is that it covers programmable messages, the computer programming of which is the essential difference between such messages and traditional paper-based documents. Such situations are intended to be covered by the Model Law, based on a consideration of the users' need for a consistent set of rules to govern a variety of communication techniques that might be used interchangeably. More generally, it may be noted that, as a matter of principle, no communication technique is excluded from the scope of the Model Law since future technical developments need to be accommodated.

9. The objectives of the Model Law are best served by the widest possible application of the Model Law. Thus, although there is provision made in the Model Law for exclusion of certain situations from the scope of articles 6, 7, 8, 11, 12, 15 and 17, an enacting State may well decide not to enact in its legislation substantial restrictions on the scope of application of the Model Law.

10. The Model Law should be regarded as a balanced and discrete set of rules, which are recommended to be enacted as a single statute. Depending on the situation in each enacting State, however, the Model Law could be implemented in various ways, either as a single statute or in several pieces of legislation (see below, para. 143).

C. Structure

11. The Model Law is divided into two parts, one dealing with electronic commerce in general and the other one dealing with electronic commerce in specific areas. It should be noted that part two of the Model Law, which deals with electronic commerce in specific areas, is composed of a chapter I only, dealing with electronic commerce as it applies to the carriage of goods. Other aspects of electronic commerce might need to be dealt with in the future, and the Model Law can be regarded as an open-ended instrument, to be complemented by future work.

12. UNCITRAL intends to continue monitoring the technical, legal and commercial developments that underline the Model Law. It might, should it regard it advisable, decide to add new model provisions to the Model Law or modify the existing ones.

D. A "framework" law to be supplemented by technical regulations

13. The Model Law is intended to provide essential procedures and principles for facilitating the use of modern techniques for recording and communicating information in various types of circumstances. However, it is a "framework" law that does not itself set forth all the rules and regulations that may be necessary to implement those techniques in an enacting State. Moreover, the Model Law is not intended to cover every aspect of the use of electronic commerce. Accordingly, an enacting State may wish to issue regulations to fill in the procedural details for procedures authorized by the Model Law and to take account of the specific, possibly changing, circumstances at play in the enacting State, without compromising the objectives of the Model Law. It is recommended that, should it decide to issue such regulation, an enacting State should give particular attention to the need to maintain the beneficial flexibility of the provisions in the Model Law.

14. It should be noted that the techniques for recording and communicating information considered in the Model Law, beyond raising matters of procedure that may need to be addressed in the implementing technical regulations, may raise certain legal questions the answers to which will not necessarily be found in the Model Law, but rather in other bodies of law. Such other bodies of law may include, for example, the applicable administrative, contract, criminal and judicial-procedure law, which the Model Law is not intended to deal with.

E. The "functional-equivalent" approach

15. The Model Law is based on the recognition that legal requirements prescribing the use of traditional paper-based documentation constitute the main obstacle to the development of modern means of communication. In the preparation of the Model Law, consideration was given to the possibility of dealing with impediments to the use of electronic commerce posed by such requirements in national laws by way of an extension of the scope of such notions as "writing," "signature" and "original," with a view to encompassing computer-based techniques. Such an approach is used in a number of existing legal instruments, e.g., article 7 of the UNCITRAL Model Law on International Commercial Arbitration and article 13 of the United Nations Convention on Contracts for the International Sale of Goods. It was observed that the Model Law should permit States to adapt their domestic legislation to developments in communications technology applicable to trade law without necessitating the wholesale removal of the paper-based requirements themselves or disturbing the

legal concepts and approaches underlying those requirements. At the same time, it was said that the electronic fulfilment of writing requirements might in some cases necessitate the development of new rules. This was due to one of many distinctions between EDI messages and paper-based documents, namely, that the latter were readable by the human eye, while the former were not so readable unless reduced to paper or displayed on a screen.

16. The Model Law thus relies on a new approach, sometimes referred to as the "functional equivalent approach," which is based on an analysis of the purposes and functions of the traditional paper-based requirement with a view to determining how those purposes or functions could be fulfilled through electronic-commerce techniques. For example, among the functions served by a paper document are the following: to provide that a document would be legible by all; to provide that a document would remain unaltered over time; to allow for the reproduction of a document so that each party would hold a copy of the same data; to allow for the authentication of data by means of a signature; and to provide that a document would be in a form acceptable to public authorities and courts. It should be noted that in respect of all of the above-mentioned functions of paper, electronic records can provide the same level of security as paper and, in most cases, a much higher degree of reliability and speed, especially with respect to the identification of the source and content of the data, provided that a number of technical and legal requirements are met. However, the adoption of the functional-equivalent approach should not result in imposing on users of electronic commerce more stringent standards of security (and the related costs) than in a paper-based environment.

17. A data message, in and of itself, cannot be regarded as an equivalent of a paper document in that it is of a different nature and does not necessarily perform all conceivable functions of a paper document. That is why the Model Law adopted a flexible standard, taking into account the various layers of existing requirements in a paper-based environment: when adopting the "functional-equivalent" approach, attention was given to the existing hierarchy of form requirements, which provides distinct levels of reliability, traceability and unalterability with respect to paper-based documents. For example, the requirement that data be presented in written form (which constitutes a "threshold requirement") is not to be confused with more stringent requirements such as "signed writing," "signed original" or "authenticated legal act."

18. The Model Law does not attempt to define a computer-based equivalent to any kind of paper document. Instead, it singles out basic functions of paper-based form requirements, with a view to providing criteria which, once they are met by data messages, enable such data messages to enjoy the same level of legal recognition as corresponding paper documents performing the same function. It should be noted that the functional-equivalent approach has been taken in articles 6 to 8 of the Model Law with respect to the concepts of "writing," "signature" and "original" but not with respect to other legal concepts dealt with in the

Model Law. For example, article 10 does not attempt to create a functional equivalent of existing storage requirements.

F. Default rules and mandatory law

19. The decision to undertake the preparation of the Model Law was based on the recognition that, in practice, solutions to most of the legal difficulties raised by the use of modern means of communication are sought within contracts. The Model Law embodies the principle of party autonomy in article 4 with respect to the provisions contained in chapter III of part one. Chapter III of part one contains a set of rules of the kind that would typically be found in agreements between parties, e.g., interchange agreements or "system rules." It should be noted that the notion of "system rules" might cover two different categories of rules, namely, general terms provided by communication networks and specific rules that might be included in those general terms to deal with bilateral relationships between originators and addressees of data messages. Article 4 (and the notion of "agreement" therein) is intended to encompass both categories of "system rules."

20. The rules contained in chapter III of part one may be used by parties as a basis for concluding such agreements. They may also be used to supplement the terms of agreements in cases of gaps or omissions in contractual stipulations. In addition, they may be regarded as setting a basic standard for situations where data messages are exchanged without a previous agreement being entered into by the communicating parties, e.g., in the context of open-networks communications.

21. The provisions contained in chapter II of part one are of a different nature. One of the main purposes of the Model Law is to facilitate the use of modern communication techniques and to provide certainty with the use of such techniques where obstacles or uncertainty resulting from statutory provisions could not be avoided by contractual stipulations. The provisions contained in chapter II may, to some extent, be regarded as a collection of exceptions to well-established rules regarding the form of legal transactions. Such well-established rules are normally of a mandatory nature since they generally reflect decisions of public policy. The provisions contained in chapter II should be regarded as stating the minimum acceptable form requirement and are, for that reason, of a mandatory nature, unless expressly stated otherwise in those provisions. The indication that such form requirements are to be regarded as the "minimum acceptable" should not, however, be construed as inviting States to establish requirements stricter than those contained in the Model Law.

G. Assistance from UNCITRAL secretariat

22. In line with its training and assistance activities, the UNCITRAL secretariat may provide technical consultations for Governments preparing legislation based on the UNCITRAL Model Law on Electronic Commerce, as it

may for Governments considering legislation based on other UNCITRAL model laws, or considering adhesion to one of the international trade law conventions prepared by UNCITRAL.

23. Further information concerning the Model Law as well as the Guide and other model laws and conventions developed by UNCITRAL, may be obtained from the secretariat at the address below. The secretariat welcomes comments concerning the Model Law and the Guide, as well as information concerning enactment of legislation based on the Model Law.

International Trade Law Branch
Office of Legal Affairs
United Nations Vienna International Centre
P.O. Box 500
A-1400, Vienna, Austria
Telephone: (43-1) 26060-4060 or 4061
Telefax: (43-1) 26060-5813 or (43-1) 2692669
Telex: 135612 uno a
E-mail: *uncitral@unov.un.or.at*
Internet Home Page: *http://www.un.or.at/uncitral*

II. ARTICLE-BY-ARTICLE REMARKS

Part one. Electronic commerce in general

Chapter I. General provisions

Article 1. Sphere of application

24. The purpose of article 1, which is to be read in conjunction with the definition of "data message" in article 2(a), is to delineate the scope of application of the Model Law. The approach used in the Model Law is to provide in principle for the coverage of all factual situations where information is generated, stored or communicated, irrespective of the medium on which such information may be affixed. It was felt during the preparation of the Model Law that exclusion of any form or medium by way of a limitation in the scope of the Model Law might result in practical difficulties and would run counter to the purpose of providing truly "media-neutral" rules. However, the focus of the Model Law is on "paperless" means of communication and, except to the extent expressly provided by the Model Law, the Model Law is not intended to alter traditional rules on paper-based communications.

25. Moreover, it was felt that the Model Law should contain an indication that its focus was on the types of situations encountered in the commercial area and that it had been prepared against the background of trade relationships. For that reason, article 1 refers to "commercial activities" and provides, in footnote ****, indications as to what is meant thereby. Such indications, which may be

particularly useful for those countries where there does not exist a discrete body of commercial law, are modelled, for reasons of consistency, on the footnote to article 1 of the UNCITRAL Model Law on International Commercial Arbitration. In certain countries, the use of footnotes in a statutory text would not be regarded as acceptable legislative practice. National authorities enacting the Model Law might thus consider the possible inclusion of the text of footnotes in the body of the Law itself.

26. The Model Law applies to all kinds of data messages that might be generated, stored or communicated, and nothing in the Model Law should prevent an enacting State from extending the scope of the Model Law to cover uses of electronic commerce outside the commercial sphere. For example, while the focus of the Model Law is not on the relationships between users of electronic commerce and public authorities, the Model Law is not intended to be inapplicable to such relationships. Footnote *** provides for alternative wordings, for possible use by enacting States that would consider it appropriate to extend the scope of the Model Law beyond the commercial sphere.

27. Some countries have special consumer protection laws that may govern certain aspects of the use of information systems. With respect to such consumer legislation, as was the case with previous UNCITRAL instruments (e.g., the UNCITRAL Model Law on International Credit Transfers), it was felt that an indication should be given that the Model Law had been drafted without special attention being given to issues that might arise in the context of consumer protection. At the same time, it was felt that there was no reason why situations involving consumers should be excluded from the scope of the Model Law by way of a general provision, particularly since the provisions of the Model Law might be found appropriate for consumer protection, depending on legislation in each enacting State. Footnote ** thus recognizes that any such consumer protection law may take precedence over the provisions in the Model Law. Legislators may wish to consider whether the piece of legislation enacting the Model Law should apply to consumers. The question of which individuals or corporate bodies would be regarded as "consumers" is left to applicable law outside the Model Law.

28. Another possible limitation of the scope of the Model Law is contained in the first footnote. In principle, the Model Law applies to both international and domestic uses of data messages. Footnote * is intended for use by enacting States that might wish to limit the applicability of the Model Law to international cases. It indicates a possible test of internationality for use by those States as a possible criterion for distinguishing international cases from domestic ones. It should be noted, however, that in some jurisdictions, particularly in federal States, considerable difficulties might arise in distinguishing international trade from domestic trade. The Model Law should not be interpreted as encouraging enacting States to limit its applicability to international cases.

29. It is recommended that application of the Model Law be made as wide as possible. Particular caution should be used in excluding the application of the

Model Law by way of a limitation of its scope to international uses of data messages, since such a limitation may be seen as not fully achieving the objectives of the Model Law. Furthermore, the variety of procedures available under the Model Law (particularly articles 6 to 8) to limit the use of data messages if necessary (e.g., for purposes of public policy) may make it less necessary to limit the scope of the Model Law. As the Model Law contains a number of articles (articles 6, 7, 8, 11, 12 , 15 and 17) that allow a degree of flexibility to enacting States to limit the scope of application of specific aspects of the Model Law, a narrowing of the scope of application of the text to international trade should not be necessary. Moreover, dividing communications in international trade into purely domestic and international parts might be difficult in practice. The legal certainty to be provided by the Model Law is necessary for both domestic and international trade, and a duality of regimes governing the use of electronic means of recording and communication of data might create a serious obstacle to the use of such means.

References[2]
A/50/17, paras. 213-219;
A/CN.9/407, paras. 37-40;
A/CN.9/406, paras. 80-85; A/CN.9/WG.IV/WP.62, article 1;
A/CN.9/390, paras. 21-43; A/CN.9/WG.IV/WP.60, article 1;
A/CN.9/387, paras. 15-28; A/CN.9/WG.IV/WP.57, article 1;
A/CN.9/373, paras. 21-25 and 29-33; A/CN.9/WG.IV/WP.55, paras. 15-20.

Article 2. Definitions

"Data message"

30. The notion of "data message" is not limited to communication but is also intended to encompass computer-generated records that are not intended for communication. Thus, the notion of "message" includes the notion of "record." However, a definition of "record" in line with the characteristic elements of "writing" in article 6 may be added in jurisdictions where that would appear to be necessary.

31. The reference to "similar means" is intended to reflect the fact that the Model Law was not intended only for application in the context of existing communication techniques but also to accommodate foreseeable technical

---

[2] Reference materials listed by symbols in this Guide belong to the following three categories of documents:

A/50/17 and A/51/17 are the reports of UNCITRAL to the General Assembly on the work of its twenty-eighth and twenty-ninth sessions, held in 1995 and 1996, respectively;

A/CN.9/. . . documents are reports and notes discussed by UNCITRAL in the context of its annual session, including reports presented by the Working Group to the Commission;

A/CN.9/WG.IV/. . . documents are working papers considered by the UNCITRAL Working Group on Electronic Commerce (formerly known as the UNCITRAL Working Group on Electronic Data Interchange) in the preparation of the Model Law.

developments. The aim of the definition of "data message" is to encompass all types of messages that are generated, stored, or communicated in essentially paperless form. For that purpose, all means of communication and storage of information that might be used to perform functions parallel to the functions performed by the means listed in the definition are intended to be covered by the reference to "similar means," although, for example, "electronic" and "optical" means of communication might not be, strictly speaking, similar. For the purposes of the Model Law, the word "similar" connotes "functionally equivalent."

32. The definition of "data message" is also intended to cover the case of revocation or amendment. A data message is presumed to have a fixed information content but it may be revoked or amended by another data message.

"Electronic Data Interchange (EDI)"

33. The definition of EDI is drawn from the definition adopted by the Working Party on Facilitation of International Trade Procedures (WP.4) of the Economic Commission for Europe, which is the United Nations body responsible for the development of UN/EDIFACT technical standards.

34. The Model Law does not settle the question whether the definition of EDI necessarily implies that EDI messages are communicated electronically from computer to computer, or whether that definition, while primarily covering situations where data messages are communicated through a telecommunications system, would also cover exceptional or incidental types of situations where data structured in the form of an EDI message would be communicated by means that do not involve telecommunications systems, for example, the case where magnetic disks containing EDI messages would be delivered to the addressee by courier. However, irrespective of whether digital data transferred manually is covered by the definition of "EDI," it should be regarded as covered by the definition of "data message" under the Model Law.

"Originator" and "Addressee"

35. In most legal systems, the notion of "person" is used to designate the subjects of rights and obligations and should be interpreted as covering both natural persons and corporate bodies or other legal entities. Data messages that are generated automatically by computers without direct human intervention are intended to be covered by subparagraph *(c)*. However, the Model Law should not be misinterpreted as allowing for a computer to be made the subject of rights and obligations. Data messages that are generated automatically by computers without direct human intervention should be regarded as "originating" from the legal entity on behalf of which the computer is operated. Questions relevant to agency that might arise in that context are to be settled under rules outside the Model Law.

36. The "addressee" under the Model Law is the person with whom the originator intends to communicate by transmitting the data message, as opposed to any person who might receive, forward or copy the data message in the course of transmission. The "originator" is the person who generated the data message even if that message was transmitted by another person. The definition of "addressee" contrasts with the definition of "originator," which is not focused on intent. It should be noted that, under the definitions of "originator" and "addressee" in the Model Law, the originator and the addressee of a given data message could be the same person, for example in the case where the data message was intended for storage by its author. However, the addressee who stores a message transmitted by an originator is not itself intended to be covered by the definition of "originator."

37. The definition of "originator" should cover not only the situation where information is generated and communicated, but also the situation where such information is generated and stored without being communicated. However, the definition of "originator" is intended to eliminate the possibility that a recipient who merely stores a data message might be regarded as an originator.

"Intermediary"

38. The focus of the Model Law is on the relationship between the originator and the addressee, and not on the relationship between either the originator or the addressee and any intermediary. However, the Model Law does not ignore the paramount importance of intermediaries in the field of electronic communications. In addition, the notion of "intermediary" is needed in the Model Law to establish the necessary distinction between originators or addressees and third parties.

39. The definition of "intermediary" is intended to cover both professional and non-professional intermediaries, i.e., any person (other than the originator and the addressee) who performs any of the functions of an intermediary. The main functions of an intermediary are listed in subparagraph *(e)*, namely receiving, transmitting or storing data messages on behalf of another person. Additional "value-added services" may be performed by network operators and other intermediaries, such as formatting, translating, recording, authenticating, certifying and preserving data messages and providing security services for electronic transactions. "Intermediary" under the Model Law is defined not as a generic category but with respect to each data message, thus recognizing that the same person could be the originator or addressee of one data message and an intermediary with respect to another data message. The Model Law, which is focused on the relationships between originators and addressees, does not, in general, deal with the rights and obligations of intermediaries.

"Information system"

40. The definition of "information system" is intended to cover the entire range of technical means used for transmitting, receiving and storing information.

For example, depending on the factual situation, the notion of "information system" could be indicating a communications network, and in other instances could include an electronic mailbox or even a telecopier. The Model Law does not address the question of whether the information system is located on the premises of the addressee or on other premises, since location of information systems is not an operative criterion under the Model Law.

References
A/51/17, paras. 116-138;
A/CN.9/407, paras. 41-52;
A/CN.9/406, paras. 132-156; A/CN.9/WG.IV/WP.62, article 2;
A/CN.9/390, paras. 44-65; A/CN.9/WG.IV/WP.60, article 2;
A/CN.9/387, paras. 29-52; A/CN.9/WG.IV/WP.57, article 2;
A/CN.9/373, paras. 11-20, 26-28 and 35-36;
A/CN.9/WG.IV/WP.55, paras. 23-26;
A/CN.9/360, paras. 29-31; A/CN.9/WG.IV/WP.53, paras. 25-33.

Article 3. Interpretation

41. Article 3 is inspired by article 7 of the United Nations Convention on Contracts for the International Sale of Goods. It is intended to provide guidance for interpretation of the Model Law by courts and other national or local authorities. The expected effect of article 3 is to limit the extent to which a uniform text, once incorporated in local legislation, would be interpreted only by reference to the concepts of local law.

42. The purpose of paragraph (1) is to draw the attention of courts and other national authorities to the fact that the provisions of the Model Law (or the provisions of the instrument implementing the Model Law), while enacted as part of domestic legislation and therefore domestic in character, should be interpreted with reference to its international origin in order to ensure uniformity in the interpretation of the Model Law in various countries.

43. As to the general principles on which the Model Law is based, the following non-exhaustive list may be considered: (1) to facilitate electronic commerce among and within nations; (2) to validate transactions entered into by means of new information technologies; (3) to promote and encourage the implementation of new information technologies; (4) to promote the uniformity of law; and (5) to support commercial practice. While the general purpose of the Model Law is to facilitate the use of electronic means of communication, it should not be construed in any way as imposing their use.

References
A/50/17, paras. 220-224;
A/CN.9/407, paras. 53-54;
A/CN.9/406, paras. 86-87; A/CN.9/WG.IV/WP.62, article 3;
A/CN.9/390, paras. 66-73; A/CN.9/WG.IV/WP.60, article 3;

A/CN.9/387, paras. 53-58; A/CN.9/WG.IV/WP.57, article 3;
A/CN.9/373, paras. 38-42; A/CN.9/WG.IV/WP.55, paras. 30-31.

Article 4. Variation by agreement

44. The decision to undertake the preparation of the Model Law was based on the recognition that, in practice, solutions to the legal difficulties raised by the use of modern means of communication are mostly sought within contracts. The Model Law is thus intended to support the principle of party autonomy. However, that principle is embodied only with respect to the provisions of the Model Law contained in chapter III of part one. The reason for such a limitation is that the provisions contained in chapter II of part one may, to some extent, be regarded as a collection of exceptions to well-established rules regarding the form of legal transactions. Such well-established rules are normally of a mandatory nature since they generally reflect decisions of public policy. An unqualified statement regarding the freedom of parties to derogate from the Model Law might thus be misinterpreted as allowing parties, through a derogation to the Model Law, to derogate from mandatory rules adopted for reasons of public policy. The provisions contained in chapter II of part one should be regarded as stating the minimum acceptable form requirement and are, for that reason, to be regarded as mandatory, unless expressly stated otherwise. The indication that such form requirements are to be regarded as the ''minimum acceptable'' should not, however, be construed as inviting States to establish requirements stricter than those contained in the Model Law.

45. Article 4 is intended to apply not only in the context of relationships between originators and addressees of data messages but also in the context of relationships involving intermediaries. Thus, the provisions of chapter III of part one could be varied either by bilateral or multilateral agreements between the parties, or by system rules agreed to by the parties. However, the text expressly limits party autonomy to rights and obligations arising as between parties so as not to suggest any implication as to the rights and obligations of third parties.

References
A/51/17, paras. 68, 90 to 93, 110, 137, 188 and 207 (article 10);
A/50/17, paras. 271-274 (article 10);
A/CN.9/407, para. 85;
A/CN.9/406, paras. 88-89; A/CN.9/WG.IV/WP.62, article 5;
A/CN.9/390, paras. 74-78; A/CN.9/WG.IV/WP.60, article 5;
A/CN.9/387, paras. 62-65; A/CN.9/WG.IV/WP.57, article 5;
A/CN.9/373, para. 37; A/CN.9/WG.IV/WP.55, paras. 27-29.

Chapter II. Application of legal requirements to data messages

Article 5. Legal recognition of data messages

46. Article 5 embodies the fundamental principle that data messages should not be discriminated against, i.e., that there should be no disparity of treatment

between data messages and paper documents. It is intended to apply notwithstanding any statutory requirements for a "writing" or an original. That fundamental principle is intended to find general application and its scope should not be limited to evidence or other matters covered in chapter II. It should be noted, however, that such a principle is not intended to override any of the requirements contained in articles 6 to 10. By stating that "information shall not be denied legal effectiveness, validity or enforceability solely on the grounds that it is in the form of a data message," article 5 merely indicates that the form in which certain information is presented or retained cannot be used as the only reason for which that information would be denied legal effectiveness, validity or enforceability. However, article 5 should not be misinterpreted as establishing the legal validity of any given data message or of any information contained therein.

References
A/51/17, paras. 92 and 97 (article 4);
A/50/17, paras. 225-227 (article 4);
A/CN.9/407, para. 55;
A/CN.9/406, paras. 91-94; A/CN.9/WG.IV/WP. 62, article 5 *bis*;
A/CN.9/390, paras. 79-87;
A/CN.9/WG.IV/WP. 60, article 5 *bis*;
A/CN.9/387, paras. 93-94.

Article 5 bis. Incorporation by reference

46-1. Article 5 *bis* was adopted by the Commission at its thirty-first session, in June 1998. It is intended to provide guidance as to how legislation aimed at facilitating the use of electronic commerce might deal with the situation where certain terms and conditions, although not stated in full but merely referred to in a data message, might need to be recognized as having the same degree of legal effectiveness as if they had been fully stated in the text of that data message. Such recognition is acceptable under the laws of many States with respect to conventional paper communications, usually with some rules of law providing safeguards, for example rules on consumer protection. The expression "incorporation by reference" is often used as a concise means of describing situations where a document refers generically to provisions which are detailed elsewhere, rather than reproducing them in full.

46-2. In an electronic environment, incorporation by reference is often regarded as essential to widespread use of electronic data interchange (EDI), electronic mail, digital certificates and other forms of electronic commerce. For example, electronic communications are typically structured in such a way that large numbers of messages are exchanged, with each message containing brief information, and relying much more frequently than paper documents on reference to information accessible elsewhere. In electronic communications, practitioners should not have imposed upon them an obligation to overload their data messages with quantities of free text when they can take advantage of extrinsic sources of information, such as databases, code lists or glossaries, by making use of abbreviations, codes and other references to such information.

46-3. Standards for incorporating data messages by reference into other data messages may also be essential to the use of public key certificates, because these certificates are generally brief records with rigidly prescribed contents that are finite in size. The trusted third party which issues the certificate, however, is likely to require the inclusion of relevant contractual terms limiting its liability. The scope, purpose and effect of a certificate in commercial practice, therefore, would be ambiguous and uncertain without external terms being incorporated by reference. This is the case especially in the context of international communications involving diverse parties who follow varied trade practices and customs.

46-4. The establishment of standards for incorporating data messages by reference into other data messages is critical to the growth of a computer-based trade infrastructure. Without the legal certainty fostered by such standards, there might be a significant risk that the application of traditional tests for determining the enforceability of terms that seek to be incorporated by reference might be ineffective when applied to corresponding electronic commerce terms because of the differences between traditional and electronic commerce mechanisms.

46-5. While electronic commerce relies heavily on the mechanism of incorporation by reference, the accessibility of the full text of the information being referred to may be considerably improved by the use of electronic communications. For example, a message may have embedded in it uniform resource locators (URLs), which direct the reader to the referenced document. Such URLs can provide "hypertext links" allowing the reader to use a pointing device (such as a mouse) to select a key word associated with a URL. The referenced text would then be displayed. In assessing the accessibility of the referenced text, factors to be considered may include: availability (hours of operation of the repository and ease of access); cost of access; integrity (verification of content, authentication of sender, and mechanism for communication error correction); and the extent to which that term is subject to later amendment (notice of updates; notice of policy of amendment).

46-6. One aim of article 5 *bis* is to facilitate incorporation by reference in an electronic context by removing the uncertainty prevailing in many jurisdictions as to whether the provisions dealing with traditional incorporation by reference are applicable to incorporation by reference in an electronic environment. However, in enacting article 5 *bis*, attention should be given to avoid introducing more restrictive requirements with respect to incorporation by reference in electronic commerce than might already apply in paper-based trade.

46-7. Another aim of the provision is to recognize that consumer-protection or other national or international law of a mandatory nature (e.g., rules protecting weaker parties in the context of contracts of adhesion) should not be interfered with. That result could also be achieved by validating incorporation by reference in an electronic environment "to the extent permitted by law," or by listing the rules of law that remain unaffected by article 5 *bis*. Article 5 *bis* is not to be interpreted as creating a specific legal regime for incorporation by reference in an electronic environment. Rather, by establishing a principle of non-discrimination, it is to be construed as making the domestic rules applicable to incorporation by reference in a paper-based environment equally applicable to incorporation by reference for the purposes of electronic commerce. For example, in a number of jurisdictions, existing rules of mandatory law only validate incorporation by reference provided that the following three conditions are met: (a) the reference clause should be inserted in the data message; (b) the document being referred to, e.g., general terms and conditions, should actually be known to the party against whom the reference document might be relied upon; and (c) the reference document should be accepted, in addition to being known, by that party.

References
A/53/17, paras. 212-221;
A/CN.9/450;
A/CN.9/446, paras. 14-24;
A/CN.9/WG.IV/WP.74;
A/52/17, paras. 248-250;
A/CN.9/437, paras. 151-155;
A/CN.9/WG.IV/WP. 71, paras 77-93;
A/51/17, paras. 222-223;
A/CN.9/421, paras. 109 and 114;
A/CN.9/WG.IV/WP.69, paras. 30, 53, 59-60 and 91;
A/CN.9/407, paras. 100-105 and 117;
A/CN.9/WG.IV/WP.66;
A/CN.9/WG.IV/WP.65;
A/CN.9/406, paras. 90 and 178-179;
A/CN.9/WG.IV/WP.55, para. 109-113;
A/CN.9/360, paras. 90-95;
A/CN.9/WG.IV/WP.53, paras. 77-78;
A/CN.9/350, paras. 95-96;
A/CN.9/333, paras. 66-68.

Article 6. Writing

47. Article 6 is intended to define the basic standard to be met by a data message in order to be considered as meeting a requirement (which may result from statute, regulation or judge-made law) that information be retained or presented "in writing" (or that the information be contained in a "document" or other paper-based instrument). It may be noted that article 6 is part of a set of three articles (articles 6, 7 and 8), which share the same structure and should be read together.

48. In the preparation of the Model Law, particular attention was paid to the functions traditionally performed by various kinds of "writings" in a paper-based environment. For example, the following non-exhaustive list indicates

reasons why national laws require the use of "writings:" (1) to ensure that there would be tangible evidence of the existence and nature of the intent of the parties to bind themselves; (2) to help the parties be aware of the consequences of their entering into a contract; (3) to provide that a document would be legible by all; (4) to provide that a document would remain unaltered over time and provide a permanent record of a transaction; (5) to allow for the reproduction of a document so that each party would hold a copy of the same data; (6) to allow for the authentication of data by means of a signature; (7) to provide that a document would be in a form acceptable to public authorities and courts; (8) to finalize the intent of the author of the "writing" and provide a record of that intent; (9) to allow for the easy storage of data in a tangible form; (10) to facilitate control and subsequent audit for accounting, tax or regulatory purposes; and (11) to bring legal rights and obligations into existence in those cases where a "writing" was required for validity purposes.

49. However, in the preparation of the Model Law, it was found that it would be inappropriate to adopt an overly comprehensive notion of the functions performed by writing. Existing requirements that data be presented in written form often combine the requirement of a "writing" with concepts distinct from writing, such as signature and original. Thus, when adopting a functional approach, attention should be given to the fact that the requirement of a "writing" should be considered as the lowest layer in a hierarchy of form requirements, which provide distinct levels of reliability, traceability and unalterability with respect to paper documents. The requirement that data be presented in written form (which can be described as a "threshold requirement") should thus not be confused with more stringent requirements such as "signed writing," "signed original" or "authenticated legal act." For example, under certain national laws, a written document that is neither dated nor signed, and the author of which either is not identified in the written document or is identified by a mere letterhead, would be regarded as a "writing" although it might be of little evidential weight in the absence of other evidence (e.g., testimony) regarding the authorship of the document. In addition, the notion of unalterability should not be considered as built into the concept of writing as an absolute requirement since a "writing" in pencil might still be considered a "writing" under certain existing legal definitions. Taking into account the way in which such issues as integrity of the data and protection against fraud are dealt with in a paper-based environment, a fraudulent document would nonetheless be regarded as a "writing." In general, notions such as "evidence" and "intent of the parties to bind themselves" are to be tied to the more general issues of reliability and authentication of the data and should not be included in the definition of a "writing."

50. The purpose of article 6 is not to establish a requirement that, in all instances, data messages should fulfill all conceivable functions of a writing. Rather than focusing upon specific functions of a "writing," for example, its evidentiary function in the context of tax law or its warning function in the context of civil law, article 6 focuses upon the basic notion of the information

being reproduced and read. That notion is expressed in article 6 in terms that were found to provide an objective criterion, namely that the information in a data message must be accessible so as to be usable for subsequent reference. The use of the word "accessible" is meant to imply that information in the form of computer data should be readable and interpretable, and that the software that might be necessary to render such information readable should be retained. The word "usable" is not intended to cover only human use but also computer processing. As to the notion of "subsequent reference," it was preferred to such notions as "durability" or "non-alterability," which would have established too harsh standards, and to such notions as "readability" or "intelligibility," which might constitute too subjective criteria.

51. The principle embodied in paragraph (3) of articles 6 and 7, and in paragraph (4) of article 8, is that an enacting State may exclude from the application of those articles certain situations to be specified in the legislation enacting the Model Law. An enacting State may wish to exclude specifically certain types of situations, depending in particular on the purpose of the formal requirement in question. One such type of situation may be the case of writing requirements intended to provide notice or warning of specific factual or legal risks, for example, requirements for warnings to be placed on certain types of products. Another specific exclusion might be considered, for example, in the context of formalities required pursuant to international treaty obligations of the enacting State (e.g., the requirement that a cheque be in writing pursuant to the Convention providing a Uniform Law for Cheques, Geneva, 1931) and other kinds of situations and areas of law that are beyond the power of the enacting State to change by means of a statute.

52. Paragraph (3) was included with a view to enhancing the acceptability of the Model Law. It recognizes that the matter of specifying exclusions should be left to enacting States, an approach that would take better account of differences in national circumstances. However, it should be noted that the objectives of the Model Law would not be achieved if paragraph (3) were used to establish blanket exceptions, and the opportunity provided by paragraph (3) in that respect should be avoided. Numerous exclusions from the scope of articles 6 to 8 would raise needless obstacles to the development of modern communication techniques, since what the Model Law contains are very fundamental principles and approaches that are expected to find general application.

References
A/51/17, paras. 180-181 and 185-187 (article 5);
A/50/17, paras. 228-241 (article 5);
A/CN.9/407, paras. 56-63;
A/CN.9/406, paras. 95-101; A/CN.9/WG.IV/WP.62, article 6;
A/CN.9/390, paras. 88-96; A/CN.9/WG.IV/WP.60, article 6;
A/CN.9/387, paras. 66-80; A/CN.9/WG.IV/WP.57, article 6;
A/CN.9/WG.IV/WP.58, annex;
A/CN.9/373, paras. 45-62; A/CN.9/WG.IV/WP.55, paras. 36-49;

A/CN.9/360, paras. 32-43; A/CN.9/WG.IV/WP.53, paras. 37-45;
A/CN.9/350, paras. 68-78;
A/CN.9/333, paras. 20-28;
A/CN.9/265, paras. 59-72.

Article 7. Signature

53. Article 7 is based on the recognition of the functions of a signature in a paper-based environment. In the preparation of the Model Law, the following functions of a signature were considered: to identify a person; to provide certainty as to the personal involvement of that person in the act of signing; to associate that person with the content of a document. It was noted that, in addition, a signature could perform a variety of functions, depending on the nature of the document that was signed. For example, a signature might attest to the intent of a party to be bound by the content of a signed contract; the intent of a person to endorse authorship of a text; the intent of a person to associate itself with the content of a document written by someone else; the fact that, and the time when, a person had been at a given place.

54. It may be noted that, alongside the traditional handwritten signature, there exist various types of procedures (e.g., stamping, perforation), sometimes also referred to as "signatures," which provide various levels of certainty. For example, in some countries, there exists a general requirement that contracts for the sale of goods above a certain amount should be "signed" in order to be enforceable. However, the concept of a signature adopted in that context is such that a stamp, perforation or even a typewritten signature or a printed letterhead might be regarded as sufficient to fulfil the signature requirement. At the other end of the spectrum, there exist requirements that combine the traditional handwritten signature with additional security procedures such as the confirmation of the signature by witnesses.

55. It might be desirable to develop functional equivalents for the various types and levels of signature requirements in existence. Such an approach would increase the level of certainty as to the degree of legal recognition that could be expected from the use of the various means of authentication used in electronic commerce practice as substitutes for "signatures." However, the notion of signature is intimately linked to the use of paper. Furthermore, any attempt to develop rules on standards and procedures to be used as substitutes for specific instances of "signatures" might create the risk of tying the legal framework provided by the Model Law to a given state of technical development.

56. With a view to ensuring that a message that was required to be authenticated should not be denied legal value for the sole reason that it was not authenticated in a manner peculiar to paper documents, article 7 adopts a comprehensive approach. It establishes the general conditions under which data messages would be regarded as authenticated with sufficient credibility and would be enforceable in the face of signature requirements which currently present barriers to electronic commerce. Article 7 focuses on the two basic

functions of a signature, namely to identify the author of a document and to confirm that the author approved the content of that document. Paragraph (1)(a) establishes the principle that, in an electronic environment, the basic legal functions of a signature are performed by way of a method that identifies the originator of a data message and confirms that the originator approved the content of that data message.

57. Paragraph (1)(b) establishes a flexible approach to the level of security to be achieved by the method of identification used under paragraph (1)(a). The method used under paragraph (1)(a) should be as reliable as is appropriate for the purpose for which the data message is generated or communicated, in the light of all the circumstances, including any agreement between the originator and the addressee of the data message.

58. In determining whether the method used under paragraph (1) is appropriate, legal, technical and commercial factors that may be taken into account include the following: (1) the sophistication of the equipment used by each of the parties; (2) the nature of their trade activity; (3) the frequency at which commercial transactions take place between the parties; (4) the kind and size of the transaction; (5) the function of signature requirements in a given statutory and regulatory environment; (6) the capability of communication systems; (7) compliance with authentication procedures set forth by intermediaries; (8) the range of authentication procedures made available by any intermediary; (9) compliance with trade customs and practice; (10) the existence of insurance coverage mechanisms against unauthorized messages; (11) the importance and the value of the information contained in the data message; (12) the availability of alternative methods of identification and the cost of implementation; (13) the degree of acceptance or non-acceptance of the method of identification in the relevant industry or field both at the time the method was agreed upon and the time when the data message was communicated; and (14) any other relevant factor.

59. Article 7 does not introduce a distinction between the situation in which users of electronic commerce are linked by a communication agreement and the situation in which parties had no prior contractual relationship regarding the use of electronic commerce. Thus, article 7 may be regarded as establishing a basic standard of authentication for data messages that might be exchanged in the absence of a prior contractual relationship and, at the same time, to provide guidance as to what might constitute an appropriate substitute for a signature if the parties used electronic communications in the context of a communication agreement. The Model Law is thus intended to provide useful guidance both in a context where national laws would leave the question of authentication of data messages entirely to the discretion of the parties and in a context where requirements for signature, which were usually set by mandatory provisions of national law, should not be made subject to alteration by agreement of the parties.

60. The notion of an "agreement between the originator and the addressee of a data message" is to be interpreted as covering not only bilateral or

multilateral agreements concluded between parties exchanging directly data messages (e.g., "trading partners agreements," "communication agreements" or "interchange agreements") but also agreements involving intermediaries such as networks (e.g., "third-party service agreements"). Agreements concluded between users of electronic commerce and networks may incorporate "system rules," i.e., administrative and technical rules and procedures to be applied when communicating data messages. However, a possible agreement between originators and addressees of data messages as to the use of a method of authentication is not conclusive evidence of whether that method is reliable or not.

61. It should be noted that, under the Model Law, the mere signing of a data message by means of a functional equivalent of a handwritten signature is not intended, in and of itself, to confer legal validity on the data message. Whether a data message that fulfilled the requirement of a signature has legal validity is to be settled under the law applicable outside the Model Law.

References
A/51/17, paras. 180-181 and 185-187 (article 6);
A/50/17, paras. 242-248 (article 6);
A/CN.9/407, paras. 64-70;
A/CN.9/406, paras. 102-105; A/CN.9/WG.IV/WP.62, article 7;
A/CN.9/390, paras. 97-109; A/CN.9/WG.IV/WP.60, article 7;
A/CN.9/387, paras. 81-90; A/CN.9/WG.IV/WP.57, article 7;
A/CN.9/WG.IV/WP.58, annex;
A/CN.9/373, paras. 63-76; A/CN.9/WG.IV/WP.55, paras. 50-63;
A/CN.9/360, paras. 71-75; A/CN.9/WG.IV/WP.53, paras. 61-66;
A/CN.9/350, paras. 86-89;
A/CN.9/333, paras. 50-59;
A/CN.9/265, paras. 49-58 and 79-80.

Article 8. Original

62. If "original" were defined as a medium on which information was fixed for the first time, it would be impossible to speak of "original" data messages, since the addressee of a data message would always receive a copy thereof. However, article 8 should be put in a different context. The notion of "original" in article 8 is useful since in practice many disputes relate to the question of originality of documents, and in electronic commerce the requirement for presentation of originals constitutes one of the main obstacles that the Model Law attempts to remove. Although in some jurisdictions the concepts of "writing," "original" and "signature" may overlap, the Model Law approaches them as three separate and distinct concepts. Article 8 is also useful in clarifying the notions of "writing" and "original," in particular in view of their importance for purposes of evidence.

63. Article 8 is pertinent to documents of title and negotiable instruments, in which the notion of uniqueness of an original is particularly relevant. However,

attention is drawn to the fact that the Model Law is not intended only to apply to documents of title and negotiable instruments, or to such areas of law where special requirements exist with respect to registration or notarization of "writings," e.g., family matters or the sale of real estate. Examples of documents that might require an "original" are trade documents such as weight certificates, agricultural certificates, quality or quantity certificates, inspection reports, insurance certificates, etc. While such documents are not negotiable or used to transfer rights or title, it is essential that they be transmitted unchanged, that is in their "original" form, so that other parties in international commerce may have confidence in their contents. In a paper-based environment, these types of documents are usually only accepted if they are "original" to lessen the chance that they be altered, which would be difficult to detect in copies. Various technical means are available to certify the contents of a data message to confirm its "originality." Without this functional equivalent of originality, the sale of goods using electronic commerce would be hampered since the issuers of such documents would be required to retransmit their data message each and every time the goods are sold, or the parties would be forced to use paper documents to supplement the electronic commerce transaction.

64. Article 8 should be regarded as stating the minimum acceptable form requirement to be met by a data message for it to be regarded as the functional equivalent of an original. The provisions of article 8 should be regarded as mandatory, to the same extent that existing provisions regarding the use of paper-based original documents would be regarded as mandatory. The indication that the form requirements stated in article 8 are to be regarded as the "minimum acceptable" should not, however, be construed as inviting States to establish requirements stricter than those contained in the Model Law.

65. Article 8 emphasizes the importance of the integrity of the information for its originality and sets out criteria to be taken into account when assessing integrity by reference to systematic recording of the information, assurance that the information was recorded without lacunae and protection of the data against alteration. It links the concept of originality to a method of authentication and puts the focus on the method of authentication to be followed in order to meet the requirement. It is based on the following elements: a simple criterion as to "integrity" of the data; a description of the elements to be taken into account in assessing the integrity; and an element of flexibility, i.e., a reference to circumstances.

66. As regards the words "the time when it was first generated in its final form" in paragraph (1)(a), it should be noted that the provision is intended to encompass the situation where information was first composed as a paper document and subsequently transferred on to a computer. In such a situation, paragraph (1)(a) is to be interpreted as requiring assurances that the information has remained complete and unaltered from the time when it was composed as a paper document onwards, and not only as from the time when it was translated into electronic form. However, where several drafts were created and stored

before the final message was composed, paragraph (1)(a) should not be misinterpreted as requiring assurance as to the integrity of the drafts.

67. Paragraph (3)(a) sets forth the criteria for assessing integrity, taking care to except necessary additions to the first (or "original") data message such as endorsements, certifications, notarizations, etc., from other alterations. As long as the contents of a data message remain complete and unaltered, necessary additions to that data message would not affect its "originality." Thus when an electronic certificate is added to the end of an "original" data message to attest to the "originality" of that data message, or when data is automatically added by computer systems at the start and the finish of a data message in order to transmit it, such additions would be considered as if they were a supplemental piece of paper with an "original" piece of paper, or the envelope and stamp used to send that "original" piece of paper.

68. As in other articles of chapter II of part one, the words "the law" in the opening phrase of article 8 are to be understood as encompassing not only statutory or regulatory law but also judicially-created law and other procedural law. In certain common law countries, where the words "the law" would normally be interpreted as referring to common law rules, as opposed to statutory requirements, it should be noted that, in the context of the Model Law, the words "the law" are intended to encompass those various sources of law. However, "the law," as used in the Model Law, is not meant to include areas of law that have not become part of the law of a State and are sometimes, somewhat imprecisely, referred to by expressions such as *"lex mercatoria"* or "law merchant."

69. Paragraph (4), as was the case with similar provisions in articles 6 and 7, was included with a view to enhancing the acceptability of the Model Law. It recognizes that the matter of specifying exclusions should be left to enacting States, an approach that would take better account of differences in national circumstances. However, it should be noted that the objectives of the Model Law would not be achieved if paragraph (4) were used to establish blanket exceptions. Numerous exclusions from the scope of articles 6 to 8 would raise needless obstacles to the development of modern communication techniques, since what the Model Law contains are very fundamental principles and approaches that are expected to find general application.

References
A/51/17, paras. 180-181 and 185-187 (article 7);
A/50/17, paras. 249-255 (article 7);
A/CN.9/407, paras. 71-79;
A/CN.9/406, paras. 106-110; A/CN.9/WG.IV/WP.62, article 8;
A/CN.9/390, paras. 110-133; A/CN.9/WG.IV/WP.60, article 8;
A/CN.9/387, paras. 91-97; A/CN.9/WG.IV/WP.57, article 8;
A/CN.9/WG.IV/WP.58, annex;
A/CN.9/373, paras. 77-96; A/CN.9/WG.IV/WP.55, paras. 64-70;

A/CN.9/360, paras. 60-70; A/CN.9/WG.IV/WP.53, paras. 56-60;
A/CN.9/350, paras. 84-85;
A/CN.9/265, paras. 43-48.

Article 9. Admissibility and evidential weight of data messages

70. The purpose of article 9 is to establish both the admissibility of data messages as evidence in legal proceedings and their evidential value. With respect to admissibility, paragraph (1), establishing that data messages should not be denied admissibility as evidence in legal proceedings on the sole ground that they are in electronic form, puts emphasis on the general principle stated in article 4 and is needed to make it expressly applicable to admissibility of evidence, an area in which particularly complex issues might arise in certain jurisdictions. The term "best evidence" is a term understood in, and necessary for, certain common law jurisdictions. However, the notion of "best evidence" could raise a great deal of uncertainty in legal systems in which such a rule is unknown. States in which the term would be regarded as meaningless and potentially misleading may wish to enact the Model Law without the reference to the "best evidence" rule contained in paragraph (1).

71. As regards the assessment of the evidential weight of a data message, paragraph (2) provides useful guidance as to how the evidential value of data messages should be assessed (e.g., depending on whether they were generated, stored or communicated in a reliable manner).

References
A/50/17, paras. 256-263 (article 8);
A/CN.9/407, paras. 80-81;
A/CN.9/406, paras. 111-113; A/CN.9/WG.IV/WP.62, article 9;
A/CN.9/390, paras. 139-143; A/CN.9/WG.IV/WP.60, article 9;
A/CN.9/387, paras. 98-109; A/CN.9/WG.IV/WP.57, article 9;
A/CN.9/WG.IV/WP.58, annex;
A/CN.9/373, paras. 97-108; A/CN.9/WG.IV/WP.55, paras. 71-81;
A/CN.9/360, paras. 44-59; A/CN.9/WG.IV/WP.53, paras. 46-55;
A/CN.9/350, paras. 79-83 and 90-91;
A/CN.9/333, paras. 29-41;
A/CN.9/265, paras. 27-48.

Article 10. Retention of data messages

72. Article 10 establishes a set of alternative rules for existing requirements regarding the storage of information (e.g., for accounting or tax purposes) that may constitute obstacles to the development of modern trade.

73. Paragraph (1) is intended to set out the conditions under which the obligation to store data messages that might exist under the applicable law would be met. Subparagraph *(a)* reproduces the conditions established under article 6 for a data message to satisfy a rule which prescribes the presentation of a

"writing." Subparagraph *(b)* emphasizes that the message does not need to be retained unaltered as long as the information stored accurately reflects the data message as it was sent. It would not be appropriate to require that information should be stored unaltered, since usually messages are decoded, compressed or converted in order to be stored.

74. Subparagraph *(c)* is intended to cover all the information that may need to be stored, which includes, apart from the message itself, certain transmittal information that may be necessary for the identification of the message. Subparagraph *(c)*, by imposing the retention of the transmittal information associated with the data message, is creating a standard that is higher than most standards existing under national laws as to the storage of paper-based communications. However, it should not be understood as imposing an obligation to retain transmittal information additional to the information contained in the data message when it was generated, stored or transmitted, or information contained in a separate data message, such as an acknowledgement of receipt. Moreover, while some transmittal information is important and has to be stored, other transmittal information can be exempted without the integrity of the data message being compromised. That is the reason why subparagraph *(c)* establishes a distinction between those elements of transmittal information that are important for the identification of the message and the very few elements of transmittal information covered in paragraph (2) (e.g., communication protocols), which are of no value with regard to the data message and which, typically, would automatically be stripped out of an incoming data message by the receiving computer before the data message actually entered the information system of the addressee.

75. In practice, storage of information, and especially storage of transmittal information, may often be carried out by someone other than the originator or the addressee, such as an intermediary. Nevertheless, it is intended that the person obligated to retain certain transmittal information cannot escape meeting that obligation simply because, for example, the communications system operated by that other person does not retain the required information. This is intended to discourage bad practice or wilful misconduct. Paragraph (3) provides that in meeting its obligations under paragraph (1), an addressee or originator may use the services of any third party, not just an intermediary.

References
A/51/17, paras. 185-187 (article 9);
A/50/17, paras. 264-270 (article 9);
A/CN.9/407, paras. 82-84;
A/CN.9/406, paras. 59-72; A/CN.9/WG.IV/WP.60, article 14;
A/CN.9/387, paras. 164-168;
A/CN.9/WG.IV/WP.57, article 14;
A/CN.9/373, paras. 123-125; A/CN.9/WG.IV/WP.55, para. 94.

Chapter III. Communication of data messages

Article 11. Formation and validity of contracts

76. Article 11 is not intended to interfere with the law on formation of contracts but rather to promote international trade by providing increased legal certainty as to the conclusion of contracts by electronic means. It deals not only with the issue of contract formation but also with the form in which an offer and an acceptance may be expressed. In certain countries, a provision along the lines of paragraph (1) might be regarded as merely stating the obvious, namely that an offer and an acceptance, as any other expression of will, can be communicated by any means, including data messages. However, the provision is needed in view of the remaining uncertainties in a considerable number of countries as to whether contracts can validly be concluded by electronic means. Such uncertainties may stem from the fact that, in certain cases, the data messages expressing offer and acceptance are generated by computers without immediate human intervention, thus raising doubts as to the expression of intent by the parties. Another reason for such uncertainties is inherent in the mode of communication and results from the absence of a paper document.

77. It may also be noted that paragraph (1) reinforces, in the context of contract formation, a principle already embodied in other articles of the Model Law, such as articles 5, 9 and 13, all of which establish the legal effectiveness of data messages. However, paragraph (1) is needed since the fact that electronic messages may have legal value as evidence and produce a number of effects, including those provided in articles 9 and 13, does not necessarily mean that they can be used for the purpose of concluding valid contracts.

78. Paragraph (1) covers not merely the cases in which both the offer and the acceptance are communicated by electronic means but also cases in which only the offer or only the acceptance is communicated electronically. As to the time and place of formation of contracts in cases where an offer or the acceptance of an offer is expressed by means of a data message, no specific rule has been included in the Model Law in order not to interfere with national law applicable to contract formation. It was felt that such a provision might exceed the aim of the Model Law, which should be limited to providing that electronic communications would achieve the same degree of legal certainty as paper-based communications. The combination of existing rules on the formation of contracts with the provisions contained in article 15 is designed to dispel uncertainty as to the time and place of formation of contracts in cases where the offer or the acceptance are exchanged electronically.

79. The words ''unless otherwise stated by the parties,'' which merely restate, in the context of contract formation, the recognition of party autonomy expressed in article 4, are intended to make it clear that the purpose of the Model Law is not to impose the use of electronic means of communication on parties who rely on the use of paper-based communication to conclude contracts. Thus,

article 11 should not be interpreted as restricting in any way party autonomy with respect to parties not involved in the use of electronic communication.

80. During the preparation of paragraph (1), it was felt that the provision might have the harmful effect of overruling otherwise applicable provisions of national law, which might prescribe specific formalities for the formation of certain contracts. Such forms include notarization and other requirements for "writings," and might respond to considerations of public policy, such as the need to protect certain parties or to warn them against specific risks. For that reason, paragraph (2) provides that an enacting State can exclude the application of paragraph (1) in certain instances to be specified in the legislation enacting the Model Law.

References
A/51/17, paras. 89-94 (article 13);
A/CN.9/407, para. 93;
A/CN.9/406, paras. 34-41; A/CN.9/WG.IV/WP.60, article 12;
A/CN.9/387, paras. 145-151; A/CN.9/WG.IV/WP.57, article 12;
A/CN.9/373, paras. 126-133; A/CN.9/WG.IV/WP.55, paras. 95-102;
A/CN.9/360, paras. 76-86; A/CN.9/WG.IV/WP.53, paras. 67-73;
A/CN.9/350, paras. 93-96;
A/CN.9/333, paras. 60-68.

Article 12. Recognition by parties of data messages

81. Article 12 was added at a late stage in the preparation of the Model Law, in recognition of the fact that article 11 was limited to dealing with data messages that were geared to the conclusion of a contract, but that the draft Model Law did not contain specific provisions on data messages that related not to the conclusion of contracts but to the performance of contractual obligations (e.g., notice of defective goods, an offer to pay, notice of place where a contract would be performed, recognition of debt). Since modern means of communication are used in a context of legal uncertainty, in the absence of specific legislation in most countries, it was felt appropriate for the Model Law not only to establish the general principle that the use of electronic communication should not be discriminated against, as expressed in article 5, but also to include specific illustrations of that principle. Contract formation is but one of the areas where such an illustration is useful and the legal validity of unilateral expressions of will, as well as other notices or statements that may be issued in the form of data messages, also needs to be mentioned.

82. As is the case with article 11, article 12 is not to impose the use of electronic means of communication but to validate such use, subject to contrary agreement by the parties. Thus, article 12 should not be used as a basis to impose on the addressee the legal consequences of a message, if the use of a non-paper-based method for its transmission comes as a surprise to the addressee.

References
A/51/17, paras. 95-99 (new article 13 *bis*).

Article 13. Attribution of data messages

83. Article 13 has its origin in article 5 of the UNCITRAL Model Law on International Credit Transfers, which defines the obligations of the sender of a payment order. Article 13 is intended to apply where there is a question as to whether a data message was really sent by the person who is indicated as being the originator. In the case of a paper-based communication the problem would arise as the result of an alleged forged signature of the purported originator. In an electronic environment, an unauthorized person may have sent the message but the authentication by code, encryption or the like would be accurate. The purpose of article 13 is not to assign responsibility. It deals rather with attribution of data messages by establishing a presumption that under certain circumstances a data message would be considered as a message of the originator, and goes on to qualify that presumption in case the addressee knew or ought to have known that the data message was not that of the originator.

84. Paragraph (1) recalls the principle that an originator is bound by a data message if it has effectively sent that message. Paragraph (2) refers to the situation where the message was sent by a person other than the originator who had the authority to act on behalf of the originator. Paragraph (2) is not intended to displace the domestic law of agency, and the question as to whether the other person did in fact and in law have the authority to act on behalf of the originator is left to the appropriate legal rules outside the Model Law.

85. Paragraph (3) deals with two kinds of situations, in which the addressee could rely on a data message as being that of the originator: firstly, situations in which the addressee properly applied an authentication procedure previously agreed to by the originator; and secondly, situations in which the data message resulted from the actions of a person who, by virtue of its relationship with the originator, had access to the originator's authentication procedures. By stating that the addressee "is entitled to regard a data as being that of the originator," paragraph (3) read in conjunction with paragraph (4)*(a)* is intended to indicate that the addressee could act on the assumption that the data message is that of the originator up to the point in time it received notice from the originator that the data message was not that of the originator, or up to the point in time when it knew or should have known that the data message was not that of the originator.

86. Under paragraph (3)*(a)*, if the addressee applies any authentication procedures previously agreed to by the originator and such application results in the proper verification of the originator as the source of the message, the message is presumed to be that of the originator. That covers not only the situation where an authentication procedure has been agreed upon by the originator and the addressee but also situations where an originator, unilaterally or as a result of an agreement with an intermediary, identified a procedure and agreed to be bound by a data message that met the requirements corresponding to that procedure.

Thus, agreements that became effective not through direct agreement between the originator and the addressee but through the participation of third-party service providers are intended to be covered by paragraph (3)*(a)*. However, it should be noted that paragraph (3)*(a)* applies only when the communication between the originator and the addressee is based on a previous agreement, but that it does not apply in an open environment.

87. The effect of paragraph (3)*(b)*, read in conjunction with paragraph (4)*(b)*, is that the originator or the addressee, as the case may be, is responsible for any unauthorized data message that can be shown to have been sent as a result of negligence of that party.

88. Paragraph (4)*(a)* should not be misinterpreted as relieving the originator from the consequences of sending a data message, with retroactive effect, irrespective of whether the addressee had acted on the assumption that the data message was that of the originator. Paragraph (4) is not intended to provide that receipt of a notice under subparagraph *(a)* would nullify the original message retroactively. Under subparagraph *(a)*, the originator is released from the binding effect of the message after the time notice is received and not before that time. Moreover, paragraph (4) should not be read as allowing the originator to avoid being bound by the data message by sending notice to the addressee under subparagraph *(a)*, in a case where the message had, in fact, been sent by the originator and the addressee properly applied agreed or reasonable authentication procedures. If the addressee can prove that the message is that of the originator, paragraph (1) would apply and not paragraph (4)*(a)*. As to the meaning of "reasonable time," the notice should be such as to give the addressee sufficient time to react. For example, in the case of just-in-time supply, the addressee should be given time to adjust its production chain.

89. With respect to paragraph (4)*(b)*, it should be noted that the Model Law could lead to the result that the addressee would be entitled to rely on a data message under paragraph (3)*(a)* if it had properly applied the agreed authentication procedures, even if it knew that the data message was not that of the originator. It was generally felt when preparing the Model Law that the risk that such a situation could arise should be accepted, in view of the need for preserving the reliability of agreed authentication procedures.

90. Paragraph (5) is intended to preclude the originator from disavowing the message once it was sent, unless the addressee knew, or should have known, that the data message was not that of the originator. In addition, paragraph (5) is intended to deal with errors in the content of the message arising from errors in transmission.

91. Paragraph (6) deals with the issue of erroneous duplication of data messages, an issue of considerable practical importance. It establishes the standard of care to be applied by the addressee to distinguish an erroneous duplicate of a data message from a separate data message.

92. Early drafts of article 13 contained an additional paragraph, expressing the principle that the attribution of authorship of a data message to the originator should not interfere with the legal consequences of that message, which should be determined by other applicable rules of national law. It was later felt that it was not necessary to express that principle in the Model Law but that it should be mentioned in this Guide.

References
A/51/17, paras. 189-194 (article 11);
A/50/17, paras. 275-303 (article 11);
A/CN.9/407, paras. 86-89;
A/CN.9/406, paras. 114-131; A/CN.9/WG.IV/WP.62, article 10;
A/CN.9/390, paras. 144-153; A/CN.9/WG.IV/WP.60, article 10;
A/CN.9/387, paras. 110-132; A/CN.9/WG.IV/WP.57, article 10;
A/CN.9/373, paras. 109-115; A/CN.9/WG.IV/WP.55, paras. 82-86.

Article 14. Acknowledgement of receipt

93. The use of functional acknowledgements is a business decision to be made by users of electronic commerce; the Model Law does not intend to impose the use of any such procedure. However, taking into account the commercial value of a system of acknowledgement of receipt and the widespread use of such systems in the context of electronic commerce, it was felt that the Model Law should address a number of legal issues arising from the use of acknowledgement procedures. It should be noted that the notion of "acknowledgement" is sometimes used to cover a variety of procedures, ranging from a mere acknowledgement of receipt of an unspecified message to an expression of agreement with the content of a specific data message. In many instances, the procedure of "acknowledgement" would parallel the system known as "return receipt requested" in postal systems. Acknowledgements of receipt may be required in a variety of instruments, e.g., in the data message itself, in bilateral or multilateral communication agreements, or in "system rules." It should be borne in mind that variety among acknowledgement procedures implies variety of the related costs. The provisions of article 14 are based on the assumption that acknowledgement procedures are to be used at the discretion of the originator. Article 14 is not intended to deal with the legal consequences that may flow from sending an acknowledgement of receipt, apart from establishing receipt of the data message. For example, where an originator sends an offer in a data message and requests acknowledgement of receipt, the acknowledgement of receipt simply evidences that the offer has been received. Whether or not sending that acknowledgement amounted to accepting the offer is not dealt with by the Model Law but by contract law outside the Model Law.

94. The purpose of paragraph (2) is to validate acknowledgement by any communication or conduct of the addressee (e.g., the shipment of the goods as an acknowledgement of receipt of a purchase order) where the originator has not agreed with the addressee that the acknowledgement should be in a particular

form. The situation where an acknowledgement has been unilaterally requested by the originator to be given in a specific form is not expressly addressed by article 14, which may entail as a possible consequence that a unilateral requirement by the originator as to the form of acknowledgements would not affect the right of the addressee to acknowledge receipt by any communication or conduct sufficient to indicate to the originator that the message had been received. Such a possible interpretation of paragraph (2) makes it particularly necessary to emphasize in the Model Law the distinction to be drawn between the effects of an acknowledgement of receipt of a data message and any communication in response to the content of that data message, a reason why paragraph (7) is needed.

95. Paragraph (3), which deals with the situation where the originator has stated that the data message is conditional on receipt of an acknowledgement, applies whether or not the originator has specified that the acknowledgement should be received by a certain time.

96. The purpose of paragraph (4) is to deal with the more common situation where an acknowledgement is requested, without any statement being made by the originator that the data message is of no effect until an acknowledgement has been received. Such a provision is needed to establish the point in time when the originator of a data message who has requested an acknowledgement of receipt is relieved from any legal implication of sending that data message if the requested acknowledgement has not been received. An example of a factual situation where a provision along the lines of paragraph (4) would be particularly useful would be that the originator of an offer to contract who has not received the requested acknowledgement from the addressee of the offer may need to know the point in time after which it is free to transfer the offer to another party. It may be noted that the provision does not create any obligation binding on the originator, but merely establishes means by which the originator, if it so wishes, can clarify its status in cases where it has not received the requested acknowledgement. It may also be noted that the provision does not create any obligation binding on the addressee of the data message, who would, in most circumstances, be free to rely or not to rely on any given data message, provided that it would bear the risk of the data message being unreliable for lack of an acknowledgement of receipt. The addressee, however, is protected since the originator who does not receive a requested acknowledgement may not automatically treat the data message as though it had never been transmitted, without giving further notice to the addressee. The procedure described under paragraph (4) is purely at the discretion of the originator. For example, where the originator sent a data message which under the agreement between the parties had to be received by a certain time, and the originator requested an acknowledgement of receipt, the addressee could not deny the legal effectiveness of the message simply by withholding the requested acknowledgement.

97. The rebuttable presumption established in paragraph (5) is needed to create certainty and would be particularly useful in the context of electronic

communication between parties that are not linked by a trading-partners agreement. The second sentence of paragraph (5) should be read in conjunction with paragraph (5) of article 13, which establishes the conditions under which, in case of an inconsistency between the text of the data message as sent and the text as received, the text as received prevails.

98. Paragraph (6) corresponds to a certain type of acknowledgement, for example, an EDIFACT message establishing that the data message received is syntactically correct, i.e., that it can be processed by the receiving computer. The reference to technical requirements, which is to be construed primarily as a reference to "data syntax" in the context of EDI communications, may be less relevant in the context of the use of other means of communication, such as telegram or telex. In addition to mere consistency with the rules of "data syntax," technical requirements set forth in applicable standards may include, for example, the use of procedures verifying the integrity of the contents of data messages.

99. Paragraph (7) is intended to dispel uncertainties that might exist as to the legal effect of an acknowledgement of receipt. For example, paragraph (7) indicates that an acknowledgement of receipt should not be confused with any communication related to the contents of the acknowledged message.

References
A/51/17, paras. 63-88 (article 12);
A/CN.9/407, paras. 90-92;
A/CN.9/406, paras. 15-33; A/CN.9/WG.IV/WP.60, article 11;
A/CN.9/387, paras. 133-144; A/CN.9/WG.IV/WP.57, article 11;
A/CN.9/373, paras. 116-122; A/CN.9/WG.IV/WP.55, paras. 87-93;
A/CN.9/360, para. 125; A/CN.9/WG.IV/WP.53, paras. 80-81;
A/CN.9/350, para. 92;
A/CN.9/333, paras. 48-49.

*Article 15. Time and place of dispatch and receipt of data
messages*

100. Article 15 results from the recognition that, for the operation of many existing rules of law, it is important to ascertain the time and place of receipt of information. The use of electronic communication techniques makes those difficult to ascertain. It is not uncommon for users of electronic commerce to communicate from one State to another without knowing the location of information systems through which communication is operated. In addition, the location of certain communication systems may change without either of the parties being aware of the change. The Model Law is thus intended to reflect the fact that the location of information systems is irrelevant and sets forth a more objective criterion, namely, the place of business of the parties. In that connection, it should be noted that article 15 is not intended to establish a conflict-of-laws rule.

101. Paragraph (1) defines the time of dispatch of a data message as the time when the data message enters an information system outside the control of the originator, which may be the information system of an intermediary or an information system of the addressee. The concept of "dispatch" refers to the commencement of the electronic transmission of the data message. Where "dispatch" already has an established meaning, article 15 is intended to supplement national rules on dispatch and not to displace them. If dispatch occurs when the data message reaches an information system of the addressee, dispatch under paragraph (1) and receipt under paragraph (2) are simultaneous, except where the data message is sent to an information system of the addressee that is not the information system designated by the addressee under paragraph (2)(a).

102. Paragraph (2), the purpose of which is to define the time of receipt of a data message, addresses the situation where the addressee unilaterally designates a specific information system for the receipt of a message (in which case the designated system may or may not be an information system of the addressee), and the data message reaches an information system of the addressee that is not the designated system. In such a situation, receipt is deemed to occur when the data message is retrieved by the addressee. By "designated information system," the Model Law is intended to cover a system that has been specifically designated by a party, for instance in the case where an offer expressly specifies the address to which acceptance should be sent. The mere indication of an electronic mail or telecopy address on a letterhead or other document should not be regarded as express designation of one or more information systems.

103. Attention is drawn to the notion of "entry" into an information system, which is used for both the definition of dispatch and that of receipt of a data message. A data message enters an information system at the time when it becomes available for processing within that information system. Whether a data message which enters an information system is intelligible or usable by the addressee is outside the purview of the Model Law. The Model Law does not intend to overrule provisions of national law under which receipt of a message may occur at the time when the message enters the sphere of the addressee, irrespective of whether the message is intelligible or usable by the addressee. Nor is the Model Law intended to run counter to trade usages, under which certain encoded messages are deemed to be received even before they are usable by, or intelligible for, the addressee. It was felt that the Model Law should not create a more stringent requirement than currently exists in a paper-based environment, where a message can be considered to be received even if it is not intelligible for the addressee or not intended to be intelligible to the addressee (e.g., where encrypted data is transmitted to a depository for the sole purpose of retention in the context of intellectual property rights protection).

104. A data message should not be considered to be dispatched if it merely reached the information system of the addressee but failed to enter it. It may be noted that the Model Law does not expressly address the question of possible malfunctioning of information systems as a basis for liability. In particular, where

the information system of the addressee does not function at all or functions improperly or, while functioning properly, cannot be entered into by the data message (e.g., in the case of a telecopier that is constantly occupied), dispatch under the Model Law does not occur. It was felt during the preparation of the Model Law that the addressee should not be placed under the burdensome obligation to maintain its information system functioning at all times by way of a general provision.

105. The purpose of paragraph (4) is to deal with the place of receipt of a data message. The principal reason for including a rule on the place of receipt of a data message is to address a circumstance characteristic of electronic commerce that might not be treated adequately under existing law, namely, that very often the information system of the addressee where the data message is received, or from which the data message is retrieved, is located in a jurisdiction other than that in which the addressee itself is located. Thus, the rationale behind the provision is to ensure that the location of an information system is not the determinant element, and that there is some reasonable connection between the addressee and what is deemed to be the place of receipt, and that that place can be readily ascertained by the originator. The Model Law does not contain specific provisions as to how the designation of an information system should be made, or whether a change could be made after such a designation by the addressee.

106. Paragraph (4), which contains a reference to the "underlying transaction," is intended to refer to both actual and contemplated underlying transactions. References to "place of business," "principal place of business" and "place of habitual residence" were adopted to bring the text in line with article 10 of the United Nations Convention on Contracts for the International Sale of Goods.

107. The effect of paragraph (4) is to introduce a distinction between the deemed place of receipt and the place actually reached by a data message at the time of its receipt under paragraph (2). That distinction is not to be interpreted as apportioning risks between the originator and the addressee in case of damage or loss of a data message between the time of its receipt under paragraph (2) and the time when it reached its place of receipt under paragraph (4). Paragraph (4) merely establishes an irrebuttable presumption regarding a legal fact, to be used where another body of law (e.g., on formation of contracts or conflict of laws) require determination of the place of receipt of a data message. However, it was felt during the preparation of the Model Law that introducing a deemed place of receipt, as distinct from the place actually reached by that data message at the time of its receipt, would be inappropriate outside the context of computerized transmissions (e.g., in the context of telegram or telex). The provision was thus limited in scope to cover only computerized transmissions of data messages. A further limitation is contained in paragraph (5), which reproduces a provision already included in articles 6, 7, 8, 11 and 12 (see above, para. 69).

References
A/51/17, paras. 100-115 (article 14);
A/CN.9/407, paras. 94-99;
A/CN.9/406, paras. 42-58; A/CN.9/WG.IV/WP.60, article 13;
A/CN.9/387, paras. 152-163; A/CN.9/WG.IV/WP.57, article 13;
A/CN.9/373, paras. 134-146; A/CN.9/WG.IV/WP.55, paras. 103-108;
A/CN.9/360, paras. 87-89; A/CN.9/WG.IV/WP.53, paras. 74-76;
A/CN.9/350, paras. 97-100;
A/CN.9/333, paras. 69-75.

Part two.   Electronic commerce in specific areas

108. As distinct from the basic rules applicable to electronic commerce in general, which appear as part one of the Model Law, part two contains rules of a more specific nature. In preparing the Model Law, the Commission agreed that such rules dealing with specific uses of electronic commerce should appear in the Model Law in a way that reflected both the specific nature of the provisions and their legal status, which should be the same as that of the general provisions contained in part one of the Model Law. While the Commission, when adopting the Model Law, only considered such specific provisions in the context of transport documents, it was agreed that such provisions should appear as chapter I of part two of the Model Law. It was felt that adopting such an open-ended structure would make it easier to add further specific provisions to the Model Law, as the need might arise, in the form of additional chapters in part two.

109. The adoption of a specific set of rules dealing with specific uses of electronic commerce, such as the use of EDI messages as substitutes for transport documents does not imply that the other provisions of the Model Law are not applicable to such documents. In particular, the provisions of part two, such as articles 16 and 17 concerning transfer of rights in goods, presuppose that the guarantees of reliability and authenticity contained in articles 6 to 8 of the Model Law are also applicable to electronic equivalents to transport documents. Part two of the Model Law does not in any way limit or restrict the field of application of the general provisions of the Model Law.

Chapter I.   Carriage of goods

110. In preparing the Model Law, the Commission noted that the carriage of goods was the context in which electronic communications were most likely to be used and in which a legal framework facilitating the use of such communications was most urgently needed. Articles 16 and 17 contain provisions that apply equally to non-negotiable transport documents and to transfer of rights in goods by way of transferable bills of lading. The principles embodied in articles 16 and 17 are applicable not only to maritime transport but also to transport of goods by other means, such as road, railroad and air transport.

742

Article 16. Actions related to contracts of carriage of goods

111. Article 16, which establishes the scope of chapter I of part two of the Model Law, is broadly drafted. It would encompass a wide variety of documents used in the context of the carriage of goods, including, for example, charter-parties. In the preparation of the Model Law, the Commission found that, by dealing comprehensively with contracts of carriage of goods, article 16 was consistent with the need to cover all transport documents, whether negotiable or non-negotiable, without excluding any specific document such as charter-parties. It was pointed out that, if an enacting State did not wish chapter I of part two to apply to a particular kind of document or contract, for example if the inclusion of such documents as charter-parties in the scope of that chapter was regarded as inappropriate under the legislation of an enacting State, that State could make use of the exclusion clause contained in paragraph (7) of article 17.

112. Article 16 is of an illustrative nature and, although the actions mentioned therein are more common in maritime trade, they are not exclusive to such type of trade and could be performed in connection with air transport or multimodal carriage of goods.

References
A/51/17, paras. 139-172 and 198-204 (draft article x);
A/CN.9/421, paras. 53-103; A/CN.9/WG.IV/WP.69, paras. 82-95;
A/50/17, paras. 307-309;
A/CN.9/407, paras. 106-118; A/CN.9/WG.IV/WP.67, annex;
A/CN.9/WG.IV/WP.66, annex II;
A/49/17, paras. 198, 199 and 201;
A/CN.9/390, paras. 155-158.

Article 17. Transport documents

113. Paragraphs (1) and (2) are derived from article 6. In the context of transport documents, it is necessary to establish not only functional equivalents of written information about the actions referred to in article 16, but also functional equivalents of the performance of such actions through the use of paper documents. Functional equivalents are particularly needed for the transfer of rights and obligations by transfer of written documents. For example, paragraphs (1) and (2) are intended to replace both the requirement for a written contract of carriage and the requirements for endorsement and transfer of possession of a bill of lading. It was felt in the preparation of the Model Law that the focus of the provision on the actions referred to in article 16 should be expressed clearly, particularly in view of the difficulties that might exist, in certain countries, for recognizing the transmission of a data message as functionally equivalent to the physical transfer of goods, or to the transfer of a document of title representing the goods.

114. The reference to "one or more data messages" in paragraphs (1), (3) and (6) is not intended to be interpreted differently from the reference to "a data

message" in the other provisions of the Model Law, which should also be understood as covering equally the situation where only one data message is generated and the situation where more than one data message is generated as support of a given piece of information. A more detailed wording was adopted in article 17 merely to reflect the fact that, in the context of transfer of rights through data messages, some of the functions traditionally performed through the single transmission of a paper bill of lading would necessarily imply the transmission of more than one data message and that such a fact, in itself, should entail no negative consequence as to the acceptability of electronic commerce in that area.

115. Paragraph (3), in combination with paragraph (4), is intended to ensure that a right can be conveyed to one person only, and that it would not be possible for more than one person at any point in time to lay claim to it. The effect of the two paragraphs is to introduce a requirement which may be referred to as the "guarantee of singularity." If procedures are made available to enable a right or obligation to be conveyed by electronic methods instead of by using a paper document, it is necessary that the guarantee of singularity be one of the essential features of such procedures. Technical security devices providing such a guarantee of singularity would almost necessarily be built into any communication system offered to the trading communities and would need to demonstrate their reliability. However, there is also a need to overcome requirements of law that the guarantee of singularity be demonstrated, for example in the case where paper documents such as bills of lading are traditionally used. A provision along the lines of paragraph (3) is thus necessary to permit the use of electronic communication instead of paper documents.

116. The words "one person and no other person" should not be interpreted as excluding situations where more than one person might jointly hold title to the goods. For example, the reference to "one person" is not intended to exclude joint ownership of rights in the goods or other rights embodied in a bill of lading.

117. The notion that a data message should be "unique" may need to be further clarified, since it may lend itself to misinterpretation. On the one hand, all data messages are necessarily unique, even if they duplicate an earlier data message, since each data message is sent at a different time from any earlier data message sent to the same person. If a data message is sent to a different person, it is even more obviously unique, even though it might be transferring the same right or obligation. Yet, all but the first transfer might be fraudulent. On the other hand, if "unique" is interpreted as referring to a data message of a unique kind, or a transfer of a unique kind, then in that sense no data message is unique, and no transfer by means of a data message is unique. Having considered the risk of such misinterpretation, the Commission decided to retain the reference to the concepts of uniqueness of the data message and uniqueness of the transfer for the purposes of article 17, in view of the fact that the notions of "uniqueness" or "singularity" of transport documents were not unknown to practitioners of transport law and users of transport documents. It was decided, however, that this

Guide should clarify that the words "a reliable method is used to render such data message or messages unique" should be interpreted as referring to the use of a reliable method to secure that data messages purporting to convey any right or obligation of a person might not be used by, or on behalf of, that person inconsistently with any other data messages by which the right or obligation was conveyed by or on behalf of that person.

118. Paragraph (5) is a necessary complement to the guarantee of singularity contained in paragraph (3). The need for security is an overriding consideration and it is essential to ensure not only that a method is used that gives reasonable assurance that the same data message is not multiplied, but also that no two media can be simultaneously used for the same purpose. Paragraph (5) addresses the fundamental need to avoid the risk of duplicate transport documents. The use of multiple forms of communication for different purposes, e.g., paper-based communications for ancillary messages and electronic communications for bills of lading, does not pose a problem. However, it is essential for the operation of any system relying on electronic equivalents of bills of lading to avoid the possibility that the same rights could at any given time be embodied both in data messages and in a paper document. Paragraph (5) also envisages the situation where a party having initially agreed to engage in electronic communications has to switch to paper communications where it later becomes unable to sustain electronic communications.

119. The reference to "terminating" the use of data messages is open to interpretation. In particular, the Model Law does not provide information as to who would effect the termination. Should an enacting State decide to provide additional information in that respect, it might wish to indicate, for example, that, since electronic commerce is usually based on the agreement of the parties, a decision to "drop down" to paper communications should also be subject to the agreement of all interested parties. Otherwise, the originator would be given the power to choose unilaterally the means of communication. Alternatively, an enacting State might wish to provide that, since paragraph (5) would have to be applied by the bearer of a bill of lading, it should be up to the bearer to decide whether it preferred to exercise its rights on the basis of a paper bill of lading or on the basis of the electronic equivalent of such a document, and to bear the costs for its decision.

120. Paragraph (5), while expressly dealing with the situation where the use of data messages is replaced by the use of a paper document, is not intended to exclude the reverse situation. The switch from data messages to a paper document should not affect any right that might exist to surrender the paper document to the issuer and start again using data messages.

121. The purpose of paragraph (6) is to deal directly with the application of certain laws to contracts for the carriage of goods by sea. For example, under the Hague and Hague-Visby Rules, a contract of carriage means a contract that is covered by a bill of lading. Use of a bill of lading or similar document of title results in the Hague and Hague-Visby Rules applying compulsorily to a contract

of carriage. Those rules would not automatically apply to contracts effected by one or more data message. Thus, a provision such as paragraph (6) is needed to ensure that the application of those rules is not excluded by the mere fact that data messages are used instead of a bill of lading in paper form. While paragraph (1) ensures that data messages are effective means for carrying out any of the actions listed in article 16, that provision does not deal with the substantive rules of law that might apply to a contract contained in, or evidenced by, data messages.

122. As to the meaning of the phrase "that rule shall not be inapplicable" in paragraph (6), a simpler way of expressing the same idea might have been to provide that rules applicable to contracts of carriage evidenced by paper documents should also apply to contracts of carriage evidenced by data messages. However, given the broad scope of application of article 17, which covers not only bills of lading but also a variety of other transport documents, such a simplified provision might have had the undesirable effect of extending the applicability of rules such as the Hamburg Rules and the Hague-Visby Rules to contracts to which such rules were never intended to apply. The Commission felt that the adopted wording was more suited to overcome the obstacle resulting from the fact that the Hague-Visby Rules and other rules compulsorily applicable to bills of lading would not automatically apply to contracts of carriage evidenced by data messages, without inadvertently extending the application of such rules to other types of contracts.

References
A/51/17, paras. 139-172 and 198-204 (draft article x);
A/CN.9/421, paras. 53-103; A/CN.9/WG.IV/WP.69, paras 82-95;
A/50/17, paras. 307-309
A/CN.9/407, paras. 106-118 A/CN.9/WG.IV/WP.67, annex;
A/CN.9/WG.IV/WP.66, annex II;
A/49/17, paras. 198, 199 and 201;
A/CN.9/390, paras. 155-158.

III. HISTORY AND BACKGROUND OF THE MODEL LAW

123. The UNCITRAL Model Law on Electronic Commerce was adopted by the United Nations Commission on International Trade Law (UNCITRAL) in 1996 in furtherance of its mandate to promote the harmonization and unification of international trade law, so as to remove unnecessary obstacles to international trade caused by inadequacies and divergences in the law affecting trade. Over the past quarter of a century, UNCITRAL, whose membership consists of States from all regions and of all levels of economic development, has implemented its mandate by formulating international conventions (the United Nations Conventions on Contracts for the International Sale of Goods, on the Limitation Period in the International Sale of Goods, on the Carriage of Goods by Sea, 1978 ("Hamburg Rules"), on the Liability of Operators of Transport Terminals in

International Trade, on International Bills of Exchange and International Promissory Notes, and on Independent Guarantees and Stand-by Letters of Credit), model laws (the UNCITRAL Model Laws on International Commercial Arbitration, on International Credit Transfers and on Procurement of Goods, Construction and Services), the UNCITRAL Arbitration Rules, the UNCITRAL Conciliation Rules, and legal guides (on construction contracts, countertrade transactions and electronic funds transfers).

124. The Model Law was prepared in response to a major change in the means by which communications are made between parties using computerized or other modern techniques in doing business (sometimes referred to as "trading partners"). The Model Law is intended to serve as a model to countries for the evaluation and modernization of certain aspects of their laws and practices in the field of commercial relationships involving the use of computerized or other modern communication techniques, and for the establishment of relevant legislation where none presently exists. The text of the Model Law, as reproduced above, is set forth in annex I to the report of UNCITRAL on the work of its twenty-ninth session.[3]

125. The Commission, at its seventeenth session (1984), considered a report of the Secretary-General entitled "Legal aspects of automatic data processing" (A/CN.9/254), which identified several legal issues relating to the legal value of computer records, the requirement of a "writing," authentication, general conditions, liability and bills of lading. The Commission took note of a report of the Working Party on Facilitation of International Trade Procedures (WP.4), which is jointly sponsored by the Economic Commission for Europe and the United Nations Conference on Trade and Development, and is responsible for the development of UN/EDIFACT standard messages. That report suggested that, since the legal problems arising in this field were essentially those of international trade law, the Commission as the core legal body in the field of international trade law appeared to be the appropriate central forum to undertake and coordinate the necessary action.[4] The Commission decided to place the subject of the legal implications of automatic data processing to the flow of international trade on its programme of work as a priority item.[5]

126. At its eighteenth session (1985), the Commission had before it a report by the Secretariat entitled "Legal value of computer records" (A/CN.9/265). That report came to the conclusion that, on a global level, there were fewer problems in the use of data stored in computers as evidence in litigation than might have been expected. It noted that a more serious legal obstacle to the use of

---

[3] *Official Records of the General Assembly, Fifty-first Session, Supplement No. 17* (A/51/17), Annex I.

[4] "Legal aspects of automatic trade data interchange" (TRADE/WP.4/R.185/Rev.1). The report submitted to the Working Party is reproduced in A/CN.9/238, annex.

[5] *Official Records of the General Assembly, Thirty-ninth Session, Supplement No. 17* (A/39/17), para. 136.

computers and computer-to-computer telecommunications in international trade arose out of requirements that documents had to be signed or be in paper form. After discussion of the report, the Commission adopted the following recommendation, which expresses some of the principles on which the Model Law is based:

"The United Nations Commission on International Trade Law,

"*Noting* that the use of automatic data processing (ADP) is about to become firmly established throughout the world in many phases of domestic and international trade as well as in administrative services,

"*Noting* also that legal rules based upon pre-ADP paper-based means of documenting international trade may create an obstacle to such use of ADP in that they lead to legal insecurity or impede the efficient use of ADP where its use is otherwise justified,

"*Noting* further with appreciation the efforts of the Council of Europe, the Customs Co-operation Council and the United Nations Economic Commission for Europe to overcome obstacles to the use of ADP in international trade arising out of these legal rules,

"*Considering* at the same time that there is no need for a unification of the rules of evidence regarding the use of computer records in international trade, in view of the experience showing that substantial differences in the rules of evidence as they apply to the paper-based system of documentation have caused so far no noticeable harm to the development of international trade,

"*Considering* also that the developments in the use of ADP are creating a desirability in a number of legal systems for an adaptation of existing legal rules to these developments, having due regard, however, to the need to encourage the employment of such ADP means that would provide the same or greater reliability as paper-based documentation,

"1. *Recommends* to Governments:

"*(a)* to review the legal rules affecting the use of computer records as evidence in litigation in order to eliminate unnecessary obstacles to their admission, to be assured that the rules are consistent with developments in technology, and to provide appropriate means for a court to evaluate the credibility of the data contained in those records;

"*(b)* to review legal requirements that certain trade transactions or trade related documents be in writing, whether the written form is a condition to the enforceability or to the validity of the transaction or document, with a view to permitting, where appropriate, the transaction or document to be recorded and transmitted in computer-readable form;

"*(c)* to review legal requirements of a handwritten signature or other paper-based method of authentication on trade related documents with a view to permitting, where appropriate, the use of electronic means of authentication;

"*(d)* to review legal requirements that documents for submission to governments be in writing and manually signed with a view to permitting, where appropriate, such documents to be submitted in computer-readable form to those administrative services which have acquired the necessary equipment and established the necessary procedures;

"2. *Recommends* to international organizations elaborating legal texts related to trade to take account of the present Recommendation in adopting such texts and, where appropriate, to consider modifying existing legal texts in line with the present Recommendation."[6]

127. That recommendation (hereinafter referred to as the "1985 UNCI-TRAL Recommendation") was endorsed by the General Assembly in resolution 40/71, paragraph 5*(b)*, of 11 December 1985 as follows:

"The General Assembly,

". . . Calls upon Governments and international organizations to take action, where appropriate, in conformity with the Commission's recommendation so as to ensure legal security in the context of the widest possible use of automated data processing in international trade; . . .".[7]

128. As was pointed out in several documents and meetings involving the international electronic commerce community, e.g. in meetings of WP. 4, there was a general feeling that, in spite of the efforts made through the 1985 UNCITRAL Recommendation, little progress had been made to achieve the removal of the mandatory requirements in national legislation regarding the use of paper and handwritten signatures. It has been suggested by the Norwegian Committee on Trade Procedures (NORPRO) in a letter to the Secretariat that "one reason for this could be that the 1985 UNCITRAL Recommendation advises on the need for legal update, but does not give any indication of how it could be done." In this vein, the Commission considered what follow-up action to the 1985 UNCITRAL Recommendation could usefully be taken so as to enhance the needed modernization of legislation. The decision by UNCITRAL to formulate model legislation on legal issues of electronic data interchange and related means of communication may be regarded as a consequence of the process that led to the adoption by the Commission of the 1985 UNCITRAL Recommendation.

129. At its twenty-first session (1988), the Commission considered a proposal to examine the need to provide for the legal principles that would apply

---

[6] *Official Records of the General Assembly, Fortieth Session, Supplement No. 17* (A/40/17), para. 360.

[7] Resolution 40/71 was reproduced in United Nations Commission on International Trade Law Yearbook, 1985, vol. XVI, Part One, D. (United Nations publication, Sales No. E.87.V.4).

to the formation of international commercial contracts by electronic means. It was noted that there existed no refined legal structure for the important and rapidly growing field of formation of contracts by electronic means and that future work in that area could help to fill a legal vacuum and to reduce uncertainties and difficulties encountered in practice. The Commission requested the Secretariat to prepare a preliminary study on the topic.[8]

130. At its twenty-third session (1990), the Commission had before it a report entitled "Preliminary study of legal issues related to the formation of contracts by electronic means" (A/CN.9/333). The report summarized work that had been undertaken in the European Communities and in the United States of America on the requirement of a "writing" as well as other issues that had been identified as arising in the formation of contracts by electronic means. The efforts to overcome some of those problems by the use of model communication agreements were also discussed.[9]

131. At its twenty-fourth session (1991), the Commission had before it a report entitled "Electronic Data Interchange" (A/CN.9/350). The report described the current activities in the various organizations involved in the legal issues of electronic data interchange (EDI) and analyzed the contents of a number of standard interchange agreements already developed or then being developed. It pointed out that such documents varied considerably according to the various needs of the different categories of users they were intended to serve and that the variety of contractual arrangements had sometimes been described as hindering the development of a satisfactory legal framework for the business use of electronic commerce. It suggested that there was a need for a general framework that would identify the issues and provide a set of legal principles and basic legal rules governing communication through electronic commerce. It concluded that such a basic framework could, to a certain extent, be created by contractual arrangements between parties to an electronic commerce relationship and that the existing contractual frameworks that were proposed to the community of users of electronic commerce were often incomplete, mutually incompatible, and inappropriate for international use since they relied to a large extent upon the structures of local law.

132. With a view to achieving the harmonization of basic rules for the promotion of electronic commerce in international trade, the report suggested that the Commission might wish to consider the desirability of preparing a standard communication agreement for use in international trade. It pointed out that work by the Commission in this field would be of particular importance since it would involve participation of all legal systems, including those of developing countries that were already or would soon be confronted with the issues of electronic commerce.

---

[8] *Official Records of the General Assembly, Forty-third Session, Supplement No. 17* (A/43/17), paras. 46 and 47, and *ibid., Forty-fourth Session, Supplement No. 17* (A/44/17), para. 289.

[9] *Ibid., Forty-fifth Session, Supplement No. 17* (A/45/17), paras. 38 to 40.

133. The Commission was agreed that the legal issues of electronic commerce would become increasingly important as the use of electronic commerce developed and that it should undertake work in that field. There was wide support for the suggestion that the Commission should undertake the preparation of a set of legal principles and basic legal rules governing communication through electronic commerce.[10] The Commission came to the conclusion that it would be premature to engage immediately in the preparation of a standard communication agreement and that it might be preferable to monitor developments in other organizations, particularly the Commission of the European Communities and the Economic Commission for Europe. It was pointed out that high-speed electronic commerce required a new examination of basic contract issues such as offer and acceptance, and that consideration should be given to legal implications of the role of central data managers in international commercial law.

134. After deliberation, the Commission decided that a session of the Working Group on International Payments would be devoted to identifying the legal issues involved and to considering possible statutory provisions, and that the Working Group would report to the Commission on the desirability and feasibility of undertaking further work such as the preparation of a standard communication agreement.[11]

135. The Working Group on International Payments, at its twenty-fourth session, recommended that the Commission should undertake work towards establishing uniform legal rules on electronic commerce. It was agreed that the goals of such work should be to facilitate the increased use of electronic commerce and to meet the need for statutory provisions to be developed in the field of electronic commerce, particularly with respect to such issues as formation of contracts; risk and liability of commercial partners and third-party service providers involved in electronic commerce relationships; extended definitions of "writing" and "original" to be used in an electronic commerce environment; and issues of negotiability and documents of title (A/CN.9/360).

136. While it was generally felt that it was desirable to seek the high degree of legal certainty and harmonization provided by the detailed provisions of a uniform law, it was also felt that care should be taken to preserve a flexible approach to some issues where legislative action might be premature or inappropriate. As an example of such an issue, it was stated that it might be fruitless to attempt to provide legislative unification of the rules on evidence that may apply to electronic commerce massaging (*ibid.*, para. 130). It was agreed

---

[10] It may be noted that the Model Law is not intended to provide a comprehensive set of rules governing all aspects of electronic commerce. The main purpose of the Model Law is to adapt existing statutory requirements so that they would no longer constitute obstacles to the use of paperless means of communication and storage of information.

[11] *Official Records of the General Assembly, Forty-sixth Session, Supplement No. 17* (A/46/17), paras. 311 to 317.

that no decision should be taken at that early stage as to the final form or the final content of the legal rules to be prepared. In line with the flexible approach to be taken, it was noted that situations might arise where the preparation of model contractual clauses would be regarded as an appropriate way of addressing specific issues (*ibid.*, para. 132).

137. The Commission, at its twenty-fifth session (1992), endorsed the recommendation contained in the report of the Working Group (*ibid.*, paras. 129-133) and entrusted the preparation of legal rules on electronic commerce (which was then referred to as "electronic data interchange" or "EDI") to the Working Group on International Payments, which it renamed the Working Group on Electronic Data Interchange.[12]

138. The Working Group devoted its twenty-fifth to twenty-eighth sessions to the preparation of legal rules applicable to "electronic data interchange (EDI) and other modern means of communication" (reports of those sessions are found in documents A/CN.9/373, 387, 390 and 406).[13]

139. The Working Group carried out its task on the basis of background working papers prepared by the Secretariat on possible issues to be included in the Model Law. Those background papers included A/CN.9/WG.IV/WP.53 (possible issues to be included in the programme of future work on the legal aspects of EDI) and A/CN.9/WG.IV/WP.55 (outline of possible uniform rules on the legal aspects of electronic data interchange). The draft articles of the Model Law were submitted by the Secretariat in documents A/CN.9/WG.IV/WP.57, 60 and 62. The Working Group also had before it a proposal by the United Kingdom of Great Britain and Northern Ireland relating to the possible contents of the draft Model Law (A/CN.9/WG.IV/WP.58).

140. The Working Group noted that, while practical solutions to the legal difficulties raised by the use of electronic commerce were often sought within contracts (A/CN.9/WG.IV/WP.53, paras. 35-36), the contractual approach to electronic commerce was developed not only because of its intrinsic advantages such as its flexibility, but also for lack of specific provisions of statutory or case law. The contractual approach was found to be limited in that it could not overcome any of the legal obstacles to the use of electronic commerce that might result from mandatory provisions of applicable statutory or case law. In that

---

[12] *Ibid., Forty-seventh Session, Supplement No. 17* (A/47/17), paras. 141 to 148.

[13] The notion of "EDI and related means of communication" as used by the Working Group is not to be construed as a reference to narrowly defined EDI under article 2(b) of the Model Law but to a variety of trade-related uses of modern communication techniques that was later referred to broadly under the rubric of "electronic commerce." The Model Law is not intended only for application in the context of existing communication techniques but rather as a set of flexible rules that should accommodate foreseeable technical developments. It should also be emphasized that the purpose of the Model Law is not only to establish rules for the movement of information communicated by means of data messages but equally to deal with the storage of information in data messages that are not intended for communication.

respect, one difficulty inherent in the use of communication agreements resulted from uncertainty as to the weight that would be carried by some contractual stipulations in case of litigation. Another limitation to the contractual approach resulted from the fact that parties to a contract could not effectively regulate the rights and obligations of third parties. At least for those parties not participating in the contractual arrangement, statutory law based on a model law or an international convention seemed to be needed (see A/CN.9/350, para. 107).

141. The Working Group considered preparing uniform rules with the aim of eliminating the legal obstacles to, and uncertainties in, the use of modern communication techniques, where effective removal of such obstacles and uncertainties could only be achieved by statutory provisions. One purpose of the uniform rules was to enable potential electronic commerce users to establish a legally secure electronic commerce relationship by way of a communication agreement within a closed network. The second purpose of the uniform rules was to support the use of electronic commerce outside such a closed network, i.e., in an open environment. However, the aim of the uniform rules was to enable, and not to impose, the use of EDI and related means of communication. Moreover, the aim of the uniform rules was not to deal with electronic commerce relationships from a technical perspective but rather to create a legal environment that would be as secure as possible, so as to facilitate the use of electronic commerce between communicating parties.

142. As to the form of the uniform rules, the Working Group was agreed that it should proceed with its work on the assumption that the uniform rules should be prepared in the form of statutory provisions. While it was agreed that the form of the text should be that of a "model law," it was felt, at first, that, owing to the special nature of the legal text being prepared, a more flexible term than "model law" needed to be found. It was observed that the title should reflect that the text contained a variety of provisions relating to existing rules scattered throughout various parts of the national laws in an enacting State. It was thus a possibility that enacting States would not incorporate the text as a whole and that the provisions of such a "model law" might not appear together in any one particular place in the national law. The text could be described, in the parlance of one legal system, as a "miscellaneous statute amendment act." The Working Group agreed that this special nature of the text would be better reflected by the use of the term "model statutory provisions." The view was also expressed that the nature and purpose of the "model statutory provisions" could be explained in an introduction or guidelines accompanying the text.

143. At its twenty-eighth session, however, the Working Group reviewed its earlier decision to formulate a legal text in the form of "model statutory provisions" (A/CN.9/390, para. 16). It was widely felt that the use of the term "model statutory provisions" might raise uncertainties as to the legal nature of the instrument. While some support was expressed for the retention of the term "model statutory provisions," the widely prevailing view was that the term "model law" should be preferred. It was widely felt that, as a result of the course

taken by the Working Group as its work progressed towards the completion of the text, the model statutory provisions could be regarded as a balanced and discrete set of rules, which could also be implemented as a whole in a single instrument (A/CN.9/406, para. 75). Depending on the situation in each enacting State, however, the Model Law could be implemented in various ways, either as a single statute or in various pieces of legislation.

144. The text of the draft Model Law as approved by the Working Group at its twenty-eighth session was sent to all Governments and to interested international organizations for comment. The comments received were reproduced in document A/CN.9/409 and Add.1-4. The text of the draft articles of the Model Law as presented to the Commission by the Working Group was contained in the annex to document A/CN.9/406.

145. At its twenty-eighth session (1995), the Commission adopted the text of articles 1 and 3 to 11 of the draft Model Law and, for lack of sufficient time, did not complete its review of the draft Model Law, which was placed on the agenda of the twenty-ninth session of the Commission.[14]

146. The Commission, at its twenty-eighth session,[15] recalled that, at its twenty-seventh session (1994), general support had been expressed in favour of a recommendation made by the Working Group that preliminary work should be undertaken on the issue of negotiability and transferability of rights in goods in a computer-based environment as soon as the preparation of the Model Law had been completed.[16] It was noted that, on that basis, a preliminary debate with respect to future work to be undertaken in the field of electronic data interchange had been held in the context of the twenty-ninth session of the Working Group (for the report on that debate, see A/CN.9/407, paras. 106-118). At that session, the Working Group also considered proposals by the International Chamber of Commerce (A/CN.9/WG.IV/WP.65) and the United Kingdom of Great Britain and Northern Ireland (A/CN.9/WG.IV/WP.66) relating to the possible inclusion in the draft Model Law of additional provisions to the effect of ensuring that certain terms and conditions that might be incorporated in a data message by means of a mere reference would be recognized as having the same degree of legal effectiveness as if they had been fully stated in the text of the data message (for the report on the discussion, see A/CN.9/407, paras. 100-105). It was agreed that the issue of incorporation by reference might need to be considered in the context of future work on negotiability and transferability of rights in goods (A/CN.9/407, para. 103). The Commission endorsed the recommendation made by the Working Group that the Secretariat should be entrusted with the preparation of a background study on negotiability and transferability of EDI transport documents, with particular emphasis on EDI maritime transport

---

[14] *Official Records of the General Assembly, Fiftieth Session, Supplement No. 17* (A/50/17), para. 306.

[15] *Ibid.*, para. 307.

[16] *Ibid., Forty-ninth Session, Supplement No. 17* (A/49/17), para. 201.

documents, taking into account the views expressed and the suggestions made at the twenty-ninth session of the Working Group.[17]

147. On the basis of the study prepared by the Secretariat (A/CN.9/WG.IV/WP.69), the Working Group, at its thirtieth session, discussed the issues of transferability of rights in the context of transport documents and approved the text of draft statutory provisions dealing with the specific issues of contracts of carriage of goods involving the use of data messages (for the report on that session, see A/CN.9/421). The text of those draft provisions as presented to the Commission by the Working Group for final review and possible addition as part II of the Model Law was contained in the annex to document A/CN.9/421.

148. In preparing the Model Law, the Working Group noted that it would be useful to provide in a commentary additional information concerning the Model Law. In particular, at the twenty-eighth session of the Working Group, during which the text of the draft Model Law was finalized for submission to the Commission, there was general support for a suggestion that the draft Model Law should be accompanied by a guide to assist States in enacting and applying the draft Model Law. The guide, much of which could be drawn from the *travaux préparatoires* of the draft Model Law, would also be helpful to users of electronic means of communication as well as to scholars in that area. The Working Group noted that, during its deliberations at that session, it had proceeded on the assumption that the draft Model Law would be accompanied by a guide. For example, the Working Group had decided in respect of a number of issues not to settle them in the draft Model Law but to address them in the guide so as to provide guidance to States enacting the draft Model Law. The Secretariat was requested to prepare a draft and submit it to the Working Group for consideration at its twenty-ninth session (A/CN.9/406, para. 177).

149. At its twenty-ninth session, the Working Group discussed the draft Guide to Enactment of the Model Law (hereinafter referred to as "the draft Guide") as set forth in a note prepared by the Secretariat (A/CN.9/WG.IV/WP.64). The Secretariat was requested to prepare a revised version of the draft Guide reflecting the decisions made by the Working Group and taking into account the various views, suggestions and concerns that had been expressed at that session. At its twenty-eighth session, the Commission placed the draft Guide to Enactment of the Model Law on the agenda of its twenty-ninth session.[18]

150. At its twenty-ninth session (1996), the Commission, after consideration of the text of the draft Model Law as revised by the drafting group, adopted the following decision at its 605th meeting, on 12 June 1996:

---

[17] *Ibid., Fiftieth Session, Supplement No. 17* (A/50/17), para. 309.
[18] *Ibid.*, para. 306.

"The United Nations Commission on International Trade Law,

"Recalling its mandate under General Assembly resolution 2205 (XXI) of 17 December 1966 to further the progressive harmonization and unification of the law of international trade, and in that respect to bear in mind the interests of all peoples, and in particular those of developing countries, in the extensive development of international trade,

"Noting that an increasing number of transactions in international trade are carried out by means of electronic data interchange and other means of communication commonly referred to as 'electronic commerce,' which involve the use of alternatives to paper-based forms of communication and storage of information,

"Recalling the recommendation on the legal value of computer records adopted by the Commission at its eighteenth session, in 1985, and paragraph 5*(b)* of General Assembly resolution 40/71 of 11 December 1985 calling upon Governments and international organizations to take action, where appropriate, in conformity with the recommendation of the Commission[19] so as to ensure legal security in the context of the widest possible use of automated data processing in international trade,

"Being of the opinion that the establishment of a model law facilitating the use of electronic commerce, and acceptable to States with different legal, social and economic systems, contributes to the development of harmonious international economic relations,

"Being convinced that the UNCITRAL Model Law on Electronic Commerce will significantly assist all States in enhancing their legislation governing the use of alternatives to paper-based forms of communication and storage of information, and in formulating such legislation where none currently exists,

> "1. *Adopts* the UNCITRAL Model Law on Electronic Commerce as it appears in annex I to the report on the current session;
> "2. *Requests* the Secretary-General to transmit the text of the UNCITRAL Model Law on Electronic Commerce, together with the Guide to Enactment of the Model Law prepared by the Secretariat, to Governments and other interested bodies;
> "3. *Recommends* that all States give favourable consideration to the UNCITRAL Model Law on Electronic Commerce when they enact or revise their laws, in view of the need for uniformity of the law applicable to alternatives to paper-based forms of communication and storage of information."[20]

---

[19] *Ibid., Fortieth Session, Supplement No. 17* (A/40/17), paras. 354-360.
[20] *Ibid., Fifty-first Session, Supplement No. 17* (A/51/17), para. 209.

PROPOSED BILL—
BUSINESS METHOD PATENT
IMPROVEMENT ACT OF 2000

106TH CONGRESS
2D SESSION

H. R. 5364 IH

IN THE HOUSE OF REPRESENTATIVES

Mr. BERMAN (for himself and Mr. BOUCHER) introduced the following bill; which was referred to the Committee on _____

A BILL

To amend title 35, United States Code, to provide for improvements in the quality of patents on certain inventions.

Be it enacted by the Senate and House of Representatives of the United States of America in Congress assembled,

SECTION 1.   SHORT TITLE.

This Act may be cited as the Business Method Patent Improvement Act of 2000.

SEC. 2.   DEFINITIONS.

Section 100 of title 35, United States Code, is amended by adding at the end the following:

(f)   The term business method means

(1)   a method of

(A) administering, managing, or otherwise operating an enterprise or organization, including a technique used in doing or conducting business; or

(B) processing financial data;

(2)   any technique used in athletics, instruction, or personal skills; and

(3)   any computer-assisted implementation of a method described in paragraph (1) or a technique described in paragraph (2).

(g)   The term business method invention means

(1)   any invention which is a business method (including any software or other apparatus); and

(2)   any invention which is comprised of any claim that is a business method.

SEC. 3. PATENTS ON BUSINESS METHOD INVENTIONS.

(a) IN GENERAL. Title 35, United States Code, is amended by inserting after chapter 31 the following new chapter:

CHAPTER 32 PATENTS ON BUSINESS METHOD INVENTIONS

Sec.
321. Business method invention determinations.
322. Opposition procedures.
323. Effect on other proceedings.
324. Burden of proof.

321. Business method invention determinations

(a) CONFIDENTIALITY. Except as provided in subsection (b), an application for a patent on a business method invention shall be kept in confidence by the Patent and Trademark Office and no information concerning the application may be given without authority of the applicant or owner unless necessary to carry out the provisions of an Act of Congress or in such special circumstances as may be determined by the Director.

(b) PUBLICATION.

(1) IN GENERAL. (A) Subject to subparagraph (E) and paragraph (2), each application for a patent on a business method invention shall be published, in accordance with procedures determined by the Director, promptly after the expiration of a period of 18 months after the earliest filing date for which a benefit is sought under this title. At the request of the applicant, an application may be published earlier than the end of that 18-month period.

(B) Within 12 months after the earliest filing date of an application for a patent under this title, the Director shall make a determination of whether any invention claimed in the application is a business method invention.

(C) After making a determination under subparagraph (B) that an invention is a business method invention, the Director shall notify the applicant of the determination and shall provide the applicant with a period of 60 days within which to respond to the determination by amending the application, withdrawing the application, or otherwise.

(D) No information concerning patent applications published under this subsection shall be made available to the public, except as the Director determines.

(E)(i) The Director shall establish procedures for making determinations under subparagraph (B), and for addressing amendments to any application that may affect the Directors determination of

759

whether the invention claimed in the application is a business method invention.

(ii) In no case shall an application that would be subject to section 122 but for this section be published later than the date that would otherwise apply to the application under section 122.

(2) EXCEPTIONS. (A) An application shall not be published under paragraph (1) if that application is

(i) no longer pending;

(ii) subject to a secrecy order under section 181 of this title;

(iii) a provisional application filed under section 111(b) of this title; or

(iv) an application for a design patent filed under chapter 16 of this title.

(B) No application for a patent shall be published under paragraph (1) if the publication or disclosure of such invention would be detrimental to the national security. The Director shall establish appropriate procedures to ensure that such applications are promptly identified and the secrecy of such inventions is maintained in accordance with chapter 17 of this title.

(3) PUBLIC PARTICIPATION. Any party shall have the opportunity to submit to the Director for the record prior art (including, but not limited to, evidence of knowledge or use, or public use or sale, under section 102), file a protest, or petition the Director to conduct a proceeding to determine whether the invention was known or used, or was in public use, or on sale, under section 102 or is obvious under section 103. The Director shall conduct such a proceeding if the petition

(i) is in writing;

(ii) is accompanied by payment of the fee set forth in section 41(a) of this title; and

(iii) sets forth in detail the basis on which the proceeding is requested.

(4) AVAILABILITY OF INFORMATION. Information submitted pursuant to paragraph (3) shall be considered during the examination of the patent application.

(5) PROVISIONAL RIGHTS. During the period of pendency of an application after publication, an applicant shall have provisional rights pursuant to section 154 of this title.

322. Opposition procedures

(a) ADMINISTRATIVE OPPOSITION PANEL.

(1) ESTABLISHMENT. The Director shall, not later than 1 year after the date of enactment of the Business Method Patent Improvement Act of 2000, establish an Administrative Opposition Panel. The Administrative Opposition Panel shall be comprised of not less than 18 administrative opposition judges, each of whom shall be an individual of competent legal knowledge and scientific ability. Upon establishment of the Administrative Opposition Panel, the Director shall publish notice of the establishment of the Panel in the Federal Register.

(2) ASSIGNMENT OF PATENT EXAMINERS TO PANEL. Patent examiners may be assigned on detail to assist the Administrative Opposition Panel in carrying out opposition proceedings under this section, except that a patent examiner may not be assigned to assist in review of a patent application examined by that patent examiner. The Director shall establish procedures by which an opposition is heard under subsection (b).

(b) OPPOSITION PROCEDURES.

(1) REQUEST FOR OPPOSITION. (A) Any person may file a request for an opposition to a patent on a business method invention on the basis of section 101, 102, 103, or 112 of this title. Such a request is valid only if the request

>(i) is made not later than 9 months after the date of issuance of the patent;
>(ii) is in writing;
>(iii) is accompanied by payment of the opposition fee set forth in section 41(a) of this title; and
>(iv) sets forth in detail the basis on which the opposition is requested.

(B) Not later than 60 days after receiving a valid request under subparagraph (A), the Director shall issue an order for an opposition proceeding to be held on the record after opportunity for a hearing, and shall promptly send a copy of the request to the owner of record of the patent. The patent owner shall be provided a reasonable period, but in no case less than 60 days after the date on which a copy of the request is given or mailed to the patent owner, within which the owner may file a statement in reply to the grounds for the request for opposition, including any amendment to the patent and new claim or claims, for consideration in the opposition proceeding. If the patent owner files such a statement, the patent owner promptly shall serve a copy of the statement on the third-party requester. Not later than 2 months after the date of such service, the third-party requester may file and have considered in the opposition proceeding a reply to the statement filed by the patent owner.

(2) CONDUCT OF OPPOSITION PROCEEDINGS. Each opposition shall be heard by one administrative opposition judge, and no party shall be permitted *ex parte* communication with the administrative opposition judge. In addition to the statements and replies set forth in paragraph (1), the administrative opposition judge may consider evidence that the judge considers relevant, including evidence that is presented in any oral testimony (including exhibits and expert testimony) in direct or cross examination, or in any deposition, affidavit, or other documentary form, whether voluntary or compelled. In any opposition proceeding, the Federal Rules of Evidence shall apply.

(3) AMENDMENTS TO PATENT CLAIMS. A patent applicant may propose to amend a patent claim or propose a new claim at any time during the opposition proceeding, except that no proposed amended or new claim enlarging the scope of a claim of the patent may be permitted at any time during an opposition proceeding under this section.

(4) DETERMINATION. Not later than 18 months after the filing of a request for an opposition under this section, the administrative opposition judge in the opposition proceeding shall determine the patentability of the subject matter of the patent, a record of the administrative opposition judges determination under this section shall be placed in the official file of the patent, and a copy promptly shall be given or mailed to the owner of record of the patent and to the third-party requester.

(5) APPEALS. Any party to the opposition may appeal a decision of the Administrative Opposition Panel under the provisions of section 134 of this title, and may seek court review under the provisions of section 141 to 145 of this title, with respect to any decision in regard to the patentability of any original or proposed amended or new claim of the patent. A patent owner may be a party to an appeal taken by a third-party requester. Any third-party requester may be a party to an appeal taken by a patent owner.

(6) CERTIFICATION OF PATENTABILITY. In an opposition proceeding under this chapter, when the time for appeal has expired or any appeal proceeding has terminated, the Director shall issue and publish a certificate canceling any claim of the patent finally determined to be unpatentable, confirming any claim of the patent determined to be patentable, and incorporating in the patent any proposed amended or new claim determined to be patentable.

(7) EFFECT OF DETERMINATION. Any proposed, amended, or new claim determined to be patentable and incorporated into a patent following an opposition proceeding shall have the same effect as that specified in section 252 of this title for reissued patents on the right of any person who made, purchased, or used within the United States, or imported into the United States, anything patented by such proposed amended or new claim, or who made substantial preparations therefor, prior to issuance of a certificate under paragraph (6) of this subsection.

323. Effect on other proceedings

(a) RIGHT TO LITIGATION. Subject to subsections (b), (c), and (d), proceedings under section 322 shall not alter or prejudice any partys right to pursue remedies under provisions of law other than this section. In the case of court proceedings, other than an appeal of a decision in an opposition proceeding under this section, the court may consider any matter independently of any opposition proceeding under this section.

(b) EFFECT OF FINAL DECISION.

(1) IN GENERAL. If a final decision has been entered against a party in a civil action arising in whole or in part under section 1338 of title 28, establishing that the party has not sustained its burden of proving the invalidity of any patent claim, or if a final decision in an *inter partes* reexamination proceeding or an opposition proceeding instituted by a third-party requester is favorable to the patentability of any original or proposed amended or new claim of the patent

(A) neither that party to the civil action, the third-party requester, nor the privies of that party or third-party requester may thereafter bring a civil action under section 1338 of title 28, or request an *inter partes* reexamination of, or an opposition to, such patent claim on the basis of issues which that party, third-party requester, or the privies of that party or third-party requester raised or could have raised in such civil action, *inter partes* reexamination proceeding, or opposition proceeding (as the case may be); and

(B) an *inter partes* reexamination or opposition requested by that party, third-party requester, or the privies of that party or third-party party requester on the basis of such issues may not thereafter be maintained by the Office.

(2) NEW EVIDENCE. Paragraph (1) does not prevent the assertion by a party to a civil action or a third-party requester of invalidity based on newly discovered prior art, or other evidence, unavailable to that party or third-party requester, as the case may be, and the Patent and Trademark Office, at the time of the civil action, *inter partes* reexamination, or opposition proceeding (as the case may be).

(c) STAY OF LITIGATION. Once an order for an opposition proceeding with respect to a patent has been issued under section 322(b)(1), any party to the proceeding may obtain a stay of any pending court proceeding (other than an appeal to the Court of Appeals for the Federal Circuit) which involves an issue of patentability of any claims of the patent which are the subject of the opposition proceeding, unless the court before which such litigation is pending determines that a stay would not serve the interests of justice.

324. Burden of proof

(a) BURDEN OF PROOF. In the case of reexamination, interference, opposition, or other legal challenge (including a civil action brought in whole or in part under section 1338 of title 28) to a patent (or an application for a patent) on a business method invention, the party producing evidence of invalidity or ineligibility shall have the burden of showing by a preponderance of the evidence the invalidity of the patent or ineligibility of the subject matter of the application.

(b) FEES. Section 41(a) of title 35, United States Code, is amended

(1) by redesignating paragraphs (7) through (15) as paragraphs (9) through (17), respectively; and

(2) by inserting after paragraph (6) the following:

(7)(A) On filing an opposition under chapter 32 to a patent on a business method invention based on prior art citations or obviousness, a fee of $200.

(B) On filing an opposition under chapter 32 to a patent on a business method invention on any other basis, a fee of $5,000.

(C) The Director may waive the payment by an individual of fees under this paragraph if such waiver is in the public interest.

(8) On filing a request for a proceeding to determine whether an invention claimed in an application was known or used, or has been in public use or on sale, under section 102, a fee of $35.

(b) CLERICAL AMENDMENT. The table of chapters for part III of title 35, United States Code, is amended by adding at the end the following:

SEC. 4.   NONOBVIOUSNESS.

Section 103 of title 35, United States Code, is amended by adding at the end the following:

(d)(1) If

(A) Subject matter within the scope of a claim addressed to a business method invention would be obtained by combining or modifying one or more prior art references, and

(B) Any of those prior art references discloses a business method which differs from what is claimed only in that the claim requires a computer technology to implement the practice of the business method invention, the invention shall be presumed obvious to a person of ordinary skill in the art at the time of the invention.

(2)(A) An applicant or patentee may rebut the presumption under paragraph (1) upon a showing by a preponderance of the evidence that the invention is not obvious to persons of ordinary skill in all relevant arts.

(B) Those areas of art which are relevant for purposes of subparagraph (A) include the field of the business method and the field of the computer implementation.

SEC. 5.   REQUIREMENT TO DISCLOSE SEARCH.

The Director of the Patent and Trademark Office shall, within 30 days after the date of enactment of this Act, publish notice of rulemaking proceedings to amend the rules of the Patent and Trademark Office to require an applicant for a patent for a business method invention to disclose in the application the extent to which the applicant searched for prior art to meet the requirements of title 35, United States Code. Such amendment shall include appropriate penalties for failure to comply with such requirement. The Director shall ensure that the amendment is implemented as promptly as possible.

SEC. 6.   CONFORMING AMENDMENTS.

(a) DEFINITIONS. Section 100(e) of title 35, United States Code, is amended by striking or *inter partes* reexamination under section 311 and inserting, *inter partes* reexamination under section 311, or an opposition under section 322.

(b) BOARD OF PATENT APPEALS AND INTERFERENCES. Section 134 of title 35, United States Code, is amended

(1) in subsection (b)

(A) by inserting "or opposition after reexamination"; and
(B) by inserting "or the Administrative Opposition Panel (as the case may be)" after administrative patent judge; and

(2) in subsection (c)

(A) by striking "proceeding and inserting reexamination proceeding or an opposition proceeding";
(B) by inserting "or the Administrative Opposition Panel (as the case may be)" after administrative patent judge; and
(C) in the last sentence, by inserting "in an *inter partes* reexamination proceeding" after requester.

(c) APPEAL TO COURT OF APPEALS. (1) Section 141 of title 35, United States Code, is amended in the second sentence by inserting after reexamination proceeding the following , and any party in an opposition proceeding, who is.

(2) Section 143 of title 35, United States Code, is amended by inserting after the third sentence the following: In any opposition

proceeding, the Administrative Opposition Panel shall submit to the court in writing the grounds for the decision of the Panel, addressing all the issues involved in the appeal.

(d) DEFENSE TO INFRINGEMENT. Section 273 of title 35, United States Code, is amended

(1) in subsection (a)

(A) by striking paragraph (3) and redesignating paragraph (4) as paragraph

(3); and

(B) in paragraphs (1) and (2) by striking ''method'' and inserting ''business method''; and

(2) in subsection (b), by striking ''method'' each place it appears and inserting ''business method.''

(e) OTHER PUBLICATION OF PATENT APPLICATIONS. Section 122 of title 35, United States Code, is amended by adding at the end the following:

(e) BUSINESS METHOD INVENTIONS. In the case of applications for business method inventions, section 321 of this title applies in lieu of this section.

SEC. 7. EFFECTIVE DATE.

(a) IN GENERAL. Subject to subsections (b), (c), and (d), this Act and the amendments made by this Act apply to

(1) any application for patent that is pending on, or that is filed on or after, the date of enactment of this Act; and

(2) any patent issued on or after the date of enactment of this Act.

(b) PENDING APPLICATIONS. In applying section 321 of title 35, United States Code, as added by section 3 of this Act, to an application for patent that is pending on the date of enactment of this Act

(1) the Director of the Patent and Trademark Office shall make the determination required by subsection (b)(1)(B) of such section 321 within 12 months after the date of enactment of this Act, or on the date specified in such section 321, whichever occurs later;

(2) subject to paragraph (3), such an application shall be published

(A) on the date specified in section 321 of title 35, United States Code, or

(B) the date on which the determination is made pursuant to paragraph (1), whichever occurs later; and

(3) in no case shall an application that would be published under section 122 of title 35, United States Code, but for the enactment of this Act, be published later than the date specified in such section 122, regardless of when the Director makes the determination under paragraph (1).

(c) PATENTS ISSUED BEFORE ESTABLISHMENT OF ADMINISTRATIVE OPPOSITION PANEL. In the case of a patent issued after the enactment of this Act but before the date on which notice of the establishment of the Administrative Opposition Panel is published under section 322(a)(1) of title 35, United States Code (as added by this Act), a request for an opposition to the patent may be filed under section 322(b)(1)(A) of title 35, United States Code (as added by this Act), notwithstanding the 9-month requirement set forth in clause (i) of that section, if the request is filed not later than 9 months after the date on which such notice is so published.

TABLE OF CASES

INDEX

A

Abstract world, 3.02[A]

Access control, 7.02[C][1]

Accounting method, 6.28

Accounting patents, 5.06[A]

Activity diagrams, 11.12[H]

Agent technology patents, 5.06[B]

Aggregation, 11.07[A], 11.12[B]

Amazon.com one-click shopping, 1.05[A], 10.10[A]

American dollar, 4.15

American Fruit Growers, Inc. v. Brogdex Company, 6.15

Ancient Greeks, 4.05

ANSI X9.17, 7.02[G][1]

Anti-fraud cash registering, 6.11

Anti-fraud restaurant menus, 6.08

Anti-mold fruit treatment, 6.15

Arabic numbering system, 3.09[A][1]

Arabic numerals, 4.09[A]

ARPANET, 9.06

Array, 12.11[A][2][a]

Arrhenius equation, 3.09[D][1]

Art of Computer Programming, The (Knuth), 12.11[A][2]

Article of manufacture, 10.10[H]

Artificial world, 3.02[A]

ASP, 7.06

Assembly line, 8.02[A][2]

Assigning budget categories to bank reports, 6.23

Assigning priorities in data processing systems, 6.24

Association, 11.07[A], 11.12[B]

Assurance of identity, 7.02[C][2]

Atomic e-business models

content provider, 5.03[A]

direct-to-customer, 5.03[B]

full-service provider, 5.03[C]

intermediary, 5.03[D]

shared infrastructure, 5.03[E]

value net integrator, 5.03[F]

virtual community, 5.03[G]

whole-of-enterprise/government, 5.03[H]

Attorney-client privilege, 12.10[C]

Auction patents, 5.06[C]

Aurelius, Marcus, 4.06

Authentication using cookies, 10.10[D]

Automated service upgrade offer acceptance system, 8.15

B

Bank account ledgers, 6.05

Bank owned life insurance (BOLI), 8.11

Barter, 4.02

BCD numbering system, 3.09[A][1]

Being Digital (Negroponte), 13.02

Benjamin Menu Card Co. v. Rand, McNally & Co., 6.08

Benson, 3.09[A]

Berardini v. Tocci, 6.12

Berman, Howard, 1.07

Berners-Lee, Tim, 9.06

Best mode requirement, 12.07

Bidding for energy, 10.10[N]

Bill of exchange, 4.09

Binary coded decimal (BCD) numbering system, 3.09[A][1]

Binary numbering system, 3.09[A][1]

Binary tree, 12.11[A][2][e]

Black box, 12.11

Block cipher, 7.02[E][1][a]